Lecture Notes in Biomathematics

Managing Editor: S. Levin

29

Kinetic Logic A Boolean Approach to the Analysis of Complex Regulatory Systems

Proceedings of the EMBO Course "Formal Analysis of Genetic Regulation", Held in Brussels, September 6–16, 1977

Edited by René Thomas

Springer-Verlag
Berlin Heidelberg New York 1979

Editorial Board
W. Bossert · H. J. Bremermann · J. D. Cowan · W. Hirsch
S. Karlin · J. B. Keller · M. Kimura · S. Levin (Managing Editor)
R. C. Lewontin · R. May · G. F. Oster · A. S. Perelson
T. Poggio · L. A. Segel

Editor
René Thomas
Department of Molecular Biology
Faculty of Sciences
University of Brussels
Rue des Chevaux, 67
B-1640 Rhode Saint Genèse, Belgium

AMS Subject Classifications (1980): 92A10, 94C10

ISBN 978-3-540-09556-9 ISBN 978-3-642-49321-8 (eBook)
DOI 10.1007/978-3-642-49321-8

Library of Congress Cataloging in Publication Data
Main entry under title:
Kinetic logic.
(Lecture notes in biomathematics ; 29)
Bibliography: p.
Includes index.
1. Genetic regulation--Mathematical models--Congresses. 2. Algebra, Boolean--Congresses. I. Thomas, René. II. European Molecular Biology Organization. III. Series.
QH450.T45 511'.32 79-21915

This work is subject to copyright. All rights are reserved, whether the whole or part of the material is concerned, specifically those of translation, reprinting, re-use of illustrations, broadcasting, reproduction by photocopying machine or similar means, and storage in data banks. Under § 54 of the German Copyright Law where copies are made for other than private use, a fee is payable to the publisher, the amount of the fee to be determined by agreement with the publisher.

© by Springer-Verlag Berlin Heidelberg 1979

2141/3140-543210

To Lewis Carroll
To Jean Florine

FOREWORD

The E M B O course on "Formal Analysis of Genetic Regulation"

A course entitled "Formal analysis of Genetic Regulation" was held at the University of Brussels from 6 to 16 September 1977 under the auspices of EMBO (European Molecular Biology Organization). As indicated by the title of the book (but not explicitly enough by the title of the course), the main emphasis was put on a dynamic analysis of systems using logical methods, that is, methods in which functions and variables take only a limited number of values - typically two. In this respect, this course was complementary to an EMBO course using continuous methods which was held some months later in Israel by Prof. Segel.

People from four very different laboratories took an active part in teaching our course in Brussels :

Drs Anne LEUSSLER and Philippe VAN HAM, from the Laboratory of Prof. Jean FLORINE (Laboratoire des Systèmes logiques et numériques, Faculté des Sciences appliquées, Université Libre de Bruxelles).

Dr Stuart KAUFFMAN (Dept. of Biochemistry and Biophysics, School of Medicine, Philadelphia).

Prof. Grégoire NICOLIS (Service de Biophysique Théorique, Faculté des Sciences, Université Libre de Bruxelles) and his temporary coworker Dr David RIGNEY (presently at the Center for Statistical Mechanics and Thermodynamics of the University of Texas at Austin, Texas).

Prof. René THOMAS (Laboratoire de Génétique, Faculté des Sciences, Université libre de Bruxelles) insured the organization (if any*) of the course as well as a substantial part of the teaching, with the help of his coworkers Jean RICHELLE, Alain GHYSEN, and secretary, Marie-Jeanne PISANESCHI.

The participants come from disciplines as diverse as pure mathematics (Prof. Riguet, from Paris), embryology (Dr Slack, from London), immunology (Dr Conti, from Geneva) and urbanism (Drs Boon and de Palma, from the Laboratory of Prof. PRIGOGINE, Brussels). A complete list of participants is in the appendix.

---

* The participants in the course will appreciate this restriction.

In order to introduce the participants to the concrete biological systems which lead some of us to develop logical methods of analysis, the first day was devoted to the description in verbal terms, that is, without formalization, of well-documented genetic circuits (Thomas) and of some salient features of developmental Biology (Kauffman).

After this, three full days were devoted to the theory and practice of combinational and sequential logics, as taught by designers of logical machines (Leussler & Van Ham). If the link with the first part of the course was not immediately obvious it became apparent when regulatory systems were described in terms of logical equations, and their dynamic behaviour analyzed (Thomas). This second part of the course might be denoted "naïve" boolean approach because at this stage one sticks to the binary character of the formalism, and in addition one remains in systems whose number of variables is low enough to allow a treatment "by hand".

The third part of the course was introduced by Nicolis and Rigney, who dealt with continuous and stochastic methods. The relationship between these methods and the logical ones was discussed by Richelle.

During the last 2-3 days, the participants held free discussions or worked in the laboratory of Prof. Florine, using the simulator "Delphin" (see Chapter VIII) and other logical devices.

The course did not actually end on the official day, as a number of participants developed original contributions after returning home. These contributions were in most but not all cases based on boolean methods.

I wish to thank very sincerely the European Molecular Biology Organization (EMBO) which sponsored this course ; the organizations which subsidized the underlying research ; Prof. Jo Bertani for his encouragement ; Prof. Jean Florine who taught me combinational and sequential logics ; the instructors of the course for their efficient teaching ; the "students" of all ages for their enthousiasm and dynamic participation ; Philippe Van Ham, Jean Richelle and Alain Ghysen for many discussions and editorial help ; Ann Roller and Oliver Doubleday, who corrected number of english mistakes in this book, but should not be considered guilty for the many mistakes added after their efforts ; my secretaries Marie-Jeanne Pisaneschi, who took part in the organization of the course and most of all Tatiana Bieliavsky, who typed most of the manuscripts produced in Brussels ; my wife and Dr. Levin for their infinite patience during the long pregnancy of this book.

LIST OF PARTICIPANTS EMBO COURSE 1977

Michael BAZIN	Queen Elizabeth College - University of London Microbiology Dept.
Françoise BOON	Université Libre de Bruxelles Chimie Physique II
Guy BURTONBOY	Université catholique de Louvain - Faculté de Médecine. Laboratoire de Virologie.
Michel CASSAN	Université de Paris - Sud - Centre d'Orsay Institut de Microbiologie
Carol CONTI	Hôpital Cantonal - Suisse Immunologie de Transplantation
Michel DE LEENER	Université Libre de Bruxelles Chimie Physique II
André de PALMA	Université Libre de Bruxelles Chimie Physique II
S.M. DUNHAM	Queen Elizabeth College - University of London Microbiology Dept.
John GUARDIOLA	Instituto Internazionale di Genetica e Biofisica - Naples.
Jacques RIGUET	Université Renè Descartes - Paris Département Mathématique
Ignace LASTERS	Vrij Universiteit Brussel Molecular Biology
Jacques LEFEVRE	Université catholique de Louvain - Faculté de Médecine Physiologie Pathologie
Claire MARTINET	Université Paris-Sud - Centre d'Orsay Microbiologie
Arthur RÖRSCH	Leiden State University - laboratory of Molecular genetics Dept. Biochemistry
Jonathan SLACK	Imperial Cancer Research Fund Mill Hill Laboratories - London
Sasa SVETINA	Institute of Biophysics Yugoslavia
Suzanne THIRY	Faculté N.D. de la Paix, Namur Belgique Département Mathématique
August VAN GOOL	University of Leuven F.A. Janssenslaboratory for Genetics
Gad YAGIL	Weizmann Institute of Science mechanism of enzyme induction Dept. Cell Biology. Israël.

INTRODUCTION

Our aims.

When one is faced with the problem of how to describe complex situations or systems, one usually considers the system as a network of elementary processes which are functionally connected. By "functional connection" it is meant that the development of one process is (positively or negatively) affected by the result of one or more of the processes considered. A network which involves feedback loops will be referred to as "regulatory system". As research progresses, the description of a system evolves and yields successive images (models), each of which has its own logical structure. The essential aims of the line of research developed in this book are : first, given the logical structure of a model, to infer the whole range of its possible dynamic behaviours (pathways) ; second, to decide which of the possible pathways will actually be followed in given conditions ; third, given experimental data and the inferred set of elements of the network, what are the simplest interactions between the elements considered, which account for the observed behaviour ?

Qualitative analysis.

Where the parameters of a system are known or can be guessed with a reasonable accuracy, a quantitative mathematical treatment can be applied. For example, the methods of analysis using sets of differential equations have been successfully used to describe complex systems, especially in the fields of theoretical physics and chemistry. However, some fields have proved less amenable to an efficient quantitative analysis, because of their enormous complexity or because the systems are inadequately documented and the impression of precision given by the quantitative treatment is illusory and misleading. In this book we endeavor to show that a qualitative analysis may be helpful in handling complex regulatory systems ; in particular, it can often extract all the essential features of a model where a quantitative description restricts the view to a limited range of values of the parameters. A leitmotiv of this book, as of the course from which it arose, is that a treatment can be qualitative and yet rigourous ; for a provocative discussion on this point, see R. Thom*, 1972.

Logical analysis.

One typical qualitative treatment is the logical (or boolean) one, which is characterized by the fact that functions and variables take only a very limited number of values (usually only two, 0 and 1).
For early applications of boolean treatments to biology see, for instance, Rashevsky (1948), Sugita (1963) and Kauffman (1969).

---

* R. Thom, the great french topologist, not to be confused with the editor of this book.

In its most elementary form (here denoted naïve boolean approach) it gives a somewhat caricatural image of the system in view of its all-or-none character . On the other hand, the description and analysis are greatly simplified. In addition, the non-linear character of regulatory systems is built into the boolean formalism (which in fact, treats each interaction as if it were infinitely non-linear ; and it is well established that non-linearities play an absolutely essential role in the operation of regulatory systems (see Nicolis & Prigogine, 1978).

In fact, man has always tended to reason in binary terms, largely because of the attractiveness of simplicity. In many fields the elementary methods of logics (the classical syllogism and sometimes sorites*) are of common use. But much less common is the application of formalisms which would allow the treatment of complex situations. Yet if simplifying assumptions can be used to understand less complex systems, then they can probably be helpful in more complex systems ; if a logical approach is useful in analyzing a small fragment of a system it seems reasonable to extend it to the complete system. One question is, of course, whether the idealization inherent in the boolean formalism respects the essential features of the qualitative behaviour of the concrete systems in which we are interested. This point is discussed by Glass & Kauffman (1973) and in this book (Chapter XVI). A second, fundamental question was : "How can one most adequately introduce time in the logical formalism ?". This question is treated in Chapters V and VI in very different perspectives. Chapter V gives the viewpoint of the builder of logical machines, whose problem is : given a set of desiderata, find the simplest and most reliable device which would behave in that way. Chapter VI deals with the question : given a logical structure, which are the possible dynamic behaviours (pathways), and what decides which pathway(s) will actually be followed ?

Translation of a model from verbal to formal terms.

Translating a model from a verbal into a formalized description usually involves simplifying assumptions, which render the formalized version at the same time more specific and less general than the verbal one ; in fact, one operates a choice between variants of the model. Some of the assumptions underlying the translation from verbal to formal are obvious, others are not ; it is absolutely essential to make all of them appear as clearly as possible. As pointed out to us by G. Kreisel, one way consists of first describing precisely in verbal terms the variant chosen, and only then operating an exact translation of the refined verbal description into the formal language. Let us consider, for instance, the statement : "the synthesis of substance α is under positive control of substance β ". Specific variants of this statement are :

*"Sorites" (Lewis Carroll) are essentially generalized syllogisms, with $n \geq 4$ terms rather than only 3 .

(I) "The rate of synthesis of α increases linearly with the concentration of substance β ".

(II) "The rate of synthesis of α increases with the concentration of substance β according to a sigmoid curve characterized by a Hill number n".

(III) "The synthesis of α requires the presence of β ".

Variant (III) can be considered an extreme case of variant (II), in which the Hill number would be infinite, that is, the sigmoid curve has become a staircase step. Each of the three specific variants of the initial statement can be expressed in algebraic terms (as a differential equation). The boolean (or logical) attitude consists of systematically reasoning in terms of step functions, which can be written very simply as logical equations. In the above-mentioned example, this amounts to choosing variant III : "The synthesis of α requires the presence of β ".
The other choices are different, sometimes more sophisticated, not necessarily better. One of our ambitions has been to establish bridges between the continuous methods using differential equations, the stochastic analysis and the boolean analysis ; to point out the major simplifying assumptions involved in the various methods and to delineate their respective domains of validity : see Chapters XIV and XVI.

Regulatory systems in biology and other fields.

The starting point of our boolean approach is the occurence, in the field of genetics, of interactive networks too complex to be conveniently described and analyzed verbally. Of course, structurally similar situations exist in other fields of biology (immunology, nervous system, embryology, ecology ...) and outside biology. Even though the very nature of the elements and the mechanisms of their interactions are unrelated in the various fields, the logical structures involved in specific systems of different fields may be similar or even identical.

Most of the examples treated in this book are genetic ones. That "kinetic logic" can be applied as such to very different fields outside biology, is shown by Chapter XVIII (Boon & de Palma), which deals with a problem of urbanism.

References

Glass, L. and Kauffman, S.A. (1973)
J. Theor. Biol. 39, 103-129.

Kauffman, S.A. (1969)
J. Theor. Biol. 22, 437

Nicolis, G. and Prigogine, I. (1978)
Self-Organization in Nonequilibrium systems (Wiley-Interscience, New York).

Rashevsky, N. (1948)
Mathematical biophysics. The University of Chicago Press.

Sugita, M. (1963)
J. Theor. Biol. 4, 179.

Thom, R. (1972)
Stabilité structurelle et morphogénèse (Benjamin, New York).

TABLE OF CONTENTS

P A R T I

Verbal description of some biological systems

CHAPTER I.

R. T H O M A S

Verbal description of two well-documented genetic circuits : the lactose system in the bacterium E. coli, and immunity in bacteriophage lambda.

In this introductory chapter , I have tried to describe things as simply as possible, yet to give (espacially at the end of section 3) an insight into the real complexity of the systems. The aim is to illustrate the biological interest of this type of circuits, and at the same time suggest that it may be worthwhile to treat them in more formal terms.

1. A short introduction about gene expression, for non - geneticists.

Proteins, many of which are enzymes, comprise one or more polypeptide chains, each a defined linear sequence of amino-acids. Genes are usually segments of double-stranded DNA which carry the information for the assembly of a specific sequence of amino acids into a polypeptide chain.

A first step in gene expression, transcription, is realized by a complex enzyme, RNA polymerase, which recognizes along the DNA molecule oriented sites called promoters. RNA polymerase sticks to a promoter, and from there it synthesizes an RNA chain complementary and antiparallel to one of the DNA strands, i.e., an RNA copy of the other DNA strand. This RNA chain is called messenger RNA (m RNA) if it codes for polypeptide chain(s).

In subsequent steps, m RNA is translated by the complex machinery (ribosomes, transfer RNAs, activation enzymes, factors, etc...) which synthesizes proteins. In eucaryotes, transcription takes place in the cell nucleus, but RNA chains undergo complex processing and transit to the cytoplasm, where translation takes place. In procaryotes there is no individualized cell nucleus.

In this book, we will often consider gene expression globally, as if proteins were the direct products of gene activity.

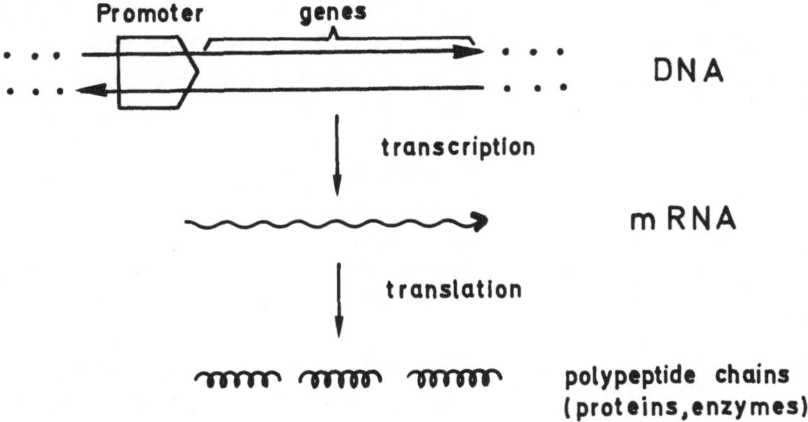

Fig.I, 1 Gene expression

Note : DNA is here drawn as two parallel lines representing its two complementary antiparallel strands. In subsequent figures, it will be drawn as a single line.

In lower organisms, one frequently finds sets of genes, adjacent on the chromosome, which are transcribed from a common promoter into a single m RNA. Such a set of genes is called an operon. The very important concept of the operon was introduced by Jacob and Monod (1961) as a result of their studies on the *lac* operon and on regulation in the bacteriophage λ.

Very often, genes in an operon have related functions. For instance, in the colibacillus *E. coli*, the *lac* operon codes for three proteins which manipulate β-galactosides and the *his* operon comprises 9 genes coding for all the enzymes of histidine anabolism. In bacteriophage λ, the so-called "late" operon codes for over 20 proteins involved in building phage heads and tails.

Different promoters have different efficiencies. In addition, their level of expression can be strongly influenced by regulatory proteins called repressors or activators depending on whether they act negatively or positively. A repressor acts at a genetic site called an operator, which, like the promotor which it may partly overlap , is located at the origin of the operon. Regulatory proteins are often allosteric proteins, i.e. they can exist in either of two conformations (typically, one active, the other inactive) depending on the presence of small molecules. For instance, transcription of the lactose operon is regulated by a repressor which is active in the absence of lactose but inactive in its presence ; lactose (or rather, a close derivative of it) is an inducer of the lactose operon. Thus, when lactose - the substrate of the enzymes coded for by the lac operon - is absent, the lac repressor is active and the enzymes are not synthesized ; if the inducer is added, the repressor is inactived and the enzymes are synthesized. In practice, the enzymes in this operon are synthesized only when their presence is useful ; this represents a remarkable and, in fact, essential, economy, as shown by the finding that mutants which have lost the control may produce, vainly, up to 25 % of their dry weight as β-galactosidase.

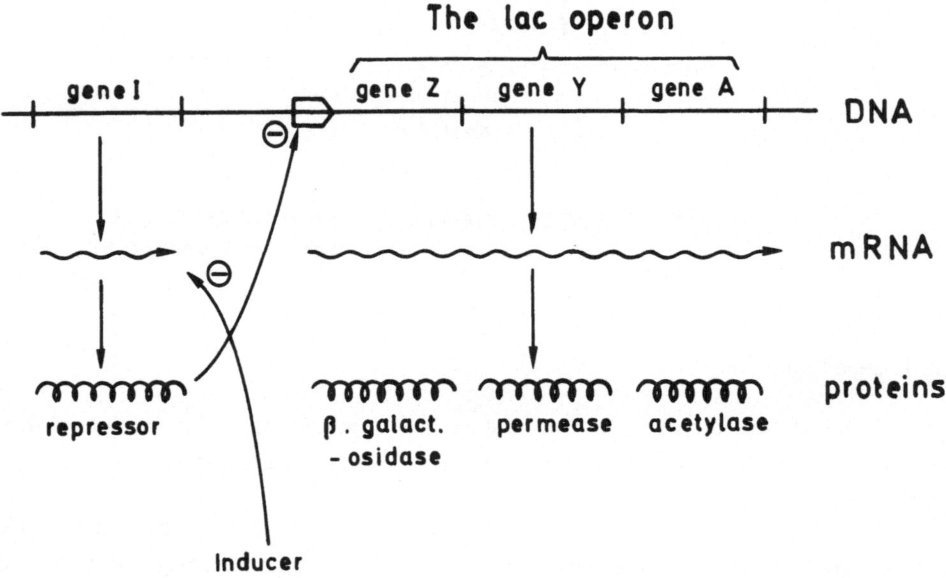

Fig.I,2 : A first scheme of the lac operon.

In contrast with this catabolic chain, which is controlled by its substrate, anabolic chains are usually controlled by their final product. In this case, the repressor is inactive unless combined with the terminal metabolite or a derivative of it, therefore called co-repressor. The economic interest of this system is also obvious : when an amino-acid, for instance, is present in sufficient amounts, it cooperates with the repressor to prevent its own synthesis, while in the absence of the amino-acid the repressor is inactive, and hence the operon is expressed.

Regulation takes place at various levels : transcription, translation, enzyme activity, etc... The most documented levels are regulation of transcription and of enzyme activity.

2. The Novick-Wiener (1957) and Cohn-Horibata (1959) experiments.

The lactose system of E. coli has been briefly described above. In this second section, additional aspects of the system will be considered. This system is also considered in Chapter III (Kauffman), XI (Nicolis), XIII (Rigney), XV (Van Ham) and XVII (Thomas).

a. The function of β-galactoside permease, the product of the second gene (Y) of the lac operon, is to concentrate external lactose inside the cells, in other words, to convert external lactose into internal lactose. It is internal lactose - or rather, a close derivative of it - which is the inducer. Thus, the system may be schematized as follows.

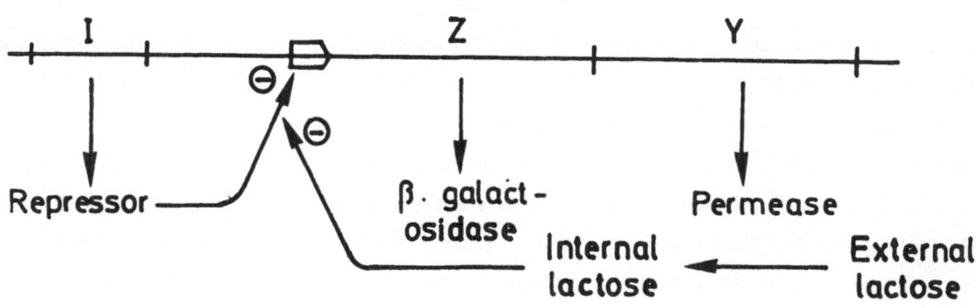

Fig.I,3 : A second scheme of the lactose operon. Additional features are taken into account, but gene expression is considered globally, without explicitely indicating m RNA.

It is seen that permease permits the occurence of internal lactose, which is necessary for the synthesis of permease. This sounds like a vicious circle, but in reality, if the external concentration of lactose is high enough there is some penetration, and induction takes place even in the absence of preexisting permease ; an internal concentration of lactose sufficient to keep the repressor inactive may be reached even with low external levels of inducer if permease is already present, but only with high external levels of inducer in the absence of preexisting permease. In other words, there are external concentrations of inducer which are sufficient to maintain induction but not to establish it.

b. The lac operon is one of the operons sensitive to an additional control called "catabolite repression" or "glucose effect". In the presence of glucose, induction tends to be prevented ; this has again a physiological interest for the cell, since it economizes the synthesis of the lac enzymes as long as there is enough of a better substrate, glucose. In practice, glucose acts as if it raised the external threshold concentration of lactose below which induction cannot be established.

c. As remarked long ago by Monod, the use of lactose itself as an inducer leads to serious difficulties in the interpretation of experimental results, because lactose serves both as the inducer of the lac operon and as a source of carbon. There are, however, analogs of lactose which are perfect inducers but are not split by β-galactosidase. Such substances (e.g. IPTG) fool the cell by inducing it to synthesize β-galactosidase and β-galactoside permease uselessly, that is, in the absence of their substrate. They are called gratuitous inducers; most studies are done with gratuitous inducers because they make it possible to analyse induction in the absence of complicating factors.

We shall now consider a typical experiment. Let us take a non-induced culture of E. coli, add to it a proper amount of gratuitous inducer and divide it into two parts, A and B. To both parts glucose is added at a concentration such that induction can be maintained but not established. To part A glucose is added immediately, whereas to part B it is added after a delay of about 15 min. Under these conditions culture A is not induced and remains indefinitely non-induced, whereas culture B is and remains indefinitely induced ; the state (induced or not) of the cultures remains unchanged even after repeated dilutions in the same medium. That the same genotype is still present in the two cultures can be shown by the fact that culture A can be induced at will by increasing the concentration of inducer or by transfer to a medium without glucose, and culture B can be de-induced at will by transfer to a medium without inducer. Thus, the same genotype, in the same actual conditions, may lead to either of two perfectly stable, heritable steady states, one in which the lac operon is turned on, the other in which it is off. The only difference between the two cultures lies in their former history. As we will see later, this bistable character is characteristic of so-called "positive" feedback loops, that is, loops with an even number of negative interactions.

As clearly stated already twenty years ago by the authors of these experiments, permease acts in an autocatalytic way since it promotes the penetration of inducer, which in turn induces permease synthesis⁺. At the concentrations of inducer and of glucose used, a cell which already contains permease (culture B) will find the external concentration of inducer sufficient to retain its induced state and transmit it to its progeny, whereas a cell without permease (culture A) will remain so because in the absence of permease the inducer concentration is too low to achieve the induced state. This leads to populations of the same genotype carrying in a inheritable way, in the same conditions, different phenotypes (epigenetic difference). This is, by the way, probably the situation of differently determined cell lines in the embryonic development of higher organisms.

+ As Cohn and Horibata remark, the same autocatalytic behaviour should be expected whenever the synthesis of an enzyme is induced by the product of its activity (as in the case of permease) rather than by its substrate (as in the case of β-galactosidase).

3. The establishment and maintenance of immunity in temperate bacteriophages.

a. Some background about temperate bacteriophages.

Bacteriophage infection typically results in the death of the infected bacterial cells, which release hundreds of bacteriophage particles identical with those used for the infection. Temperate bacteriophages can behave in this way (lytic development), or they can establish with the infected cell a permanent association (lysogenization) which usually involves the physical integration of the viral DNA in the continuity of the bacterial DNA. The resulting bacterial line is called a lysogenic bacterium (or lysogen) and the integrated phage DNA is called prophage. With very few exceptions, lysogens are immune towards infection with the phage they carry as a prophage, and most of its mutants. This immunity is due to a repressor protein coded by a viral gene (called cI in the case of the extensively studied phage λ). By a mechanism described below, the repressor produced by the prophage prevents the expression of essentially all the genes of an infecting λ phage, thus ensuring immunity. The same mechanism retroacts on the prophage itself, whose expression is paralyzed except for the cI region ; the prophage is nevertheless perpetuated , because it has become part of the bacterial chromosome and it is passively replicated as such.

The lytic development of λ depends to a great extent on a sequence of positive controls, in which genes N and Q play a major role. It begins with the expression of two so-called "early" operons (comprizing genes N - int and cro -Q : see Fig.1,4). But the extension of transcription beyond genes N and cro requires the product of gene N. Among the genes of the cro - Q operon one finds genes 0 and P , involved in DNA replication, and Q . The product of gene Q promotes a massive transcription of the so-called "late" operon, which comprize genes involved in cell lysis (S,R) and in the production of phage heads (A→F) and tails (G→J).

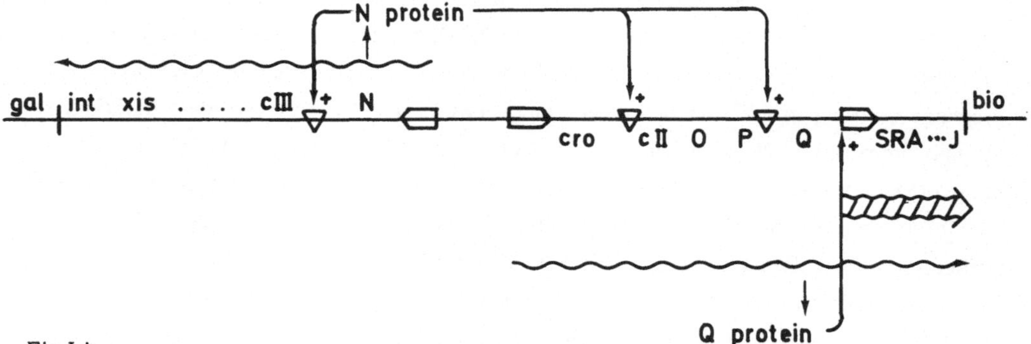

Fig.I,4

A schematic genetic map of prophage λ, showing the positive control of lytic development. Gal and bio are bacterial gene clusters. The other genes (int, ... J) are prophage genes. Wavey lines represent mRNA. The arrows with a + symbol represent positive regulation.

The repressor responsible for <u>immunity</u> acts by preventing the initiation of transcription at the level of the two early promoters :

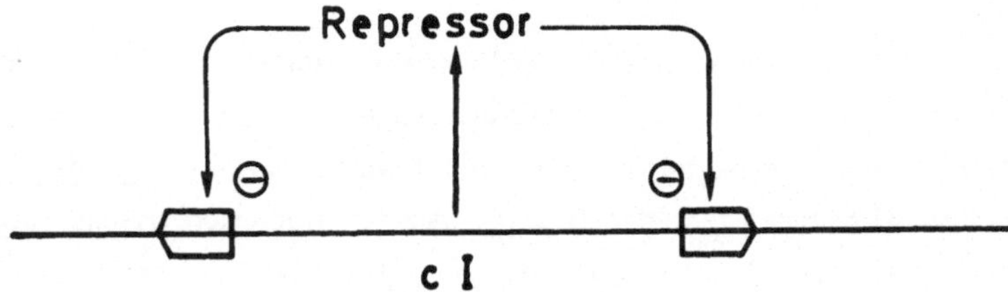

As one can infer from Fig.I,4, this results, directly on indirectly, in a block of the whole viral genome, except the small region comprizing gene cI itself. In addition, the product of gene cro exerts a negative control at the same sites at the classical repressor (cI product).

When phage λ infects a bacterium, a choice is offered between the lytic and the lysogenic pathways. In order to lysogenize , the phage must fulfill two conditions : integration of its DNA (the viral genome) into the bacterial DNA (bacterial genome) ; and the establishment of immunity, so that the letal functions of the viral genome are off. These two conditions are not independent, as functions involved in integration are under the negative control of immunity. This aspect will be treated later (see chapter XVII, section 3) . In this first examination of the control of lysogenization I shall deal only with one of the conditions, the establishment immunity.

As the λ repressor can block virtually all the other genes of the phage, one might have thought that the success or failure of the establishment of immunity results from a simple race between gene cI and the lytic functions. However, it turned out that gene cI is itself under a complex control, as shown by Eisen et al (1970) .

The mutant λcI_{857} produces a reversibly thermosensitive repressor ; cI_{857} lysogens grow at low temperatures (up to 36°C) but at high temperatures (about 40°C) the repressor is inactivated, immunity is thereby lifted, and the cells lyse. Eisen <u>at al</u> used a prophage carrying in addition two amber mutations in gene N. Although most functions are strongly impaired by N mutations, heat induction of cI_{857}NN lysogens kills the cells with great efficiency. This is because N^- phage can still replicate its DNA (albeit at a low rate), and <u>in situ</u> replication of the prophage kills the bacterium by interfering with bacterial replication ; the cell might escape this fate if the prophage were able to excise, but this function is efficiently blocked by the N mutations. After heat induction of this strain, the rare (10^{-6} or so) surviving cells all carry an additional mutation which results in a block of replication. Similar strains can also be obtained by appropriate crosses.

We have now lysogens (cI_{857}NN repl$^-$), which can live at low temperature, with immunity, but also survive at high temperature, without immunity.

Let us start with a culture of cI_{857}NN repl$^-$ at low temperature, with immunity present. If we shift to high temperature, the repressor immediately denatures and immunity is lost. If we return quickly to low temperature, the repressor renatures and immunity is restored.

If, however, the system is left for some hours at high temperature before returning to low temperature, immunity is not restored immediately. This suggests that at high temperature not only is the repressor inactivated but in addition its synthesis is blocked and as the cells grow at high temperature the repressor synthesized previously is progressively diluted out. The simplest way to account for the possibility that inactivation of the repressor results in a block of its synthesis is to postulate, as Eisen et al did, that the synthesis of the repressor in its active form exerts a positive control on its own synthesis, in other words, that the synthesis of the repressor is autocatalytic. It is now confirmed (and in addition understood in molecular terms) that at moderate concentrations the repressor indeed exerts a positive control on its own synthesis.

As mentioned above, lowering the temperature after a long period at high temperature does not result in an immediate recovery of immunity. An hour or so after the return to low temperature, two possible situations occur : in some strains immunity is never recovered,in others, immunity is eventually recovered. The result depends on the state of a gene called cro . If the phage cannot express gene cro (either because it carries a cro$^-$ mutation or because the promoter of the operon comprising gene cro is inactive) then immunity is finally recovered, otherwise it is not. In short, this result is due to the fact that gene cro exerts a negative control on the expression of gene cI ; this point will be analyzed in more detail later in Chapter XVII, sections 1 and 2.

The important point in these experiments is that a cell can live indefinitely in the same conditions in either of two alternative states : at low temperature, a (cI_{857}NN repl$^-$,cro$^-$) strain can live indefinitely with or without immunity, depending on whether its ancesters have always lived at low temperature, or have been subjected to a stay at high temperature. The situation is reminiscent of that described above (this Chapter, section 2) in the case of the lac system, and again it is interpreted in terms of the autocatalytic production of a cell component : in the simplified system studied here the repressor is necessary for its own synthesis, and consequently, either it is absent, and it will remain absent indefinitely, or it is present, and more of it will be synthesized permanently.

If one examines more normal situations, in which gene N is normal etc... , one finds that there are in fact two distinct mechanisms for the expression of gene cI :

- an establishment mechanism, which leads to a massive but transient synthesis of repressor
- a maintenance mechanism which operates at a lower rate but which is permanent in established lysogens.

These two mechanisms use distinct promoters, and they are subject to distinct controls (see Fig.1,5):

Establishment

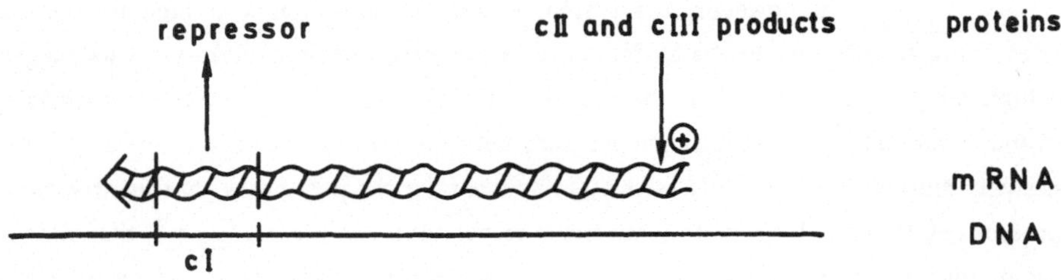

Maintenance

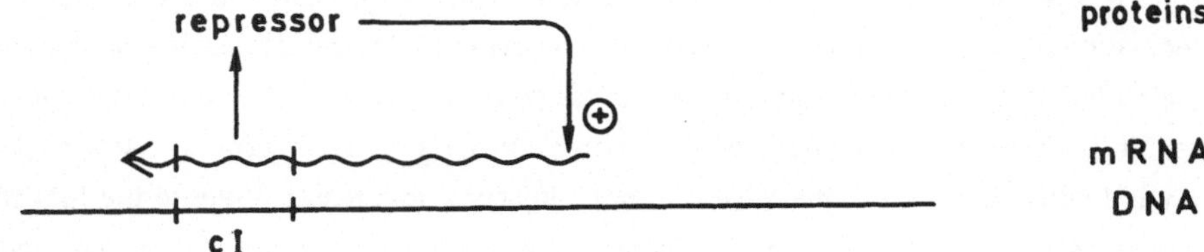

Fig.I,5 : Establishment and maintenance of immunity.

To outline the situation at infection, immunity develops in a fraction of the cells through the establishment mechanism. But as soon as immunity is established, this mechanism is turned off, since genes cII and cIII are under negative control of immunity. In counterpart, as soon as immunity is established, the maintenance mechanism is turned on, and remains active indefinitely, in view of the positive control exerted by the repressor on its own synthesis via this mechanism.

This seems simple. In fact, the situation is quite complex ; genes cII and cIII, involved in the establishment of mechanism, are themselve under positive control of gene N (see Fig.I,4) and under negative control of cI and of cro ; gene N is under negative control of cI and cro ; cro is under negative control of cI and itself ...

A characteristic feature of this system is the presence of multiple feedback loops , of which Fig.I,6 gives a simplified idea.

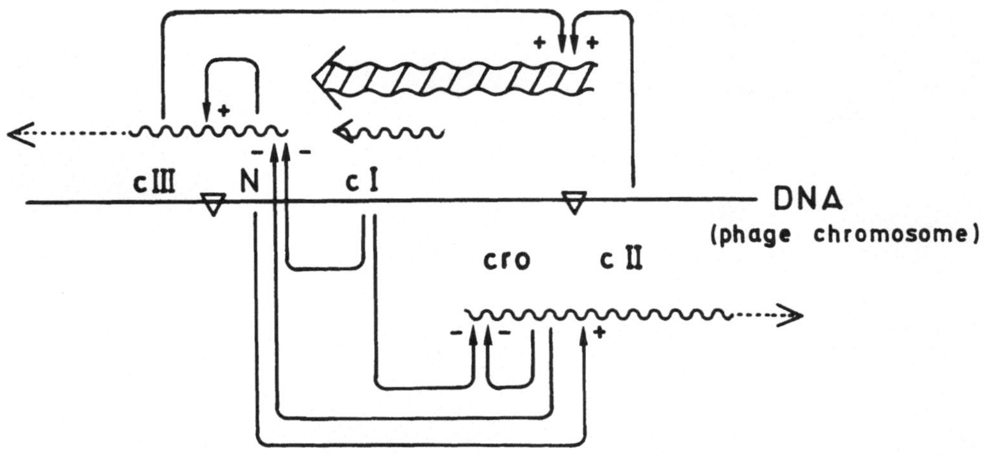

Fig I,6 The wavey lines represent the mRNAs transcribed from the phage chromosome (DNA). The arrows $\xrightarrow{+}$ and $\xrightarrow{-}$ refer to positive and negative controls, respectively, exerted at the level of transcription. The symbols ▾ represent terminator sequences ; at their level transcription stops unless the product of gene N is present.

Having studied this type of situation for some time I was led in desperation to develop a formal analysis. Some aspects of the control of immunity will be treated later in this course in formal terms (Chapter XVII, sections 1, 2, 3 and 4).

References

Cohn, M. & Horibata, K. (1959)
J. Bact. 78, 601.

Eisen, H., Pereira da Silva, L. & Brachet,P. (1970)
Proc. Natl. Acad. Sci. U.S. 66, 855.

Jacob, F. and Monod, J. (1961)
J. Mol.Biol. 3, 318.

Novick, A. & Wiener, M. (1957)
Proc. Natl. Acad. Sci. U.S. 43, 553.

CHAPTER II

THE MOLAR BEHAVIOR OF CELLS IN DEVELOPMENT

Stuart A. Kauffman

University of Pennsylvania
Department of Biochemistry and Biophysics
School of Medicine
Philadelphia, PA I9104

Molar Behaviors of Cells in Development

Metazoan development poses among the most severe epistimological problems confronted by scientists. The aim of this chapter is to sketch these issues; in the next chapter I will discuss two complementary, but still naive approaches to some of the major problems raised.

It is now well established that the diverse cell types in a metazoan almost always differ by virtue of differential expression of genes and their products, rather than by selective loss, or amplification, of genetic material from distinct cell types during ontogeny. Among the kinds of evidence which support constancy of genetic material are these: Constancy in the karyotypes of most or all cell types in an organism; presence of the same amount and kinds of DNA sequences in different cell types of an organism; the capacity of nuclei derived from cells at relatively late stages of embryogenesis, to support normal development after injection into enucleate eggs; the capacity of single differentiated cells of some adult plants to generate a complete new plant; a large number of metaplasias, in which cells on one developmental pathway give rise to those on a separate developmental pathway. At the same time, a very large body of evidence establishes that the patterns of gene activity differ in different cell types.

These results pose the core problem of developmental biology, since if cells differ by virtue of the patterns of activities of genes and their products, then in order to understand development, we must inevitably understand how the system of genes and their products are able to coordinate their patterns of activity to generate an adult from a zygote. The number of structural genes in a higher metazoan is unknown. Based on the typical size of proteins, mammalian cells contain enough DNA to code for well over 10^6 distinct proteins. Even if attention is restricted to unique sequences of DNA, and even if the bulk of that DNA and repetitive DNA were to serve regulatory roles, there remains enough DNA to code easily for 10^4 to 10^5 proteins. Even this crude estimate states the epistimological problem. If there are on the order of 10^5 structural genes, and about the same number of regulatory functions, then our task consists in understanding how that integrated regulatory system directs development.

It is worth trying to see why this task is more formidable than, for example, the characterization of biochemical pathways in intermediate metabolism. The progressive conversion of substrate to product occurs by simple modifications of the substrate at each step. This imposes an inherent order on the sequence of conversions, which falls out readily as the idea of a pathway of conversions. We easily understand what we need to explain: the pathway, the control points regulating flux down that pathway, the mechanisms of catalysis, and the mechanisms regulating the catalytic rates.

Differentiation appears dissimilar to intermediate metabolism in fundamental conceptual respects. First, there is no parallel to the sequential modification of substrate to product in the more complicated gene regulatory systems worked out in bacteria and phage. The molecules mediating repression and activation of specific genes or their products need bear no known structural similarity to the products of the genes which are regulated. The richness of allosteric interactions, and DNA, RNA protein interactions, appears to allow essentially arbitrary control connections to exist among the regulated processes. Thus, although it is sometimes possible to make educated guesses, from the functional requirements of a cell, how the control interactions will be realized, the molecular mechanisms mediating those reactions are not constrained by the kind of steric similarity relating competitive inhibitor of an enzyme to the normal substrate of that enzyme.

Second, we really have no clear idea of what constitutes a "cell type". The morphological criteria used by histologists to classify cells is adequate to show that marked differences exist between cells derived from different tissues.

If one attempts to make clear what the term "cell type" means at the molecular level, then it seems that we properly think of a cell type as a rather precisely constrained pattern of activities of the entire genome and its products, and we think of different cell types as having rather sharply different patterns of activities. This presumption mirrors first, the histological observation of sharp differences between cell properties, with rather few "intermediate" cases, and the more contemporary measurements of actual differences in gene activities in different cell types. But at present, I think it is correct to say that we have no clear understanding of how constrained such a pattern must be to constitute "modulation" within a cell type rather than differentiation of one cell type to another, nor as yet do we have clear criteria to decide whether this distinction is merely a semantic one, or reflects differences in the mechanisms underlying differentiation and modulation. Furthermore, we do not know whether the constrained pattern of gene activities which constitute a specific cell type, reflects the integrated activity of a <u>single</u> coupled system with about 10^5 components, whether that system is fractured into a number of essentially independent subsystems, whether a relatively few nodal control loci dominate the dynamical activity of the entire system as a coupled whole, and so on. In short, we have at present, virtually no idea how these 10^5 processes are conjointly regulated, even if we are beginning to have some idea of the molecular means which mediate that regulation.

The epistemological problem lies in understanding how this integrated system works. The most successful exemplar of the analysis of systems comprised of many components is in the reduction of classical thermodynamics to statistical mechanics. In statistical mechanics, it is legitimate to assume that each particle in, say, a gas, obeys the same Newtonian laws. It has been this identity, coupled with the existence of conserved properties, such as the total energy of the system, which has allowed statistical mechanics to be built. In a coupled genetic system, the regulatory rule governing the activity of one locus will in general differ from the rule at another locus, and the specific regulatory "inputs" to one locus differ from those at another locus. Inevitably, we are confronted with a system having 10^5 components which regulate one another's activities in fashions we must seek to unravel.

In our search, we might adopt a "tops down", or a "bottoms up" effort. In the latter case we would hope by brute force, to discover each control point, its molecular nature, and its direct functional coupling to those loci which either regulated it, or were regulated by it. In such a manner we would gradually build up the picture of the dynamical behavior of the integrated system. In a tops down approach, we would attempt to understand the molar behaviors of the system as a guide to guessing what features in the underlying control machinery must exist to account for the molar behaviors we observe. It seems obvious that both should be pursued simultaneously. I want now to describe a number of "molar" properties of development in metazoans which might be clues to the organization of the underlying control machinery.

The zygote of a typical higher metazoan undergoes on the order of 30 to 50 further divisions in generating the adult. During the course of those division, differentiation ensues. The criteria by which we might count the number of distinct cell types in an organism are unclear, and would yield a higher number were finer criteria used. One set of criteria is that used by histologists in classifying cell types. On that basis, it is fair to say that a mammal has on the order of a few hundred cell types, while a hydra has on the order at most a few tens of cell types. The number of cell types of which a given genome is capable is a perfectly reasonable molar property of that genome. If one does use the histologists criteria, and plots the number of cell types against the complexity of its DNA, then roughly speaking, the number of cell types increases more slowly than the complexity of the DNA. In one such plot, the number of cell types increases crudely as a square root function of the complexity of the DNA (1). The implications of this observation include the deduction that adding the next cell type requires proportionally more genome as the number of cell types increases. This is not at all trivial, since we can easily imagine control couplings among genes and their products which would allow the number of cell types to increase as the number of genes squared or cubed.

A second important molar behavior exhibited by cells is the occurrence of pathways of differentiation. This implies that any cell type can change directly into rather few cell types. And in fact, it is a central property of ontogeny among all metazoans, that any one cell type differentiates directly into only a few other cell types. Again, this is not a trivial observation, since we could readily design gene regulatory systems which allowed one cell type to differentiate directly into hundreds of other cell types. Rather, it is likely to reflect some deep properties of the underlying genomic system.

A third general molar property of cell types is that they appear to be rather sharply distinct from one another, with few intermediates except, perhaps, along pathways of differentiation. Again this is not trivial, since we could construct genetic regulatory systems in which different "cell types" might grade indefinitely into one another.

Fourth, being a specific "cell type" appears often to be a stable, clonally heritable property of the cell in question. This implies that the processes which specify and maintain that cell types are stable with respect to cell division, and to some range of perturbing stimuli which impinge on each cell due to noise in the control system. A given cell type, therefore is often a regenerable, and stable constellation of gene activities.

Fifth, inductive stimuli are capable of triggering specific pathways of differentiation in stem cells for those pathways, or intertissue interactions during normal entogeny.

Sixth, in a variety of circumstances, cells on one developmental pathway can "switch" to developmental pathways they would not normally follow. Such metaplasia is widespread, and will be discussed further below. For example, regenerating newt neural retina cells derive directly from cells which were formerly pigment cells (2); similarly blood lymphocytes can first transform into phagocytic macrophages, then into collagen secreting fibroblasts(3). Cell types which transform into one another during the normal course of ontogeny along developmental pathways, and those which transform into one another through metaplasia are particularly interesting because they suggest that the pair of cell types which can so transform are in a fundamental sense, functional neighbors. Their developmental programs must somehow be more similar than pairs of cell types which cannot directly interconvert. In general, the range of cell types to which any one cell type can give rise during normal ontogeny and via metaplasia is very limited, suggesting that, in some sense, each program has only a few near neighbors.

Seventh, cell interaction in an integrated tissue appears to be requisite for the maintenance of some aspects of differentiation. Cells isolated from their normal contacts in an integrated tissue are frequently said to "dedifferentiate", although the precise criteria to say so seem insecure.

Among these molar properties, patterns of metaplasia are particularly interesting. Among the most striking of metaplasias are the homeotic mutants known among insects. For example, Drosophila melanogaster is a holometabolous insect with egg, larva, pupa and adult stages. The ectoderm of the adult derives during metamorphosis form the terminal maturation and eversion of specialized pools of 10,000 to 40,000 cells found in the larva called imaginal discs (4). Each is determined early in embryogenesis to form a particular part of the adult epidermis; wing, first leg, eye-antenna, genital, etc. Homeosis refers in general to the replacement of one normal body part by another. In Drosophila, a large number of mutants are known which effect such replacements (5). For example, Nasobemia converts portions of the antenna into portions of the second leg, eyeless-Opthalmoptera converts eye to wing tissue, tumorous head converts head to genital tissue. Homeotic mutants include dominant and recessives, chromosome rearrangements and apparent point mutants. The implication of such mutants is that single loci are capable of diverting tissue from one to another developmental pathway. Three other forms of metaplasia are well evidenced in insects and arthopods in general. Perturbations during the development of wild type Drosophila and other insects can "phenocopy" known homeotic mutants. For example, heat or ether shocks in the first three hours of development can transform the metathorax into the mesothorax, a condition mimicing the bithorax mutants (6). In insects and other arthropods, if an antenna is removed, a leg occasionally regenerates in its stead; a process called homeotic regeneration (5). Finally, in Drosophila melanogaster, culture of imaginal discs determined to form one adult cuticular structure sometimes results in the "transdetermination" of that disc tissue to a new state giving rise to a different adult ectodermal structure (7). For example, genital disc fragments cultured in adult abdomen then subjected to metamorphosis by injection into larvae about to enter metamorphosis often yields adult antenna or leg tissue rather than adult genital tissue.

Homeotic mutants and transdetermination in Drosophila offer important clues to the organization of the "decision machinery" which mediates developmental commitments. Homeotic mutants exibit a number of general features. First, a single homeotic mutant generally converts a given tissue in only a single direction. Second, a single homeotic mutant often affects more than one tissue. For example, Nasobemia converts both antenna to mesothoracic leg, and simultaneously can convert eye to wing; almost all of these coordinated conversions can be called "parallel" in the sense that the mutant transforms tissue A to B, and C to D. Thus, except for a few cases, homeotic mutants transform any given tissue in only one direction, but often transform two or more tissues simultaneously in a coordinated fashion. Third, the tissues which transform into one another are not always neighbors on the fate map (8) of these structure on the Drosophila egg. For example, tumorous head (9) converts head tissue, which lies near the anterior end of the egg on the fate map, to genital tissue, which lies near the posterior pole of the egg on the fate map. Therefore, lack of geometric proximity on the fate map does not insure functional "distance" of developmental programs. However, many neighbors on the fate map do transform to one another in some homeotic mutants, suggesting among other possibilities that proximity may generally imply similarity of developmental programs. Fourth, homeotic mutants present in combinations yield the predicted combinations of transformations, as if each acted independently. For example, postbithorax converts the posterior half of the metathoracic haltere into the posterior half of the mesothoracic wing (10). The mutant engrailed converts the posterior wing into a mirror image anterior half (11). A fly homozygous in both postbithorax and engrailed, has the posterior half of the halter converted into anterior wing tissue (12). The implications of these observations will be discussed below.

Transdetermination in Drosophila exhibits a number of equally important molar properties. First, any one tissue can transdetermine directly into more than one, but only two or three other tissue types. Second, sequences of transdetermination exist, for example, from genital to antenna to wing to mesothorax. Thus, there are allowed and forbidded one step transitions, and sequences of transitions separating tissues which cannot directly reach one another. There is, in addition, a global orientation toward mesothorax, in the sense that transitions toward it are always more probable than transitions away from it. Figure 1 shows the observed patterns of transdetermination among the major imaginal discs (13).

Consider two alternative hypothesis. First, imagine that the different tissue states "genital, leg, antenna, wing, etc." corresponded to different states of a single coupled genetic system with as many distinct stable patterns of gene activity as tissue types. Such a hypothesis makes no predictions about which tissues transform to one another, about the existence of allowed and forbidden one step transitions, about sequences of transdeterminations, or about a global orientation toward thorax. Restrict attention for the moment to the transitions from genital to antenna or leg, and from antenna or leg to wing. The hypothesis of a gene system with four alternative patterns of gene activity 0,1,2,3, such that 0 = genital, 1 = antenna, 2 = wing, and 3 = leg, makes no predictions about transdetermination. Alternatively, suppose that that two independent genetic systems exist, each with two alternative patterns of gene activities, 0 and 1. Suppose the combination of states 00 = genital, 01 = leg, 10 = antenna and 11 = wing. Immediately, the genital to leg and leg to wing transitions are 1 step transitions while the genital to wing transition requires two steps. Imagine that the 1 state is more stable than the 0 state of each genetic system, then the 0 to 1 transition is more probable than the reverse, and one obtains the global orientation towards wing. This simple model does make predictions about transdetermination frequencies. For example, the antenna to wing transition should be more frequent than the antenna to leg conversion, since both involve changing the 0 in the second genetic system in antenna to a 1, while the conversion of antenna to leg requires the additional alteration of the 1 state of the first system in

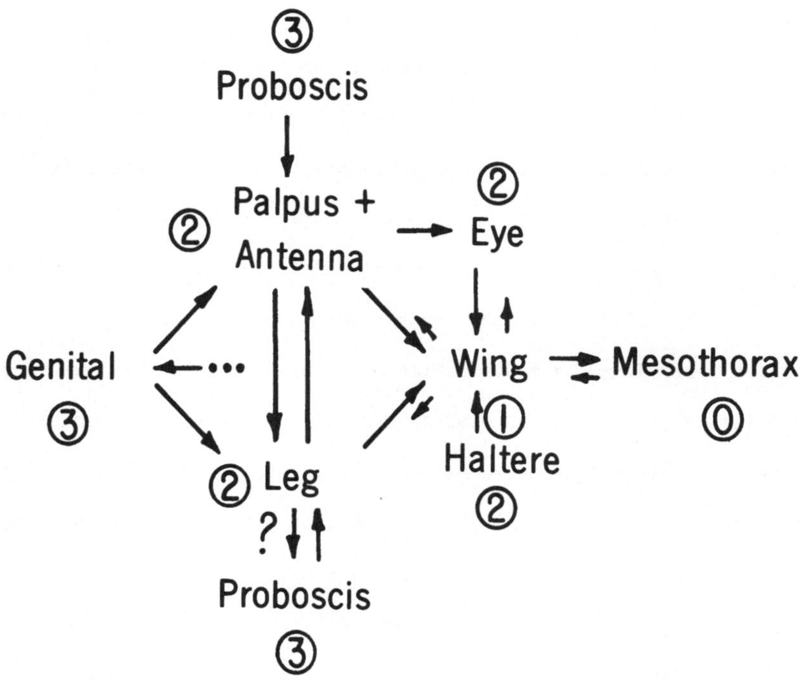

Figure 1. Transdeterminations among the imaginal discs of _Drosophila melanogaster_. Lengths of arrows represent relative frequencies of transdetermination. Encircled numbers indicate the minimum numbers of transdetermination steps needed before the disc can reach the determined state of mesothorax (13).

antenna to a 0.

The simplest way to account for sequences of transdeterminations which diverge and converge, is to imagine that the "determined state" in each imaginal disc is encoded by a particular combination of states of several essentially independent subsystems.

In Figure 2, I show such a combinatorial model for the determined state of the different imaginal disc tissues, on the fate map of the Drosophila egg. The model was derived from a modification of a Turing like mechanisms to generate positional information in the developing egg (14), but that other model is not of importance to this discussion. Rather, the relative success of the combinatorial model is important. In Table 1, I show a number of predictions of relative transdetermination frequencies from the model, all similar to that described above. All but two appear to be true. The same combinatorial model accounts for the qualitative features of transdetermination described above. If any genetic system among the four independent "switches" can change state in transdetermination, then any tissue should transdetermine to more than one tissue, but sequences and a global orientation toward the combination reflecting the most stable state of each subsystem should be seen.

The same combinatorial model explains the major features of the metaplasia seen in homeotic mutants. First, if any homeotic mutant affects a single gene system, then any homeotic mutant should transform one tissue in only one direction. That direction should be among those seen in transdeterminations from that tissue. Second, one state of one gene system often occurs in more than one tissue, hence a homeotic mutant altering a 0 to 1 on a given switch, might be expected to yield coordinated parallel transformations of A to B and C to D. Third, the model in Figure 2 shows that areas which are distant on the fate map can have similar combinations of gene system states. For example, antenna and genital differ only in the state of the first "switch gene system". Thus homeotic mutants might

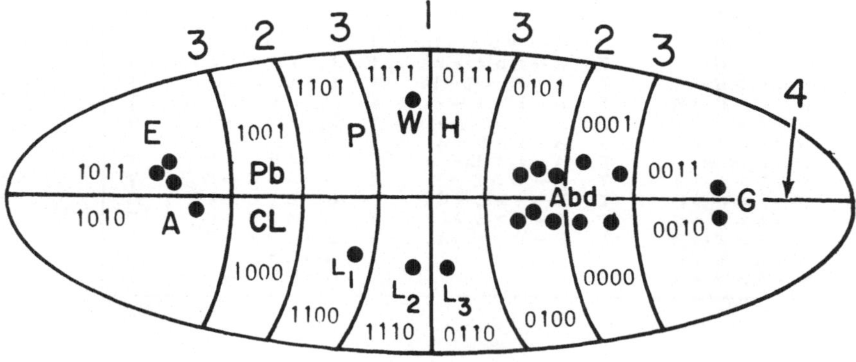

Figure 2. The fate map of the blastoderm. A-antenna; E, eye; Pb, proboscis; Cl, clypeo-labrum; P, prothrax; W, wing and mesothorax; H, haltere; Abd, abdominal segments; G, genital. Assignment of clypeo-labrum to a position below proboscis is tentative. The numbered lines are a hypothetical (14) sequence of compartmental boundaries subdividing the progressing blastoderm. The combinations of four 1 and 0 digits, (1010), (1111), etc. show a proposed combinatorial epigenetic code word in each compartment, specifying the determined state of each major imaginal disc (14).

Prediction	Status	Prediction	Status	Prediction	Status
$H \to W > H \to A$	T	$A \to W > A \to H$	T	$L \to W > L \to E$	T
$H \to W > H \to L_{1,2}$	T	$A \to L > A \to W$	F	$L_{1,2} \to W > L_{1,2} \to H$	T
$H \to W > H \to E$	T	$A \to Pb > G \to Pb$	?	$L > A > L \to E$	T
$H \to W > H \to Pb$	T	$A \to E > A \to W$	F	$L_{1,2} \to A > L_{1,2} \to G$	T
$W \to A > H \to A$	T	$A \to G > L_{1,2} \to G$	T	$L_2 \to G > L_3 \to A$	?
$W \to E > H \to E$	T	$A \to E > E \to A$	T	$L_1 \to Pb > L_1 \to G$	?
$W \to L_{1,2} > H \to L_{1,2}$	T	$A \to L_2 > L_2 \to A$	?T	$G \to A > G \to Pb$	T
$W \to L > W \to A$	T	$E \to W > E \to H$	T	$G \to A > G \to W$	T
$W \to L > W \to G$	T	$E \to A > E \to G$	T	$G \to L_{2,3} > G \to W$	?T
$W \to A > W \to G$	T	$E \to A > E \to L$	T	$G \to A > G \to L_{1,2}$	?T
$W \to E > W \to Pb$	T	$E \to W > E \to L$	T	$G \to A > A \to G$	T
$W \to E > W \to G$	T			$G \to L > L \to G$	T
$W \to E > W \to A$	?			$G \to H > G \to W$	?F

Table 1. Predicted relative transdetermination frequencies derived from the chemical wave model applied to the blastoderm. $L_{1,2} \to A > L_{1,2} \to G$ means the model predicts transdetermination from the first or second leg to antenna is greater than to genital. Abbreviations are explained in the legend of Figure 2 (14).

Mutant	Symbol	Transformation			Coordination	Code Change	Switches Required
Antennapedia[1]	Antp	antenna	→	leg 2	-	1010→1110	1
Pointed wing	Pw	antenna	→	wing	-	1010→1111	2
Nasobemia	Ns	antenna	→	leg 2	parallel	1010→1110	1
		eye	→	wing		1011→1111	1
dachsous	ds	tarsus	→	arista	-	1110→1010	1
Opthalmoptera[2]	Opt^G	eye	→	wing	-	1011→1111	1
Hexaptera	Hx	prothorax	→	mesothorax	-	1101→1111	1
podoptera	pod	wing	→	leg		1111→1110	1
tetraltera[3]	tet	wing	→	haltere	-	1111→0111	1
Contrabithorax	Cbx	wing	→	haltere	parallel	1111→0111	1
		leg 2	→	leg 3		1110→0110	1
Ultrabithorax	Ubx	haltere	→	wing	parallel	0111→1111	1
		leg 3	→	leg 2		0110→1110	1
tumorous head	tuh1,3	eye	→	genital	parallel	1011→0011	1
		antenna	→	genital	divergent	1010→0010	1
		antenna	→	leg		1010→1110	1*
lethal(3)III-10	1(3)III-10	haltere	→	wing	parallel	0111→1111	1
lethal(3)XVI-18	1(3)XVI-18	genital	→	antenna	divergent	0010→1010	1
		genital	→	leg		0010→0110	1*
lethal(3)703	1(3)703	antenna	→	leg	parallel	1010→1110	1
lethal(3)1803R	1(3)1803R	genital	→	leg	divergent	0010→0110	1
		genital	→	antenna	parallel	0010→1010	1*
		haltere	→	wing		0111→1111	1
proboscipedia	pb	proboscis	→	antenna	divergent	1000→1010	1
		proboscis	→	leg		1000→1100	1*
extrasexcombs[4]	ecs	leg 2	→	leg 1	convergent	1110→1100	1
		leg 3	→	leg 1		0110→1100	2
Polycomb	Pc	antenna	→	leg 2		1010→1110	1
lethal(4)29	1(4)29	leg 2	→	leg 1	convergent	1110→1100	1*
		leg 3	→	leg 1		0110→1100	2*

Table 2. Observed homeotic transformation, and the code changes required for the code scheme in Figure 2. A set of homeotic mutants causing the same transformation is represented by one member (4,5): 1) Antennapedia, Antennapedix, aristapedia, aristatarsia; 2) Opthalmoptera, opthalmoptera, eyes-reduced; 3) tetraltera, Metaplasia, Haltere mimic; 4) extrasex combs, Extrasexcomb, reduplicated sex comb, sparse arista. Transformations with 1* and 2* require one additional switch to account for other transformations of that homeotic (14).

be expected to cause conversions between distant areas on the fate map, as well as between neighboring domains.

The specific model in Figure 2 actually accounts for most major homeotic mutants. In the model, each tissue is characterized by a unique combination of switch states in a kind of epigenetic code word written as a binary code with four digits. Any specific homeotic mutant might cause a transformation requiring 1,2,3 or 4 digits to alter state. On average, one would expect 2 digits to be required to change states for a four digit bit code applied at random to the different tissues. The model in Figure 2 accounts for the observed homeotic transitions among tissues discriminated by the code words of the model with little more than 1 switch per transition, Table 2. Further evidence supporting this model will not be presented here.

From the point of view of a general discussion of the molar features of ontogeny, the discussion of homeotic mutants and transdetermination in _Drosophila_ finds its importance in the very real possibility that the committed state in each tissue can be thought of as being specified by a small number of essentially _independent_ subsystems, each with only _two_ alternative states, such that the combination of their states encodes the determined state in each tissue. Thus, this combinatorial model offers the start of one account for some of the molar behaviors described.

The major problem is to attempt to link the molar behaviors we can describe, with the underlying behaviors of a genetic system with 10^4 to 10^5 components. Even were the genetic system to fall apart into several independent subsystems, each is likely to have many components. Can we begin to make some guesses about the functional organization of such coupled systems based on an amalgam of rational design principles, on our current knowledge of the coupling between genes in bacteria and phage, and on statistical arguments about kinds of components we surmise exist in eukaryotic cells?

I think it is now beginning to be possible to do so. In order to begin, however, it is necessary to have some formal language to describe the dynamical coupling of activities of very many components. For a variety of reasons, it is convenient to use the idealization that any gene is either active or inactive, its product either present or absent. This is a severe idealization, since it is well known that genes can function at various intermediate rates of activity, and their product levels can be continuously modulated. But the systems we wish to consider are so complex, that examination of a dynamical "skeleton" is probably the only practical first step. Insights gained thereby can later be used to study the consequences of relaxing the idealizations. This approach will be discussed in the next chapter.

1. Kauffman, S.A. (1969). J. Theoret. Biol. 22 ,437.

2. Stone, L.S. (1950). Anat. Record 106, 89.

3. Petrakis, N.L., Davis, M. and Lucia, S.P. (1961). Blood 17, 109.

4. Gehring, W. and Nothiger, R. (1973). In Developmental Systems II, Insects" p.161 (S. Counce and C.H. Waddington, Eds.) Academic Press, New York.

5. Ouweneel, W. J. (1976). In Adv. in Genetics 18, 179.

6. Gloor, H. (1947). Rev. Scisse Zool. 54, 637.

7. Hadorn, E. (1966). in "Major Problems in Developmental Biology" (M. Locke, Ed.) p. 85. Academic Press, New York.

8. Hotta, Y. and Benza, S. (1973). " Genetic Mechanism of Development" (F. Ruddle, Ed.) p. 129, Academic Press, New York.

9. Postlethwait, J. H., Bryant, P. J. and Schubiger, G. Develop. Biol. 29, 237.

10. Lewis, E. (1964). Symp. Soc. Develop. Biol. 23, 231.

11. Lawrence, P. D. and Morata, G. (1976). Develop. Biol. 50, 321.

12. Garcia-Bellido, A. (1975). In "Cell Patterning" CIBA Foundation Symposium 29, p. 161.

13. Kauffman, S.A. (1973). Science 181, 310.

14. Kauffman, S.A., Shymko, R., Trabert, K. (1978). Science 199, 259.

CHAPTER III

ASSESSING THE PROBABLE REGULATORY STRUCTURES AND DYNAMICS OF THE METAZOAN GENOME

Stuart A. Kauffman

University of Pennsylvania
Department of Biochemistry and Biophysics
School of Medicine
Philadelphia, PA 19104

In the previous chapter I sketched several molar behaviors of cells during metazoan ontogeny. The purpose of the present chapter is to discuss an approach to relating the observed molar behaviors to the behaviors of the underlying coupled system of genes and products which regulate one anothers activities. We cannot in the near future expect to work through the control connections among so many components. However, we can hope to characterize a number of local, small scale properties of the organization of cellular control systems, such as specifying the typical number of molecular variables controlling any given process and specifying the ways variations in the regulating processes affect the controlled process. Specification of such small scale, local properties should be useful in two ways: (1) The local properties form the basis for hypotheses about the organization of larger control circuits; (2) The implications of the small scale properties for the large scale dynamical behavior of cellular control systems can be assessed. Systematic use of such local characteristics for both these purposes can be made by constructing a set of all the possible large control systems, each member of which is built using only those small scale properties. This set, or ensemble, represents the class of hypotheses about the total architecture of cellular control systems implied by known small scale properties of the organization. Examination of the typical, or average "wiring diagram" of ensemble members will allow hypotheses about the most probable kinds of intermediate and large control circuits which may be expected from small scale properties we already know. Examination of the typical large scale dynamical behaviors of the ensemble members will allow us to assess the most probable large scale behaviors of cellular control systems having the known small scale properties. The primary purpose in characterizing small scale properties of cell control systems and constructing an ensemble of possible control systems is to examine the implications of known small scale features for probable large scale properties, rather than directly to help learn about molecular mechanisms or the small scale properties themselves.

In this chapter, I present evidence for three claims: (1) Nearly all known regulated genes and processes are controlled according to a very small, similar and simple class of rules; (2) This small scale property appears sufficient to account for the known molar dynamical behaviors of the integrated genetic control systems described above; (3) Restriction to this class of control rules predicts the existence of simple but powerful intermediate size regulatory circuits with properties useful for directing differentiation, and allows a reasonable first guess as to the overall functional organization of a metazoan's regulatory machinery.

The Lac Operator

Consider the *lac* operator of *E. coli*. If bound by repressor (Jacob and Monod, 1963) it prevents transcription of the adjacent structural genes, Z, Y, A. Binding of the operator by repressor is controlled by the presence of repressor molecules, and lactose, the inducer. A transgalactosidation derivative of lactose, allo-lactose (Zubay and Chambers, 1971; Burstein, Cohn, Kepes and Monod, 1965; Muller-Hill, Rikenberg and Wallenfels, 1964) binds to a site on the repressor molecule, weakening the repressor operator bond and removing the repressor. Binding of the operator is dependent on the concentrations of repressor molecules and allo-lactose. In the absence of allo-lactose, binding of one operator saturates as repressor concentration increases. At a fixed maximal level of about twelve repressor molecules per cell (Bretscher, 1967), binding of lactose derivative to repressor saturates as lactose level increases (Riggs, Newby and Bourgeois, 1970).

To understand the behavior of the operator locus, it is convenient to examine its response to saturating, and minimum concentrations of its controlling molecular variables. In those cases the operator can be bound only if repressor is present in saturating levels, and no allo-lactose is present. The operator cannot be bound if there is either (1) no repressor or (2) saturating levels of allo-lactose. The striking feature of this process is that each control variable, acting alone, can determine one of the two possible states of the operator, regardless of the concentration of the other regulatory molecule. Absence of repressor, or high levels of allo-lactose, each alone, assures the operator cannot bind repressor. Both must be coordinated to assure the operator <u>is</u> bound. I aim to show it is typical of regulated genes and processes that at least one control variable has a state which determines the outcome of the regulated processes regardless of the states of other regulatory variables. This property defines a class of control rules I term canalizing functions.[1]

Canalizing Functions

The simplest way to describe the behavior of a regulated locus like an operator at saturating and minimal levels of its control variables, is to consider it an on-off device and use logical (Boolean) algebra. Let "Op = 1" mean repressor is bound to the operator, "Op = 0" mean the operator is unbound, "repressor = 1" mean repressor is present in saturating levels and "repressor = 0" mean repressor is absent. The behavior of the lac operator can be described by a table listing all possible combinations of the states of the molecular variables controlling it, and its subsequent response. Any such table is a Boolean function.

TABLE I

The lac operator, a canalizing function

Allo-lactose	Repressor	Op
0	0	0
0	1	1
1	0	0
1	1	0

Boolean functions may be classified by the minimum number of controlling variables whose states must be specified in order to determine one specific state of the regulated processes. For example, one variable suffices to determine one state of the function, shown in Table I, governing the lac operator; repressor = 0 determines Op = 0, regardless of the state of allo-lactose, the other control variable. Also, specifying allo-lactose = 1 by itself determines Op = 0 regardless of the state of repressor. Determining the state Op = 1 requires coordination of both variables, allo-lactose must equal 0 and repressor must equal 1.

I shall call a function canalizing if at least _one_ control variable has _one_ state which, by itself, can determine _one_ state of the regulated processes. In this sense, the lac operator is canalizing, since allo-lactose = 1 determines Op = 0 regardless of the state of repressor. Only one state of a canalizing variable, the _canalizing state_, guarantees that the recipient goes to a specific state, the other state of that variable guarantees nothing. Allo-lactose = 1 guarantees the operator is unbound, allo-lactose = 0 guarantees nothing for the state of the operator still depends on the state of the repressor. The _canalized state_ of a process governed by a canalizing function is that state which can be determined by a single regulatory variable; for the lac operator the canalized state is Op = 0. A process controlled by K variables may have up to K canalizing variables. The lac operator has two canalizing variables since either allo-lactose = 1 alone, or repressor = 0 alone guarantees Op = 0. By contrast, consider a hypothetical structural gene (ST) with three distinct adjacent operators, where repression of transcription requires binding of any two or more operators. Since determining either the state 0 or the state 1 for transcription would require specifying the states of at least two control variables this function is not canalizing. Consider two hypothetical promoters regulating

TABLE II

A non-canalizing function

Op_1	Op_2	Op_3	ST
0	0	0	1
0	0	1	1
0	1	0	1
0	1	1	0
1	0	0	1
1	0	1	0
1	1	0	0
1	1	1	0

one structural gene (St), in which transcription occurs if either, but not both, promoters are bound by polymerase (Table III). This Exclusive Or function is not canalized by either control variable, for no state of either promoter alone can determine whether transcription occurs or fails.

TABLE III

P_1	P_2	St
0	0	0
0	1	1
1	0	1
1	1	0

A non-canalizing function

Boolean functions are a convenient idealization. Gene activities and other biosynthetic processes are more realistically described by Michaelis Menten or cooperative sigmoidal binding curves. These binding curves are typically monotonically increasing or decreasing functions of one or more controlling variables, and bounded due to saturation. The concept of canalization generalizes naturally to these functions (Kauffman, 1970; Newman and Rice, 1971). For simplicity, the Boolean notation will be used below.

Fig. 1 (Page 48)

Bacterial and Phage Genes are Governed by Canalizing Functions

I shall consider the direct control variables for a structural gene in phage or bacteria to be "cis" acting sites within the same chromosome such as promoter, operator and termination sites. These cis regulatory sites are usually directly controlled by diffusible "trans" acting substances such as repressors, inducers, Rho, etc.

The proximate regulating loci for the lac structural genes in *E. coli* are the promoter and the operator (Zubay and Chambers, 1971). The operon structure is POZYA, in which P = promoter, 0 = operator, Z, Y, A are the structural genes and the five loci occur in the order POZYA on the chromosome. Let 'P = 1' represent polymerase bound to promoter, and 'Op = 1' represent operator blocked. Transcription occurs only if P = 1, Op = 0, Table IV.

TABLE IV

P	Op	ZYA
0	0	0
0	1	0
1	0	1
1	1	0

(2)

This is the Not If Boolean function in which both control variables are canalizing, since P = 0 or Op = 1 each alone determines (ZYA) = 0. Note that P = 1 or Op = 0 alone is unable to determine a state of transcription. Only one state of a canalizing control canalizes the behavior of the target.

The lac operator, as described above, can be canalized to Op = 0 by each of its two control variables.

Binding of polymerase at the lac promoter is regulated by polymerase core and sigma factor (Zubay and Chambers, 1971). In addition, like other catabolite repressible operons, cyclic AMP and the catabolite gene activation protein, CAP, are required for activation (De Crombrugghe _et al._, 1971). Binding of repressor at the operator (Zubay and Chambers, 1971) may sterically hinder binding of polymerase to the promoter. All five control variables are canalizing; absence of each of the first four, or presence of bound operator, alone determines polymerase fails to bind to promoter.

The probable sequence of genes of the arabinose operon on _E. coli's_ chromosome is COIBAD (Zubay, Gielow and Englesberg, 1971). B, A, and D are structural genes, 0 is an operator, and I probably functions as a promoter. C product, regulator protein, probably can exist in two forms, P_1 and P_2 which attach respectively to 0 and I. The complex P_1 at 0 inhibits transcription of BAD. P_2 at I is required for transcription. L-arabinose is a specific effector, probably binding to the regulatory protein and stabilizing P_2 over P_1. Activation requires CAP and cyclic AMP. CAP probably interacts with a locus in or near I.

Like lac, the arabinose structural genes are governed according to the canalizing Not If function. Each cis control variable alone can determine transcription fails, Table V.

TABLE V

0	I	BAD
0	0	0
0	1	1
1	0	0
1	1	0

The Ara 0 locus is controlled by L-arabinose and C protein. Let Op = 1 mean the 0 locus is bound. The function, Not If, is canalized by both molecular controls, since C = 0, or L-arbinose = 1 each suffices to determine Op = 0, Table VI.

TABLE VI

C	L-arabinose	Op
0	0	0
0	1	0
1	0	1
1	1	0

The presumptive molecules controlling the I locus are polymerase core, sigma factor, CAP, cyclic AMP, C protein and L-arabinose. All six variables are canalizing, since absence of each determines that polymerase is not bound at I.

Bacteriophage lambda's right operator, 0_r, is regulated by lambda repressor, and some metabolic signal induced by UV irradiation and other stimuli (Ptashne, 1971; Ptashne, 1967). which appears to render the repressor unable to bind 0_r. The function is Not If, canalized by each variable to the state $0_r = 0$.

Lambda's left operator, 0_1, is more complex. Lambda repressor, the product of gene C_1, binds to it and is removed, presumably by the same substance (X) which frees 0_r during lytic induction. However, the product of lambda gene _tof_ represses leftward transcription by binding at 0_1 even in the presence of X (Szybalski _et al_., 1970; Eisen and Ptashne, 1971; Kumar, Calef and Szybalski, 1970). 0_1 is governed by the following function:

TABLE VII

tof	C_1	X	0_1
0	0	0	0
0	0	1	0
0	1	0	1
0	1	1	0
1	0	0	1
1	0	1	1
1	1	0	1
1	1	1	1

TABLE VIII

C_1	X	0_r	tof
0	0	0	1
0	1	0	1
1	0	1	0
1	1	0	1

This function has a canalized state, $0_1 = 1$, and tof is a canalizing control variable, since tof = 1 determines $0_1 = 1$ regardless of the states of C_1 or X. Neither remaining control variable is canalizing because no state of C_1 or X alone can determine $0_1 = 1$.

Processes governed according to canalizing functions are not limited to direct gene activity. In *E. coli*, the arginine operon may be regulated by blocking translation (Vogel, McLellan, Hiroven, and Vogel, 1971). The exact nature of the repressing complex is not yet known, but it is clear that translation block requires the Arg R product as aporepressor, and arginine as coreproessor. Absence of either control molecule vetoes the block. Histidine (Brenner and Ames, 1971) may also involve translational control and appears governed by a canalizing function.

Since canalizing functions become a very small fraction of the possible control rules as the number of controlling variables increases above 3 (Kauffman, 1970), and Table II and Table III consider possible regulated genes which are not controlled by canalizing functions, the assertion that regulated genes are commonly controlled according to canalizing functions is not trivially true.

The examples given were chosen only because data concerning them are clear. Numerous examples can be found in the literature on bacterial and phage gene regulation (Metabolic Pathways, 5: 1971; Umbarger, 1969; The Bacteriophage Lambda, 1971). Difference in molecular processes between bacteria and higher cells weaken generalization based on detailed molecular mechanisms. However, functional properties of control organization may be less dependent on their molecular embodiment. While examples of canalizing functions were drawn from phage and bacteria, the property of canalization depends only on the properties of monotonicity and saturation in the binding curves of ligands controlling activity of catalytic elements. Langmuir and sigmoidal binding curves are common and canalization may occur at many distances from direct gene activity. It appears difficult to find many examples of controlled metabolic and genetic processes which are not regulated according to canalizing functions. For the theory I shall develop, it is sufficient if the majority, but not necessarily all, processes with more than one control variable utilize canalizing functions.

Large Scale Dynamics

The purpose in trying to establish the class of phenomenological control rules governing gene activities is to characterize small scale properties of the organization of cell control systems. Such local properties at best constrain hypotheses about large scale organization to the enormous class of possible large control systems built consistent with the local properties. The implications of these small scale properties for the large scale dynamical behaviors of control systems can be assessed by constructing an ensemble of all the possible large control systems in which each member is built using only the local properties, and asking whether the ensemble possesses "typical", or expected large scale dynamic properties which occur in the vast majority of the ensemble, but which do not occur in systems built without these local constraints.

The examination of regulated bacterial processes above seems to warrant two conclusions: most genes are probably regulated according to canalizing functions; most are directly controlled by rather few, one to six, other processes. The later implies that the mean connectivity of the control system is low.

To assess the implications of these local properties, one would ideally use realistic Michaelis Menten, sigmoidal, or other continuous kinetic equations to express the kinds of canalizing functions found, and explore the behavior of the ensemble of systems built using them. The ideal is not directly approachable, for no adequate techniques exist to examine large systems of such non-linear differential equations. However, the question may be approached by substituting the idealization of a regulated process like gene activity as a binary, on-off device. Very large systems of such binary genes are easily studied by simulation.

To discover the typical dynamical behavior of this ensemble of systems requires examining the behavior of systems sampled at random from that ensemble. This may be done by building control systems in which each model gene is assigned at random one to five or six other genes as its control variables, and each gene is assigned at random one of the possible canalizing functions on its control variables. Once built, any such network is fixed in structure and is a random sample from the ensemble of possible control systems. The dynamical behaviors of members may be compared to typical members of an ensemble of control systems built without using canalized functions. Since I have reported results of such studies elsewhere (Kauffman, 1970; Kauffman, 1969*a*; Kauffman, 1969*b*; Kauffman, 1971; Glass and Kauffman, 1972; Glass and Kauffman, 1978), I will only summarize.

In a network of 10,000 binary genes, each governed by a *non*-canalizing function of many other model genes, the following typically occurs:

1. The net has $2^N = 2^{10,000}$ ($\cong 10^{3000}$) distinct states comprised by each possible combination of gene activities for the N genes. When released from an initial state, the net settles into and cycles repeatedly through a recurrent set of about $2^{N/2} = 2^{5000} \cong 10^{1500}$ states.
2. The net has about $10,000/e \cong 3700$ such recurrent patterns of behaviors. The system must settle into one of these.
3. A minimal perturbation, defined as reversing the state of a single model gene, is almost certain to move the system from its current pattern of behavior to some other dynamic pattern.
4. The various possible minimal perturbations to any one pattern can cause the system to jump from that pattern of behavior directly to very many of the other 3700 behaviors.

In sharp contrast, a model genetic net with 10,000 "genes" in which a majority of about 60% or more have one or more canalizing control variables, typically has the following properties:

1. When released from an initial state, the net settles into and cycles repeatedly in a recurrent pattern through about $\sqrt{N} = \sqrt{10,000} = 100$ states out of its potential $2^{10,000} \cong 10^{3000}$ states. Behavior is thus enormously restricted.
2. When released from any initial state, the system must settle into one of about $\sqrt{N} = \sqrt{10,000} = 100$ such recurrent patterns, each comprised of a distinct set of about 100 states through which the net cycles.
3. For about 90% of all minimal perturbations, the system returns to the recurrent pattern of behavior from which it was perturbed, exhibiting homeostasis.
4. The set of all minimal perturbations can cause the net to jump directly from any one recurrent dynamic pattern to only five or six others of the 100 possible behavior patterns. There is a local topology of neighboring behavior patterns.

Enormously restricted, orderly dynamic behavior occurs in virtually any large net of binary genes built using canalizing functions. If the typical large scale dynamic behaviors of this class of systems parallels some large scale behaviors of metazoan cells, those behaviors may be explicable as consequences of these simple local properties.

Typical large scale properties which occur in almost all ensemble members must be insensitive to details of network construction. The large scale properties of cells which are candidates to parallel such average, structurally insensitive, properties of this ensemble of control systems must be those which depend on general features of control systems, not their detailed architecture. These are likely to be properties which are universal, and occur in all or most organisms, whose diversity of detailed control systems precludes common behaviors depending upon those diverse details. The natural biological observables which arise are therefore unlike the small isolatable fragments of cellular systems capable of reasonably complete description which are usually studied. In early attempts to link small scale properties of cell control systems with their large scale dynamic behaviors, the suggestion is to consider those large scale properties which depend the least on detailed construction, for there hope of explanation with incomplete knowledge is best.

The following large scale properties of cells seem universal and appear to parallel those of the ensemble of control systems built using canalizing functions of few variables:

A. The temporal pattern of gene activities corresponding to one cell type in any organism must be enormously restricted in comparison to the potential combinations of gene activities, to a small number of states, or combinations of genes activities through which the cell "modulates" in its ongoing activity. A vast number of possible gene control systems are incapable of this restriction, in particular, large gene control systems built with <u>non</u>-canalizing functions. However, temporal gene activity patterns, each with enormous restriction to 100 out of 10^{3000} states for a 10,000 gene net, are expected if canalizing functions are used.

B. Any organism possesses a particular number of stable, distinct cell types. The number of cell types in an organism increases with the number of its genes and complexity of its DNA. Such a correlation should reflect general features of control organization in all organisms. Estimates of numbers of cell types and of genes are diffiuclt. Previously (Kauffman, 1969<u>a</u>) I presented evidence that the number of cell types in an organism increases as roughly a square root function of the haploid DNA content of its cells, from about 1 or 2 cell types for bacteria, to roughly 10^2 for man. Were I to have overestimated the number of genes in higher cells, by not allowing for redundant DNA, by an order of magnitude, it would still appear that the number of cell types increases as the number of genes to a <u>fractional</u> power. A stable distinct cell type can be interpreted as a distinct, steady or perhaps cyclic pattern of behavior into which an entire cellular control system settles. If the gene system has 100 distinct patterns of gene activity, then each corresponds to the ongoing activity of one cell type; for example, one activity pattern corresponds to cardiac muscle, another distinct pattern to lymphocyte, etc. It is easy to build control systems in which the number of cell types increases as the number of genes to a power greater than 1, or even 2. Constraint to the class of canalizing functions yields an ensemble of control systems in which the number of alternate recurrent behavior patterns, or cell types, increases roughly as a square root function of the number of model genes.

C. Cell types and model cell types exhibit homeostasis, remaining the same cell type in the face of a wide variety of perturbations. This property does not occur if non-canalizing functions are used.

D. In virtually all developing systems, no cell type differentiates directly into more than two to five or six other cell types, although it may indirectly develop into many by repeated branching differentiation. Control systems using non-canalizing functions are able to pass from any cell type to a large number of other cell types with minimal perturbations or signals. By contrast, in control systems using canalizing functions, typically one cell type can be induced by small perturbations or signals to differentiate to only a few neighboring cell types, although most perturbations leave it the same cell type.

In asserting that these dynamic properties of randomly constructed model systems parallel those of cells, I am not claiming that control systems evolved over 2 billion years are random. Rather, with respect to these particular global dynamic properties, cell control systems may be typical of the ensemble studied by random sampling from that ensemble.

Although the similarities in behavior between control systems using binary switching variables and homologous systems using realistic continuous equations are not straightforward, there are indications (Glass and Kauffman, 1973; Kauffman, 1970; Newman and Rice, 1971) that the highly orderly dynamic behaviors of the ensemble of control systems studied using Boolean canalizing functions also occur when continuous canalizing functions are used.

The Design of Large Control Circuits: Extended Forcing Structures

The ensemble of control systems built using currently known small scale properties is a conceptual tool which allows both study of the implications of those properties for large scale behavior, and, by examination of average properties of the "wiring diagrams" of ensemble members, yields hypotheses about the most probable kinds of intermediate sized circuits to expect as consequences of the known local properties. Restriction to use of canalizing function with few controlling variables makes probable the existence of simple and powerful intermediate sized circuits, which I call extended forcing structures.

The simplest forcing structure to picture is a familiar model which consists in many hierarchically arranged genes, each controlled by several others and activated if any one of its controlling variables is active. If any gene is active, that alone suffices to activate all the genes it directly controls whether other variables controlling those genes are active or inactive. The active state of a gene propagates directly or indirectly to all members of the hierarchy below it. The inactive state of a gene is not guaranteed to propagate. The hierarchal batteries of genes proposed by Britten and Davidson (Britten and Davidson, 1969; Britten and Davidson, 1971)[3] are this kind of forcing structure.

Although such positive control cascade derepression circuits have not yet been found, they are members of a more general class of control systems using positive and negative control of which examples are known. The important properties of the cascade derepression hierarchy are: (1) that each gene is regulated according to a canalizing function,(in this case each gene is activated if any one gene controlling it is active); (2) the canalized state of each gene,(in this case it is the active state which can be determined by one control variable alone), is also the state of that gene which canalizes the genes it in turn directly controls, regardless of the states of other variables controlling those genes. Taken together, these properties assure that the canalizing state,(here the active state), propagates in the hierarchy, but the non-canalizing (here, inactive) state may not propagate.

These properties define a transitive relation between two regulated processes like gene activity I term forcing: Process A forces Process B if: (1) Process A is canalized by one or more control variables; (2) Process A is a canalizing control variable of Process B; (3) the canalized state of A is the state of A which canalizes B. For example, the lac promoter has five canalizing controlling variables which canalize it to P = 0. Lac promoter is a canalizing control variable of the structural genes, and P = 0 is the state of P which determines that transcription fails. Therefore, the lac promoter forces the lac structural genes P = 0 forces (ZYA) = 0.

Two processes canalized by any or all control variables can also be connected so neither forces the other. The lac operator is itself canalized by lactose or repressor to the state Op = 0. The operator canalizes the structural genes, since Op = 1 determines transcription fails. However, the canalized state of Op is 0, not 1. Since the canalized state of Op is not the state which canalizes the structural genes, the operator does not force the structural genes.

By definition, the relation 'Process A forces Process B' is transitive, so that if B also forces C, then A forces C indirectly through B. In this way, extended forcing structures may be constructed. The forcing structures may or may not contain forcing loops.

The following are the major characteristics of extended forcing structures:

(Figure 1).

1. A canalizing state introduced to any element in the structure canalizes all descendent elements, since the canalized state propagates deterministically to all descendent members regardless of the states of any other control variables. A non-canalized state is not guaranteed to propagate. In Figure 1, if A is in its canalizing state, 1, it determines that B goes to state 0, whatever the state of B's other controlling variable, L. B = 0 in turn determines that C goes to 0 whatever the state of M, C's other controlling variable. C = 0 determines D = 1 and E = 1. E = 1 forces A = 1, completing the forcing loop. If A is in its non-canalizing state, 0, it is unable by itself to determine the next state of B. The non-canalizing state is not guaranteed to propagate in the structure.
2. The canalized state may differ at different points in the structure. A = 1 forces B = 0, etc.
3. The canalized state of a process is identical for all its canalizing control variables, therefore extended forcing structures can have redundancy of control.
4. Any forcing loop is a positive feedback loop with a maximum of two steady states. Any loop has a stable steady state when each element is in its canalized state. In that state, the loop is insensitive to all external regulatory events. Some forcing loops also have a metastable steady state with each element in its non-canalized state. In that state, the loop is sensitive to external control variables. In Figure 1, A = 1, B= 0, C = 0, E = 1 is the insensitive forced state. No states of X, L, M, or Q, additional variables controlling loop members, can alter the forced state of the loop; A = 0, B = 1, C = 1, E = 0 is the complementary

FORCING STRUCTURE

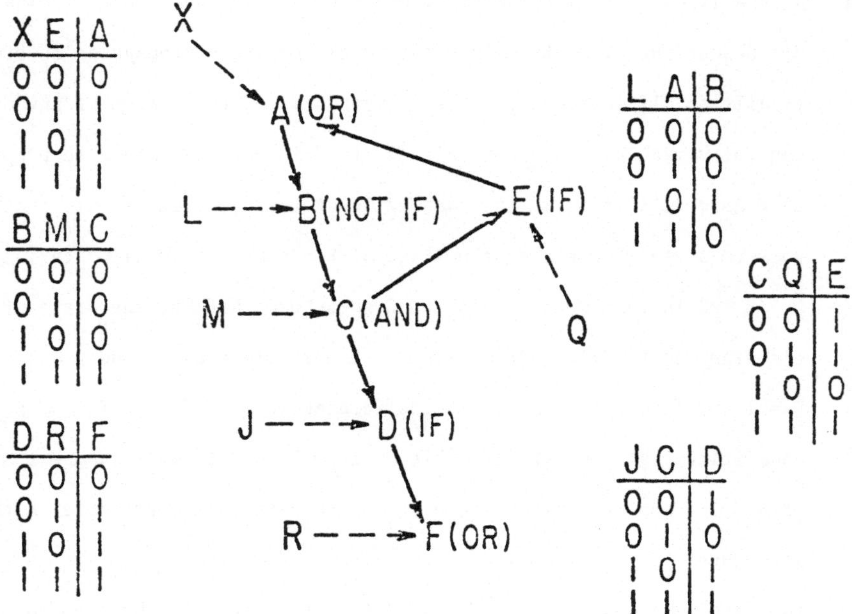

STABLE STEADY STATE

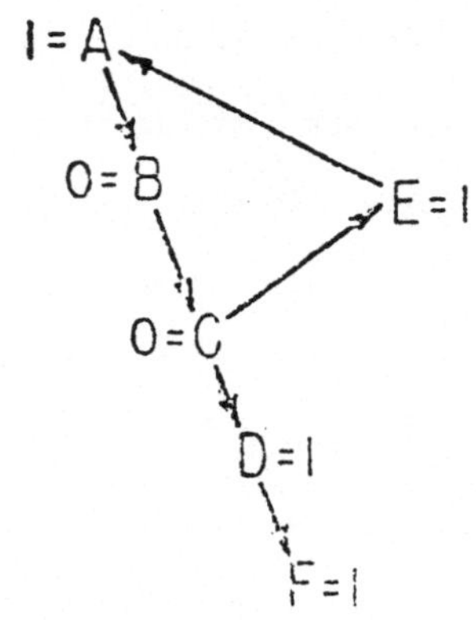

Fig 1

sensitive metastable state, which can be altered by appropriate states of X, L, M, or Q.

5. If a forcing loop is fixed in its forced steady state, all members of its descendent forcing structure become fixed in their forced states.
6. Distinct forcing structures may be coupled to one another through a variety of non-forcing control connections allowing subtle and flexible control relations between forcing structures. One forcing structure is sensitive to another only through those of its components which are not in their canalized state.

The power of forcing structures, in which any one gene has one state capable of determining the subsequent states of all descendent members of the structures, are to be contrasted with those obtained if non-canalizing functions are used. By definition, in a non-canalizing function of two or more variables, no single controlling variable alone can determine any state of the target process. To determine one state of the target requires specifying the states of at least two controlling variables; to specify that each of those two is in its correct state requires specifying for each the states of at least two of its controlling variables, and so on. For example, suppose every gene in Figure 1 realized the non-canalized _Exclusive Or_ function (Table III). Then no state of any single variable alone could determine any subsequent behavior in the structure. Further, the loop ABCE possesses no steady state insensitive to external variable.

These behaviors of forcing structures and non-forcing circuits are independent of the fact that each process in such a structure may itself be a nearly irreversible. Repressor is removed by the binding of allo-lactose, which weakens the repressor operator bond. Such loci could be regulated, as is the lac operator, according to a canalizing function, or a non-canalizing function like _Exclusive Or_, shown in Table III.

The occurrence of extended forcing structures is not merely a hypothesis. An analysis of the regulatory system in bacteriophage lambda has shown such a system (Kauffman, 1974), which will not be described further here.

Statistical Features of Gene Control Systems Based on Ensemble Averages

A major use of the ensemble approach is to try to estimate the major features of the functional organization of metazoan regulatory systems which would occur if cells were typical members of the ensemble built using a specified set of local properties. The two local properties of low connectivity and use of canalizing functions does in fact predict major organizational features. Imagine the control system in a metazoan, with its 10^4 or so loci, imagine each regulatory input from locus A to B as a black arrow from A to B. Some subset of these regulatory connections are forcing connections. Color each forcing connection as a red arrow. Then the genomic system consists in a very complex graph of points connected by black or red arrows. The connected sets of red arrows are all the extended forcing structures in the network. Using percolation theory, it is possible to predict the character of these extended forcing structures as a function of the number of red forcing connections present in the net. Suppose there are N loci which regulate one another, and M of the regulatory connections are red forcing connections. Let M red arrows be assigned at random to pairs of points representing the N regulated loci. If M is much less than N, then the vast majority of loci will not be connected by red arrows, but only by black ones. Occasionally, a pair will be connected by a red arrow. As M increases toward N, at first small two and three loci structures connected by red arrows will form, then rather longer connected sequences of 4, 5, or 6 loci will form, along with red forcing loops having a few members. As M increases about to N, when $M \approx N$, a kind of crystallization occurs in which the majority of loci are joined into an enormous graph of points connected by red arrows which ramifies throughout the entire system. Isolated pockets of loci remain which are not yet part of the now dominant single connected forcing structure. These pockets are largely isolated from one another in the sense that in any pocket rather few loci are connected only by black arrows representing non-forcing connections, while inputs from outside the pocket come from loci which are themselves members of the extended forcing structure. If the total number of regulatory inputs per locus, K, is small, with mean 2, reflecting overall low connectivity in the net, then for any member of such a pocket, some of its inputs derive from outside the pocket, and only 1 or more rarely 2 will derive from other members of the pocket of loci itself (Kauffman, 1970).

The behavior of such a system is easily understood. The ramified red forcing structure, which encompasses the majority of elements in the net, will fall to its overall forced state at each loci. This result derives from a simple theorem. If two forcing loops having numbers of loci which are prime with respect to one another, P_1, P_2, e.g. 3 and 4, are joined by a common component, then if any locus is ever in its forced value, the two loops will fall to the forced state in a maximum of $P_1 \cdot P_2$ time moments (Ibid). An enormous forcing structure will typically contain many loops which are relatively prime, and the entire extended structure will fall to its forced state. This is seen in simulations of such nets, for after starting at an arbitrary initial state, the number of loci which continue to change values dwindles roughly exponentially during the approach to a limit cycle until about .70 of the N elements are fixed either in state 1 or 0. The same .70 N elements are fixed in the identical 1 or 0 states in all state cycles of the system. The (Kauffman, 1969) different state cycles of the net differ, therefore, due to the behaviors of those elements in rather isolated pockets which are not members of the ramified forcing structure. These pockets are isolated in the sense that although members may have direct and indirect control connections from outside the pocket, they are isolated by the constancy of the forcing structure, in its fixed state, such that the further dynamical behavior of each pocket is determined internally. Thus, the dominant functional consequences of an enorous ramified forcing structure in a net with thousands of loci are three: 1. Many elements, which are members of that forcing structure, fall to a fixed state and return stably to that fixed state if perturbed. 2. The consequent "walls of constancy" provided by the forcing structure, isolate pockets of loci from one another which are then essentially functionally independent. The further dynamics reflects the behavior of a number of functionally independent subsystems. 3. Since many, and on average, most of the control inputs to members of one pocket come from members of the forcing structure, the average functional connectivity within a pocket is less than K, the average connectivity for the whole net. If the mean value of K is 2, for example, and .70 N elements are members of the forcing structure, then the average connectivity within a pocket is about 1.3, rather than 2, since .7L of the 2L inputs to the L members of the pocket come from the forcing structure. Therefore, for K about 2, the isolated pockets consist largely of control subsystems which are either one input loops and tails, or sparsely cross connected sets of such loop and tail structures.

The behavior of single input loops of boolean elements is simple to undertand. Any element of such a loop either affirms or negates its single input. If the number of loci which negate their inputs is even, then the loop has two alternate steady states. For example, if no element negates its input, then the loop has two steady states, with all elements 1, or all 0. In addition, in a precisely synchronous net, each such loop has a number of modes of oscillation. For example, if no element negates its input, a six element loop can have a single 1 propagating around the loop, two 1 values separated by various numbers of elements around the loop, 3 1 values, etc. However, the two steady states are absorbing states. In simulations of such 1 input loops where the net is no longer synchronous, (R. Thomas, personal communication) but a given element may remain in the 1 or 0 state for an interval of time chosen from some distribution, one input loops with an even number of negative interactions ("positive loops") typically fall to one or the other steady state. Thus, if the idealization of synchrony is relaxed, one input positive loops typically are subsystems with only two alternate asymptotic states. For a large variety of choices of sigmoidal and Michaelis Menton kinetics, it is also true that such loops typically fall to one or another of two alternative steady states, (Glass and Kauffman, 1972, 1973).

Single input loops in synchronous nets of binary elements, in which an odd number of elements negate their inputs, oscillate with a maximum period of 2 times the number of elements in the loop. Loss of synchronly does not alter the occurrence of oscillating behavior. In continuous analogues to the binary variable system, the capacity of such negative feedback loops to oscillate depends upon both the number of elements in the loop, and the precise character of the continuous kinetics. For example, if only one element is characterized by sigmoidal kinetics, and the rest by Michaelis Menten kinetics, then there is a reciprocal relation between the number of components in the loop and the Hill coefficient of the sigmoidal component needed to achieve oscillations. If the loop is too short, or N too small, the negative feedback loop has a single stable steady state.

These results, therefore, suggest that in a large system of coupled processes regulated by Michaelis Menten and sigmoidal type kinetics, the net will typically contain an enormous ramified forcing structure, with isolated pockets of functionally independent subsystems. Many such subsystems will typically have either two alternative steady states; will have a single steady state or will oscillate. Other such subsystems will be more complex than isolated 1 input loops and dependent 1 input tails, and will have some components having two or more input variables which are left free by the forcing structure to alter their states.

Functional Implications

These properties represent a first guess at the functional architecture of complex genetic systems based on the implications of two small scale properties: low connectivity and use of canalizing functions. Nevertheless, the expected large scale organization predicted by these alone already begins to suggest obvious functional correlates. The diverse cell types of any organism must maintain the activity of a very large number of genes in order to accomplish "housekeeping" functions common to all cells. The active elements of the ramified forcing structure provide an easy way to accomplish this major task, since that state of each locus is active in all "cell types" of the model genome. Any cell type is stable to most perturbations. This property is exemplified in the typical members of the ensemble which was examined. The stability results both from the stability of the fixed forced state of the ramified forcing structure, and the stability of the steady states or oscillation patterns of the relative isolated subsystems.

Analysis of Drosophila development, among other systems, has begun to suggest that developmental commitments can be viewed as combinatorial. As described in the previous chapter, the determined state of each distinct imaginal disc appears as if it might be characterized by a unique combination of binary alternative states of several independent genetic subsystems, such that wing is (1111), leg (1010), etc. Here each 1 or 0 represents one or another alternate stable state of a gene subsystem of unknown complexity. The success of that model suggests that these subsystems behave rather independently. Typical members of the ensemble of large gene systems which were analyzed have pockets of functionally isolated 1 input loops which typically have two alternate steady states. The differential action of these loops accounts in large part, for the different model "cell types" in such a genome. Such pockets provide one way a gene system could achieve subsystems which behave with relative independence and have only two alternative steady states.

The isolated pockets of loci with alternative states are therefore the nodes which carry differentiation in typical membersof this ensemble. During the course of normal development, the commitments taken in any area of the embryo must be coordinated with those taken at other locations and stages of development. This, in turn requires that the subsystems which carry commitments cannot be fully isolated at least during the laying down of commitments, although once a commitment is taken, the subsystem maintaining that commitment might become functionally isolated from the coordinated processes which trigger the initial commitment. In this respect, the <u>inactive</u> members of the ramified forcing structure are of particular interest. These elements are inactive on all state cycles, that is, on all model cell types. They therefore only play a role in the dynamical behavior of the system during transients leading to assymptotic state cycles. In particular, when members of the ramified forcing structure are <u>not</u> in their fixed forced value, then it is both the case that they themselves are functionally varying inputs to members of the isolated pockets which are only isolated by the <u>fixed</u> state of the forcing structure, and also true that cross talk between different pockets can occur as well. Thus, such inactive members of the forcing structures forced state, are well suited to serve as critical control variables transiently affecting the isolated pockets, or allowing interaction betweeen those pockets, during the course of commitment. After commitment is taken, such pockets carrying commitments can become functionally isolated from their triggering stimuli.

Inevitably, such functional assignments are speculation. They are, nevertheless, speculation informed by a reasonable guess at the overall functional architecture of a metazoan genomic system.

<u>Conclusion</u>

There are only two ways we shall come to explain large scale properties of cells. Either all properties of interest will be consequences of circuitry sufficiently simple for detailed characterization, or we shall have to use incomplete knowledge and provide statistical explanations. The ensembles of control systems constructed using any currently known small scale properties of the organization should prove useful tools both to predict the organization of large circuits, and to explain large scale behaviors in the face of incomplete knowledge.

Although the use of ensembles of systems and their average properties has been successful in statistical mechanics, the procedures I am suggesting are less familiar in cell biology. The notion that a large scale property of cell control systems, such as the number of cell types into which any one cell type can directly differentiate, is an observable deserving explanation seems premature, precisely because it is probably not explicable from the properties of any small number of genes or processes. Further, the explanation offered seems uncertain. To assert that any single large scale behavior is a consequence of cells being typical members of some ensemble of systems may be false, since cells might be atypical members of the ensemble and the large scale property might be due to other mechanisms. However, reasonably strong evidence that cells are typical members of an ensemble would be provided if many distinct large scale properties of cells simultaneously parallel typical properties of ensemble members. The observed restriction of regulated genes and processes to canalizing functions of few controlling variables is an excessively simple small scale property which already may predict simultaneously four or five large scale dynamic behaviors of cells, as well as the existence of extended forcing structures and functionally isolated subsystems. Extensions of this early theory consist in discovering additional large scale properties of cells which are insensitive to details of construction and additional small scale properties through which to explain them. The possibility of an adequate theory of large scale properties of cells as local properties are better understood seems good, even should it prove impossible to characterize control systems completely.

FOOTNOTES

1. In previous publications I called "canalizing" functions "forcible" functions.
2. The input state with promoter and operator simultaneously bound may be sterically impossible (Zubay and Chambers, 1971). If state 0 is assigned to transcription for that forbidden input state, the function is Not If, with two canalizing variables; if state 1 is assigned, transcription becomes a trivially canalizing function of the promoter alone.
3. The control systems suggested by Britten and Davidson (1969) are a specific type of extended forcing structure. They make the interesting suggestion that redundant DNA serves as cis regulatory sites for structural genes, and also as templates for synthesis of diffusible regulatory molecules. Repetitive sequences give multiplexing of regulatory variables. All genes are presumed to be non-specifically repressed by histones, and to be activated by any specific input. The hierarchical batteries of genes created are extended forcing structures, for every gene realizes the canalizing OR function on all specific control variables, and the canalized state of each gene--active--canalizes its targets. In their model all specific connections are forcing. This property is false in bacteria and phage. Lac operator does not force lac structural genes. There is no reason, however, why their concept of hierarchial batteries cannot be widened to the more general notion of extended forcing structures, for which redundant DNA might provide some of the regulatory circuitry.

REFERENCES

1. Asburner, M. (1971). Nature New Biology 230, 222.
2. "The Bacteriophage Lambda". (1971). (A. D. Hershey, Ed.) Cold Spring Harbor Laboratory.
3. Berendes, H. D. (1965). Developmental Biology 11, 371.
4. Brenner, M. and Ames, B. N. (1971). In "Metabolic Pathways" 5 (H. J. Vogel, Ed.) p. 349. Academi Press, New York.
5. Bretscher, M. S. (1967). Nature 217, 509.
6. Britten, R. J. and Davidson, E. H. (1969). Science 165, 349.
7. Britten, R. J. and Davidson, E. H. (1971). J. Theoret. Biol. 32, 123.
8. Burstein, C., Cohn, M., Kepes, A. and Monod, J. (1965). Biochim. Biophys. Acta 95, 634.
9. Clever, U. (1966). Developmental Biology 14, 421.
10. De Crombrugghe, B., Chen, B., Anderson, W., Nissley, P., Gottesman, M., and Pastan, I. (1971). Nature New Biology 231, 139.
11. Echols, H. (1971). In "The Bacteriophage Lambda". (A. D. Hershey, Ed.) p. 247, Cold Spring Harbor Laboratory.
12. Eisen, H. and Ptashne, M. (1971). In "The Bacteriophage Lambda". (A. D. Hershey, Ed.) p. 239. Cold Spring Harbor Laboratory.
13. Gehring, W., Mindek, G. and Hadorn, E. (1968). J. Embryol. Morph. 20, 307.
14. Gehring, W. (1967). Developmental Biology 16, 438.
15. Glass, L. and Kauffman, S. A. (1972). J. Theoret. Biol. 34, 219.
16. Glass, L. and Kauffman, S. A. (1973). J. Theoret. Biol. 39, 103.
17. Hadorn, E. (1977). Developmental Biology 31, 424.
18. Heinemann, S. F. and Spiegleman, W. G. (1970). Proc. Nat. Acad. Sci. U.S.A. 67, 1122.
19. Jacob, F. and Monod, J. (1963). "21st Symp. Soc. Study of Development and Growth". Academic Press, London.

20. Kauffman, S. A. (1969a). J. Theoret. Biol. 22, 437.

21. Kauffman, S. A. (1969b). Nature 224, 177.

22. Kauffman, S. A. (1971). In "Current Topics in Developmental Biology" 6, p. 145. Academic Press, New York.

23. Kauffman, S. A. (1973). Science 181, 310.

24. Kauffman, S. K. (1974). J. Theoret. Biol. 44, 167.

25. Kumar, S., Calef, E. and Szybalski, W. (1970). Cold Spring Harbor Symp. Quant. Biol. 35, 331.

26. "Metabolic Pathways" 5. (1971). (H. J. Vogel, Ed.) Academic Press, New York.

27. Muller-Hill, B., Rikenberg, H. V. and Wallenfels, K. (1964). J. Mol. Biol. 10, 303.

28. Newman S. and Rice, S. (1971). Proc. Nat. Acad. Sci. U.S.A. 68, 92.

29. Neubauer, Z. and Calef. E. (1970). J. Mol. Biol. 51, 1.

30. Ohno, S. (1971). Nature 234, 134.

31. Ptashne, M. (1971). In "The Bacteriophage Lambda". (A. D. Hershey, Ed.) Cold Spring Harbor Laboratory, p. 221.

32. Ptashne, M. (1967). Nature 214, 232.

33. Reichardt, L. and Kaiser, A. D. (1971). Proc. Nat. Acad. Sci. U.S.A. 68, 2185.

34. Riggs, A. D., Newby, R. F. and Bourgeois, S. J. (1970). J. Mol. Biol. 51, 303.

35. Roberts, J. W. (1969). Nature 224, 1168.

36. Szybalski, W., Bovre, K., Fiandt, M., Hayes, S., Hradecna, Z., Kumar, S., Lozeron, H.A., Nijkamp, H. J. J. and Stevens, W. F. (1970). Cold Spring Harbor Symp. Quant. Biol. 35, 341.

37. Thomas, R. (1971). In "The Bacteriophage Lambda". (A. D. Hershey, Ed.) Cold Spring Harbor Laboratory. p. 211.

38. Tobler, H. (1966). J. Embryol. Exp. Morph. 16, 609.

39. Umbarger, H. E. (1969). Annual Review of Biochemistry 38, 323.

40. Vogel, R., McLellan, W., Hiroven, A. and Vogel, H. (1971). In "Metabolic Pathways" 5. (H. J. Vogel, Ed.) p. 463. Academic Press, New York.

41. Zubay, G. and Chambers, D. A. (1971). "Regulating the lac operaon" In "Metabolic Pathways" 5 (H. J. Vogel, Ed.) Academic Press, New York.

42. Zubay, G., Gielow, L. and Englasberg, E. (1971). Nature New Biology 233, 164.

PART II

Naive boolean analysis

CHAPTER IV.

Combinational systems.

A. LEUSSLER & P. VAN HAM.+

IV.1. Introduction.

Boolean (or logical) algebras use variables which can have only a limited number of values. The purpose of this chapter is to give scientists involved in systems modelling the background needed to use the boolean approach in their own field of research. For rigorous and complete mathematical developments the reader should refer to the bibliography. In this book, we use the term "boolean" in its limited, but most common sense, meaning "binary". In binary algebra, the variables can have only two values, 0 and 1 , or F (false) and T (true), or L (low) and H (high).

"That man is speaking", "that relay is off", "that substance is present at a concentration exceeding a given threshold"... one can associate with each of these statements a binary variable which has the value 1 if the statement is true, 0 if not. Boolean functions can be used to describe the relations between boolean variables. For instance, in an electronic circuit comprising on-off switches (inputs) and lamps (outputs) which lit or not depending on the state of the switches, one can associate a boolean variable with each switch and a boolean function with each lamp.

Consider a circuit having n binary inputs : $x_1, x_2, \ldots, x_n$ and m outputs $Z_1, Z_2, \ldots, Z_m$ (fig.1.) . A circuit will be called a combinational circuit if the outputs can be expressed as m boolean functions of the input variables $x_1, x_2, \ldots, x_n$, that is :

$$Z_1 = f_1 (x_1, x_2, \ldots, x_n)$$
$$Z_2 = f_2 (x_1, x_2, \ldots, x_n)$$
$$\vdots$$
$$Z_m = f_m (x_1, x_2, \ldots, x_n)$$

This means that, in these circuits, for each input combination there is a unique and well-defined output combination.

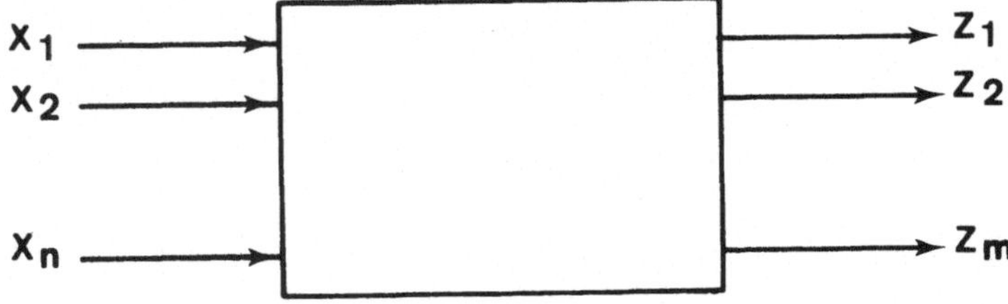

Fig.IV.1. Combinational Circuit.

+ Service des systèmes logiques et numériques, Faculté des Sciences Appliquées, Université libre de Bruxelles.

IV.2. Basic binary operations.

Boolean variables can be subjected to various operations, one of which, the NOT operation, can apply to a single input variable, while the other operations apply to two or more input variables.

As shown in table 1, the function

Z = NOT a, denoted $\bar{a}$, is the logical complement. It has the value 1 when a has the value 0 , and the value 0 when a has the value 1 .

Z = a OR b, denoted a + b , is the inclusive or, also called logical sum. It has the value 1 iff (that is, if, and only if) a or b or both have the value 1* .

Z = a AND b, denoted a.b, or simply ab , is the logical product. It has the value 1 iff a and b both have the value 1.

Other operations are NOR, NAND and EXCLUSIVE OR :

Z = NOR (a,b) = NOT (aORb), denoted $\overline{a + b}$, is the logical complement of a sum. It has the value 1 when a and b have the value 0 .

Z = NAND (a,b) = NOT (aANDb), denoted $\overline{ab}$, is the logical complement of a product. It has the value 1 when a or b or both have the value 0 .

Z = a EXCLUSIVE OR b, denoted $a \oplus b$, also called modulo 2-sum, has the value 1 when either a or b but not both have the value 1 . For n-variables operations, the output of the exclusive or is 1 iff an odd number of inputs is equal to 1 .

	NOT			OR	AND	NOR	NAND	EXCLUSIVE OR
		Symbol used in electronic schemes						
a	$\bar{a}$	a	b	a+b	a.b	$\overline{a+b}$	$\overline{a.b}$	$a \oplus b$
0	1	0	0	0	0	1	1	0
1	0	0	1	1	0	0	1	1
		1	1	1	1	0	0	0
		1	0	1	0	0	1	1

Table 1 : Summary of the basic boolean operations.

* The logical sum is equivalent in the set theory to the union ($\vee$ or $\cup$)and the logical product to the conjunction ($\wedge$ or $\cap$).

Any boolean function can be written in terms of the single operation NOR or in terms of the single operation NAND, which for this reason are called universal operations. Any boolean function can also be implemented with basic sets of binary operations : EXCLUSIVE OR, AND ; EXCLUSIVE OR, OR ; NOT, OR ; NOT, AND .

However, in this book, we will write the boolean functions with the three operations :

NOT, AND, OR .

Note that, for example, a (b+c) is read "a AND (b OR c)" rather than "a times (b plus c)".

IV.3. Implementation of a boolean function.

We would like to show now how a situation, initially described in verbal terms, can be formalized by means of boolean expressions.

Let us suppose that I have to go out and do not want to get wet should it rain. Among the various possibilities, I consider only whether it rains and whether I have my umbrella with me. (I assume that, if I have it with me, I shall use it when necessary). In order to formalize this little problem, we introduce two boolean variables (considered as inputs).

a=1 if it rains

a=0 if not

b=1 if I have my umbrella

b=0 if I do not have my umbrella

and a function F (considered as output)

F=1 if I do not get wet

F=0 if I do

There are four combinations of values of the two variables :

a=1 , b=1 , also written ab=1, or simply ab, or 11 : it rains and I have my umbrella

a=0 , b=1 , also written $\bar{a}b$=1, or $\bar{a}b$, or 01 : it does not rain and I have my umbrella

a=0 , b=0 , also written $\bar{a}\bar{b}$=1, or $\bar{a}\bar{b}$, or 00 : it does not rain and I do not have my umbrella

a=1 , b=0 , also written $a\bar{b}$=1, or $a\bar{b}$, or 10 : it rains and I do not have my umbrella

Clearly, I shall not get wet in the first three cases, but I shall get wet in the last case. Thus, $F = 1$ iff ab OR $\bar{a}b$ OR $\bar{a}\bar{b}$, that is, $F = ab + \bar{a}b + \bar{a}\bar{b}$

(and $F = 0$ iff $a\bar{b}$; thus $\overline{F} = a\bar{b}$).

In this very simple case, it is immediately obvious that the conditions for not getting wet can be expressed in a simpler way ; I shall not get wet if it does not rain ($\bar{a}$) OR if I have my umbrella (b) :

$$F = \bar{a} + b$$

Note that in this expression, $\bar{a}$ means " $\bar{a}$ whatever the value of variable b", and b means "b whatever the value of a".

For more complex situations common sense is insufficient for finding the simplest expression (which is in general not unique). We shall therefore consider below how a boolean expression can be simplified systematically. Among various possibilities, we will almost always use here Karnaugh (or Veitch) maps (see IV.5.).

IV.4. Simplification rules in Boolean Algebra.

The interested reader will find in the appendix of this chapter a number of relations used in Boolean Algebra.

Some relations used in Boolean algebra are apparently identical with relations used in classical algebra (although the symbols + and . have different meanings). For instance :

$a + 0 = a$ means that the expression a OR zero has the same value as a (1 if $a = 1$, 0 if $a = 0$).

$a \cdot 0 = 0$

$a \cdot 1 = a$

The laws of commutativity, associativity, and distributivity are also expressed in the same way in boolean and in classical algebra (see the appendix).

Other relations are clearly different from those used in classical algebra. For instance :

$a + 1 = 1$, means that the expression a OR one has the value 1 whatever the value of a.

$a + a = a$

$a \cdot a = a$

etc...

Finally, the operation NOT has properties with no apparent correspondance in classical algebra . For instance $(\bar{\bar{a}}) = a$, which means that NOT (NOT a) has the same logical value as a .

$a + \bar{a} = 1$

$a \cdot \bar{a} = 0$

Other expressions include the very important "de Morgan laws" :

$\overline{a + b + c} + \ldots = \bar{a} \cdot \bar{b} \cdot \bar{c} \ldots$

$\overline{a \cdot b \cdot c} \cdot \ldots = \bar{a} + \bar{b} + \bar{c} \ldots$

A proof of the de Morgan laws (in the case of two variables) is that for each combination of values of the variables a and b , the values of the expressions $\overline{a + b}$ and $\bar{a} \cdot \bar{b}$ are the same (and similarly for $\overline{ab}$ and $\bar{a} + \bar{b}$ (Table 2)).

a	b	$\bar{a}$	$\bar{b}$	a+b	$\overline{a+b}$	$\bar{a}.\bar{b}$	a.b	$\overline{a.b}$	$\bar{a}+\bar{b}$
0	0	1	1	0	1	1	0	1	1
0	1	1	0	1	0	0	0	1	1
1	1	0	0	1	0	0	1	0	0
1	0	0	1	1	0	0	0	1	1

Table 2 : Demonstration of the de Morgan lows in the case of two variables.

IV.5. Different representations of boolean functions.

Different methods for presenting input-output requirements are used. Truth tables and Karnaugh (or Veitch) maps are described below.

IV.5.1. Truth tables.

In the preceeding section we have given a verification of the de Morgan laws. We used a table in which the two columns to the left provide the input states, that is, the combinations of values of the variables ; in the present case, there are only two variables (a and b), and there are thus $2^2 = 4$ combinations of values of the variables (00, 01, 11, 10) ; for n variables, there would be 2^n such combinations, and thus 2^n rows in the table. The other columns, which constitute the table proper, correspond each to a function ($\bar{a}$, $\bar{b}$, a + b , etc ...) , whose value is given for each input state. This is a truth table .
The umbrella problem can also be represented by a truth table (Table 3) :

a b	F
0 0	1
0 1	1
1 1	1
1 0	0

Table 3. The truth table for the umbrella problem (see § IV .3) .

As already shown, function F has the value 1 for three out of the 4 input states and it can be written as the sum of these terms.

IV.5.2. Karnaugh or Veitch maps.

A Karnaugh or Veitch map is simply a reorganized configuration of the truth table (see table 4a). A Karnaugh map for a function with two variables (the umbrella example) is shown in table 4b.

	0	1	a
0			
1			
b			

4 a)

F	0	1	a
0	1	0	
1	1	1	
b			

4 b)

Table 4 a) Structure of a Karnaugh map with two variables.

Columns 0 and 1 correspond to the two values of variable a , rows 0 and 1 correspond to the two values of variable b . The four empty squares thus correspond to the four combinations of values of variables a and b . Each square is filled with a 0 or a 1 depending on the function (there are thus $2^4 = 16$ different functions of two binary variables).

b) The Karnaugh map for the umbrella example (§ IV.3) in which $F = \bar{a} + b$

We can extend the Karnaugh map to a greater number of inputs, as shown in table 5.

Table 5a) Karnaugh map for a function F_2 with three input variables ;

$F_2 = a\bar{b} + \bar{c}$

F_2	00	01	11	10	a b
0	1	1	1	1	
1	0	0	0	1	
c					

Table 5b) Karnaugh map for a function F_3 with four input variables ;

$F_3 = ab\bar{c} + \bar{b}c\bar{d} + ad$

F_3	00	01	11	10	a b
00	0	0	1	0	
01	0	0	1	1	
11	0	0	1	1	
10	1	0	0	1	
cd					

Table 5 c) Karnaugh map for a function F_4 of five variables ;

$F_4 = ab\bar{e} + \bar{b}c + b\bar{c}\bar{d}$

F_4	e = 0				e = 1				
	00	01	11	10	00	01	11	10	a b
00	0	1	1	0	0	1	1	0	
01	0	0	1	0	0	0	0	0	
11	1	0	1	1	1	0	0	1	
10	1	0	1	1	1	0	0	1	
cd									

We can see that 2 variables lead to $2^2 = 4$ squares, 3 variables to $2^3 = 8$ squares, 4 variables to $2^4 = 16$ squares, etc : each time another input variable is added, the surface of the Karnaugh map doubles.

Notice that the entries are not written in the sequence corresponding to the increasing binary numbers (00, 01, 10, 11). Rather, we use the sequence 00, 01, 11, 10, so that translation from one row (or column) to the next modifies the value of a single variable. Two columns or rows which differ by the value of a single variable are called adjacent. Note that the first and last rows (columns) are also adjacent, as if the table were turned into a closed surface like a tore (fig.2) (see for instance Mc Cluskey, 1962, 1965 ; Florine, 1964, 1969).

Fig.IV.2. Adjacency in a Karnaugh map.

This sequence (also called a Gray code) is very useful for simplifying boolean functions, as will become apparent below. Table 5c in fact consists of two sub tables, both using the Gray code. Alternately, one could have used a single table of eight columns, corresponding to the combinations of values of variables a, b and e in the Gray sequence.

000, 001, 011, 010, 110, 111, 101, 100

IV.5.3. Cubical representation.

Another geometric representation of a boolean function is obtained by mapping a boolean function of n variables on a n-dimensional unit cube (see for instance Miller, 1965 ; Kuntzmann, 1965).

1 variable corresponds to a 1-cube, which is a line with 2 vertices (0,1) as shown in fig.3a ; 2 variables correspond to a 2-cube, which has 4 vertices (00, 01, 11, 10) (fig. 3b) .

Notice that this figure can be obtained by projecting the 1-cube (putting a second line underneath) and prefixing the first 1-cube with value 0 and the second 1-cube with value 1. Higher dimensional cubes can be obtained by a similar projection procedure, as can be seen in figures 3c and 3d for a 3-cube and a 4-cube with eight and sixteen vertices, respectively. Each vertex corresponds to a square in the Karnaugh map.

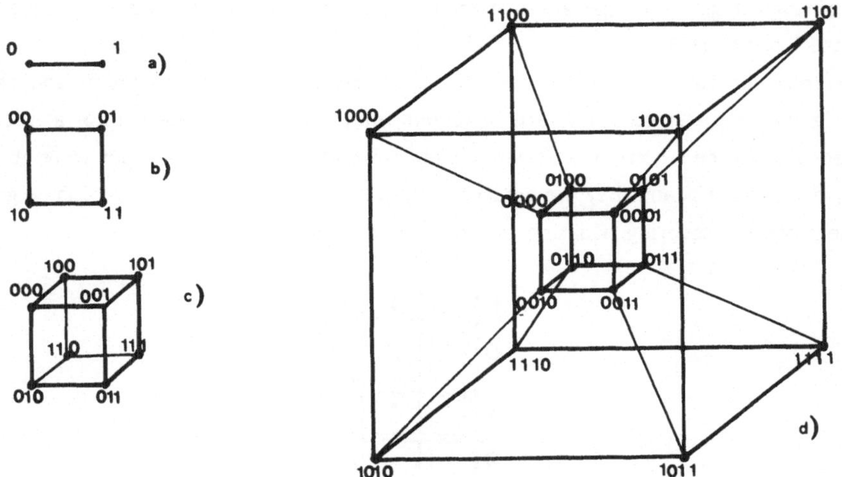

Fig.IV.3. Cubical representation of a) 1-cube ; b) 2-cube ; c) 3-cube and d) 4-cube

IV.6. Simplification of boolean functions using a Karnaugh map : prime implicants.

As seen above, a truth table or a Karnaugh map can be constructed from a boolean function, thus showing explicitly the value of the function for each input state. But these tables, and especially the Karnaugh maps, can also be used to find a simpler expression of the function ; thus, instead of deriving a map from an algebraic expression of a function, we want now to derive from the map a simpler algebraic expression of the function.

In short, the method consists of finding an expression, as simple as possible, which covers all the "1" of the table and none of the "0".

A boolean function is often expressed as a sum of products. We want to simplify each function by minimizing the number of variables (factors) explicitly present in each product, and the number of products in the sum. This can be achieved algebraically by using simplification rules (see section IV.4. and appendix).

For example, since $a + \bar{a} = 1$, any two products which differ by the value of only one variable can be combined into a single product with one less variable. For instance :

$\bar{a}b + \bar{a}\bar{b} = \bar{a}\,(b + \bar{b})$ (Distributivity)

$= \bar{a}\,.\,1$

$= \bar{a}$

and similarly,

$ab + \bar{a}b = (a + \bar{a})\,b = 1\,.\,b = b$

In a Karnaugh map, these operations amount to joining together two adjacent squares for which the function is 1 :

for instance, in Table 7, the squares corresponding to the input states $\bar{a}b$ and $\bar{a}\bar{b}$ both give F a value 1 ; since they are adjacent they can be grouped into a rectangle corresponding to the situation "$\bar{a}$, whatever the value of b". Similarily, input state ab and $\bar{a}b$ both give F a value 1, and they are adjacent ; thus, the corresponding squares can be grouped into the rectangle covering situation "b, whatever the value of a".

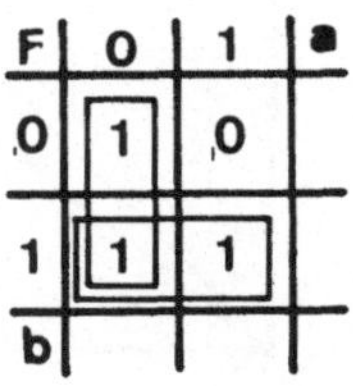

Table 7. Combining adjacent squares.

Thus the operations $ab + \bar{a}b = b$ and $\bar{a}b + \bar{a}\bar{b} = \bar{a}$ are materialized by the boxes (2-squares surfaces) which isolate, respectively, states b and $\bar{a}$ of the input variables.

It can be seen, from table 7, that $F = \bar{a}\bar{b} + \bar{a}b + ab$ is the same as the simpler expression $F = \bar{a} + b$ .One might, however, feel uneasy because of the redundancy of this expression : the input state $\bar{a}b$ is (implicitly) present twice, once in the term $\bar{a}$, once in the term b . In fact, the redundancy expresses the fact that since the conditions $\bar{a}$ and b are each sufficient to give the value 1 to F , there are two reasons ($\bar{a}$ and b) why the term $\bar{a}b$ has the value 1 . The redundancy could be avoided by $F = \bar{a} + ab$, or $F = \bar{a}\bar{b} + b$, but these are both uselessly longer than $F = \bar{a} + b$

Let us consider now three expressions using four variables : (see table 8).

$F_1 = \bar{a}\bar{b}\bar{c}\bar{d} + \bar{a}\bar{b}\bar{c}d + \bar{a}b\bar{c}\bar{d} + \bar{a}b\bar{c}d$

$= \bar{a}\,\bar{b}\bar{c}(\bar{d} + d) + \bar{a}b\,\bar{c}(\bar{d} + d)$ Two pairs of squares are grouped in Table 8a (dashed lines), etc

$= \bar{a}\bar{b}\bar{c} + \bar{a}b\bar{c}$

$= \bar{a}(\bar{b} + b)\bar{c}$

$= \bar{a}\bar{c}$

$F_2 \quad \bar{a}\bar{b}\bar{c}\bar{d} + \bar{a}b\bar{c}\bar{d} + ab\bar{c}\bar{d} + a\bar{b}\bar{c}\bar{d}$

$= \bar{c}\bar{d}\left[\bar{a}(\bar{b}+b) + a(b + \bar{b})\right]$

$= \bar{c}\bar{d}(\bar{a} + a)$

$= \bar{c}\bar{d}$

$F_3 = \bar{a}\bar{b}\bar{c}\bar{d} + a\bar{b}\bar{c}\bar{d} + \bar{a}\bar{b}c\bar{d} + a\bar{b}c\bar{d}$

$= (\bar{a} + a)\bar{b}\bar{c}\bar{d} + (\bar{a} + a)\bar{b}c\bar{d}$

$= \bar{b}(\bar{c} + c)\bar{d}$

$= \bar{b}\bar{d}$

All these simplifications are seen, in Tables 8a, b and c. Similary, the eight adjacent squares of Table 8d can be grouped to give $F_4 = \bar{d}$, and those of Table 8e can be grouped to give the term $\bar{b}\,d$; thus, the 5-variable function F_5 can be expressed as $\bar{b}\,d$. Notice that fusions will necessarily include a number of squares equal to a power of 2, since the repetitive application of the rule $\bar{a} + a = 1$ leads to double the surface of the term each time it is used.

Table 8a

F_1	00	01	11	10 ab
00	1	1	•	•
01	1	1	•	•
11	•	•	•	•
10	•	•	•	•
cd		$F_1 = \bar{a}\bar{c}$		

Table 8b

F_2	00	01	11	10 ab
00	1	1	1	1
01	•	•	•	•
11	•	•	•	•
10	•	•	•	•
cd		$F_2 = \bar{c}\bar{d}$		

Table 8c

F_3	00	01	11	10 ab
00	1	•	•	1
01	•	•	•	•
11	•	•	•	•
10	1	•	•	1
cd		$F_3 = \bar{b}\bar{d}$		

Table 8d

F_4	00	01	11	10 ab
00	1	1	1	1
01	•	•	•	•
11	•	•	•	•
10	1	1	1	1
cd		$F_4 = \bar{d}$		

Table 8e

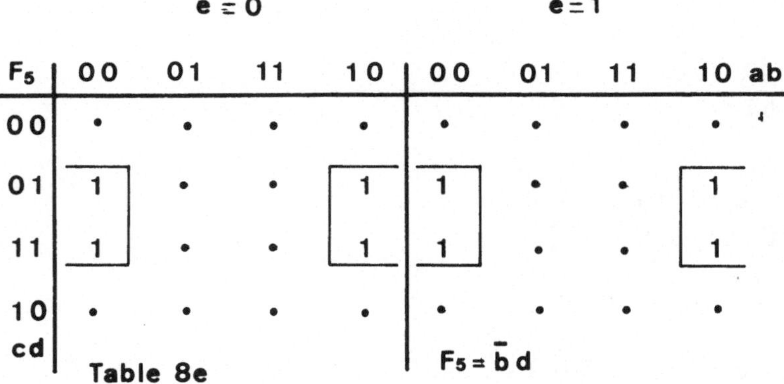

F_5	e = 0: 00	01	11	10	e = 1: 00	01	11	10 ab
00	•	•	•	•	•	•	•	•
01	1	•	•	1	1	•	•	1
11	1	•	•	1	1	•	•	1
10	•	•	•	•	•	•	•	•
cd								

Table 8e $F_5 = \bar{b}\,d$

Table 8 a, b, c, d, e : Different surfaces obtained by grouping adjacent 1-squares.

A term (or a surface of adjacent squares) which cannot be combined with any other is called a prime implicant. For instance, in function F_1 , $\bar{a}\,\bar{c}$, which results from the grouping of $\bar{a}\,\bar{b}\,c$ and $a\,b\,c$, but which cannot be fused itself with any other term, is a prime implicant.

A boolean function can always be written as the sum of its prime implicants. For example function F_6 (Table 9a) has the following prime implicants : $\bar{a}\,c$, $\bar{a}\,b\,d$, $a\,\bar{b}\,d$, $a\,\bar{c}\,d$, $b\,\bar{c}\,d$ and $\bar{b}\,c\,d$. Thus, one can write :

$$F_6 = \bar{a}\,c + \bar{a}\,b\,d + a\,\bar{b}\,d + a\,\bar{c}\,d + b\,\bar{c}\,d + \bar{b}\,c\,d$$

However, it can be seen that not all the prime implicants are necessary to describe the function, that is, to cover every square of the table containing a value of 1 and none of the squares containing a value of 0.

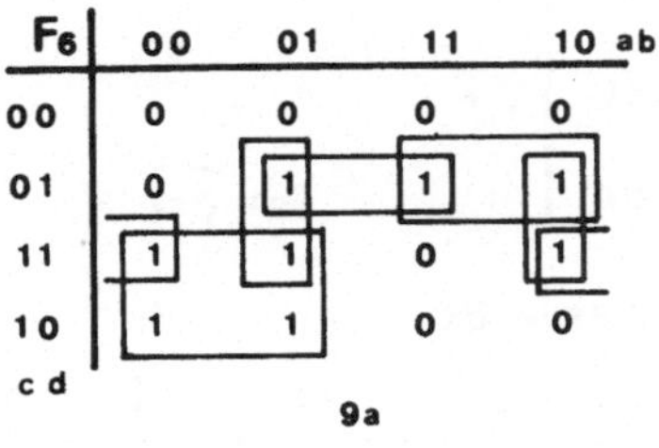

F_6	00	01	11	10 ab
00	0	0	0	0
01	0	1	1	1
11	1	1	0	1
10	1	1	0	0
c d				

9a

F_6	00	01	11	10 ab
00	0	0	0	0
01	0	1	1	1
11	1	1	0	1
10	1	1	0	0
c d				

9b

F_6	00	01	11	10 ab
00	0	0	0	0
01	0	1	1	1
11	1	1	0	1
10	1	1	0	0
c d				

9c

Tables 9a), b) and c)

For instance, the two states ($\bar{a}\ b\ \bar{c}\ d$ and $a\ b\ \bar{c}\ d$) which constitute the prime implicant $b\ \bar{c}\ d$ are each present in another prime implicant ($\bar{a}\ b\ d$ and $a\ \bar{c}\ d$ respectively). Similarly, $a\ \bar{b}\ d$ is redundant with ($a\ \bar{c}\ d + \bar{b}\ c\ d$). Thus, the prime implicants $b\ \bar{c}\ d$ and $a\ \bar{b}\ d$ can be dropped, and the expression becomes :

$$F_6 = \bar{a}\ c + \bar{a}\ b\ d + a\ \bar{c}\ d + \bar{b}\ c\ d \text{ (see Table 9b).}$$

But an even simpler expression of F_6 can be obtained by another choice of prime implicants (see Table 9c).

$$F_6 = \bar{a}\ c + b\ \bar{c}\ d + a\ \bar{b}\ d$$

These three prime implicants constitute the <u>minimal covering</u> of the function F_6. One way to find this minimal covering consists in first identifying the so-called <u>essential prime implicants</u> if any. One calls a prime implicant essential if it comprizes at least one "1" not present in any other prime implicant. For instance, in the example of Table 10a there are three "1" (labelled 1^+) which are each represented in only one prime implicant ; these essential prime implicants obviously have to be taken into account in <u>any</u> covering of function F_7 .

F_7 cd \ ab	00	01	11	10
00	0	0	0	0
01	1*	1	1	1
11	0	1	1	1*
10	1*	1	0	0

10a

F_7 cd \ ab	00	01	11	10
00	0	0	0	0
01	1	1	1	1
11	0	1	1	1
10	1	1	0	0

10b

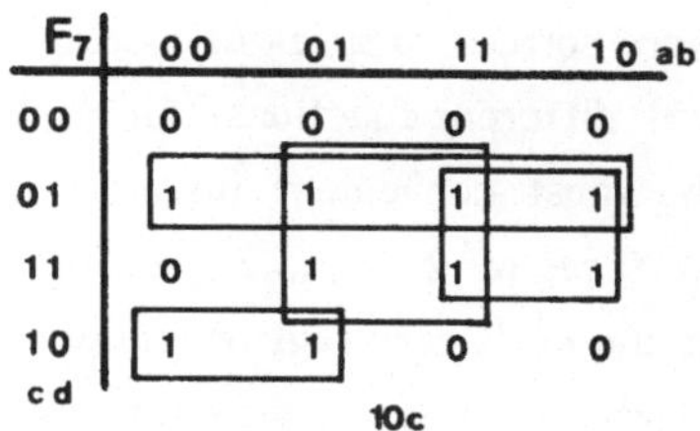

F_7 cd \ ab	00	01	11	10
00	0	0	0	0
01	1	1	1	1
11	0	1	1	1
10	1	1	0	0

10c

Tables 10a), b) and c)

One can see, however, (Table 10b) that in the present case the essential prime implicants do not suffice to define the function : state $\bar{a}\,b\,c\,d$ is missing. The most economical way to take it into account is to consider, in addition to the essential prime implicants, the largest (and thus, the simplest) prime implicant(s) which include $\bar{a}\ b\ c\ d$, in this case, $b\ d$ (Table 10c).

Thus, the simplest expression of F_7 is :

$$F_7 = a\ d + b\ d + \bar{c}\ d + \bar{a}\ c\ \bar{d}$$

In many cases, the situation is not so straightforward, and there exist two or more equally optimal solutions.

The technique based on the Karnaugh map is satisfactory for functions with up to five or six variables, but different techniques must be used for a larger number of variables. One is the Quine-Mc Cluskey minimization chart method. A specialized logical machine giving the optimal covering for incompletely specified (see next paragraph) functions of up to 8 variables has been conceived and constructed by Florine (1964). There now exist computer programs which can deal with more variables. (see chapter IX)

IV.7. Incompletely specified functions.

The boolean functions considered above are completely specified , that is, their value is defined for each input state. One often deals, however, with situations such that, for certain input states, it is indifferent whether a function takes the value 1 or 0. For instance, for some values of the input variables the requirements of a logical machine may be fulfilled for either value of one of the functions ; or, there are combinations of values of the input variables which correspond to physically impossible situations+.

In such cases, one speaks of incompletely specified functions and one uses, in addition to 1 and 0 , a third value, denoted don't care. On a Karnaugh map, the don't care input states are indicated by a dash in the corresponding squares. Each dash can be replaced by either a 0 or a 1 as desired, and a map with n dashes can thus account for 2^n different functions.

Rather than speak of incompletely defined functions, it would be more correct to speak of incompletely specified requirements or maps, which are compatible with different functions. Since all of these functions fulfill the requirements, one can choose the most convenient (usually the simplest) one. In practice, one extends the domains (surfaces of 2 , 4 , ...2^n squares, see 4.6) containing values 1 by giving a value 1 to some of the don't care regions, chosen to obtain the smallest possible number of prime implicants, which should be as simple (as large on the map) as possible.

+ for instance, let a , b and c be positive numbers ; the statements $a < b$, $b < c$ and $a < c$ can be represented by logical variables m , n , p respectively, such that $m = 1$ if $a < b$, $m = 0$ if not etc... If $a < b$ and $b < c$, necessarily $a < c$; thus m . n implies p , in other words the situation $m\ n\ \bar{p}$ will never occur and it has no importance whether a function of m , n and p has the value 1 or the value 0 for this input state.

For instance, in Table 11a, there are 5 dashes. Each one can be replaced by a 0 or a 1 .

If we take into account only the "1" present in Table 11a, the function choosen will be :

$$F = \bar{a}\,\bar{b}\,d + a\,b\,\bar{c}\,\bar{d}\ .$$

But if we replace by "1" the two dashes in column 01 and the dash corresponding to input state 1101 , the prime implicant $\bar{a}\,\bar{b}\,d$ simplifies into $\bar{a}\,d$ (that is, $\bar{a}\,d$, whatever the value of b) and $a\,b\,\bar{c}\,\bar{d}$ simplifies into $a\,b\,\bar{c}$ (that is, $a\,b\,\bar{c}$, whatever the value of d). This leads to function F8 (Table 11b) which comprises the prime implicants $\bar{a}\,d$ and a b c . $F_8 = \bar{a}\,d + a\,b\,\bar{c}$ is the simplest function which fulfills the requirements of Table 11a ; it is notably simpler than the function originally proposed ($F = \bar{a}\,\bar{b}\,d + a\,b\,\bar{c}\,\bar{d}$) .

cd \ ab	00	01	11	10
00	0	0	1	0
01	1	–	–	0
11	1	–	–	0
10	0	0	0	–

11a

F8 cd \ ab	00	01	11	10
00	0	0	1	0
01	1	1	1	0
11	1	1	0	0
10	0	0	0	0

11b

Table 11. An incompletely specified map (11a) with five don't cares, compatible with $2^5 = 32$ different functions. The simplest one, in which three dashes have been replaced by 1 and two by 0, is given in 11 b ; the prime implicants are $\bar{a}\,d$ and $a\,b\,\bar{c}$ and we write $F_8 = \bar{a}\,d + a\,b\,\bar{c}$.

IV.8. Applications.

IV.8.1. Binary adder.

A binary adder is a device which gives the arithmetic (not logical !) sum of binary numbers.

In the same way that a decimal number, say 84 , can be decomposed as :

$$84 = 80 + 4$$
$$= 8.10^1 + 4.10^0 ,$$

the equivalent binary number, 1 0 1 0 1 0 0 (84 in base 2 or modulo 2), which comprizes 7 binary digits (or bits) can be decomposed[+] :

$$1\,0\,1\,0\,1\,0\,0 = 1.2^6 + 0.2^5 + 1.2^4 + 0.2^3 + 1.2^2 + 0.2^1 + 0.2^0$$

Let us now add two numbers, A and B , of 4 bits each. We will have an input state of 8 variables, 4 for each number (fig.IV.4.). Since the greatest binary number of 4 bits is 1111 (the binary equivalent of 15) , $A \leqslant 15$, $B \leqslant 15$ and $S=A+B \leqslant 30$. The sum may thus comprise up to 5 binary digits, and the output will thus require 5 binary functions S_0 S_1 S_2 S_3 S_4

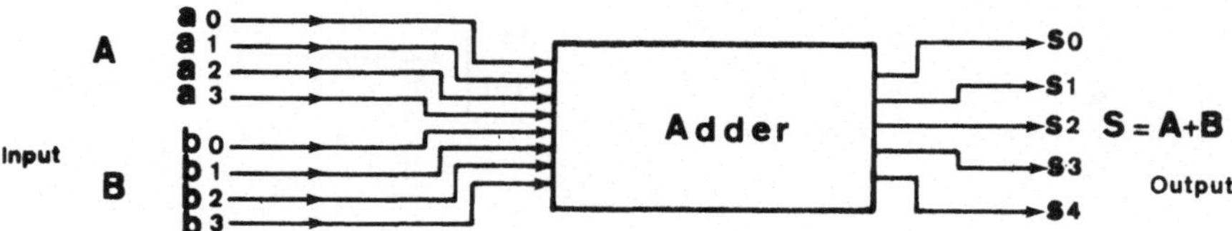

Fig.IV.4 . A parallel adder.

Thus, at first view, the simple problem of adding two binary numbers of four digits each would require the construction of a device with 8 input variables and 5 output functions. The elaboration of such a device is beyond the possibilities of a Karnaugh map !

In fact, instead of this so-called parallel treatment, one uses an iterative treatment, in which an elementary operation is used repeatedly.

The addition of two decimal numbers and their binary equivalents are shown in Fig.IV.5. Note that the digits of rank 0,1 , ...i (counting from the right) have a weight $10^0, 10^1$, $...10^i$ or 2^0, 2^1 , $...2^i$ in the decimal and binary notations, respectively.

---

* More generally, a binary number of n digits : a_{n-1} ... a_i ... a_2 a_1 a_0 (in which the general term a_i is 1 or 0)can be decomposed :

$$a = a_{n-1}\, 2^{n-1} + \dots + a_i\, 2^i + \dots + a_2\, 2^2 + a_1\, 2^1 + a_0\, 2^0 = \sum_{i=0}^{i=n-1} a_i\, 2^i$$

Each elementary operation consists of adding the digits a_i and b_i, and c_i which is the carry over from the preceding column. The number so obtained is decomposed into a sum S_i and a carry over C_i, which is used as the input variable c_{i+1} in the next elementary addition.

rank	2	1	0	
weight	10^2	10^1	10^0	
	1	0	(0)	carry
	0	8	4	
	0	9	4	
	1	7	8	
	0	1	0	

	rank	7	6	5	4	3	2	1	0
	weight	2^7	2^6	2^5	2^4	2^3	2^2	2^1	2^0
c_i		1	0	1	1	1	0	0	(0)
a_i			1	0	1	0	1	0	0
b_i			1	0	1	1	1	1	0
S_i		1	0	1	1	0	0	1	0
C_i		0	1	0	1	1	1	0	0

Fig.IV.5. Addition of two decimal numbers (84 + 94) and their binary equivalents.

Table 12a and 12b show, respectively, the values of S_i and C_i as functions of a_i, b_i and c_i ; for instance, if $a_i = 1$, $b_i = 1$ and $c_i = 0$, $S_i = 0$ $C_i = 1$, etc ...

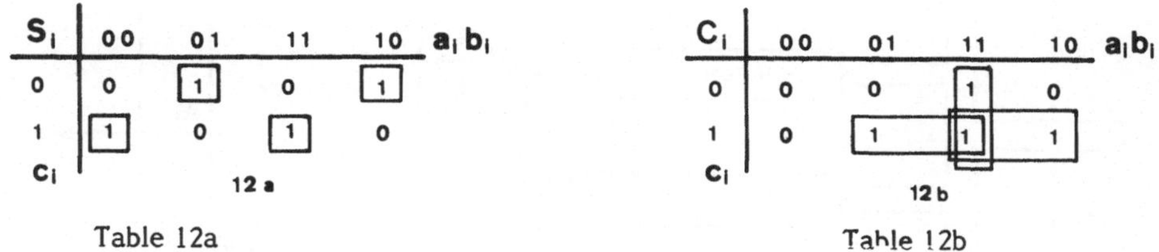

S_i \ c_i	00	01	11	10
0	0	1	0	1
1	1	0	1	0

12 a ($a_i b_i$)

C_i \ c_i	00	01	11	10
0	0	0	1	0
1	0	1	1	1

12 b ($a_i b_i$)

Table 12a Table 12b

Tables 12a and b. Karnaugh maps of S_i and C_i as functions of a_i, b_i and c_i .

As shown in Tables 12 a and b ,

$$S_i = \bar{a}_i \bar{b}_i c_i + \bar{a}_i b_i \bar{c}_i + a_i b_i c_i + a_i \bar{b}_i \bar{c}_i$$

$$C_i = a_i b_i + b_i c_i + a_i c_i$$

If one constructs and connects together in the proper way (see Fig.IV.6) n such circuits, one can add binary numbers of n bits.

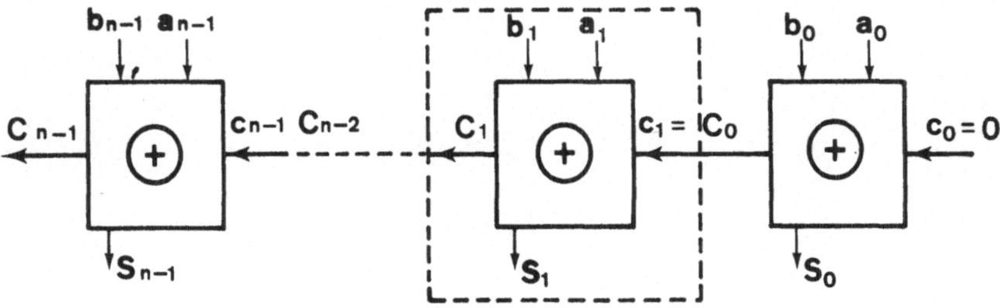

Fig.IV.6. A schematic representation of an iterative (also called cascadable) adder.

IV.8.2. Majority functions.

Let us suppose we have four identical logical circuits. Normally, they give the same output values if they have the same input state. We want to design a "fault tolerant" circuit, so that if one of the four circuits fails, we will obtain the correct output from the three other circuits (fig.IV.7). This can be done by a majority function : the final output will be 0 if there are a majority of 0 values in the output configurations and it will be 1 if there are a majority of 1 values. The other cases can be considered as don't cares ; this attitude is justified because if a failure of one of the circuits is exceptional and all four circuits operate independantly it will almost never happen that two circuits give an output 1 and two give an output 0 . This attitude is implemented in the truth table (Table 13) and in the Karnaugh map (Table 14) ; the simplest function consistent with this map is $M_1 = a\,b + c\,d$ (in which the input states 1 1 0 0 and 0 0 1 1 result in $M = 1$ while the other input states with two 0 and two 1 result in $M = 0$).

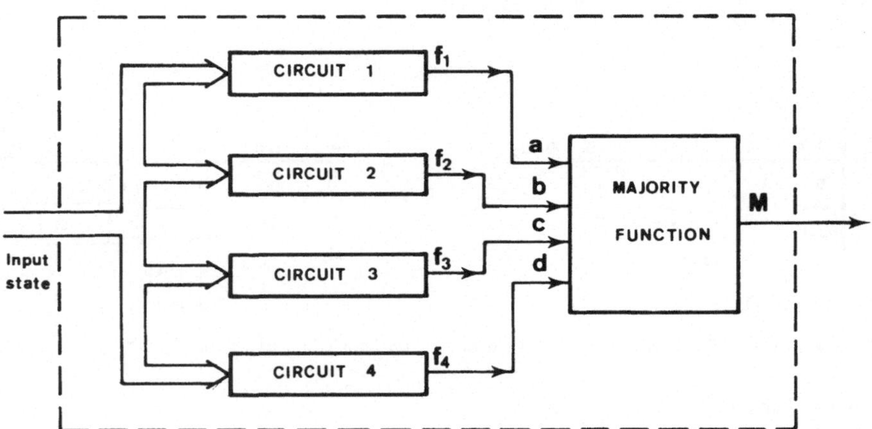

Fig.IV.7 . Fault tolerant circuit using majority function (schematic configuration).

Another attitude is that values 0 override values 1 (table 15) ; the solution is then :

$$M_2 = a\,b\,c + a\,b\,d + a\,c\,d + b\,c\,d$$

If values 1 override values 0 , we have Table 16a, which gives :

$$M_3 = a\,b + c\,d + a\,c + a\,d + b\,c + b\,d$$

Note that in this particular case, finding the prime implicants of the complementary function $\overline{M}_3$ (Table 16b) is easier :

$$\overline{M}_3 = \bar{a}\,\bar{b}\,\bar{c} + \bar{a}\,\bar{b}\,\bar{d} + \bar{a}\,\bar{c}\,\bar{d} + \bar{b}\,\bar{c}\,\bar{d}$$

Then M_3 will be found by complementing $\overline{M}_3$:

$$M_3 = \overline{\bar{a}\,\bar{b}\,\bar{c} + \bar{a}\,\bar{b}\,\bar{d} + \bar{a}\,\bar{c}\,\bar{d} + \bar{b}\,\bar{c}\,\bar{d}}$$

13

a	b	c	d	M
0	0	0	0	0
0	0	0	1	0
0	0	1	0	0
0	0	1	1	-
0	1	0	0	0
0	1	0	1	-
0	1	1	0	-
0	1	1	1	1
1	0	0	0	0
1	0	0	1	-
1	0	1	0	-
1	0	1	1	1
1	1	0	0	-
1	1	0	1	1
1	1	1	0	1
1	1	1	1	1

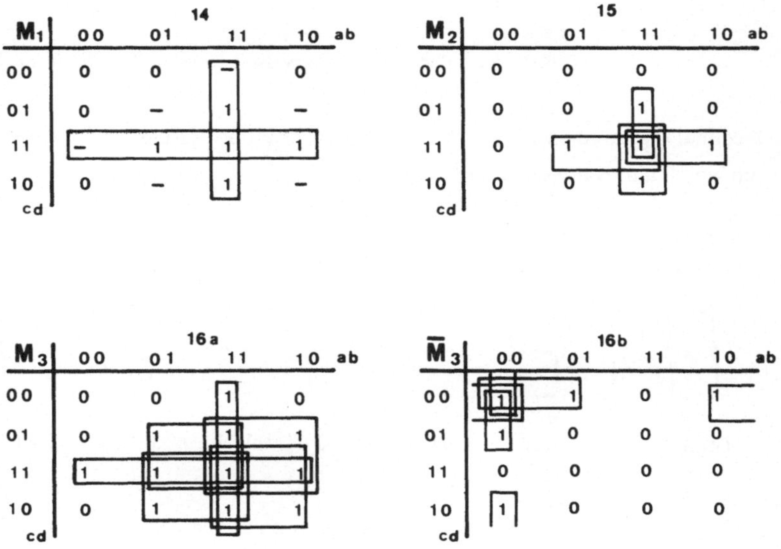

14

M_1 (cd \ ab)	00	01	11	10
00	0	0	–	0
01	0	–	1	–
11	–	1	1	1
10	0	–	1	–

15

M_2 (cd \ ab)	00	01	11	10
00	0	0	0	0
01	0	0	1	0
11	0	1	1	1
10	0	0	1	0

16 a

M_3 (cd \ ab)	00	01	11	10
00	0	0	1	0
01	0	1	1	1
11	1	1	1	1
10	0	1	1	1

16b

$\overline{M}_3$ (cd \ ab)	00	01	11	10
00	1	1	0	1
01	1	0	0	0
11	0	0	0	0
10	1	0	0	0

Tables 13 - 16. A truth table and four Karnaugh maps showing various ways to treat the problem of the majority function.

IV.8.3. Binary coded decimal to seven-segment decoder.

Although digital electronic machines operate on a binary basis, it is more convenient for numerical results to be visualized (on the front pannel of a pocket calculator, for example) in decimal form. This is commonly accomplished by composing the numbers from seven segments arranged in the form of number 8 . Each of the seven segments (a , b , c , d , e , f , g : see fig.IV. 8) is a lamp (light-emitting diode) which can be on or off.

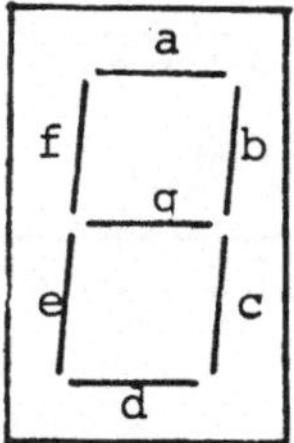

Fig.IV. 8. The seven lamps used to visualize decimal digits.

Decimal digits 0 , 1 , 2 , 3 , 4 , 5 , 6 , 7 , 8 or 9 , are visualized by such a display depending on which lamps are on (fig.IV. 9) .

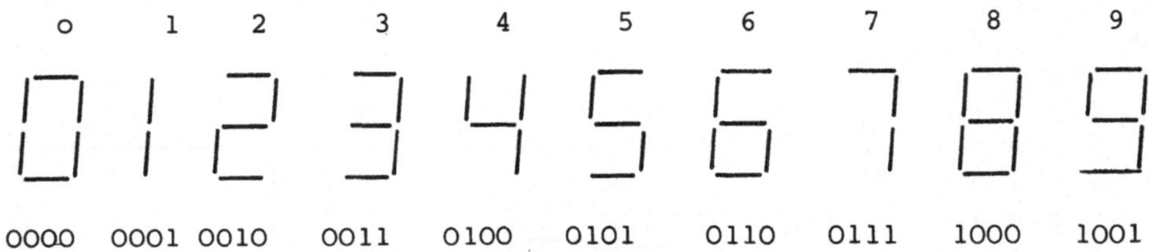

Fig.IV.9. Visualization of the decimal digits.

On the other hand, inside a logical machine the numbers 0 , ...9 are encoded as combinations of values of four binary variables X_0 , X_1 , X_2 , X_3 (0 0 0 0 for the decimal digit 0 ; 0 0 0 1 for the decimal digit 1 ; 0 0 1 0 for the decimal digit 2 , etc...). The problem thus consists in finding seven functions (a , ...g) of the four binary variables (X_0 , ... X_3) such that for each binary number the proper decimal digit is visualized.

We must thus write seven Karnaugh maps, one for each function a, b, ... g .

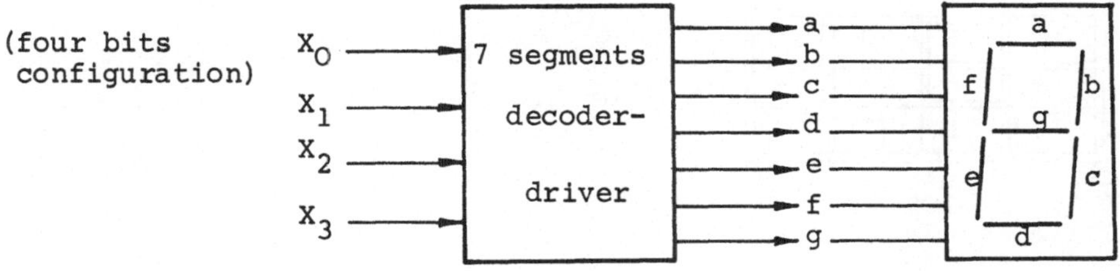

Fig.IV. 10 Schematic diagram of the decoder B C D 7 segment.

For instance, (see fig.IV,9) lamp a must be on (function a must be 1) for the decimal digits 0, 2, 3, 5, 6, 7, 8 and 9 ,which correspond to the binary numbers 0000 , 0010, 0011, 0101, 0110, 0111, 1000 and 1001, that is, to the input states $\bar{X}_0 \bar{X}_1 \bar{X}_2 \bar{X}_3$, $\bar{X}_0 \bar{X}_1 X_2 \bar{X}_3$, etc ..., respectively .

Of course, of the sixteen (16) possible combinations of values of the four input variables, we use only ten (10) , the binary equivalents of the decimal digits ; the six other input states (1 0 1 0 to 1 1 1 1 inclusively) do not code for individual decimal digits and are thus considered as don't care conditions.

Thus, the Karnaugh map for functions a to g are given in Tables 17.

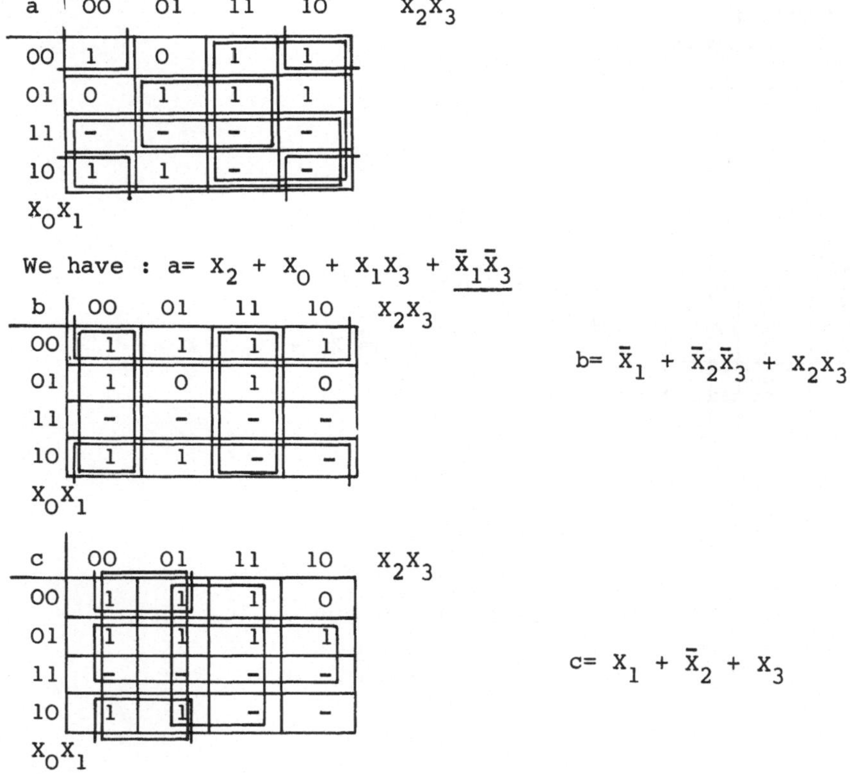

a (X_0X_1 \ X_2X_3)	00	01	11	10
00	1	0	1	1
01	0	1	1	1
11	-	-	-	-
10	1	1	-	-

We have : $a = X_2 + X_0 + X_1X_3 + \bar{X}_1\bar{X}_3$

b (X_0X_1 \ X_2X_3)	00	01	11	10
00	1	1	1	1
01	1	0	1	0
11	-	-	-	-
10	1	1	-	-

$b = \bar{X}_1 + \bar{X}_2\bar{X}_3 + X_2X_3$

c (X_0X_1 \ X_2X_3)	00	01	11	10
00	1	1	1	0
01	1	1	1	1
11	-	-	-	-
10	1	1	-	-

$c = X_1 + \bar{X}_2 + X_3$

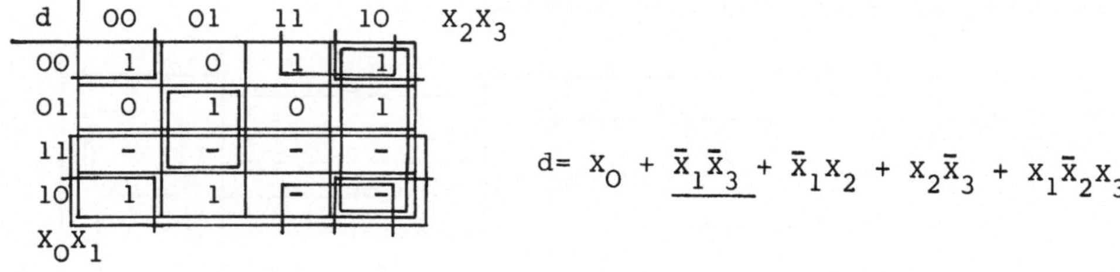

d (X_0X_1 \ X_2X_3)	00	01	11	10
00	1	0	1	1
01	0	1	0	1
11	-	-	-	-
10	1	1	-	-

$$d = X_0 + \underline{\bar{X}_1\bar{X}_3} + \bar{X}_1X_2 + X_2\bar{X}_3 + X_1\bar{X}_2X_3$$

e (X_0X_1 \ X_2X_3)	00	01	11	10
00	1	0	0	1
01	0	0	0	1
11	-	-	-	-
10	1	0	-	-

$$e = \underline{\bar{X}_1\bar{X}_3} + X_2X_3$$

f (X_0X_1 \ X_2X_3)	00	01	11	10
00	1	0	0	0
01	0	0	1	0
11	-	-	-	-
10	1	1	-	-

$$f = X_0 + \bar{X}_1\bar{X}_2\bar{X}_3 + X_1X_2X_3$$

g (X_0X_1 \ X_2X_3)	00	01	11	10
00	0	0	1	1
01	1	1	1	1
11	-	-	-	-
10	1	1	-	-

$$g = X_0 + X_1 + X_2$$

Table 17, a to g : The Karnaugh maps for functions a to g .

Fig.IV.11. shows the electronic diagram of function a (built with AND, OR and NOT circuits).

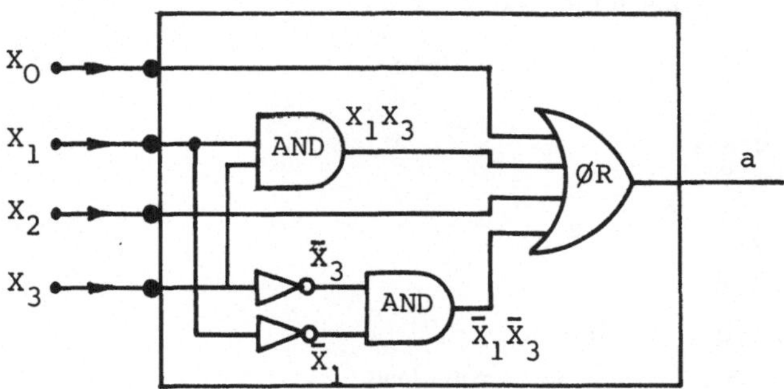

Fig.IV.11. Schematic electronic diagram of function "a".

Notice that there exist integrated circuits (Medium Scale Integration) which implement functions such as a, b, ..., g and many others, and which are contained in a single little box about 20 x 5 x 3 mm.

APPENDIX

I. Properties similar to classical algebra.

$a + 0 = a$

$a \cdot 0 = 0$

$a \cdot 1 = a$

$a + b = b + a$
$a \cdot b = b \cdot a$ commutative laws

$a + (b + c) = (a + b) + c$
$a \cdot (b \cdot c) = (a \cdot b) \cdot c$ associative laws

$a \cdot (b + c) = (a \cdot b) + (a \cdot c)$
$(a + b) \cdot (c + d) = (a \cdot c) + (a \cdot d) + (b \cdot c) + (b \cdot d)$ distributive laws.

II. Properties different from classical algebra.

$a + 1 = 1$ (1)

$a + a = a$ (2)
$a \cdot a = a$ (3) idempotent laws

$a + (a \cdot b) = a$ (4)

$a \cdot (a + b) = a$ (5)

$(a + b) \cdot (a + c) = a + bc$ (6)

III. Properties peculiar to boolean algebra

$\bar{\bar{a}} = a$ involution law (1)

$a + \bar{a} = 1$ complementarity laws (2)

$a \cdot \bar{a} = 0$ (3)

$(a + b) \cdot (a + \bar{b}) = a$ (4)

$a \cdot b + a \cdot \bar{b} = a$ (5)

$a \cdot \bar{b} + b = a + b$ (6)

$(a + \bar{b}) \cdot b = a \cdot b$ (7)

$a \cdot b + \bar{a} \cdot c + b \cdot c = a \cdot b + \bar{a} \cdot c$ (8)

$\overline{a + b + c} = \bar{a} \cdot \bar{b} \cdot \bar{c}$ dualization laws (9)

or

$\overline{a \cdot b \cdot c} = \bar{a} + \bar{b} + \bar{c}$ de Morgan's laws (10)

Groupe I properties will not be proved as they are similar to ordinary algebra but are applied to binary-valued variables. In group II, equations 1, 2 and 3 are deduced from the definition of the OR and AND operations (Table 1).

Proof II.4 $a + (a \cdot b) = a\,(1 + b) = a \cdot 1 = a$

Proof II.5. $a \cdot (a + b) = aa + ab = a + ab = a\,(1 + b) = a \cdot 1 = a$

Proof II.6. $(a + b) \cdot (a + c) = aa + ac + ba + bc = a\,(1 + b + c) + bc$

$= a \cdot 1 + bc = a + bc$

In group III, equations 1, 2 and 3 are also proved from the definition of the basic operations.

Proof III.4. $(a + b) \cdot (a + \bar{b}) = aa + a\bar{b} + ba + b\bar{b}$

$= a\,(1 + b + \bar{b}) = a \cdot 1 = a$

Proof III.5. $a \cdot b + a \cdot \bar{b} = a\,(b + \bar{b}) = a \cdot 1 = a$

Demonstration of III.6. is rather tricky. As $a + \bar{a} = 1$, we may multiply any variable by $(a + \bar{a})$ without changing the function.

And since $a\,b + a\,b = a\,b$ we may also add one $a\,b$ term in the OR function.

So :

Proof III.6. $a \cdot \bar{b} + b = a\bar{b} + b\,(a + \bar{a}) = a\bar{b} + ab + \bar{a}b + ab$

$= a\,(b + \bar{b}) + b\,(\bar{a} + a) = a + b$

This property is used to find the final equation in the umbrella example of section 4.3 .

In fact, all these relations are much easier to find using Karnaugh maps.

BIBLIOGRAPHY

E.J. Mc CLUSKEY 1962 & T.C. BARTEE — "A Survey of switching circuit theory" (Mc. Graw Hill).

E.J. Mc CLUSKEY 1965 — "Introduction to the theory of switching circuits" (Mc. Graw Hill).

R.F. MILLER 1965 — "Switching Theory : Vol.1. Combinational Circuits
Vol.2. Sequential Circuits and Machines.
(J. Wiley & Sons).

J. FLORINE 1964 — "La Synthèse des Machines Logiques et Son automatisation". (Dunod)

J. FLORINE 1969 — "Automatismes à Séquences etCommandes Numériques". (Dunod) .

ACKNOWLEDGMENTS

The authors cordially thank Professor René THOMAS, editor, for many helpful discussions and for his contribution to the manuscript.

CHAPTER V.

Sequential Systems.

BY A. LEUSSLER & P. VAN HAM+.

V.1. Introduction.

We would like first to remind the reader that Chapters IV and V are written from the point of view of the designer of logical machines. The problems treated are thus usually of the following type : given a desired behaviour (desiderata) find a logical structure, as simple as possible, which will behave in that way. Extension of the concepts and methods described here to the analysis of models will be dealt with in subsequent Chapters.

In the problems treated in Chapter IV ("combinational" problems), the value of each output function was entirely determined by the combination of values of the input variables (the input state of the system). We will now deal with problems (sequential

+ Laboratoire des systèmes logiques et numériques , Faculté des Sciences appliquées , Université libre de Bruxelles.

problems) in which there are functions whose value depends not only on the present input state, but also on former input states ; more specifically the value of these functions depends on a sequence of input states. In combinational problems one directly writes a Karnaugh map giving the values of the functions for each input state. In sequential problems, input sequences must be somehow memorized. The strategy used (see Miller, 1965 ; Florine 1964 , 1969) will be described first in a rather general way, in order to introduce the notions of internal functions, variables and states ; this short introduction may seem somewhat abstract, but it will be immediately followed by simple concrete examples.

The general idea (see Fig.1.) is to introduce additional functions and variables, the so-called internal (or memorization or secondary) functions (Y_1 ,... Y_n) and variables (y_1 ,... y_n) such that the system can be described in terms of logical equations relating the output (Z_1 ,... Z_r) and internal (Y_1 ,... Y_n) functions with the input (x_1 ,...x_m) and internal (y_1 ,...y_n) variables. The sequential problem is thus formally transformed into a combinational problem, but more variables and functions are involved. As regards the combinations of values of the variable, one has now to distinguish : an input state (that is, the state of the input variables), an internal state (the state of the internal variables), and a total state. Only from the total state of the variables can one compute the values of all the functions.

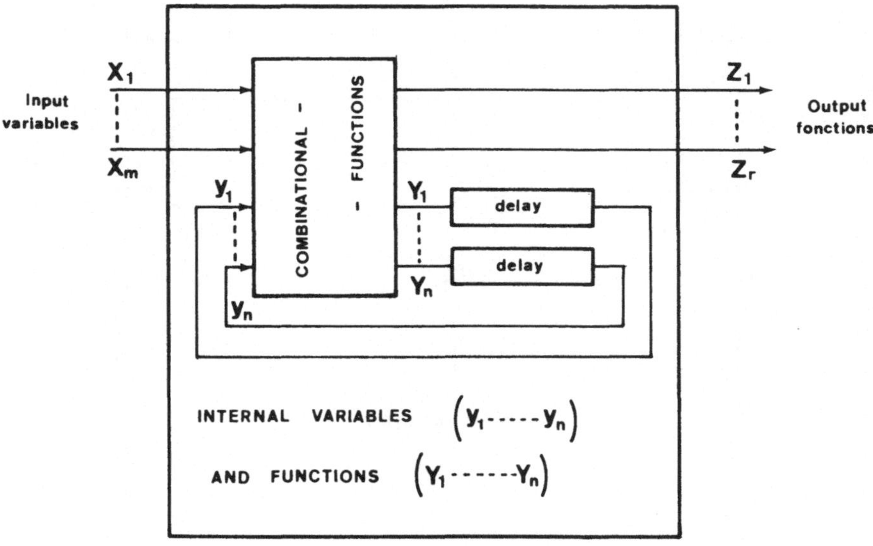

Fig.V.1. A sequential circuit.

In practice, we start with a description in terms of input sequences and output values, and at this stage the internal situation is still a black box. We will show (§ V.2.2. and V.2.3.) how to proceed in order to obtain a set of logical equations relating the output and internal functions to the input and internal variables in such a way that the desiderata are fulfilled. It must be pointed out that this can be done in various ways , but of course we want the black box (internal system) to be as simple as possible.

Before proceeding to the examples let us describe the relation between internal functions and variables.

Each internal function Y_i is associated with an internal variable y_i such that in stationary situations (here called <u>stable states</u>) the logical values of Y_i and y_i are the same, but when the value of the function Y_i changes, the value of its associated variable y_i changes only after a small time delay Δt (Fig. V.2.) which may be different for a switching-on (Δt_ε) and for a switching-off (Δt_δ) of Y_i .

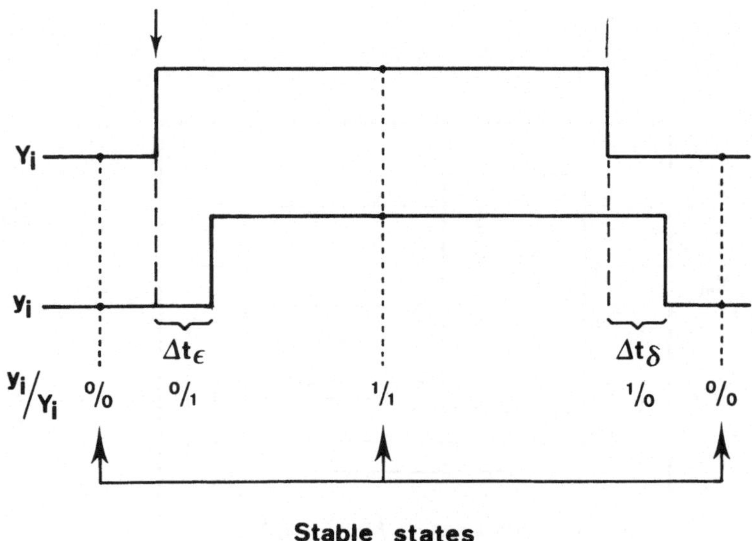

Fig.V.2. The relation between an internal function Y_i and the associated internal variable y_i .

Thus, the variable y_i serves as a memory of a preceeding value of function Y_i ; this is why one often calls y_i ,...y_n memorization variables. Note that when a system is in a stable state, it will not depart from it unless one changes the value of an appropriate input variable.

V.2. First example : a table lamp with a button switch.

Let us consider as a first example of sequential circuit a table lamp with a button switch. The lamp is switched on by pressing the button ; however, one does not have to maintain the pressure on the button, for the light remains on after one has released the pressure. To switch the light off, one pushes again on the button, but the light remains off after releasing the pressure. In order to formalize the description of this system, we associate with the button an input variable a, which has the value 1 when it is pushed down and 0 when it is released, and with the lamp an output function F , which has the value 1 when the light is on, the value 0 when it is off. The sequence is (Fig.V.3) :

Fig.V.3. Input - Output requirements in the "button switch" example.

Thus, whether $a = 0$ or $a = 1$, one can have $F = 0$ or $F = 1$: knowing the value of the input variable is not sufficient to know the value of function F . The four states described in Fig.V.3 are stable states, since in order to proceed from one to another one has to change the value of the input variable.

V.2.1. State diagram

The desired behaviour can be described by a state diagram, which is an oriented graph. Each node of the graph represents a stable state of the system and each arrow represents the transition between two states. At this stage all one knows is the sequence of the situations, each characterized by the values of the input variables and output functions.

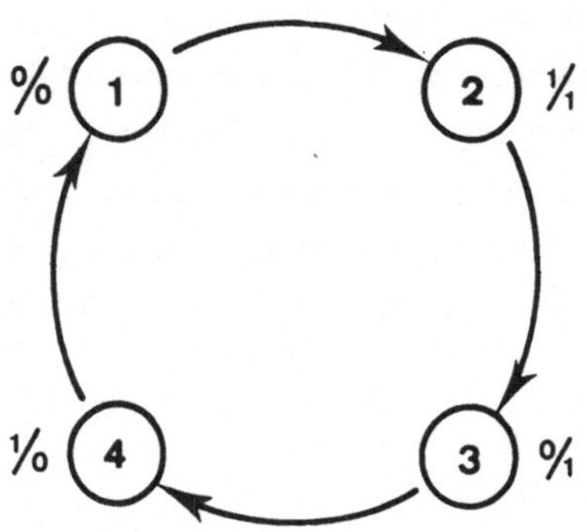

Fig.V.4 gives the state diagram of the table lamp. The four situations described in the preceeding section are arbitrarily labelled 1 , 2 , 3 , and 4 and the couples of digits 0 / 0 etc, refer to the corresponding values of the input variable and of the output function, respectively (a/F).

Fig.V.4. The state diagram for the table lamp example.

V.2.2. Flow table.

Our goal is to obtain a set of logical equations relating output and internal functions to the values of input and internal variables ; but at the present stage of the description internal functions and variables are probably felt by the reader as mere abstractions. Internal states begin to appear in a more concrete way in the flow table, where there is a column for each of the input states[+] and a row for each of the internal states. The stable states are represented by circled numbers, as in the state diagram. However, the transitions (which were symbolized by arrows in the state diagram) are now explicitely considered to involve an unstable state, in which internal functions have already changed their value, but these new values are not yet adopted by the corresponding internal variables (see Fig.V.2). This amounts to decomposing the transition into two steps.

+ There is an additional column in which the values of the output functions are mentioned.

First, one changes the value of an input variable (in the present case, the input variable a) ;

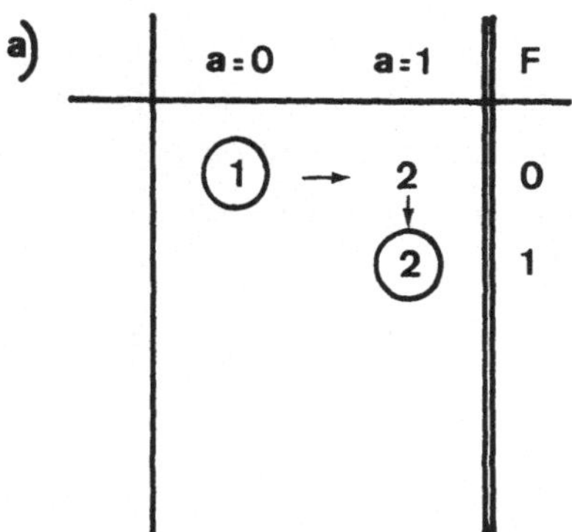

Table 1a : The beginning of the construction of a flow table.

this is seen on the flow table (Table 1a) as a shift from column to column. The state thus reached is unstable insofar as internal functions may have changed but the corresponding internal variables have not yet adopted the new values. This unstable state is designated by a number (uncircled) corresponding to its address, that is, to the stable state which will be reached next+.

Second, as the delay Δt(see Fig.V.2) has elapsed, the system proceeds to the next stable state ; this spontaneous change is seen on the flow table (Table 1a) as a change from one to another row. Thus, just as the columns correspond to the various input states the rows++ correspond to the various internal states of the variables.

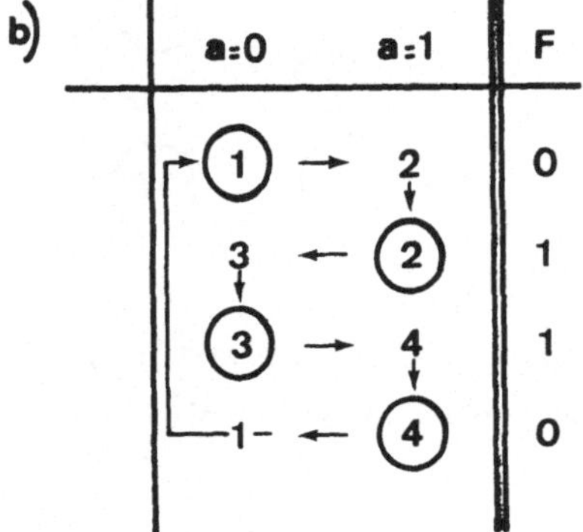

Table 1b : The flow table for the table lamp example.

In the table lamp example, state 1, in which a = 0 and F = 0 , is a stable state and will not change unless one changes the value of the input variable (a = 1) by pushing the button.
This volontary change is seen on the table (Table 1a) as a horizontal shift from column (a = 0) to column (a = 1) .

+ As will be seen below (this chapter, section V,3 and especially chapter VI, there may be more than one, and sometimes many unstable states between two stable states.

++As will be seen in the treatment of another example, one may often simplify the flow tables by condensing rows ; this results in the suppression of non essential transient states, or, which is equivalent, of non essential internal states.

It leads to an unstable state 2, so designated because it will next proceed to the stable state 2 (a = 1 , F = 1) ; this second, spontaneous step, is seen as a jump from the first to the second row (see Table 1a). The same reasoning applies to the transitions from 2 to 3, from 3 to 4 and from 4 to 1.

As just seen, the changes from row to row correspond to changes in the values of the internal variables. Our next task is to assign to each row in the flow table a combination of values of the internal variables. This assignment (or coding) is arbitrary, although there are some constraints (see section V.2.3).

V.2.3. Assignment of the internal variables

As the rows of the flow table (Table 1) correspond to different internal states, one can assign to each of them a combination of values of the internal variables. In our case, there are four (2^2) rows ; in order to code them we thus need two internal variables, y_1 and y_2. Let us assign to states 1, 2, 3 and 4 the combinations of values of y_1 y_2, 0 0, 0 1, 1 1 and 1 0, respectively.

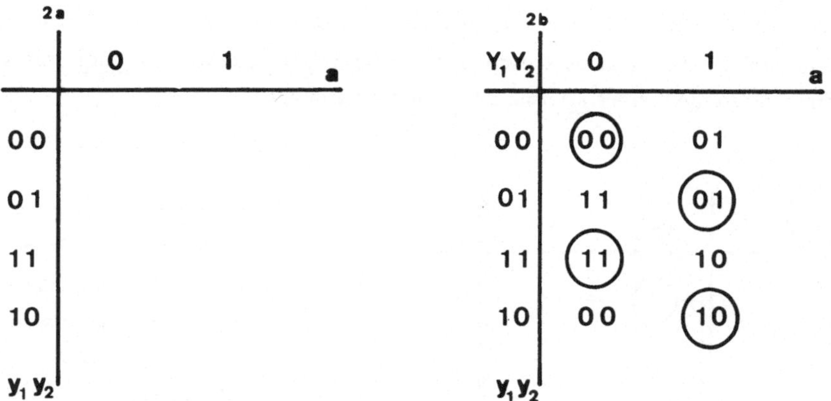

2a

$y_1 y_2$ \ a	0	1
00		
01		
11		
10		

2b

$y_1 y_2$ \ a ($Y_1 Y_2$)	0	1
00	00	01
01	11	01
11	11	10
10	00	10

Table 2 a) The skeleton of the table. b) The complete table ; the names (Y_1 , Y_2) of the internal functions are given in the N-W corner and their values for each state of the variables in the corresponding place of the table.

One can now construct a table (Table 2a) in which the columns correspond to the values of the input variable a and the rows to the values of the internal variables y_1 and y_2. This table can be filled (Table 2b) with the values of the internal functions Y_1 and Y_2 for each combination of values of the input and internal variables. We know that in the stable states (circled in the table) Y_1 and Y_2 have the same value as y_1 and y_2, respectively (see Fig.V.2.) and that in the transient states the values of Y_1 and Y_2 are the next values of y_1 and y_2.

Thus, in Table 2, the symbols 1 , 2 , 3 , 4 ,which represented the stable states in the flow table, are replaced by the corresponding values of Y_1 and Y_2 , 00 , 01 , 11 , 10 , respectively. As regards the transient states 1 , 2 , 3 , 4 , functions Y_1 and Y_2 already have their new values, although y_1 and y_2 still have the old values of Y_1 and Y_2 . For instance, starting from 00 , as a is switched from 0 to 1 (horizontal shift), the values of Y_1 Y_2 immediately shift to 01 , but the values of y_1 and y_2 are still 00 . After a short time, y_1 and y_2 adopt the new values of Y_1 and Y_2 respectively (spontaneous vertical shift in the table), thus leading to stable state 01 ; and so on. Note that Table 2a is in fact a Karnaugh (or Veitch) map. The only difference with the maps described in Chapter IV is that the values of the functions are related not only to input (columns) but also to internal (rows) variables. This illustrates the statement (section V.1) that the treatment just described formally transforms a sequential problem into a combinational one.

We can now find, with the methods of Chapter IV, the equations giving Y_1 and Y_2 as functions of the input (a) and internal (y_1 and y_2) variables ; it suffices to split Table 2 in order to separate the values of Y_1 and Y_2 and try to extract the minimal set of prime implicants covering the functions (Table 3).

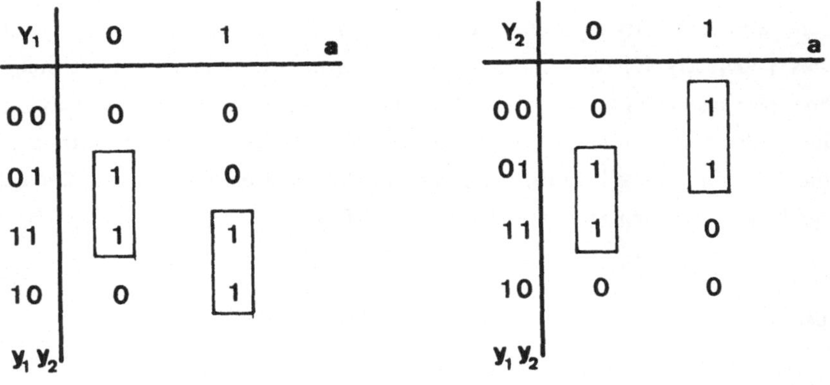

Y_1 ($y_1 y_2$ \ a)	0	1
00	0	0
01	1	0
11	1	1
10	0	1

Y_2 ($y_1 y_2$ \ a)	0	1
00	0	1
01	1	1
11	1	0
10	0	0

Table 3. The values of Y_1 and Y_2 as functions of the input and internal variables.

Thus $Y_1 = a\,y_1 + \bar{a}\,y_2$, and $Y_2 = a\,\bar{y}_1 + \bar{a}\,y_2$

One now wants to find the equation for the output function F. In Table 1[b], we find the values of F corresponding to the stable states 1 , 2 , 3 and 4 . This gives the incomplete Table 4a. How can we arrive at Table 4b ?

a

F ($y_1 y_2$)	0	1
00	0	
01		1
11	1	
10		0

b

F ($y_1 y_2$)	0	1
00	0	–
01	1	1
11	1	–
10	0	0

Table 4. The values of F as a function of the input and internal variables.

The first transition is from F = 0 to F = 1 .

We do not care whether, during the small period Δt, F still has the value 0 or already has the value 1 ; thus we write that the corresponding value of F is don't care (—). In the second transition (from state 2 to state 3) F has the same value , 1 , in both stable states ; obviously, F must hold the value 1 during the transition period (transient state 3). In the third transition (from state 3 to state 4) , F switches from 1 to 0 ; the transient value is don't care (—) . In the last transition from state 4 to state 1 , F keeps the value 0 ; thus F must hold the value 0 during the transient state 1 . One can see in Table 4b that the simplest expression of F is $F = y_2$

Thus,

$$F = y_2$$

$$Y_1 = a\,y_1 + \bar{a}\,y_2$$

$$Y_2 = a\,\bar{y}_1 + \bar{a}\,y_2$$

If one wires up a circuit based on these equations, it will have the behaviour expected for our table lamp.

V.3. Second example : the simple minded cat.

Here we design an arbitrary behaviour, which could be that of, say, a simple-minded cat - a very simple minded cat indeed - and we want to construct a logical machine which would simulate this behaviour. The rules arbitrarily adopted are as follows. Only three types of information are considered : the cat sees a mouse to its left, the cat sees a mouse to its right, the mouse is caught. Accordingly, we introduce three input variables, ℓ (a mouse to the left), r (a mouse to the right) and c (the mouse is caught). We do not consider the appearance of a mouse in front of the cat or the simultaneous appearance of two mice, one to the left, one to the right. Moreover, we consider that once the cat has seen a mouse, its behaviour will not be influenced by another mouse until it has caught the first one (c = 1) and eaten (or released) it (c back to 0). There are three output functions : GL , GR , and T . When the cat sees a mouse to its left (ℓ = 1) it turns to the left (GL = 1) and when this is achieved there is no more a mouse to its left, so ℓ = 0 . The same for r and GR . Once the cat has performed its left (or right) turn, it trails the mouse, and it is interested only in this mouse (T = 1) ; this behaviour is not affected by the appearance of a new mouse to the left (ℓ = 1) or to the right (r = 1). Its interest persists until the mouse has been caught (c = 1) and released (or eaten) (c = 0 again).

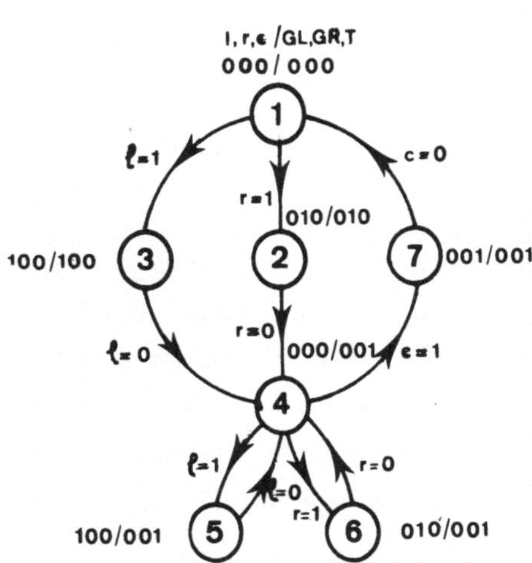

Fig.V.5. The state diagram for the simple minded cat machine.

Fig.V.5. gives the state diagram of the "simple minded cat machine", in which one distinguishes 7 stable states ; the values of the input variables and output functions (ℓ r c / GL GR T) are given for each of these states.

Table 5 gives the flow table for this system ; the states are shown in relation to the sequences of changes of the input variables, and the corresponding values of the output functions GL , GR and T are given in the additional column to the right of the table.

	c=0				c=1				
ℓr	00	01	11	10	00	01	11	10	GL,GR,T
	(1)	2	–	3	–	–	–	–	000
	4	(2)	–	–	–	–	–	–	010
	4	–	–	(3)	–	–	–	–	100
	(4)	5	–	6	7	–	–	–	001
	4	(5)	–	–	–	–	–	–	001
	4	–	–	(6)	–	–	–	–	001
	1	–	–	–	(7)	–	–	–	001

From the stable state 1 ,which may be viewed as the initial state, the system is sensitive to two events (ℓ = 1 or r = 1), represented by horizontal shifts to the columns 10 and 01 , respectively . This leads to the stable states 3 or 2, through the transient states 3 or 2 , respectively; and so on.

Table 5. The flow table for the simple minded cat machine.

As already alluded to at the end of section 1, the number of rows in such a table can often be reduced. It is beyond the scope of this chapter to explain in detail how this is done, but a complete analysis can be found in Florine (1964; 1969). Let us simply mention here that the simplification amounts to the suppression of non-essential transient states, and that in practice one can merge together rows if the stable states are in different columns and the unstable states are the same*.

For instance, in the present case, one can merge together rows 1 and 7 , rows 2 and 3 , and rows 4 , 5 and 6 . This leads to table 6 **

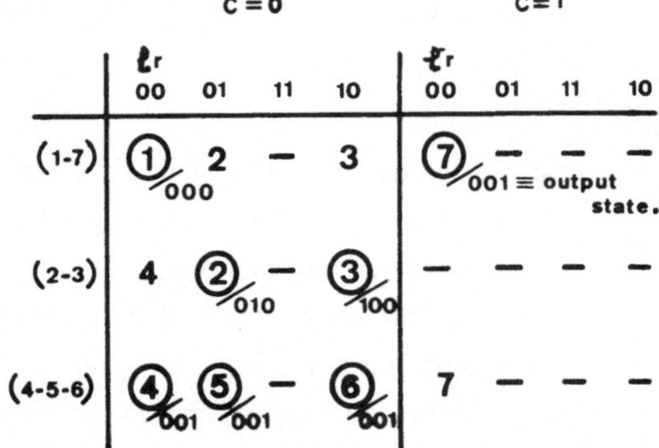

	c=0 ℓr 00	01	11	10	c=1 ℓr 00	01	11	10
(1-7)	(1)/000	2	–	3	(7)/001 ≡ output state.	–	–	–
(2-3)	4	(2)/010	–	(3)/100	–	–	–	–
(4-5-6)	(4)/001	(5)/001	–	(6)/001	7	–	–	–

Usually, and in particular in the present case, there are several ways to reduce the table. As the number of rows is now reduced to 3 , the system can be handled with 2 internal variables , y_1 and y_2 (which might code for 2^2 = 4 rows).

Table 6. A reduced flow table for the "simple minded cat" example.

---

* More precisely, in order to merge two rows, the symbols present in a single column can be merged as follows : a stable state, say (2) and either the corresponding unstable state (in this case, 2) or a don't care yield the stable state ; an unstable state and a don't care give the unstable state ; two unstable states say 4 , can be merged if they are the same. In short, two rows can be merged together if there is no contradiction between them ; moreover the stable states remain stable during the merging operation.

** Notice that after merging it is no longer possible to give an output state value (GL, GR, T) for each row. In fact, a specific output state is now attached to each stable internal state and we may thus have different output values in a single row. For instance 1 has the output value (000) and 7 (001) in Table 6 .

If we arbitrarily assign to the three lines of the flow table the combinations of values 00, 01 and 11 of y_1 and y_2 , the values of Y_1 and Y_2 as functions of the input and internal variables are given in table 7.

Y_1,Y_2	C=0 ℓr 00	01	11	10	C=1 ℓr 00	01	11	10
00	(00)	01	–	01	(00)	–	–	–
01	11	(01)	–	(01)	–	–	–	–
11	(11)	(11)	–	(11)	10	–	–	–
10	–	–	–	–	00	–	–	–
y_1,y_2								

Table 7 . The values of Y_1 and Y_2 as functions of the input and internal variables, in the case of the simple minded cat example.

As regards the stable states, there is no problem ; in each case the values of Y_1 and Y_2 are the same as those of y_1 and y_2 , respectively. Stable states located in the same row have the same values of Y_1 and Y_2 , but they can be distinguished from each other because their input state is different.

As for the transient states, one can usually take the combination of Y_1 and Y_2 corresponding to the state to which the system is adressed.

For instance, if from state 1 (state 00 in column 000) one wants to move to state 3 (01 in column 100) one first operates the horizontal shift from column 000 to column 100 (corresponding to the move : $\ell = 0 \rightarrow \ell = 1$, r and c unchanged) ; this leads to the transient state 3 in which y_1 and y_2 still have their original value (00 , since the system is still in row 00) but Y_1 and Y_2 already have their new values, corresponding to state 3 ; thus in this transient state $Y_1Y_2 = 01$.

There is a complication for the transition from state ④ to state ⑦. As in state 4 the values of Y_1 and Y_2 are 11, and in state 7 they are 00, one would in principle impose on the transient state 7 the values 00, and write these digits in column (ℓr = 00, c = 1), row 11. However the shift from $y_1 y_2 / Y_1 Y_2$ = 11/00 to $y_1 y_2$ / $Y_1 Y_2$ = 00 / 00 would require an <u>exactly simultaneous</u> switching of y_1, and y_2 from 0 to 1. This can be obtained by special methods (synchronous machine with clocking). If this is not used the system will usually transit through 01 / 00 or 10 / 00 (thus, through the rows 01 or 10) depending on whether the turnoff delay Δt has been smaller for y_2 or for y_1. This is called a <u>race</u>. In some cases the race is in practice unimportant, since from either one or the other "transient state" the system will go to the correct final address. However, in many cases, the system will reach a different final stable state, depending on the result of the race. In such cases one deals with so-called <u>critical races</u>; these are carefully avoided, as the outcome depends on small features of the memory organs in an unpredictable way. In the present case, the race may be avoided simply by going first from line 11 to line 10, and then from line 10 to line 00. The pathway is thus 11 / 11 (stable state ④) → 11 / 10 → 10 / 00 → 00 / 00 (stable state ⑦); in other words, one uses two transient states to proceed from ④ to ⑦ (see table 7). Table 8, a and b, differs from table 7 only by the fact that the values of Y_1 and Y_2 are given separately, thus making it possible to find the prime implicants easier.

Y_1 / $y_1 y_2$	ℓr 00 (c=0)	01	11	10	ℓr 00 (c=1)	01	11	10
00	0	0	-	0	0	-	-	-
01	1	0	-	0	-	-	-	-
11	1	1	-	1	1	-	-	-
10	-	-	-	-	0	-	-	-

Table 8a

Y_2 / $y_1 y_2$	ℓr 00 (c=0)	01	11	10	ℓr 00 (c=1)	01	11	10
00	0	1	-	1	0	-	-	-
01	1	1	-	1	-	-	-	-
11	1	1	-	1	0	-	-	-
10	-	-	-	-	0	-	-	-

Table 8b

This gives : $Y_1 = y_1 y_2 + r y_2$ and $Y_2 = y_2 \bar{c} + \ell + r$

The equations of functions GL , GR and T can be found from the following tables 9 a, b, c :

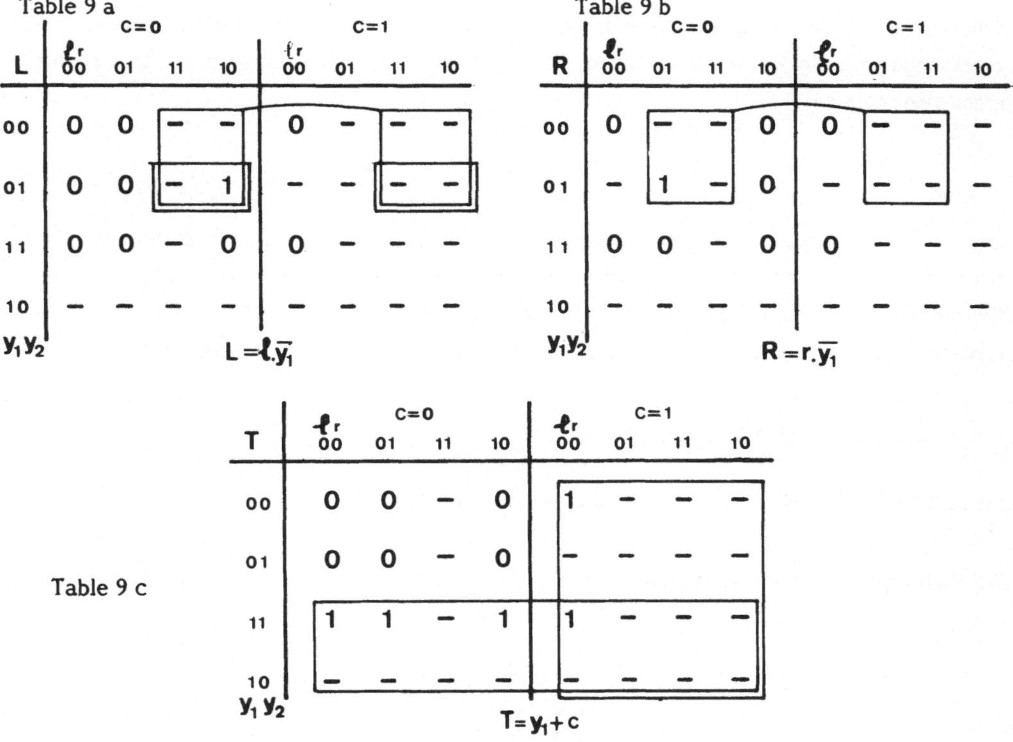

Table 9 a

L ($y_1 y_2$ \ ℓr)	C=0: 00	01	11	10	C=1: 00	01	11	10
00	0	0	-	-	0	-	-	-
01	0	0	-	1	-	-	-	-
11	0	0	-	0	0	-	-	-
10	-	-	-	-	-	-	-	-

$L = \ell . \overline{y_1}$

Table 9 b

R ($y_1 y_2$ \ ℓr)	C=0: 00	01	11	10	C=1: 00	01	11	10
00	0	-	-	0	0	-	-	-
01	-	1	-	0	-	-	-	-
11	0	0	-	0	0	-	-	-
10	-	-	-	-	-	-	-	-

$R = r . \overline{y_1}$

Table 9 c

T ($y_1 y_2$ \ ℓr)	C=0: 00	01	11	10	C=1: 00	01	11	10
00	0	0	-	0	1	-	-	-
01	0	0	-	0	-	-	-	-
11	1	1	-	1	1	-	-	-
10	-	-	-	-	-	-	-	-

$T = y_1 + c$

V.4. Third example : the detector of rotational direction.

V.4.1.Desiderata. An apparatus comprises a rotating disk with alternate opaque and transparent sectors. We wish to detect at any time the direction of rotation of the disk. For this purpose , we project on the disc a beam of light and place behind it two photodiodes as shown in Fig.V.6..

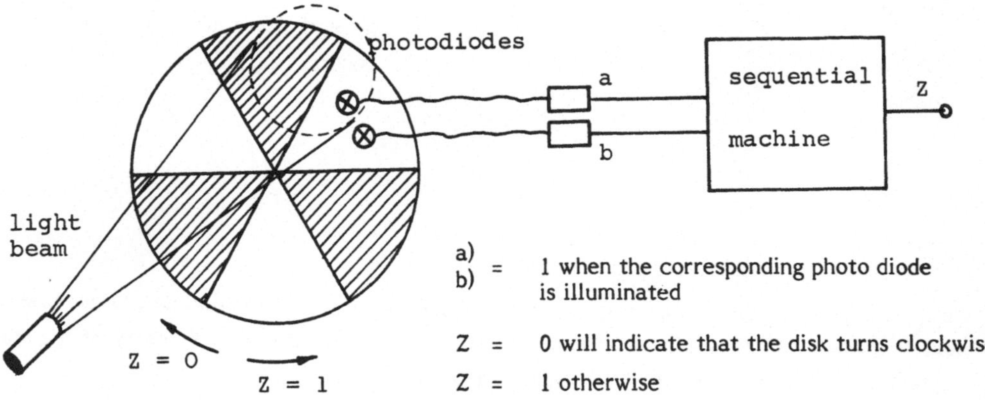

a)
b) = 1 when the corresponding photo diode is illuminated

Z = 0 will indicate that the disk turns clockwise

Z = 1 otherwise

Fig.V.6.

Input variables a and b are associated with the two diodes (their value being 1 when the corresponding diode is illuminated) . We will construct a machine which computes an output function Z of the variables a and b, such that Z = 0 , when the sequences of a and b correspond to a clockwise rotation, and Z = 1 if these sequences correspond to a counter clockwise rotation.

Of course, there is a time lag in the detection of a change in the direction of rotation. This time decreases as the number of sectors increases. However, there is a limit to decreasing the sectors' area : the distance between the photodiodes must be sufficient to distinguish separately for each of them the sequences of illuminated and non-illuminated situations.

V.4.2. Elaboration of the flow-table .

Initially let us suppose that a, b = 0 0 and Z = 0 (state 00/0).

We shall represent that situation by state 1 .

present state	ab 00	01	11	10	Z
1	1	?	?	?	0
		Next state			

Table 10 .

In table 10 we have supposed that (00/0) is a stable state, that is : the next state will be equal to the present state until an input change occurs.

Now, we have 3 possibilities corresponding to 3 input changes. We can eliminate 11 because a and b cannot vary simultaneously from 00 to 11 by construction. This entry of the state table is thus filled by a "don't care" condition (Table 11).

To store the input sequences 00 , 01 and 00 , 10 we create two new states : 2 , 3 . They will also be stable if no other input change occurs.

	ab 00	01	11	10	Z
1	(1)	2	-	3	0
2	?	(2)	?	?	1
3	?	?	?	(3)	0

Table 11. In this table we have circled the stable states.

The value of the output Z must be equal to 1 in state 2 and to 0 in state 3 because in the first case, the photodiode b is illuminated before photodiode a (00 ⟶ 01 : the disk is turning counter-clockwise and Z = 1) and, in the second the photodiode a is illuminated first (00 ⟶ 10 : the disk is turning clockwise and Z = 0).

We see now that we have 6 new possibilities.

For example, from state (2), if we have ab = 01 ⟶ 00 , then the disk has turned clockwise again because b is no longer illuminated and Z must be 0 . We can recognize this state as state 1 and the sequence 00 ⟶ 01 ⟶ 00 returns to state 1 . We may write a 1 as the next state of 2 in the 00 column. (Table 12)

Now, if we follow a similar reasonning for state (3), the sequence 00 ⟶ 10 ⟶ 00 does not return to state 1 . In fact, when ab goes from 10 to 00 and a is no longer illuminated, then the disk has changed its rotating direction and Z becomes 1 . We store the sequence 00 ⟶ 10 ⟶ 00 by saying that the next state of state 3 is, for example, 4 in column 00 . At each step it is only important to see if a sequence requires a new state so that it will be distinguished from all the others.

So, the sequence 00 , 00 , 00 , ... , 00 has the same final state as 00 , 00 , 01 , 00 , 00 , ... , 00 but not to 00 , 00 , 10 , 00 , 00 ,... , 00 .

The table becomes :

	ab 00	01	11	10	Z
1	(1)	2	-	3	0
2	1	(2)	5	-	1
3	4	-	6	(3)	0
4	(4)	2	-	3	1
5	?	?	(5)	?	?
6	?	?	(6)	?	?
7	?	(7)	?	?	?
8	?	?	?	(8)	?

Table 12.

Finally, the state table includes 8 rows :

	ab 00	01	11	10	Z
1	(1)	2	-	3	0
2	1	(2)	5	-	1
3	4	-	6	(3)	0
4	(4)	2	-	3	1
5	-	7	(5)	8	1
6	-	7	(6)	8	0
7	1	(7)	5	-	0
8	4	-	6	(8)	1

Table 13.

From the rules mentioned at the end of section V.3 , we know that table 13 can be condensed by merging rows as follows :

1 - 2 , 3 - 4 , 5 - 7 , 6 - 8

Now the table becomes : Table 14.

	00	01	11	10	ab
1-2	①/ Z=0	②/ Z=1	5	3	
3-4	④/ 1	2	6	③/ 0	
5-7	1	⑦/ 0	⑤/ 1	8	
6-8	4	7	⑥/ 0	⑧/ 1	

Table 14.

We may now rename the different states :

1 - 2 = A ; 3 - 4 = B , 5 - 7 = C ; 6 - 8 = D

	00	01	11	10
A	Ⓐ/ 0	Ⓐ/ 1	C	B
B	Ⓑ/ 1	A	D	Ⓑ/ 0
C	A	Ⓒ/ 0	Ⓒ/ 1	D
D	B	C	Ⓓ/ 0	Ⓓ/ 1

Table 15.

The last table shows that only four different states are necessary to recognize the equivalence classes of input sequences. Moreover, the output function Z depends not only on the internal state but also on the input state : we have different values of Z in the same row.

Now, to synthetize the logical machine simulating the desiderata included in the final table we must assign a boolean code to each state. Suppose we choose : A = 00, B = 01 , C = 11 , D = 10

Y_1Y_2 \ ab	00	01	11	10
A ≡ 00	(00)	(00)	11 *	01
B ≡ 01	(01)	00	10 *	(01)
C ≡ 11	00 *	(11)	(11)	10
D ≡ 10	01 *	11	(10)	(10)
y_1y_2				

Table 16.

With this assignment, we can see that four cases of critical races appear * . From the designer point of vue, critical races must be eliminated. Let us change the assignment :

A = 00 ; C = 01 ; D = 11 ; B = 10

Y_1Y_2 \ ab	00	01	11	10
A ≡ 00	(00)	(00)	01	10
C ≡ 01	00	(01)	(01)	11
D ≡ 11	10	01	(11)	(11)
B ≡ 10	(10)	00	11	(10)
y_1y_2				

Table 17.

The last table does not contain any races between secondary variables.

The analysis and synthesis of Y_1, Y_2, Z give :

Y_1 \ ab	00	01	11	10
00	0	0	0	1
01	0	0	0	1
11	1	0	1	1
10	1	0	1	1
y_1y_2				

$$Y_1 = a\bar{b} + ay_1 + \bar{b}y_1$$

$$= a\bar{b} + (a+\bar{b})\ y_1$$

Table 18.

Y_2	00	01	11	10	ab
00	0	0	1	0	
01	0	1	1	1	
11	0	1	1	1	
10	0	0	1	0	
y_1y_2					

$$Y_2 = ab + ay_2 + by_2$$
$$= ab + (a+b)\ y_2$$

Table 19.

Z	00	01	11	10	ab
00	0	1	1	0	
01	0	0	1	1	
11	1	0	0	1	
10	1	0	1	0	
y_1y_2					

$$Z = b\bar{y}_1\bar{y}_2 + a\bar{y}_1y_2 + a\bar{b}y_2 + ab\bar{y}_2 + \bar{a}\bar{b}y_1$$

Table 20.

The following figure shows the final logical diagram.

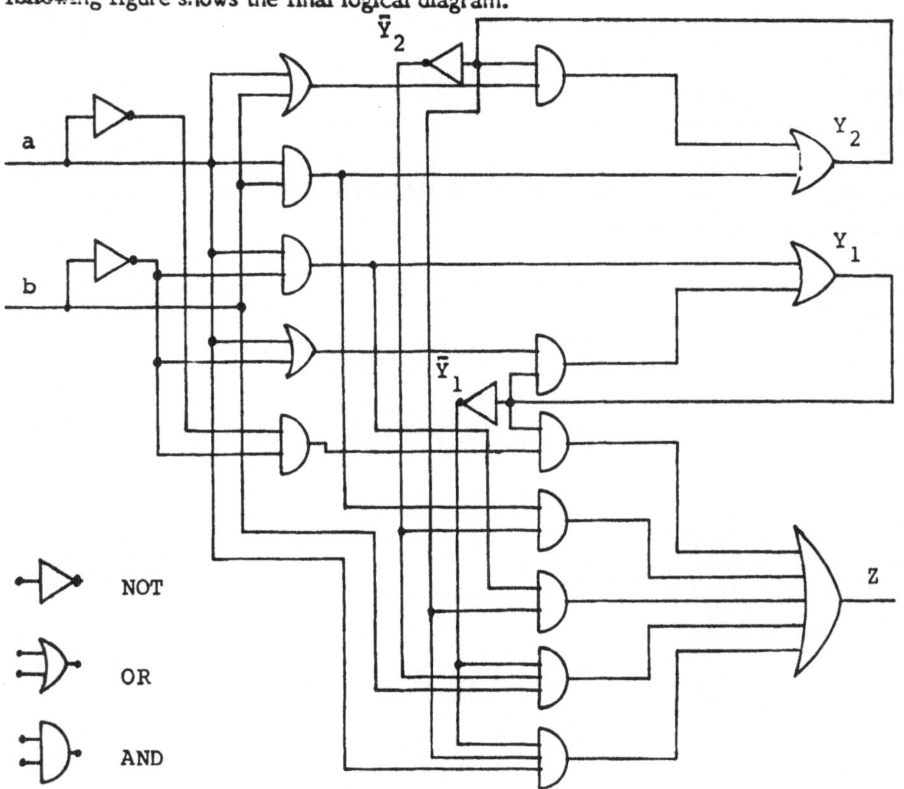

Figure V.7.

BIBLIOGRAPHY

E.J. Mc CLUSKEY 1962 & T.C. BARTEE — " A Survey of switching circuit theory" (Mc. Graw Hill).

E.J. Mc CLUSKEY 1965 — "Introduction to the theory of switching circuits" (Mc. Graw Hill).

R.F. MILLER1965 — "Switching Theory : Vol.1. Combinational Circuits
Vol.2. Sequential Circuits and Machines.
(J. Wiley & Sons).

J. FLORINE 1964 — "La Synthèse des Machines Logiques et Son automatisation". (Dunod).

J. FLORINE 1969 — "Automatismes à Séquences et Commandes Numériques". (Dunod)

ACKNOWLEDGMENTS

The authors cordially thank Professor René THOMAS, editor, for many helpful discussions and for his contribution to the manuscript.

CHAPTER VI.

Kinetic logic : a boolean analysis of the dynamic behaviour of control circuits.

R. THOMAS.

In this central chapter I show how a system amenable to boolean description can be analyzed in terms of a set of logical equations. Each equation relates, for any time, the values of a function a, b, c,...(associated with the state, on or off, of a gene, of a chemical reaction, etc...), to the values of input variables, and of memorization variables α, β, γ,...(associated with the presence or absence of the product of a gene, of a chemical reaction, etc...). Time is present in a similar way as in differential equations ; in fact, in our logical equations a, b, c,... play essentially the role of the time derivatives of α, β, γ. From the set of logical equations describing a system, one can derive its stable steady states (if any), the pathways (temporal sequences of states) and the conditions determining which pathway will be followed.

Chapters IV and V have introduced the reader to a type of problem which can usually be formulated as follows : given a set of desiderata, find a set of logical equations, as simple as possible, symbolizing a logical machine which will realize these desiderata in a reliable way. Although such a process might be used by scientists as one way to derive a model from experimental facts, a somewhat different situation is more frequent : a set of experimental data has suggested a model through a complex pathway involving both reasoning and intuition, and we want to analyze the predictions of the model in a rigourous way. I shall show that the conceptual process used by engineers to conceive a logical machine can function in the reverse direction as a tool in the analysis of the dynamic behaviours of a model. However, for reasons described below, one has to introduce time in a more sophisticated way than just giving the state of the system at time $t + \Delta t$ as function of its state at time t .

VI.1. The viewpoint of the designer of logical machines : from the desiderata to the logical equations.

In sequential problems (see chapter V), the value of a function Y_i may depend not only on the present values of the input variables x_1,... x_n, but also on previous values of functions Y_1,... Y_m . One way to treat such problems is to formulate the state of the system at time $t + 1$ as a function of its state at time t , or the state at time t as a function of the state at time $t - \Delta t$.

$$Y_i(t) = f_i\left[x_1(t),\ldots x_n(t)\;;\;Y_1(t-\Delta t),\ldots Y_m(t-\Delta t)\right]$$

values of the input variables at time t — values of functions at time $t - \Delta t$

A remarkable subtlety (see Florine, 1964 and 1973 ; this book, chapter V, by A. Leussler and P. Van Ham) consists in associating with a function Y_i a so-called memorization variable (or internal variable) y_i, defined as having at time t the value that the corresponding function Y_i had at time $t - \Delta t$:

$$y_i(t) = Y_i(t - \Delta t)$$

Equation (1) thus becomes :

$$Y_i(t) = f_i\left[x_1(t),\ldots x_n(t)\;;\;y_1(t),\ldots y_m(t)\right]$$

values of the input variables at time t — values of the internal variables at time t

and the sequential problem is now formally reduced to a combinational problem, since each function $Y_i(t)$ is given as a function of the values of input and memorization variables at this same time t .

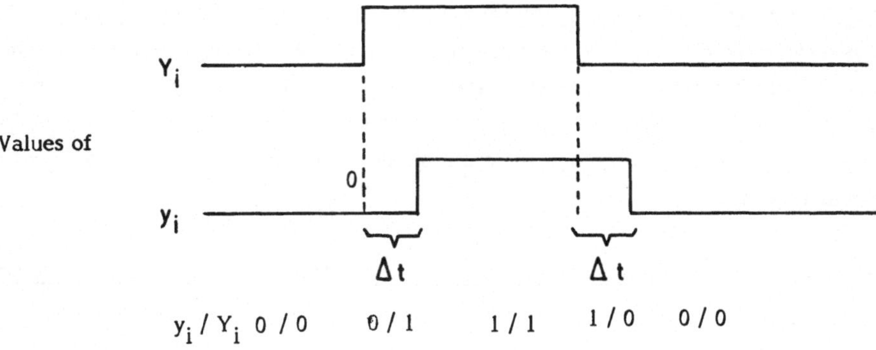

Fig.VI.0. The engineer's classical relationship between a function Y_i and its memorization variable y_i . A change of the value of Y_i is memorized by the system because in stationary situations Y_i and y_i have the same value, but when the value of Y_i changes y_i, the memorization variable keeps the former value of Y_i for a short time.

Suppose now that at time t a single signal switches on two functions, Y_1 and Y_2. The values $y_1,y_2/\,Y_1,Y_2$, which were initially 0 0 / 0 0 , thus shift to 0 0 / 1 1 . In principle, both memorization variables will jump together to the value 1 at time $t + \Delta t$.

But in practice the delays $\Delta_1 t$ and $\Delta_2 t$, corresponding to functions Y_1 and Y_2, may not be exactly equal, depending on uncontrolled differences in memorization organs ; in this case either y_1 or y_2 would switch first, and this in an unpredictable way. This is called a <u>race</u> , and where the alternatives eventually lead to different final states, a <u>critical race</u> . Critical races are carefully avoided by logic machine builders, because they lead to different final states depending on uncontrolled factors.

VI.2. <u>The reverse process, from the logical equations to the behaviour of the system : principles and difficulties.</u>

When one tries to conceive a logical machine which would realize a set of desiderata, one proceeds from these desiderata to a state table , (see Chap. IV and V) and from this table one derives a set of logical equations which contain the necessary and sufficient information to wire up the logical machine.

Instead, one may ask : given a system in which the interactions between the elements can be adequately described by a set of logical equations, can one predict the dynamic behaviour of the system ? In this case, the conceptual process described in the proceeding section may be used in the reverse direction : first one describes the structure of the system in terms of logical equations ; from them one builds a state table ; from this (see below) one derives the temporal sequences of states (pathways) and the conditions which determine which pathway will be followed.

In a first phase, one tried to relate the state of a system at time t + 1 to its state at time t . For instance, the table and graph below are redrawn from one of the very first papers in which boolean algebra is applied to biology (Kauffman, 1969). Functions a, b, c refer to the state, on or off, of three genes ; their values at time t + 1 is given as a function of their values at time t . With this quantization of time initially used, two variables whose values are committed to switch will switch in a synchronous way. For instance, state 010 is followed by state 001 (a change involving the two variables b and c), rather than by a choice between 000 (variable b has commuted first) and 011 (variable c has commuted first). With such conventions, two or more states can converge to a common next state, but the system cannot "bifurcate" from one state to two or more distinct next states.

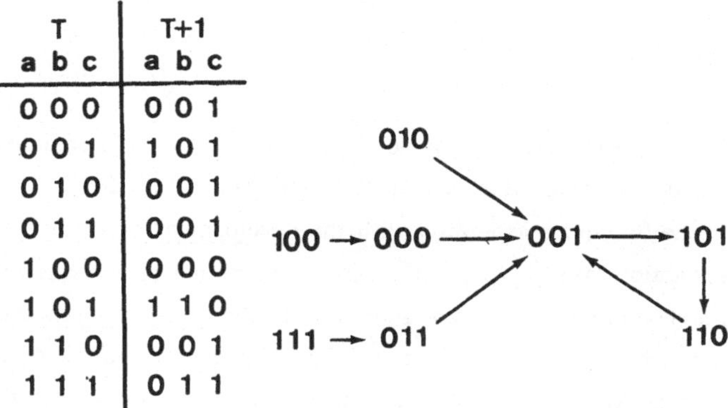

T a b c	T+1 a b c
0 0 0	0 0 1
0 0 1	1 0 1
0 1 0	0 0 1
0 1 1	0 0 1
1 0 0	0 0 0
1 0 1	1 1 0
1 1 0	0 0 1
1 1 1	0 1 1

Table 1. A system treated in synchronous terms (redrawn from Kaufman, 1969).

Reasoning in terms of memorization variables related to functions by a time shift Δt does not help much here. Of course, one could decide to associate with each function Y_i a characteritic time delay :

$$y_i(t) = Y_i(t - \Delta_i t)$$

One might even generalize by somehow expressing that $\Delta_i t$ may be different depending on whether y_i is on the point of switching on or off. This is a so called asynchronous treatment. But this does not help much as long as the time delays are perfectly arbitrary .

VI.3. A new definition of the relationship between functions and their associated memorization variable : involvement of time in non-arbitrary time delays.

(Thomas, 1973 ; Thomas & Van Ham, 1974).

The above mentioned difficulties can be overcome by adopting a different relationship between a function and its associated memorization variable. Instead of defining a memorization variable in terms of an arbitrary time shift Δt applied to the function, we use as a function and its associated memorization variable two related but qualitatively different things.For instance,in chemistry,one can use as a function the occurence of a reaction and as its associated memorization variable the presence or absence at a significant concentration of the product of the reaction. Clearly the function and its associated memorization variable are related to different things - the speed of the reaction and the concentration of the product of the reaction, respectively. To take a genetic example, let S be a gene, s its state of expression ($s = 1$ if the gene is on, $s = 0$ if it is off) and σ the presence of an efficient concentration of the gene product (assuming that there is a threshold concentration of this gene product, $\sigma = 1$ if the concentration reaches or exceeds this threshold, $\sigma = 0$ if it is below the threshold) :

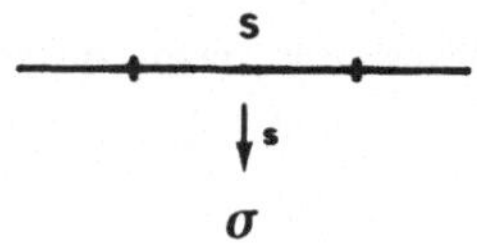

We take s as a function and σ as its associated memorization variable. It must be clear that s and σ are distinct ; function s is related to the expression of a gene (or, more generally, with the development of a process), which has the dimensions of a reaction rate $\frac{dx}{dt}$; the associated memorization variable σ is related to the presence of a gene product (or, more generally, to the result of a process), which has the dimensions of a concentration (x). (Thomas,1973 ; Thomas & Van Ham, 1974). In other words a function such as s describes a flux, its memorization variable σ describes a stock.

The memorization variable is now related to its function by delays which are no longer arbitrary, and which can be (and usually are)

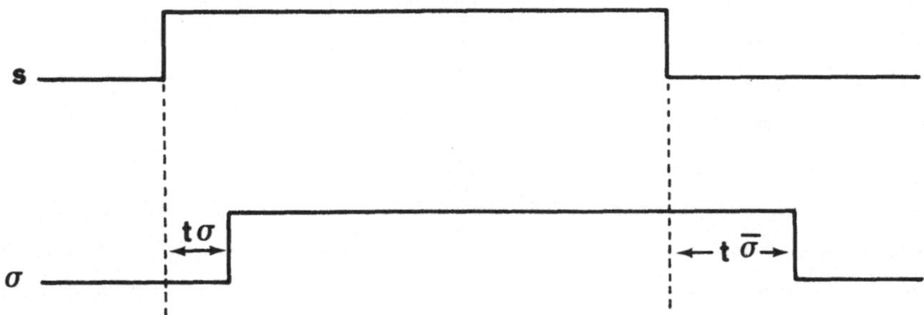

different for the shift up and for the shift down. In concrete terms, in the case of gene expression, t_{σ} is the time delay between the instant the gene is switched on and the appearance of a significant concentration of the gene product at the proper site ; this includes as an absolute minimum the times required for transcription and translation, usually some more time to allow a sufficient accumulation of the product, and in many cases diffusion to an appropriate site. The "off" delay, $t_{\bar{\sigma}}$, is the time lapse between the instant a gene is switched off and the effective disappearance of the gene product, that is, the moment when the concentration of the product at its site of action has dropped below its effective level. The delay $t_{\bar{\sigma}}$ is a complex function of various elements, of which a crucial one is the lifetime of the gene product. As certain regulatory proteins have a structure such that they are destroyed extremely rapidly in vivo (the half life of the N protein of bacteriophage λ is of the order of 2 minutes) and others are extremely stable (the λ repressor may persist for days in the absence of re-synthesis) the off delays are extremely variable, whereas the on delays are frequently of the order of some minutes.

In this new optic, the dynamic behaviour of systems is completely different from that described in section 2 of this chapter. Let us come back to the example taken from Kauffman's 1969 paper. Instead of describing how the values of a, b and c at time t + 1 can be derived from the values at time t, we now relate the value of functions a, b, c (which deal with the state, on or off, of genes A, B, C) to the values of the memorization variables α, β , γ (which deal with the presence or absence of the corresponding gene products) : $a = \bar{\beta} . \gamma$; $b = \alpha . \gamma$; $c = \bar{\alpha} + \beta$. The graphs showing the possible temporal evolutions of this system can be found in appendix n°1.

At first view one might object that chronology has disappeared from the equations, because they provide, at any time , the values of a, b, c ... as functions of the values of the input variables and of the memorization variables α ,β , γ , ... at the same time. However, except for its discontinuous character this description is very similar to that of a set of differential equations. Differential equations relate, for any time, the value of a time derivative to the values of the variables. In our logical equations a, b, c, ... play essentially the role of the time derivatives of α, β, γ ... . The exact relationship between the boolean functions a, b, c, ... and continuous time derivatives will be analyzed elsewhere (Thomas & Richelle,in preparation). As will be seen below, time reappears explicitly at the level of the graphs showing the pathways (temporal sequences of states) and of the analysis of the conditions determining which pathway is used. An important feature of these graphs is their "non-Markovian" character : in many cases, the issue of a choice depends on events which have taken place 2, 3, ... n steps ahead.

Another type of asynchronous treatment has been used by Glass (1975). In this work, he still relates boolean functions at time $t + \Delta t$ with the values of other* functions at time t, but with variable values of Δt. In this and subsequent work Glass systematically uses the n-cube representation of states. More recently, Glass (1977) has adopted a very interesting formalism using boolean functions built in differential equations.

VI.4. Temporal analysis of control circuits

a) equations, matrices (states tables) and identification of the stable states.

We can now analyse the dynamic behaviour of simple control circuits. As first example, let us consider a system comprising three genes, A, R1 , R2+, which interact as follows : α, the product of gene A , is required for the expression of gene R1 ; $\rho 1$, the product of gene R1 , exerts a negative control on the expression of gene R2 , whose product $\rho 2$ exerts a negative control on the expression of gene A . In addition, the product $\rho 1$ is reversibly inactivated at high temperature (T).

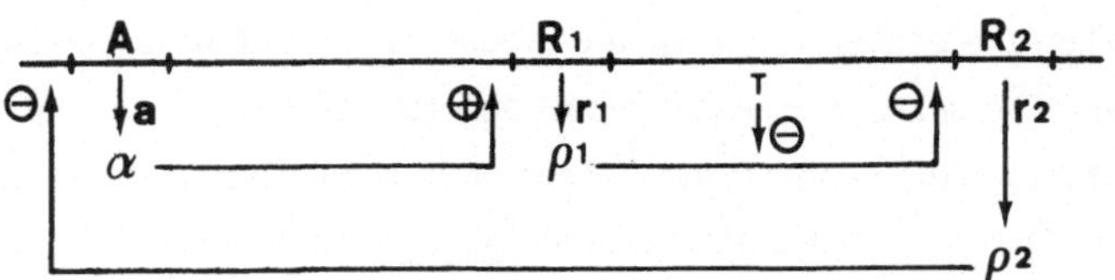

Fig.VI.1. A simple positive feedback loop (see also chapter VII)

This system can be described in terms of logical equations as follows :

$$a = A \cdot \overline{\rho}2$$
$$r1 = R1 \cdot \alpha$$
$$r2 = R2 \cdot (\overline{\rho 1} + T)$$

The meaning of the logical connectives . , + etc ... is given in Chapter IV.

The third equation, for example, means that the conditions for the expression (r2) of the third gene are :

1) the presence in the system of a genetically normal specimen of the gene (that is, R2) and

2) the absence of the product $\rho 1$ (that is, $\overline{\rho 1}$) or a temperature such that this product, if present, would be inactive (T).

---

* He excludes self-input.

\+ A stands for activator, R for repressor.

It is seen that the equations use four input variables, A , R1 , R2 and T . A complete state table would thus comprise 16 (2^4) columns, corresponding to the various combinations of values of these variables. Here, I shall only consider the most interesting case, in which each of the three genes is present as a functional allele (A , R1 and R2 = 1). The equations will thus reduce to :

$$a = \overline{\rho 2}$$
$$r1 = \alpha$$
$$r2 = \overline{\rho 1} + T \;.$$

and the state table, to two columns (T = 0 , T = 1) as there is only one input variable left. Before presenting and analyzing the matrix, it may be useful to show a simplified representation of the system :

α
\+
−
ρ1
ρ2
OR
T

In this representation, one shows only the variables (here, α, $\rho 1$, $\rho 2$ and T) with arrows which specify how they interact with each other. It is essential to stress that $\alpha \xrightarrow{+} \rho 1$ does not mean that substance α is transformed into substance $\rho 1$; rather, α is necessary for the synthesis of $\rho 1$. The matrix, which can be directly read from the equations, is :

	T = 0	T = 1
α, ρ_1, ρ_2	a, r_1, r_2	a, r_1, r_2
0 0 0	101	101
0 0 1	(001)	(001)
0 1 1	000	001
0 1 0	100	101
1 1 0	(1 10)	111
1 1 1	0 10	011
101	011	011
100	111	111

Table 2. The state table for the system $a = \overline{\rho 2}$, $r1 = \alpha$, $r2 = \overline{\rho 1} + T$

For the analysis of the state table, we must remember that the column to the left, outside the matrix, gives the states of the memorization variables, this is, in our case, it tells which gene products are present and which are absent. The columns in the matrix tell the corresponding states of the functions, that is, in our case, they tell which genes are on. For instance, in the state 0 0 0 of the memorization variables, none of the gene products is present. The corresponding values of functions a, r1 , r2 , are given by

1 , 0 , 1 , respectively (see the matrix) ; in other words all three gene products are absent, but genes A and R2 are on, so that their products are being made. One deals thus with an unstable state, since sooner or later one of the gene products, α or ρ 2 , will reach an efficient concentration. For state 0 0 1 of the memorization variables (products), the state of the functions (genes) is also 0 0 1 . This situation will thus not change spontaneously. We are dealing with a stable state (circled in the matrix : see Chapter V). This is not an equilibrium state, since the gene R2 is on and product ρ 2 is continuously produced ; it is in fact a stable steady state.

Inspection of the matrix shows that, at low temperature, the system has two stable states, one (0 0 1) in which gene R2 is on and the others are off, one (1 1 0) in which genes A and R1 are on and gene R2 is off. All the other states are transient ones. How exactly the system will finally reach one of the stable states will be examined below. At this stage, I just want to recall the interest of systems which may, in identical conditions, be stabilized in either of two distinct states. The right column of the matrix shows that at high temperature the system has only one stable state (0 0 1) .

Even before a detailed study of the pathways (temporal sequences of state) it may be interesting to point out that this simple system exhibits a phenomenon of hysteresis. Suppose that the system is at low temperature in the stable state 1 1 0 and we switch to high temperature . The system will eventually stabilize in the only stable state permitted at high temperature, 0 0 1 . If one switches back to low temperature, the system will remain in state 0 0 1 rather than return to its initial state 1 1 0 , because 0 0 1 is also stable at low temperature. The pathway is shown below.

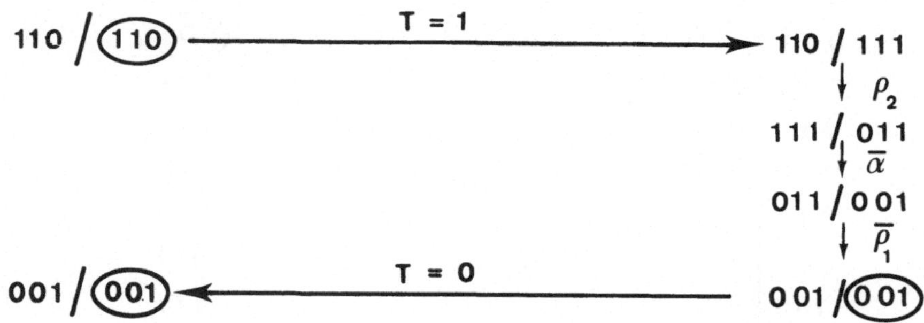

In this notation a state is described by the values of the memorization variables followed, after a slash, by the values of the corresponding functions (here : α, ρ1, ρ 2 / a, r1 , r2). A more compact notation consists in giving the values of the memorization variables surmounted by a dash where the value is different from that of the corresponding function (in which case the value of the memorization variable must change). The same pathway is now written as follows.

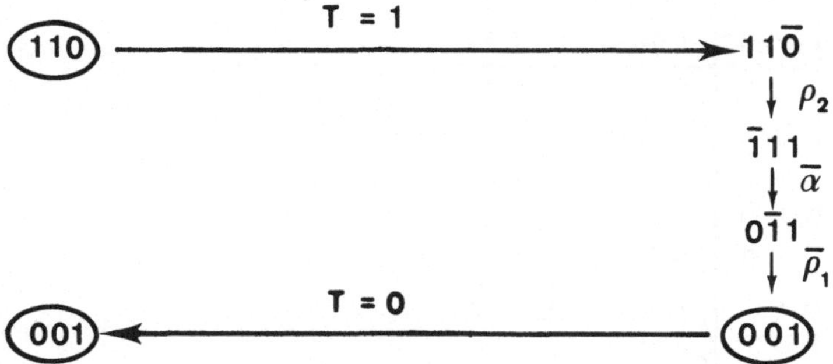

I shall now describe a second system which, although superficially similar to the system discussed above, in fact behaves quite differently. This system also comprises three genes, R , A1 and A2 , functionally connected as follows :

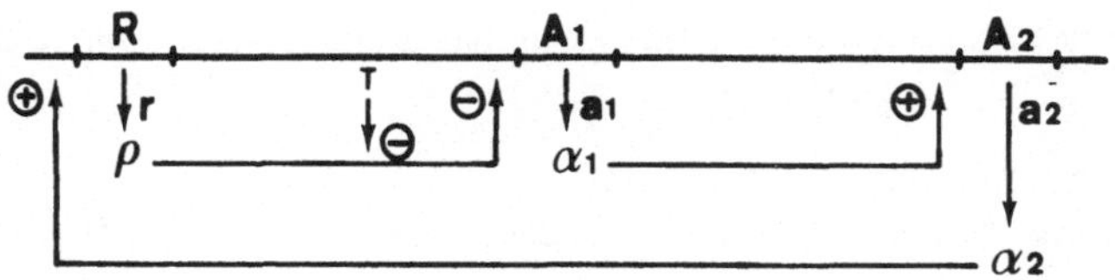

Fig.VI.2. A simple negative feedback loop, (see also Chapter VII).

As in the first circuit, one of the gene products (ρ) is reversibly inactivated at high temperature.

The equations are :

$$r = R.\alpha 2$$
$$a1 = A1.(\bar{\rho} + T)$$
$$a2 = A2.\alpha 1$$

In the particular case in which all three genes are genetically normal (R, A1 and A2 = 1), the equations become :

$$r = \alpha 2$$
$$a1 = \bar{\rho} + T$$
$$a2 = \alpha 1$$

The simplified picture of the system is :

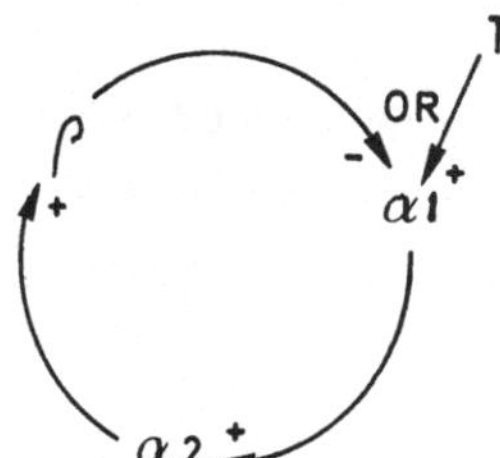

ρ,α_1,α_2	T=0 r,a1,a2	T=1 r,a1,a2
0 0 0	0 1 0	0 1 0
0 0 1	1 1 0	1 1 0
0 1 1	1 1 1	1 1 1
0 1 0	0 1 1	0 1 1
1 1 0	0 0 1	0 1 1
1 1 1	1 0 1	(1 1 1)
1 0 1	1 0 0	1 1 0
1 0 0	0 0 0	0 1 0

Table 3. The state table for the system

$$r = \alpha 2$$
$$a1 = \bar{\rho} + T$$
$$a2 = \alpha 1$$

As one can see from the state table, the boolean analysis does not predict any stable steady state for this system at low temperature ; as we shall see below, the system is supposed to fall into, and follow indefinitively, a cycle comprising 6 boolean states. However at high temperature, the system has a stable state, which, in this case, will be reached sooner or later (see below).

b) The pathways (sequences of states).

Let us examine more closely the first system described in section 4a, (Fig.VI.1.and Table 2) beginning with the situation T = 1 (that is, at high temperature).

α,ρ_1,ρ_2	a,r1,r2
$\bar{0}$ 0 $\bar{0}$	1 0 1
0 0 1	(0 0 1)
0 $\bar{1}$ 1	0 0 1
$\bar{0}$ $\bar{1}$ $\bar{0}$	1 0 1
1 1 $\bar{0}$	1 1 1
$\bar{1}$ 1 1	0 1 1
$\bar{1}$ $\bar{0}$ 1	0 1 1
1 $\bar{0}$ $\bar{0}$	1 1 1

This state table repeats part of the table 2 but it is completed by adding the dashes to indicate which values of the memorization variables will change. When a state is reached which calls for more than one change, it is assumed that one of the changes occurs first. The commitment of the system for the other changes may be cancelled or not, depending on the case. For instance, if one proceeds from $\bar{0}\,\bar{1}\,\bar{0}$ to $1\,1\,\bar{0}$ (that is, the call for synthesizing α is realized), one sees that the call for synthesizing $\rho 1$ is cancelled, while the call for synthesizing $\rho 2$ still holds. For more detail see Van Ham (1975).

Arbitrarily, we choose $\bar{0}\,0\,\bar{0}$ as the initial state. Functions a and r2 are both on, and consequently products α and $\rho 2$, initially absent, are being synthesized. Depending on whether α or $\rho 2$ first reaches the value 1 , the next step will be $1\,\bar{0}\,\bar{0}$ or (0 0 1) :

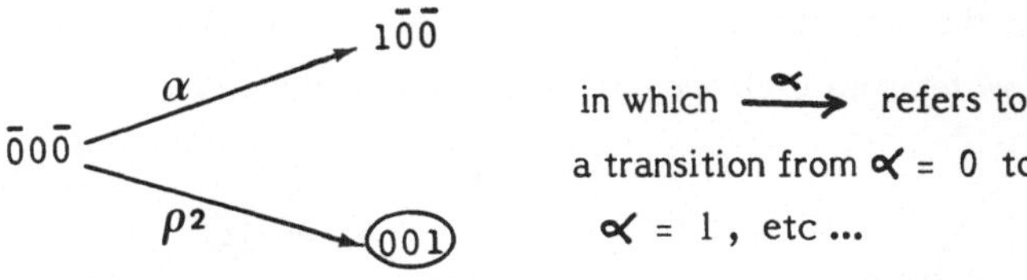

(001) is stable, but $1\bar{0}\bar{0}$ is not. From $1\bar{0}\bar{0}$ one can proceed to $11\bar{0}$ or to $\bar{1}\bar{0}1$ depending on whether $\rho 1$ or $\rho 2$ first reaches the value 1 , and so on :

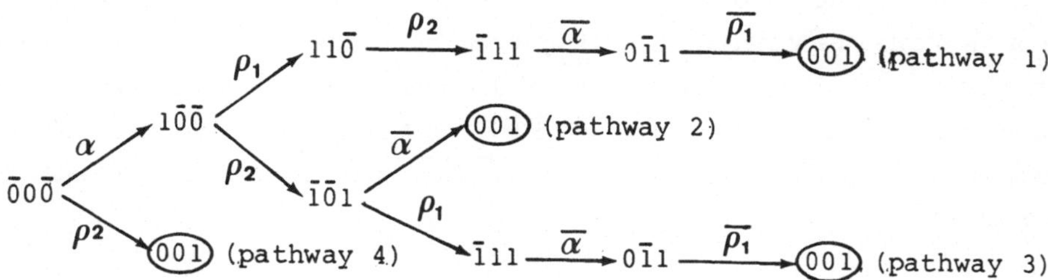

Thus, from the initial state $\bar{0}0\bar{0}$*, the system may follow either of various pathways, but in all cases it will finally reach the stable steady state 0 0 1.

It is often convenient, especially in more complex cases, to represent the states by their octal translation (which is of course the same as the decimal translation for binary numbers of three bits or less).

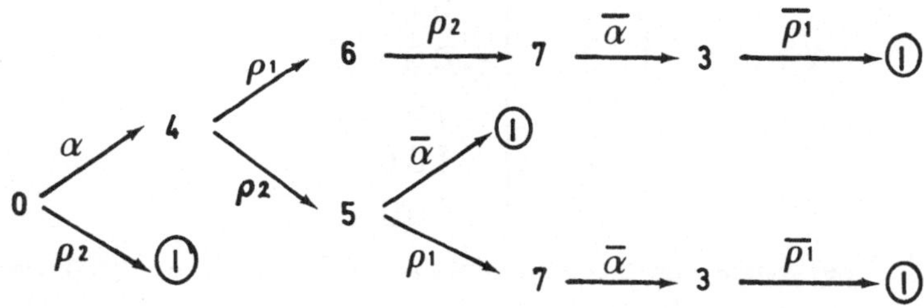

* In principle, one can proceed from any state. A more complete graph is obtained in this system if one chooses $\bar{0}\bar{1}\bar{0}$ as the initial state. One can easily show that this state can be reached from none of the other states whereas all the other states can be reached from it.

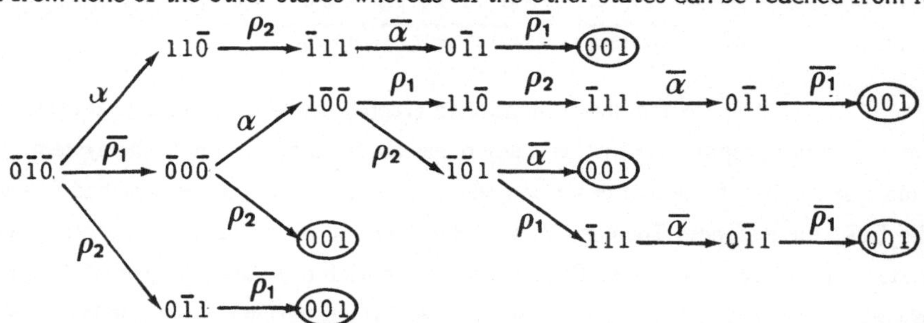

In the other situation of the first system (Fig.VI.1. and Table 2), column T = 0, that is, at low temperature), one obtains the following graph :

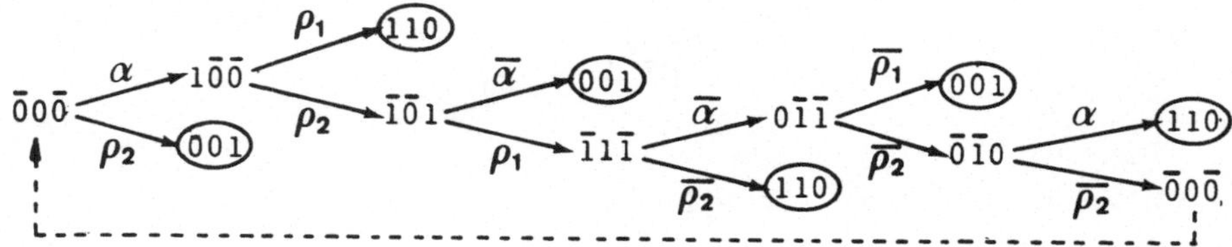

There are six transitory states. From any of these states, the system can proceed either to one of the stable steady states, (001) or (110) depending on the case, or to another transitory state ; but sooner or later it will reach one of the stable states. As will be seen in appendix n°2, the cycle $\bar{0}0\bar{0} \rightarrow 1\bar{0}\bar{0} \rightarrow \bar{1}\bar{0}1 \rightarrow \bar{1}1\bar{1} \rightarrow 0\bar{1}\bar{1} \rightarrow \bar{0}\bar{1}0$ might be followed in principle, but the conditions are irrealistic ; this situation corresponds to the unstable steady state described in the continuous analysis.

As we have seen already, the second system described above (Fig. VI,2 ; table 3) has no stable steady state at low temperature :

$\alpha, \rho 1, \rho 2$	r, a1, a 2
$0\bar{0}0$	0 1 0
$\bar{0}\bar{0}\bar{1}$	1 1 0
$\bar{0}11$	1 1 1
$01\bar{0}$	01 1
$\bar{1}\bar{1}\bar{0}$	001
$1\bar{1}1$	101
$10\bar{1}$	100
$\bar{1}00$	000

From state $\bar{0}\bar{0}\bar{1}$ the system can proceed to states $10\bar{1}$, $\bar{0}11$ or $0\bar{0}0$, depending on whether the event α, $\rho 1$ or $\overline{\rho 2}$ takes place first ; similarily, from state $\bar{1}\bar{1}\bar{0}$ the system can proceed to states $01\bar{0}$, $\bar{1}00$ or $1\bar{1}1$ depending on whether $\bar{\alpha}$, $\overline{\rho 1}$, or $\rho 2$ takes place first. But once the second state is reached there is no choice any more other than to follow a cycle :

$$0\bar{0}0 \rightarrow 01\bar{0} \rightarrow \bar{0}11 \rightarrow 1\bar{1}1 \rightarrow 10\bar{1} \rightarrow \bar{1}00$$

Thus the final state of this system is a permanent oscillation.

In the above mentioned graphs, the same state appears several times. One may also use graphs with just as many nodes as there are states in the system. In three-variable systems like those just described, each of the 2^3 states can be ascribed to a vertex of a cube (see, for instance, Thomas & Van Ham, 1974). More generally, one can ascribe to each state a vertex of a square (2-cube) for a 2-variable system, of a cube (3-cube) for a three-variable system and of a hypercube (n-cube) for a n-variable system (Glass , 1975).

In this representation, the graphs corresponding to the first (Fig VI,1, Table 2) and second systems (Fig.VI,2, Table 3) in the case of T = 0 (low temperature) are respectively :

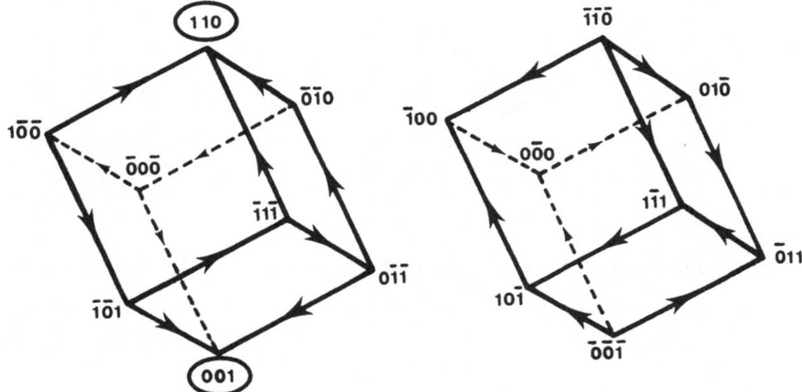

In the first case (the cube to the left), as already described, the system may for some time follow the cycle of the six transitory states, but sooner or later it will fall into one of the vertices corresponding to the stable steady states (0 0 1) or (1 1 0).

In the second case (the cube to the right) there is no stable steady state. Sooner or later the system reaches a permanent situation which is seen in our formalism as a stable cycle .

$$(0\bar{0}0 \to 01\bar{0} \to \bar{0}11 \to 1\bar{1}1 \to 10\bar{1} \to \bar{1}00$$

or, in short, 0 - 2 - 3 - 7 - 5 - 4)

c) <u>The decision as to which pathway is followed.</u>

Let us go back to the first example, (see Fig.VI.1. and Table 2, in section 4a), this time at high temperature, taking $\bar{0}0\bar{0}$ as the initial state . In this particular case, all the pathways lead to a common stable state. It may nevertheless be interesting to determine the conditions which lead to the various pathways.

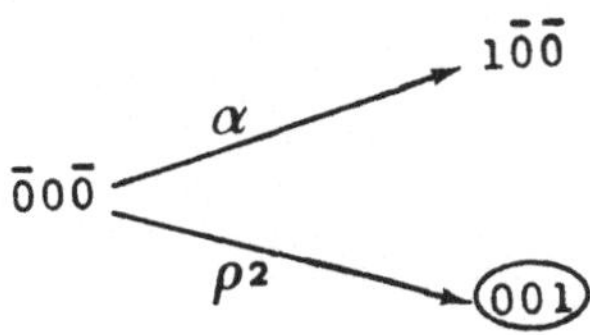

Whether the system proceeds to state $1\bar{0}\bar{0}$ or stabilizes in state (0 0 1) depends on the relative lengths of the time delays t_{α} and $t_{\rho 2}$; $t_{\alpha} < t_{\rho 2}$ implies the transition $\bar{0}0\bar{0} \xrightarrow{\alpha} 1\bar{0}\bar{0}$ whereas $t_{\alpha} > t_{\rho 2}$ implies the transition $\bar{0}0\bar{0} \xrightarrow[\rho 2]{} 001$*.

* An improbable exact equality between t_{α} and $t_{\rho 2}$ would lead to the double transition $\bar{0}0\bar{0} \Rightarrow \bar{1}\bar{0}1$. This possibility is not treated here. However, should it take place, it would automatically be taken into account in the simulations.

From state $1\,\bar{0}\,\bar{0}$, the system may proceed to $1\,1\,\bar{0}$ or to $\bar{1}\,\bar{0}\,1$:

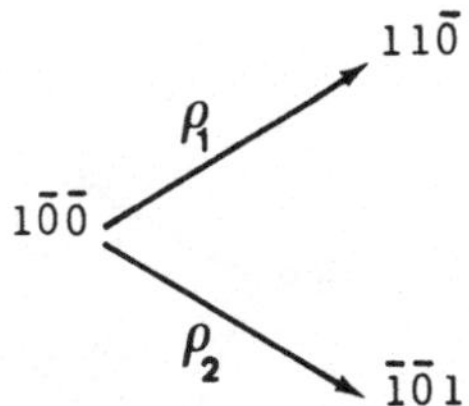

One might be tempted at first view to think that the decision depends on the relative lengths of $t_{\rho 1}$ and $t_{\rho 2}$. However, gene R1 is switched on only when the system reaches state $1\,\bar{0}\,\bar{0}$, whereas gene R2 is already on during the transition $\bar{0}\,0\,\bar{0} \rightarrow 1\,\bar{0}\,0$, that is, the period t_{α} .

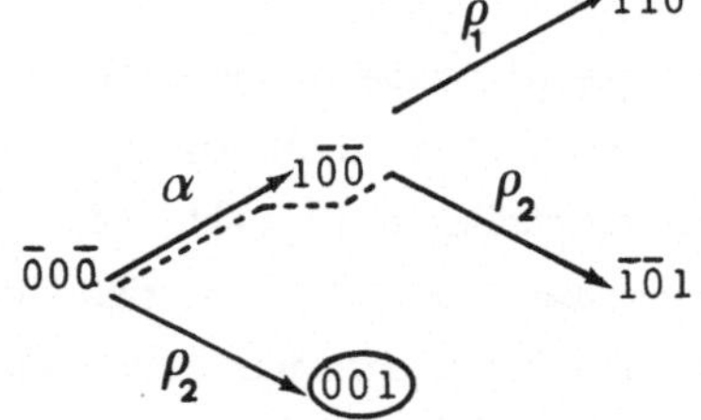

Thus, the decision to proceed to state $1\,1\,\bar{0}$ or $\bar{1}\,\bar{0}\,1$ depends on the relative lengths of $t_{\alpha} + t_{\rho 1}$ and $t_{\rho 2}$ (+ is here the arithmetic sum).

For the sake of clarity, we use a symbolism in which $\bullet\!\xrightarrow{\rho 1}$ means that the synthesis of product $\rho\,1$ has just been switched on, $\angle\!\xrightarrow{\rho 2}$ means that the synthesis of product $\rho\,2$ was already on during the preceeding period and $\alpha\!\xrightarrow{\rho 2}$, that the product $\rho\,2$ was already synthesized during the two preceeding periods. The graph becomes :

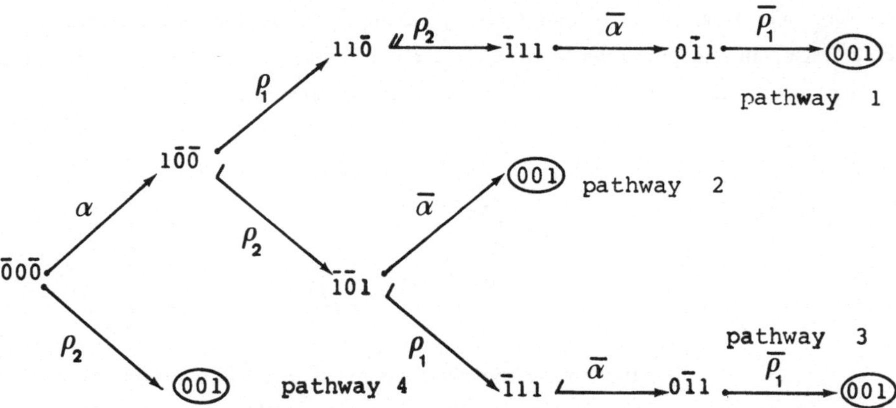

In this graph there are three "bifurcations". As seen above, at the level of state $\bar{0}\,0\,\bar{0}$ the decision depends on t_{α} vs $t_{\rho 2}$, at the level of state $1\,\bar{0}\,\bar{0}$ it depends on $(t_{\alpha} + t_{\rho 1})$ vs $t_{\rho 2}$ and at the level of state $\bar{1}\,\bar{0}\,1$ it depends on $(t_{\rho 2} + t_{\bar{\alpha}})$ vs $(t_{\alpha} + t_{\rho 1})$.

One way to handle the situation is to associate a logical variable with each of the three inequalities :

$m \equiv t_{\alpha} < t_{\rho 2}$ (this means that $m = 1$ if the statement "$t_{\alpha} < t_{\rho 2}$" is true, $m = 0$ if the statement is false)

$$n \equiv (t_{\alpha} + t_{\rho 1}) < t_{\rho 2}$$

$$p \equiv (t_{\rho 2} + t_{\bar{\alpha}}) < (t_{\alpha} + t_{\rho 1})$$

Clearly, the conditions which lead to the varions pathways are :

for pathway 1 : $m.n$
2 : $m.\bar{n}.p$
3 : $m.\bar{n}.\bar{p}$
4 : $\bar{m}$

But the expression of these conditions can be simplified as follows. Variables m, n and p are not independent of each other. If $t_{\rho 2} < t_{\alpha}$, that is, $\bar{m}$, one has necessarily $t_{\rho 2} < (t_{\alpha} + t_{\rho 1})$, that is, $\bar{n}$. In other words, $\bar{m}$ implies $\bar{n}$, and the situation $\bar{m}.n$ will never be found. Similarily, $n \rightarrow \bar{p}$, and the situation $n.p$ will never be found. For this reason, the matrix showing which pathway is followed, as a function of the values of m , n and p , will involve "don't care" states (—) corresponding to these impossible situations :

NO OF THE PATHWAY (p \ m , n)	00	01	11	10
0	4	—	1	3
1	4	—	—	2

This allows one to simplify (see chapter IV) the expressions. The conditions become :

for pathway 1 : n
for pathway 2 : $m.p$
for pathway 3 : $m.\bar{n}.\bar{p}$
for pathway 4 : $\bar{m}$

Thus, the conditions which lead to pathway 2 are reduced to $m.p$, that is, $t_{\alpha} < t_{\rho 2}$ and $(t_{\rho 2} + t_{\bar{\alpha}}) < (t_{\alpha} + t_{\rho 1})$.

In more complex systems, the proceedure just described can bring about much more extensive simplifications. From a more general viewpoint, it is interesting to know not only <u>that</u> the pathway chosen by the system depends of the relative values of the delays or linear combinations of them, but exactly <u>which</u> conditions are involved in the choice. Usually, only part of the time delays appear in these conditions.

d) Further caracterization of the states.

In transient states, one or more variables are committed to change their value. Which variable (if more then one is involved) will actually commute, and when exactly the change will take place, may depend on events one, two, three, ... n stages removed from the decision. Even though two stages are represented by the same binary set, the situation can be significantly different. For instance, in the 3rd graph of page 124 , state 1 1 1 /0 1 1 (or $\bar{1}$ 1 1) appears twice. In the first case, the local sequence is ... 1 1 $\bar{0}$ $\xrightarrow{e2}$ $\bar{1}$ 1 1 $\xrightarrow{\bar{\alpha}}$ 0 $\bar{1}$ 1 ... , in which $\xrightarrow{\bar{\alpha}}$ means that gene A has been switched off at the initiation of state $\bar{1}$ 1 1 .
In the second case, the local sequence is ... $\bar{1}$ $\bar{0}$ 1 $\xrightarrow{e1}$ $\bar{1}$ 1 1 $\xrightarrow{\bar{\alpha}}$ 0 $\bar{1}$ 1 ... in which $\xrightarrow{\bar{\alpha}}$ expresses that gene A has been switched off already at the preceeding stage. Thus, if the delay $t_{\bar{\alpha}}$ is essentially the same in both cases the transition from $\bar{1}$ 1 1 to $\bar{0}$ 1 1 will take more time in the first case than in the second case. In the situation described in appendix n° 2, with all delays identical except t_{α} , the system goes repeatedly through the state labelled $\bar{0}$ 0 $\bar{0}$, but the exact situation is slightly different each time as regards the times remaining before commutation, until the system finally leaves the cycle and jumps to one of the two stable states. These situations can be analyzed in detail, either on paper or, when they are too complex, with the automatized methods of Van Ham (Chapter IX).

e) Statistical aspect.

Let us consider the graph of the pathways derived from a logical structure. If we do not know anything about the values of the time delays, all the pathways are possible. At the other extreme, if we assign a precise value to each delay, the system will choose one well-defined pathway. In practice, however, a given delay will not have exactly the same value in different cells of a population ; at the level of the population, a delay cannot be described by a fixed value but rather by an average value and a distribution function. If the choice between two pathways depends on the relative values of, say t_{α} and t_{e2} (this chapter, section 4, b), and the average value of t_{α} is smaller than that of t_{e2} , the system will usually choose the transition corresponding to the delay t_{α} (in this case, the transition $\bar{0}$ 0 $\bar{0}$ $\longrightarrow$ 1 $\bar{0}$ $\bar{0}$) rather than the transition $\bar{0}$ 0 $\bar{0}$ $\longrightarrow$ (0 0 1). However, depending on the difference between $\langle t_{\alpha} \rangle$ and $\langle t_{e2} \rangle$ and on the exact distributions, a fraction of the cells may choose the alternative transition.

This attitude is applied to a concrete case in chapter XVII, section 3, using a program developped by P. Van Ham (see chapter IX). Another important generalization consists of treating the value of each delay as a function of the state of the system (Van Ham, in preparation).

APPENDIX n° 1 .

In section 2, I used as an example (see Table 1) a system described in synchronous terms by Kaufman (1969). As remarked in section 3, the system can be described very simply in terms of logical equations :

$$a = \bar{\beta} \cdot \gamma$$
$$b = \alpha \cdot \gamma$$
$$c = \bar{\alpha} + \beta$$

Exactly equal delays would lead to the graph of section 2. If however one takes into account that the delays have usually different values, one gets the following graph :

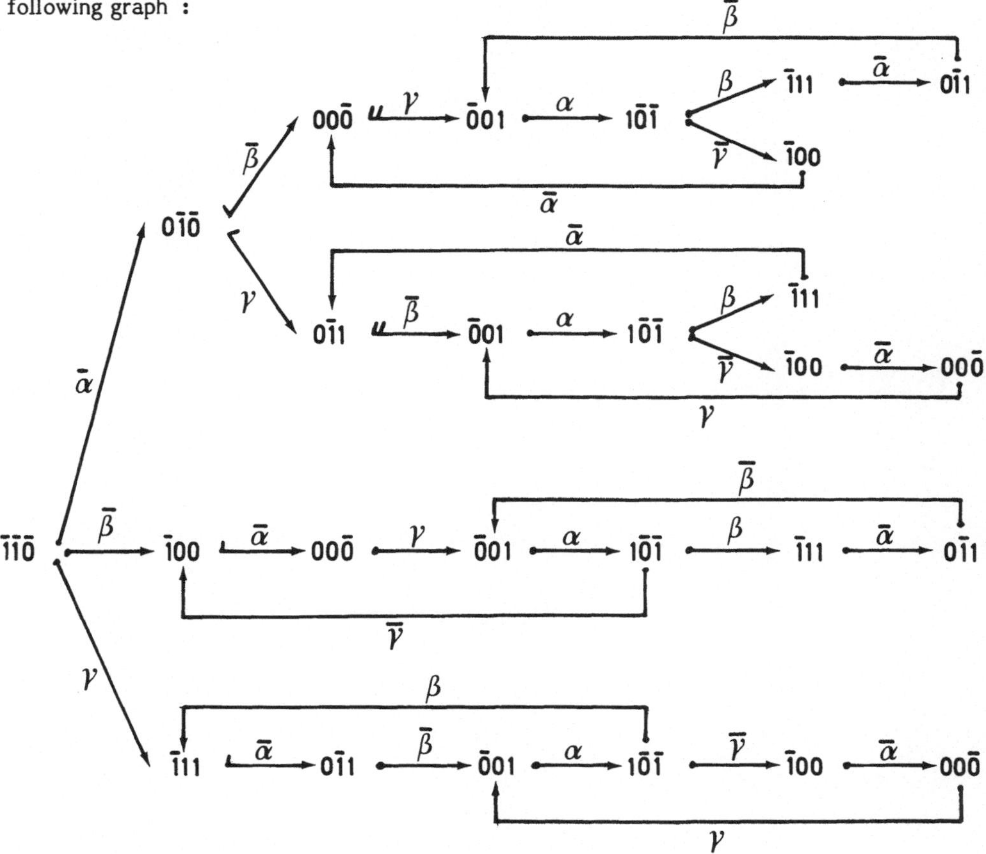

As one can see, there are no stable states, but there are two cycles : 0 - 1 - 5 - 4 and 1 - 5 - 7 - 3 which involve, respectively the transitions γ-α-$\bar{\gamma}$-$\bar{\alpha}$ and α-β-$\bar{\alpha}$-$\bar{\beta}$. The technique used is decribed in section 4, b and c .

The cubic representation of this system is :

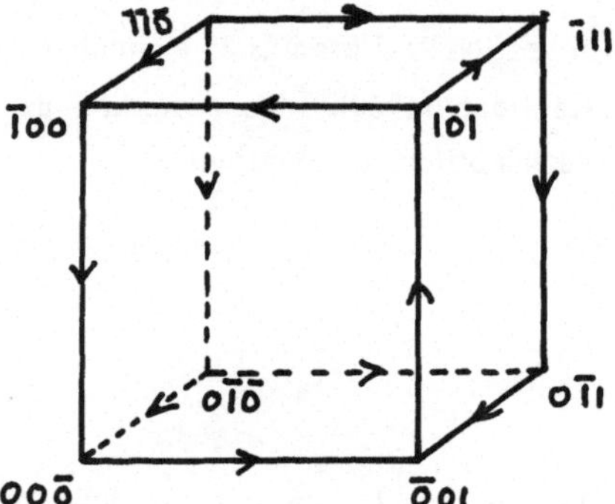

APPENDIX n° 2 .

As mentioned in this Chapter, section 4a, the positive loop described by the equations.

$$a = \overline{\rho 2}$$
$$r1 = \alpha$$
$$r2 = \overline{\rho 1}$$

(the system of Fig.VI.1. and table 2, with T = 0) provides a choice between the two stable states (0 0 1) and (1 1 0) , and also an unstable situation which is seen in the boolean formalism as a 6-membered cycle. This situation cannot persist in fact, as it requires extremely particular conditions in the lengths of the time delays and any fluctuation would lead the system to one of the stable steady states. It may nevertheless be interesting to give a brief analysis of the drastic conditions which would be required to remain in this cycle .

The graph of the pathways (starting arbitrarily from state $\bar{0}\,0\,\bar{0}$) is as follows :

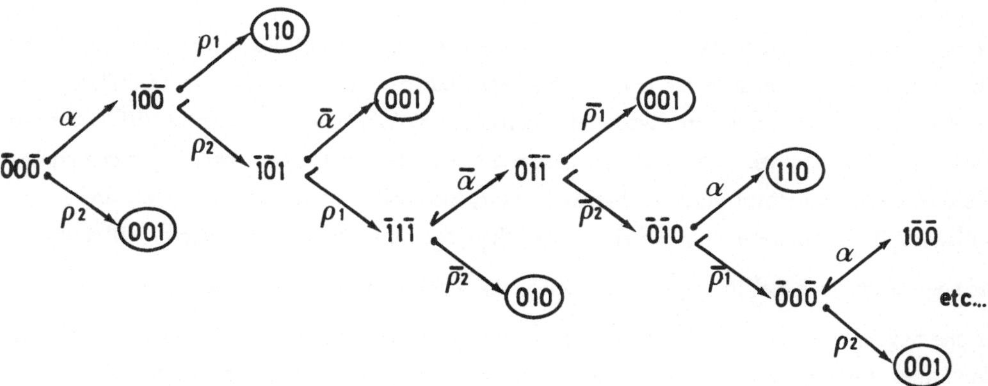

Note that for the initial state and for the seventh (13th , 19th ...) transient states one uses the same binary description ($\bar{0}\,0\,\bar{0}$, or, more explicitly 0 0 0 / 1 0 1), which means that in each of these cases the products α, $\rho 1$ and $\rho 2$ are absent but that genes A and R2 are on. However, the situations are not exactly the same, since in the first case both delays t_{α} and $t_{\rho 2}$ have been initiated together at the initial time ($\bar{0}\,0\,\bar{0}$ $\overset{\alpha}{\rightarrow}$ $\overset{\rho 2}{\rightarrow}$
while after one turn the synthesis of α and of $\rho 2$ are initiated at different times : when the system reaches the state $\bar{0}\,0\,\bar{0}$ for $\rho 2$ but already at the preceeding step ($\bar{0}$ 1 0) for α.

The conditions to enter or remain in the cycle (that is, to proceed from $\bar{0}\,0\,\bar{0}$ back to $\bar{0}\,0\,\bar{0}$ through five intermediate steps) can be described as follows.

If $m \equiv t_{\alpha} < t_{\rho 2}$

$n \equiv (t_{\alpha} + t_{\rho 1}) < t_{\rho 2}$

$p \equiv (t_{\rho 2} + t_{\bar{\alpha}}) < (t_{\alpha} + t_{\rho 1})$

$q \equiv (t_{\rho 2} + t_{\bar{\alpha}}) < (t_{\alpha} + t_{\rho 1} + t_{\bar{\rho} 2})$

$r \equiv (t_{\rho 2} + t_{\bar{\alpha}} + t_{\bar{\rho} 1}) < (t_{\alpha} + t_{\rho 1} + t_{\bar{\rho} 2})$

$s \equiv (2\, t_{\alpha} + t_{\rho 1} + t_{\bar{\rho} 2}) < (t_{\rho 2} + t_{\bar{\alpha}} + t_{\bar{\rho} 1})$

the conditions to <u>enter</u> the cycle are :

$$m\ \bar{n}\ \bar{p}\ q\ \bar{r}\ s, \quad \text{that is}$$

$$t_{\alpha} < t_{\rho 2} < (t_{\alpha} + t_{\rho 1}) < (t_{\rho 2} + t_{\bar{\alpha}}) < (t_{\alpha} + t_{\rho 1} + t_{\bar{\rho} 2}) < (t_{\rho 2} + t_{\bar{\alpha}} + t_{\bar{\rho} 1}) < (2t_{\alpha} + t_{\bar{\rho} 2} + t_{\rho 1})$$

The condition to <u>remain</u> in the cycle is simply :

$$(t_{\alpha} + t_{\rho 1} + t_{\bar{\rho} 2}) = (t_{\rho 2} + t_{\bar{\alpha}} + t_{\bar{\rho} 1}) \quad (1)$$

Note that in this case there is a contradiction between the conditions to enter the cycle and the condition to remain in the cycle. This behaviour can be simulated without difficulty on "Delphine", the simulator conceived and constructed by Van Ham (see Chapter VIII). One can easily see, for instance, that if all the delays are exactly equal except one (t_{α} if one choses $\bar{0}\,0\,\bar{0}$ as the initial state) which is slightly shorter, the conditions to enter the cycle are fulfilled. If the value or t_{α} remains lower than the other delays, the system will follow a number of cycles equal to $\dfrac{t_n}{t_n - t_{\alpha}}$ (t_n being the lenghth of the other time delays).

If one begins with these values of the delays and increment t_{α} in order to render it equal to the other time delays once the system has reached state $\bar{0}\,\bar{1}\,0$, the system will cycle indefinitely.

Finally, if one starts with all the time delays exactly equal, the system will perform the cycle

depending on the initial state (the dubble arrows refer to the fact that two variables switch at a time). More generally, the first cycle is followed whenever $t_{\alpha} = t_{\rho 2}$, $t_{\bar{\alpha}} = t_{\rho 1}$, $t_{\bar{\rho} 1} = t_{\bar{\rho} 2}$, and the initial state is $\bar{0}\,0\,\bar{0}$, $\bar{1}\,\bar{0}\,1$ or $0\,\bar{1}\,\bar{1}$. Similarily, the other cycle is found if $t_{\rho 1} = t_{\rho 2}$, $t_{\bar{\alpha}} = t_{\bar{\rho} 2}$, $t_{\alpha} = t_{\bar{\rho} 1}$, and the initial state is $1\,\bar{0}\,\bar{0}$, $\bar{1}\,1\,\bar{1}$ or $\bar{0}\,\bar{1}\,0$.

CHAPTER VII.

The dynamic behaviour of boolean systems comprising feedback loops.

R. THOMAS.

In the absence of non-linear interactions, systems comprising feedback loops display the same type of "trivial" behaviour as systems without loops ; they tend towards a single stable steady state. If one or more interaction is sufficiently non-linear, looped systems can display a "non trivial" behaviour such as multiple steady states or sustained oscillations (see Nicolis and Prigogine, 1978 ; Othmer, 1976 ; for a more detailed bibliography, see chapter XIV by Richelle). Non trivial behaviours can also be found if the system combines a mild non linearity with appropriate time delays (see Richelle, 1977 ; this book, chapter XIV). There are reasons to think that this borderline between the trivial and non trivial behaviour once overcome, the qualitative behaviour is not further modified by further increasing the non-linearity of the interactions. This was shown very clearly in a particular case by Glass and Kauffman (1972). The boolean approximation is an extreme attitude, which treats each interaction as if it were extremely non-linear (step function). In fact, if one chooses to describe a system in boolean terms, one stands on the non-trivial side of the borderline ; this is why, according to the boolean description, a positive loop generates multiple steady states and a negative loop results in sustained oscillations (see below).

I would like to emphasize that, strictly speaking the analysis given below is valid only for systems in which the interactions can be described by step functions. However, I assume that it can be applied more generally to looped systems which are sufficiently non-linear or involve sufficient delays to be on the "non-trivial" side of the behavioural borderline ; this is apparently the case for many regulatory systems in genetic and other fields.

In this chapter I describe systems of increasing complexity and I try to find general rules. As described here (and in Thomas, 1978), they are conjectures rather than theorems ; however, for a number of them a formal demonstration can be found in Van Ham & Lasters (1978). The formalism used is described in chapter VI. Generally A, a, α (or B, b, β, ...) represent, respectively, a gene, its state of expression and the presence of its product.

1) In systems made of a linear, open (chain) or closed (loop) sequence of control elements, the final states depend essentially on the parity of the number of negative control elements. A chain or a loop will be qualified positive or negative depending on whether it comprizes an even or an odd number of negative elements.

2) For an open sequence of control elements (chain), the final state is a single stable steady state. Let us examine, for instance, the system defined by the equations ($a = 1$; $b = \bar{\alpha}$; $c = \beta$; $d = \gamma$; $e = \bar{\delta}$, $z = \bar{\varepsilon}$). This system, which can be represented :

$$\alpha \xrightarrow{-} \beta \xrightarrow{+} \gamma \xrightarrow{+} \delta \xrightarrow{-} \varepsilon \xrightarrow{-} \zeta$$

finally reaches a stable state in which α is present, β absent because its synthesis is prevented by α , γ absent because β is necessary for its synthesis etc ... ; the stable state can be represented (1 0 0 0 1 0). One deals here with a negative chain (odd number of negative elements) ; the last element, which is the final product of the control chain is "off" in the final state. Conversedly, if one deals with a positive chain (even number of negative elements) the last element of the control chain is "on" in the final state. In any case, the system can always be "reduced" by dropping a positive control or by replacing a pair of negative controls by a positive one, or "extended" by the reverse operations ; the effect is indifferent at the level of the final state, on or off, of the last element.

3) Positive loops (loops with an even number of negative steps) can accomodate two stable steady states (and one unstable cycle which is probably of little pratical interest*).

Five simple examples of positive loops are given in Fig.1. The first case mentioned is typical autocatalysis ; α is required for its own synthesis. If the requirement is stringent, clearly either α is absent and it will remain absent indefinitely, or it is present and it will continue to be synthesized, and remain present indefinitely. In the second case, there are also two stable steady states, one with both α and β absent, one with both α and β present. In the third case, there are again two stable steady states, with either α or β present. Fig.1 also shows two positive loops with three elements. One of them has been described in more detail in chapter VI.

---

* see Chapter VI, appendix 2.

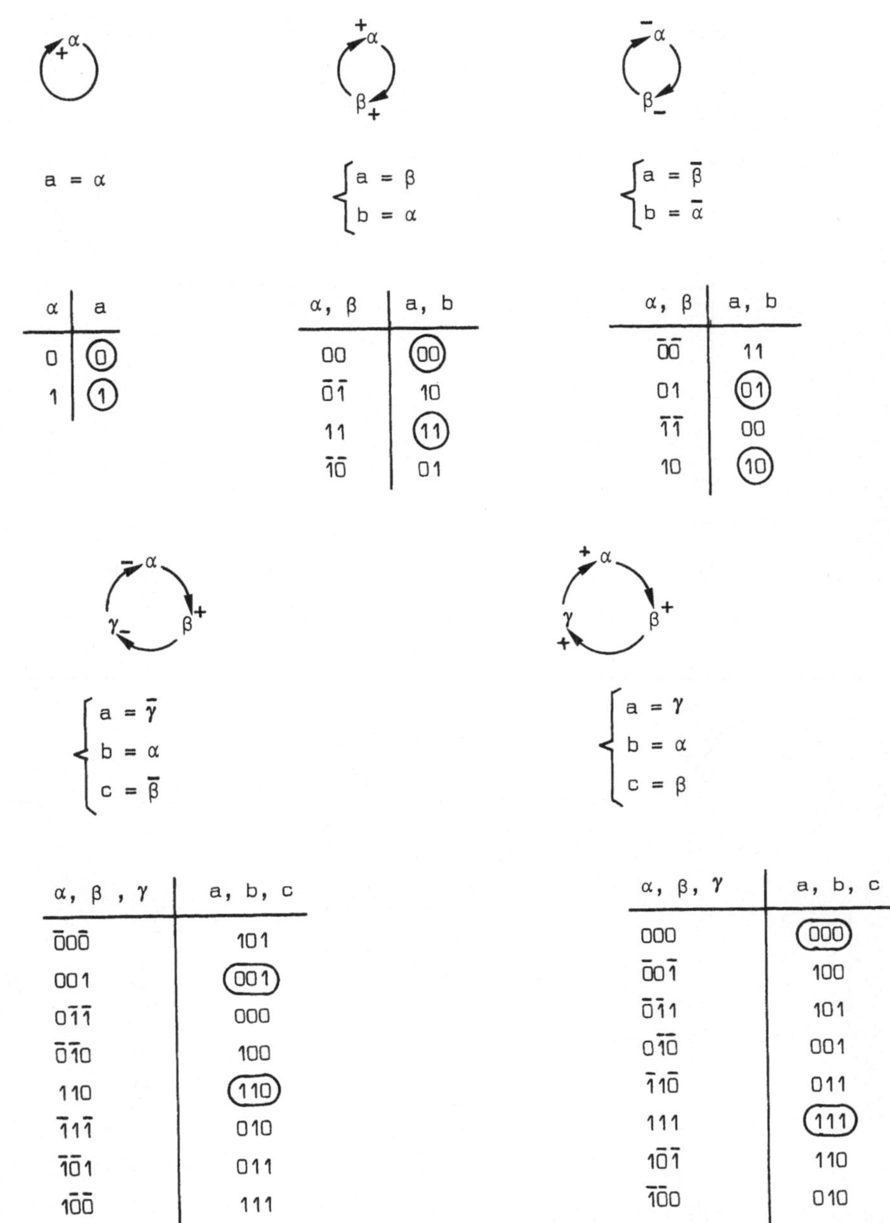

$a = \alpha$

$\begin{cases} a = \beta \\ b = \alpha \end{cases}$

$\begin{cases} a = \bar{\beta} \\ b = \bar{\alpha} \end{cases}$

α	a
0	(0)
1	(1)

α, β	a, b
00	(00)
$\bar{0}\bar{1}$	10
11	(11)
$\bar{1}\bar{0}$	01

α, β	a, b
$\bar{0}\bar{0}$	11
01	(01)
$\bar{1}\bar{1}$	00
10	(10)

$\begin{cases} a = \bar{\gamma} \\ b = \alpha \\ c = \bar{\beta} \end{cases}$

$\begin{cases} a = \gamma \\ b = \alpha \\ c = \beta \end{cases}$

α, β, γ	a, b, c
$\bar{0}0\bar{0}$	101
001	(001)
$0\bar{1}\bar{1}$	000
$\bar{0}\bar{1}0$	100
110	(110)
$\bar{1}1\bar{1}$	010
$\bar{1}\bar{0}1$	011
$1\bar{0}\bar{0}$	111

α, β, γ	a, b, c
000	(000)
$\bar{0}0\bar{1}$	100
$\bar{0}\bar{1}1$	101
$0\bar{1}\bar{0}$	001
$\bar{1}1\bar{0}$	011
111	(111)
$1\bar{0}\bar{1}$	110
$\bar{1}\bar{0}0$	010

Figure 1. Simple examples of positive feedback loops. For the meaning of the matrices, see chapter VI.

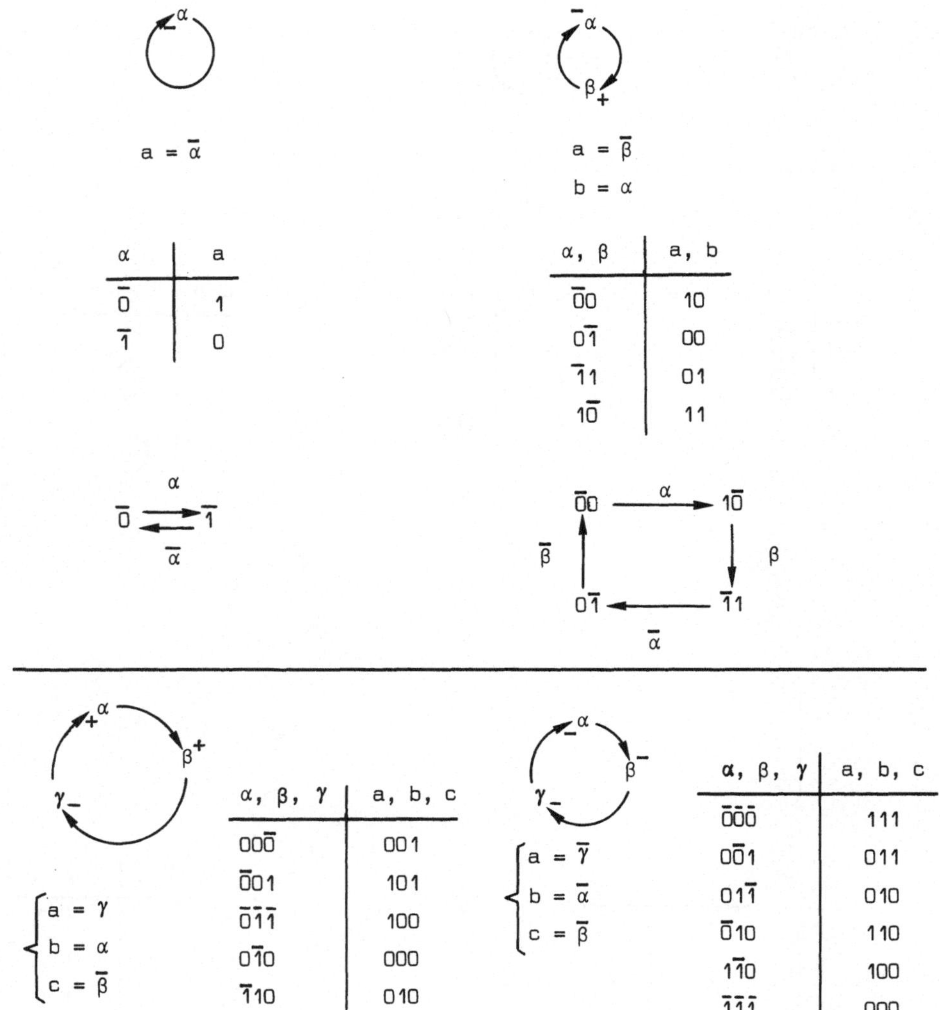

α	a
$\bar{0}$	1
$\bar{1}$	0

α, β	a, b
$\bar{0}0$	10
$0\bar{1}$	00
$\bar{1}1$	01
$1\bar{0}$	11

α, β, γ	a, b, c
$00\bar{0}$	001
$\bar{0}01$	101
$\bar{0}\bar{1}\bar{1}$	100
$0\bar{1}0$	000
$\bar{1}10$	010
$11\bar{1}$	110
$1\bar{0}1$	111
$\bar{1}\bar{0}\bar{0}$	011

α, β, γ	a, b, c
$\bar{0}\bar{0}\bar{0}$	111
$0\bar{0}1$	011
$01\bar{1}$	010
$\bar{0}10$	110
$1\bar{1}0$	100
$\bar{1}\bar{1}\bar{1}$	000
$\bar{1}01$	001
$10\bar{0}$	101

Figure 2. Four simple negative loops.

4) Negative loops (loops with an odd number of negative steps) treated with the boolean approximation yield sustained oscillations (a stable cycle). Four examples are shown in Fig.2. The final states are, respectively, the cycles 0 - 1 , 0 - 2 - 3 - 1 , 0 - 1 - 5 - 7 - 6 - 2 and 1 - 3 - 2 - 6 - 4 - 5 . Note that in these cycles each state has only one possible next state (for instance in the 4th example $0\,\bar{0}\,1$ is necessarily followed by $0\,1\,\bar{1}$ etc...) ; they are thus stable cycles.

5) Loops with a graft.

One can graft on a loop a positive or a negative control chain, with a AND or a OR connection. Some grafts disrupt the loop, thus leading to a single stable steady state ; other grafts do not modify the essential behaviour of the loops.

a) Positive chains with a OR connection, and negative chains with a AND connection are disruptive grafts. An example is found in chapter VI (p. 115). In this particular case, the graft is an input variable, and its value can thus be put at will at 0 (in which case the graft is inefficient) or at 1 . This system oscillates at low temperature (T = 0) but it tends to a stable steady state at high temperature (T = 1) because whatever the value of ρ, $a1 = \bar{\rho} + T = 1$; if one lowers again the temperature, the system again oscillates.
In other cases, the graft corresponds to a regulatory gene which is itself not regulated (in genetic jargon, it is expressed "constitutively") ; as long as the product of this gene is absent, the graft is not effective, but as soon as the product is present, the loop is disrupted, and the system tends (this time irreversibly) to a stable steady state.
This can be generalized to a graft which is itself a chain of regulatory genes.

b) Positive chains with a AND connection, and negative chains with a OR connection are non-disruptive grafts.An example is found in chapter XVII, p.360 ; in spite of the chain grafted on the positive loop the system still has the choice between two stable steady states.

6. Tangent loops.

Two loops can be independant (see for instance section 8 a) or interconnected. In this case they can have no common element ("coupled" loops : see for instance section 8 b) or at least one common element ("conjugated" loops) ; I suggest to call them "tangent" if then have only one common element and "secant" if then have two or more elements in common. This terminology is more precise than that used in Thomas (1978).

a) Two tangent positive loops.

These systems behave essentially like a single positive loop as they admit two stable steady states (see Fig.3).

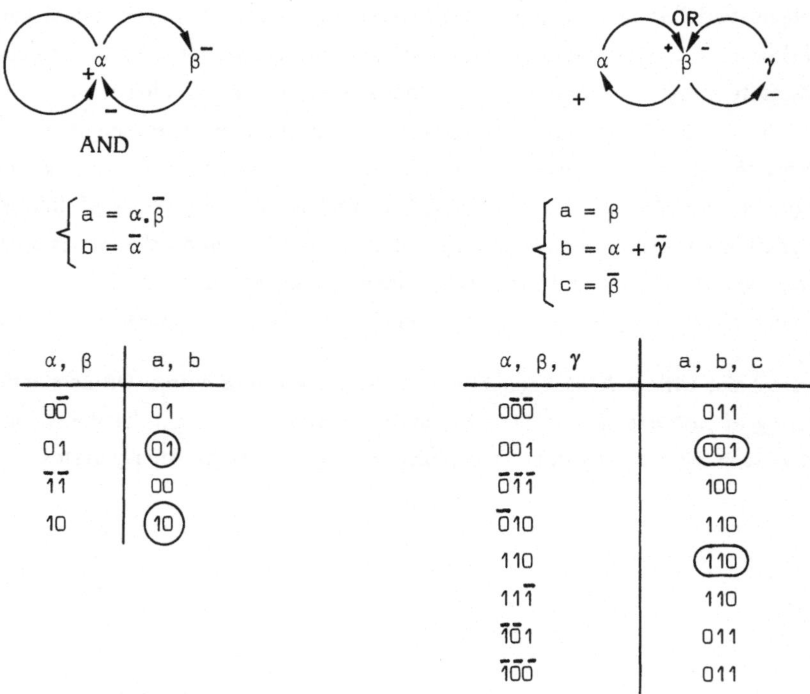

$$\begin{cases} a = \alpha . \bar{\beta} \\ b = \bar{\alpha} \end{cases}$$

$$\begin{cases} a = \beta \\ b = \alpha + \bar{\gamma} \\ c = \bar{\beta} \end{cases}$$

α, β	a, b
$0\bar{0}$	01
01	01
$\bar{1}\bar{1}$	00
10	10

α, β, γ	a, b, c
$0\bar{0}\bar{0}$	011
001	001
$\bar{0}\bar{1}\bar{1}$	100
$\bar{0}10$	110
110	110
$11\bar{1}$	110
$\bar{1}\bar{0}1$	011
$\bar{1}\bar{0}\bar{0}$	011

Figure 3 . Systems comprising two tangent positive loops.

b) Two tangent negative loops.

These systems resemble a single negative loop insofar as there is no stable steady state. However, instead of a single, stable cycle, there may be several cycles. For instance, the system described in Fig.4 can follow five different elementary* cycles (see the legend).

$$\begin{cases} a = \beta \\ b = \bar{\alpha} + \gamma \\ c = \bar{\beta} \end{cases}$$

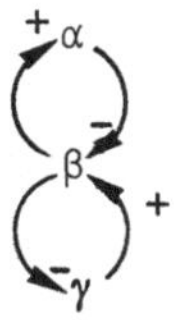

α, β, γ	a, b, c
$0\bar{0}\bar{0}$	011
$0\bar{0}1$	011
$\bar{0}1\bar{1}$	110
$\bar{0}10$	110
$1\bar{1}0$	100
$11\bar{1}$	110
$\bar{1}\bar{0}1$	011
$\bar{1}0\bar{0}$	001

Fig.4. A system formed by two conjugated negative loops. The cycles (described by the decimal or octal equivalents of the binary states) are :

0 - 2 - 6 - 4 ,
4 - 5 - 7 - 6 ,
0 - 1 - 3 - 7 - 6 - 4 - 5 ,
1 - 3 - 2 - 6 - 4 - 5 ,
1 - 3 - 7 - 6 - 4 - 5 .

---

* When a state is common to two cycles, the system might follow first one of the cycles, then the other. One would deal with a "compound" cycle, built from two "elementary" cycles.

c) <u>Two tangent loops, one positive, one negative.</u>

As compared with all the preceeding ones, these systems display a qualitatively new behaviour ; they offer a choice between a stable steady state and cycle(s). For instance, the system described in Fig.5 can choose between the stable state (1 1 0) and the cycle $0\bar{0}1 \xrightarrow{\beta} \bar{0}1\bar{1} \xrightarrow{\alpha} 1\bar{1}\bar{1} \xrightarrow{\bar{\alpha}} \bar{1}01$ (with return $\bar{1}01 \xrightarrow{\bar{\beta}} 0\bar{0}1$) (or, in short, 1 - 3 - 7 - 5).

$$\begin{cases} a = \beta \\ b = \bar{\alpha} + \bar{\gamma} \\ c = \bar{\beta} \end{cases}$$

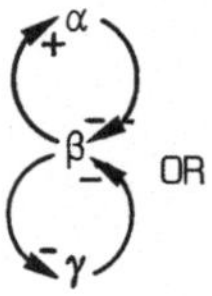

α, β, γ	a, b, c
$0\bar{0}\bar{0}$	011
$0\bar{0}1$	011
$\bar{0}1\bar{1}$	110
$\bar{0}10$	110
110	(110)
$1\bar{1}\bar{1}$	100
$\bar{1}01$	001
$\bar{1}\bar{0}\bar{0}$	011

Fig.5. A system formed by two tangent loops, one positive, one negative. There are two possible final states : the stable state (1 1 0) and the cycle 1 - 3 - 7 - 5 .

8) <u>Other systems.</u>

So far, rules have been obtained for the dynamic behaviour of the following types of systems : open chains (linear or branched) loops, loops with a linear chain(s) grafted, two tangent loops (defined as two loops with <u>one</u> common element (see Thomas 1978). A number of other logical structures have been analyzed, but not in a systematic way. I wish to emphasize again the fact that very simple logical structures can lead to quite complex behaviours. For instance :

a) systems comprising two or more <u>independant loops</u>. One of the simplest ones comprises two negative loops of one element ($a = \bar{\alpha}$; $b = \bar{\beta}$). Each element pulses with its characteristic period ; if these periods had no common measure, the behaviour of the system would be aperiodic.

b) systems comprising <u>coupled</u> loops, that is, loops which are interconnected but have no element in common. For instance :

($a = \bar{\alpha}$; $b = \bar{\alpha} + \bar{\beta}$)

If the time delays t_{α} and $t_{\bar{\alpha}}$ are much longer than the time delays of β, the second loop will go through alternate periods of rest (when $\alpha = 0$) and of oscillation (when $\alpha = 1$).

c) systems comprising <u>more than two tangent loops</u>. Generalizations are not immediate. For instance, among the systems in which two elements form three tangent loops ($a = \bar{\alpha}\bar{\beta}$; $b = \bar{\alpha}\bar{\beta}$)admits <u>no</u> stable state, ($a = \bar{\alpha}\beta$; $b = \alpha\bar{\beta}$) admits <u>one</u>, ($a = \alpha\beta$; $b = \alpha\beta$) admits <u>two</u> and ($a = \alpha\bar{\beta}$; $b = \bar{\alpha}\beta$)admits <u>three</u>.

d) systems comprising <u>secant loops</u> have not been analyzed systematically so far. It must be pointed out that the number of loops may be much higher than the number of elements of the system. For instance, the system ($a = \alpha . \bar{\beta} . \gamma$; $b = \alpha . \beta . \bar{\gamma}$; $c = \bar{\alpha} . \beta . \gamma$) comprises, as far as I can see, 8 loops (4 positive loops and 4 negative loops). A rather complex example of secant loops is found in the formal analysis of the lactose operon <u>(chapter XVII, section 5).</u>

<u>Can one apply the boolean formalism to the analysis of chemical reactions</u> ?

As emphasized several times in this book, the formalism developped in Chapter VI is efficient for the analysis of systems whose elements interact with each other <u>without being used up in the process</u>. This is the case for the systems in which most of the elements taken into consideration act catalytically on the synthesis of one another.

However, when the elements considered explicitly in the models are substances which react with and transform into one another ("chemical" systems), the naïve boolean formalism comes up against serious difficulties. For instance, how can one take into account the precise stoichiometry of the reaction ? And is there any interest to treat such systems in purely qualitative terms ?

In spite of this rather pessimistic <u>a priori</u> view, I have investigated the possibility of applying boolean analysis to chemical systems with feedback loops. One reason was to try understanding an apparent paradox. On the one hand, we have seen that according to the boolean analysis a positive loop permits multiple steady states but does not generate by itself stable oscillations ; on the other hand, the occurence of sustained oscillations is often accounted for in terms of models whose salient feature is precisely a positive loop (see for instance Goldbeter, 1975). As will be seen below, the apparent discrepancy is now clearly understood.

The first problem is : how can one describe in boolean terms the fact that the "positive control" exerted by substance α on the formation of substance β consists of the transformation of α into β (the situation most usually found in chemical systems) rather than in a catalytic effect (as in the regulatory systems considered so far in this book) ?

Let us take as an example a system of two reactions $\Rightarrow \alpha \overset{E}{\Rightarrow} \beta$ (with a "+" feedback arrow from β to the first reaction), in which the conversion of α into β requires enzyme E, and the production of α from a source requires the presence of β as catalyst. At first view, this is a pure positive loop, in which α takes part in the synthesis of β and β takes part in the synthesis of α. However, the situation is complicated by the fact that α is used up in the formation of β. How can one describe this in boolean terms ? In accordance withe the naïve boolean attitude, I shall reason as if each of the reactions ($\Rightarrow \alpha$) and ($\alpha \Rightarrow \beta$) were either on or off. The condition for reaction ($\Rightarrow \alpha$) is the presence of β as a catalyst ; this can be written : $(\Rightarrow \alpha) = \beta$. The conditions for reaction ($\alpha \Rightarrow \beta$) are the presences of α and of enzyme E ; this can be written : $(\alpha \Rightarrow \beta) = \alpha\,.E$.

Let a be the boolean function associated with the resultant production of α ; assuming that when reaction ($\alpha \Rightarrow \beta$) is on it is faster than reaction ($\Rightarrow \alpha$), one can write :

$$a = (\Rightarrow \alpha) \cdot \overline{(\alpha \Rightarrow \beta)}$$

(in which ($\Rightarrow \alpha$) means that the reaction ($\Rightarrow \alpha$) is on and $\overline{(\alpha \Rightarrow \beta)}$, that reaction ($\alpha \Rightarrow \beta$) is off). Replacing these expressions by their value, it comes :

$$\begin{aligned} a &= \beta \cdot \overline{\alpha E} \\ &= \beta(\bar{\alpha} + \bar{E}) \end{aligned} \qquad \text{(I)}$$

Similarily,

$$\begin{aligned} b &= (\alpha \Rightarrow \beta) \\ &= \alpha \cdot E \end{aligned} \qquad \text{(II)}$$

The presence of $\bar{\alpha}$ in equation (I) shows that formally the system comprizes a negative loop $\overset{-}{\circlearrowleft}\alpha$. A comparison of the initial scheme (Fig.VII,6a) with the graph of the logical structure of the same system (Fig.VII,6b), derived from the logical equations (I) and (II) , shows how a system which looks at first view as a simple positive loop may in fact comprize a hidden negative loop, due to the fact that α is used up in the formation of β.

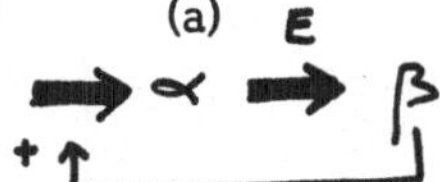

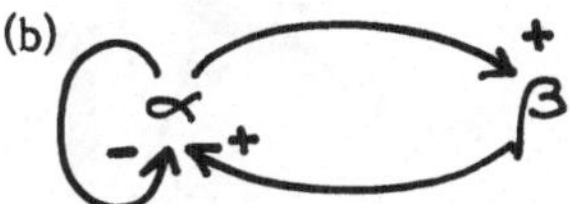

Fig.VII,6 (a) Classical scheme of the system, in which $\Rightarrow$ symbolyzes the conversion of substances into on another, and $\xrightarrow{+}$ a catalytic effect.

(b) A graph of the logical structure of the same system, as derived from equations

$a = \beta(\bar{\alpha} + \bar{E})$

$b = \alpha \cdot E$

One sees that the system comprizes a "hidden" negative loop. Only the internal variables α and β are represented.

Note that the negative loop revealed by the examination of the logical equations is seen in the corresponding differential equations as a negative term $-f(\alpha)$ in the equation of $\frac{d\alpha}{dt}$ and as a positive term $+f(\alpha)$ in the equation of $\frac{d\beta}{dt}$*.

---

* The negative terms representing the degradation of a substance into unspecified products are not treated as explicit interactions in our formalism ; they are accounted for at the level of the "decay delays".

In order to account for experimentally observed oscillations in a system, Goldbeter (1975) proposed a model (see Fig.VII, 7,a) whose salient feature is a positive loop (a cross catalysis between substances β and γ). As shown in the first part of this chapter, the boolean analysis indicates thant an isolated positive loop provides a choice between two stable steady states, but does not by itself generate oscillations. Yet, as shown by Goldbeter in terms of continuous analysis, his system undergoes sustained oscillations for appropriate values of the parameters. What seemed at first view a serious discrepancy between the boolean and continuous methods is easily understood once one realizes that the logical structure of the system in fact comprize hidden negative loops, due to the fact that susbstance α is used up in the production of β and γ , and γ in the production of β . The footnote to the next page gives the description of the model in terms of differential equations (Goldbeter 1975) and a translation into boolean terms. Fig.VII,7 gives a comparison between the initial representation of the model, in which only the positive loop is apparent, (a), and a graph of its logical structure (b), in which the various interactions, including the hidden negative loops and mentioned ; these interactions can be seen in the differential and in the logical equations.

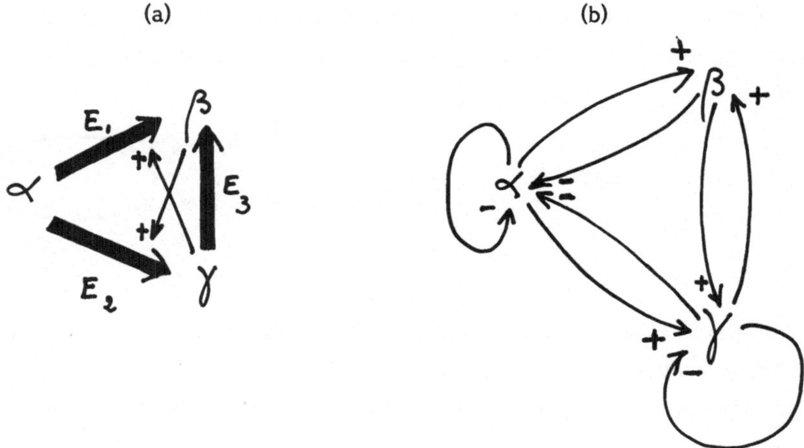

Fig.VII,7. Two representations of Goldbeter's model (1975).

a) A scheme in which the transformation of substances into one another is symbolized ➡ , and ⟶+ represents the catalytic effect of a substance on a reaction.

b) The graph of the logical structure of the system ; the various interactions are indicated by arrows, and one sees the negative loops arising from the transformation of α into β and γ and of γ into β. One has not tried to show in this graph that γ takes part in the synthesis of β both as a catalyst (of reaction α ➡ β)and as a substrate (of reaction γ ➡ β), but this is taken into account in the equations.

From the boolean equations (and also from the differential equations given in the footnote*
one can see that the system involves four negative interactions : in addition to the negative retroactions exerted by α and by γ on themselves, β exerts a negative effect on α as it catalyzes its conversion into γ, and γ exerts a negative effect on α as it catalyzes its conversion into β. One can find in the graph (Fig.VII,7b), four negative loops : a negative loop of α on itself, a negative loop of γ on itself, a loop $\alpha \xrightarrow{+} \beta \xrightarrow{+} \gamma \xrightarrow{-} \alpha$ and a loop $\alpha \xrightarrow{+} \gamma \xrightarrow{+} \beta \xrightarrow{-} \alpha$.

* The model of Goldbeter (1975), as expressed by himself in terms of differential equations :

$$\dot{\alpha} = \underbrace{\nu_1}_{\substack{\text{rate of}\\ \text{injection}\\ \text{of } \alpha}} - \underbrace{\frac{\varepsilon_1 \alpha (1+\alpha)(1+\gamma)^2}{L_1 + (1+\alpha)^2(1+\gamma)^2}}_{\substack{\text{loss of } \alpha \text{ due to its}\\ \text{conversion into } \beta\\ \text{(catalyzed by } \gamma)}} - \underbrace{\frac{\varepsilon_2 \alpha (1+\alpha)(1+\beta)^2}{L_2 + (1+\alpha)^2 + (1+\beta)^2}}_{\substack{\text{loss of } \alpha \text{ due to its}\\ \text{conversion into } \gamma\\ \text{(catalyzed by } \beta)}}$$

$$\dot{\beta} = \underbrace{\frac{\varepsilon_1 \alpha (1+\alpha)(1+\gamma)^2}{L_1 + (1+\alpha)^2 (1+\gamma)^2}}_{\substack{\text{same as the second term}\\ \text{of } \dot{\alpha} \text{ (production of } \beta\\ \text{from } \alpha, \text{ catalyzed by } \gamma)}} - \underbrace{k_1 \beta}_{\substack{\text{degradation}\\ \text{of } \beta \text{ into}\\ \text{unspecified}\\ \text{products}}} + \underbrace{k_2 \gamma}_{\substack{\text{production}\\ \text{of } \beta \text{ from } \gamma}}$$

$$\dot{\gamma} = \underbrace{\frac{\varepsilon_2 1 (1+\alpha)(1+\beta)^2}{L_2 + (1+\alpha)^2 (1+\beta)^2}}_{\substack{\text{same as the third term of}\\ \dot{\alpha} \text{ (production of } \gamma \text{ from}\\ \alpha, \text{ catalyzed by } \beta)}} - \underbrace{k_2 \gamma}_{\substack{\text{loss of } \gamma \text{ due to its}\\ \text{conversion into } \beta \text{ (de-}\\ \text{gradation of } \gamma \text{ into un-}\\ \text{specified products is \underline{not} considered)}}}$$

One boolean translation is :

$$a = \nu_1 + \nu_0 \, (\bar{\alpha} + \overline{\varepsilon_1 \gamma} \, . \, \overline{\varepsilon_2 \beta}) \qquad \text{(III)}$$

$$b = \gamma \, (\alpha \, \varepsilon_1 + \varepsilon_3) \qquad \text{(IV)}$$

$$c = \alpha \, \beta \, \varepsilon_2 \, (\bar{\gamma} + \bar{\varepsilon}_3) \qquad \text{(V)}$$

in which α, β, γ are boolean variables associated with the <u>concentrations</u> of products α, β, γ ; a, b, c are the boolean functions associated with the <u>fluxes</u> of these substances (as already mentioned in chapter VI, functions a, b, c are essentially the logical equivalents of the time derivatives $\dot{\alpha}, \dot{\beta}, \dot{\gamma}$). One uses three increasing injection rates of α ($\bar{\nu}_0$, $\nu_0 \bar{\nu}_1$, ν_1) such that at the lower and higher levels a = 0 and a = 1 , respectively, whatever the other conditions, but at the intermediate level ($\nu_0 \bar{\nu}_1$), a = 1 only if neither reaction ($\alpha \Rightarrow \beta$) nor reaction ($\alpha \Rightarrow \gamma$) is on.

The input variables ν_0, ν_1, ε_1, ε_2 and ε_3 are "boolean parameters" ; the symbols have been chosen partly in the hope of facilitating the comparison with the continuous analysis.

One way to analyze the situation consists of first looking how the system of logical equations simplifies for various combinations of values of the "boolean parameters" : see, for instance, tables 1 and 2 .

$\varepsilon_1 = 1 , \ \varepsilon_3 = 1$

	$\gamma_0 = 0$	$\gamma_0 = 1,\ \gamma_1 = 0$	$\gamma_1 = 1$
$\varepsilon_2 = 1$	1 $a = 0$ $b = \gamma$ $c = \alpha\beta\bar{\gamma}$	2 $a = \bar{\alpha} + \bar{\beta} . \bar{\gamma}$ $b = \gamma$ $c = \alpha\beta\bar{\gamma}$	3 $a = 1$ $b = \gamma$ $c = \alpha\beta\bar{\gamma}$
$\varepsilon_2 = 0$	4 $a = 0$ $b = \gamma$ $c = 0$	5 $a = \bar{\alpha} + \bar{\gamma}$ $b = \gamma$ $c = 0$	6 $a = 1$ $b = \gamma$ $c = 0$

Table I. The sets of logical equations giving a, b and c for three increasing rates of injection of α ($\bar{\gamma}_0$, $\gamma_0\bar{\gamma}_1$, γ_1) and two levels ($\bar{\varepsilon}_2$, ε_2) of the activity of enzyme E_2 .
Variables ε_1 and ε_3 have the value 1 .

$\varepsilon_2 = 1 , \ \varepsilon_3 = 1$

	γ_0	$\gamma_0\ \gamma_1$	γ_1
$\varepsilon_1 = 1$	1 $a = 0$ $b = \gamma$ $c = \alpha\beta\bar{\gamma}$	2 $a = \bar{\alpha} + \bar{\beta} . \bar{\gamma}$ $b = \gamma$ $c = \alpha\beta\bar{\gamma}$	3 $a = 1$ $b = \gamma$ $c = \alpha\beta\bar{\gamma}$
$\varepsilon_1 = 0$	7 $a = 0$ $b = \bar{\gamma}$ $c = \alpha\beta\bar{\gamma}$	8 $a = \bar{\alpha} + \bar{\beta}$ $b = \gamma$ $c = \alpha\beta\bar{\gamma}$	9 $a = 1$ $b = \gamma$ $c = \alpha\beta\bar{\gamma}$

Table 2. The sets of equations giving a, b and c for three increasing rates of injection of α ($\bar{\gamma}_0$, $\gamma_0\bar{\gamma}_1$, γ_1) and two levels ($\bar{\varepsilon}_1$, ε_1) of the activity of enzyme E_1 . The input variables ε_2 and ε_3 have value 1 .

The various situations of tables 1 and 2 can be (and have been) analyzed with the method described in Chapter VI, which tells :

- to number and nature of the stable steady state(s).
- whether the logical structure considered can generate oscillations at all, and if it does, whether these oscillations are permanent or only transient
- which are the possible pathways (sequences of boolean states) and which conditions determine the pathway actually followed.
- what is the domain of stability of the cycles, if any (that is, for which values of the delays one remains in the cycle).

In brief, the logical structure described by equations III, IV and V :

- precludes any oscillations in some conditions (boxes 2 and 6 in table 1 and 2).
- allows for transient oscillations in other cases (boxes 1, 5 and 7, tables 1 and 2).
- allows for stable oscillations in cases 2, 8, 3 and 9 (tables 1 and 2).

If one wants to apply these forecasts to a typically chemical system, one must remember that these systems do not involve absolute time delays of an appreciable length ; in these conditions no oscillations at all are supposed to be generated by negative loops comprising less than two elements, and no sustained oscillations by negative loops comprising less than three elements (for a discussion of these problems, see Chapter XIV by J. Richelle). Taking this element into account, we expect stable oscillations to occur actually only in situations corresponding to boxes 2 and 8 (tables 1 and 2), the only ones in which the logical structure comprizes negative loop(s) with three elements. The agreement with the treatment of Goldbeter is rather striking ; both analyses indicate the occurence of stable oscillations for intermediate values of the injection rate of α, and these oscillations are predicted even in the absence of enzyme E_1 ($\varepsilon_1 = 0$) but not in the absence of enzyme E_2 ($\varepsilon_2 = 0$). The boolean analysis also shows that in the absence of reaction ($\gamma \rightarrow \beta$) ($\varepsilon_3 = 0$, not shown in tables 1 and 2) the system can accomodate multiple steady states ; this is also found by continuous analysis (Goldbeter, personal communications).

In fact, a substantial part of the above-mentioned information can be obtained directly from the equations (or from the graph of the logical structure) with the help of the "rules" given at the beginning of this chapter. One interesting exercise consists of "cutting" one by one each of the interactions, in order to check which ones are really essential in order to obtain this or this dynamic behaviour. The general impression which emerges is that the presence of a negative loop in the logical structure of a system is a necessary, but not sufficient condition for the occurence of oscillations, and that the presence of a positive loop is a necessary, but not sufficient condition for the occurence of multiple steady states.

References

Glass, L. & Kauffman, S.A. (1978)
J. Theor. Biol. 34, 219.

Goldbeter,A. (1975)
Nature, 253, 540.

Nicolis, G. & Prigogine, I. (1978)
Self-Organization in Nonequilibrium systems.
Whiley-Interscience, New York.

Othmer, H.G., (1976)
J. Math. Biol. 3, 53-78.

Richelle, J. (1977)
Bull. Cl. Sc. Acad.R.Belg. 63, 534.

Thomas, R. (1978)
J. Theor. Biol. 73, 631.

Van Ham, P. & Lasters, I. (1978)
J. Theor. Biol. 72, 269.

CHAPTER VIII.

Delphin : a logical machine with incrementable phase delays.

Philippe VAN HAM[*]

VIII.1. Introduction.

The purpose of this chapter is to describe a specialized electronic hardware tool for simulation of logical models.
The electronic technology makes it possible to construct the basic logical functions (AND, OR, NOT, EXOR) so that the designer of a logical machine can materialize its logical equations in an electronic circuit. For didactic and research purposes, a great number of logical simulator has been constructed to test the new sequential structures or calculators conceived by the designers.
Besides some software tools (see chapter on simulation programs) for general purpose computers, it is very useful for the model maker to also have the possibility of wiring up the logical equations directly on simulators.

However, from the model maker point of vue, the simulator must comprise not only basic logical functions but also the delays with adjustable duration. (See Fig.1.) because they are an important part of the model.

That is the reason why we have conceived and constructed a set of 24 delays on a simulator called DELPHIN with some additional capabilities listed below :

1.1. Turn-on and turn-off delays may, of course, be different, and each of them is specified by three decimal digits.
1.2. Each delay may be initialized in a previously chosen state of the input and output.
1.3. Any delay may be enabled to work alone or synchronously with any other (this property is transitive).
1.3. Each delay is incrementable independently for turn-on delay or turn-off delay. The incrementation is controlled by two additional inputs, one for the choice between incrementation and decrementation, one for the command itself. The value of a delay may therefore vary with time between 000 and 999.
1.5. The user has a decimal visualization of the current value of each delay.
1.6. The user may choose the increment in the set :
1 , 2 , 3 , 4 , 5 , 6 , 7 , 8 , 9 , 10 ,20 , 100

---

* Service des systèmes logiques et numériques - Université libre de Bruxelles.

VIII.2. Functional aspects of DELPHIN.

VIII.2.1. Input - Output requirements.

The delays are inserted in the feedback paths of a sequential logical machine simulating a boolean model.

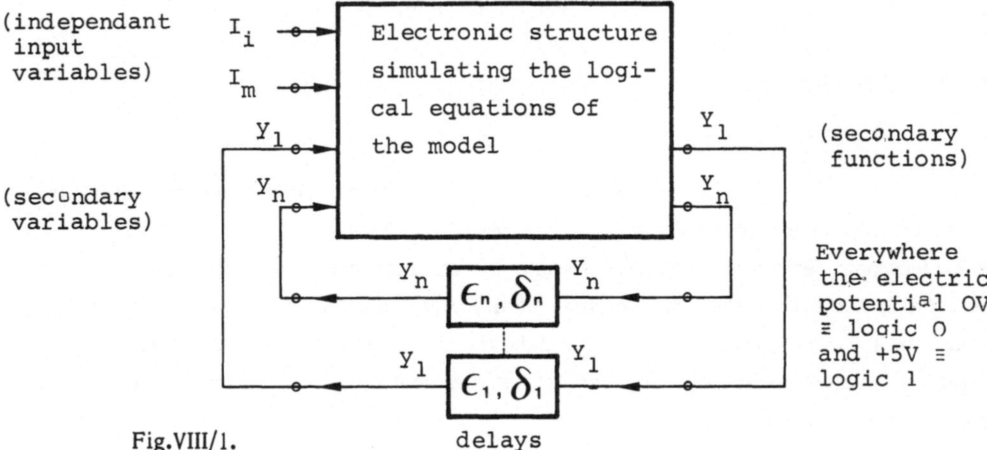

Fig.VIII/1.

One can see in fig.VIII/1 that the input of a delay is a secondary function and its output a secondary variable.

The dynamic requirements are summarized in fig.VIII/2. This figure shows the turn-on and -off delays (ϵ, δ) but also two cases of annihilation, that is, where Y (t) takes a new value before y (t) has switched after the typical delay. That corresponds to an inertial delay behaviour (Van Ham, 1975).

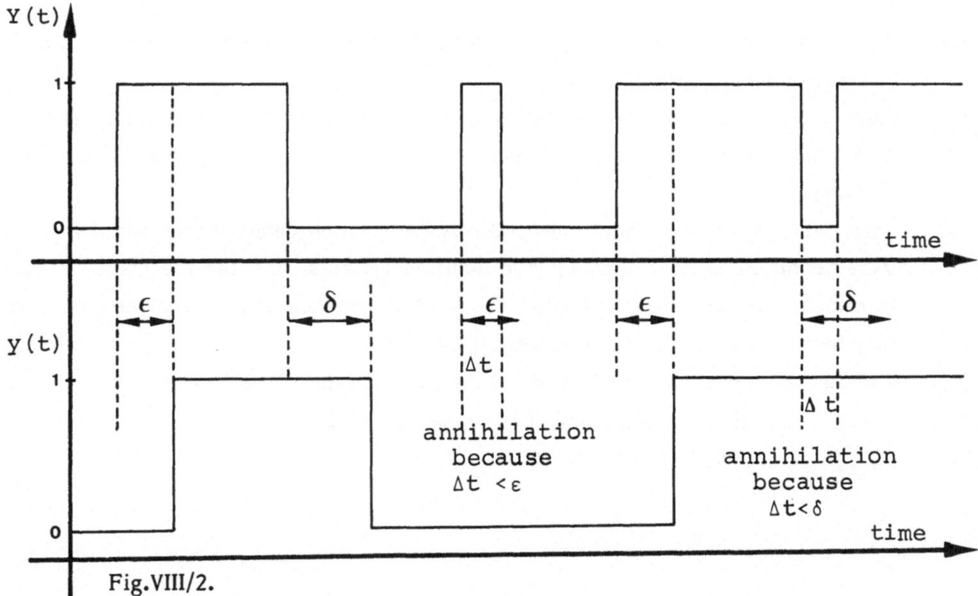

Fig.VIII/2.

VIII.2.2. External aspect of the front panel for one delay.

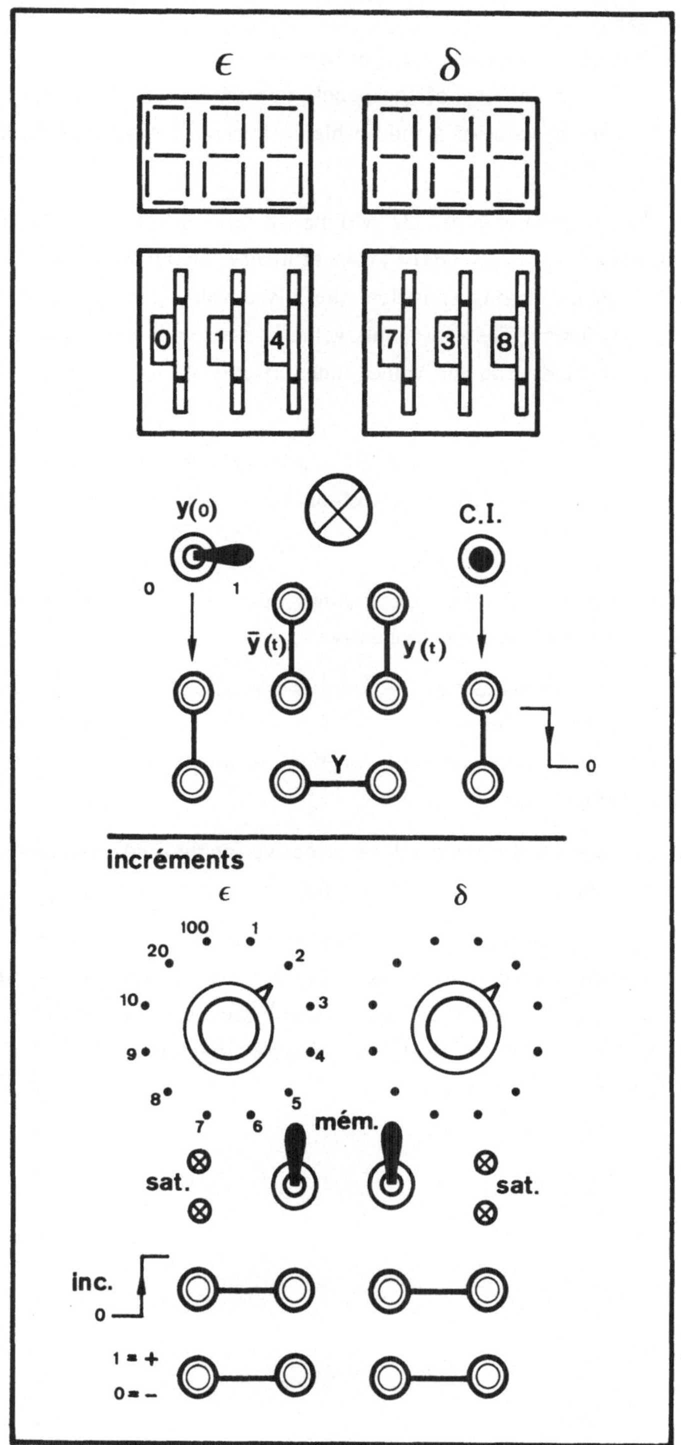

Fig.VIII/3.

Figure VIII.3. shows what a delay looks like to the user.
Two triplets of decade-switches make it possible to choose the initial value of ε and δ between 000 and 999 .
The unit of time may be choosen by a selector between 1 μs and 10 sec.

At the top of the panel, two triplets of 7-segment displays give a visualization of the current value of ε and δ which is initially the same as those of the decade switches.

Under the decade switches, a three position switch makes it possible to preset the output of the delay (y(t)) at an arbitrary chosen value (y (0) = 0 or y (0) = 1). When the switch is in the medium position, the initial value is made equal to the electric potential of a special input. Therefore, this value can be made dependent on the output of another sequential machine. The y(0) values make it possible to choose the initial state in a model simulation.

Another control input (C.I.) starts the hole system at any arbitrary time. This input can be activated by a push-button or another special external electrical input.

This input may be wired with its equivalent on any other delays, so when one of them is reset to 0 , all wired delays are started synchronously.

The bottom of the panel is specialized for variable delays. Selectors permit the increment to be chosen in unit of time base . One input (inc.), when it switches on, starts the internal incrementation routine, another input (1 = + , 0 = -) selects incrementation or decrementation.

Of course, the new value of the delay will be effective for the next variation of Y , that is, if the system is performing a time delay this last one will not be changed.

Moreover, there are two saturation levels : 000 and 999.
A decrementation acting on the current value 000 will give 000 and an incrementation acting on 999 will give 999 . Light emitting diodes (sat.) are used to indicate such saturation (999 : red light, 000 green light). It is also possible to store or not the first time the saturation occurs by a switch (mem).

VIII.2.3. Delphin : Functional diagram.

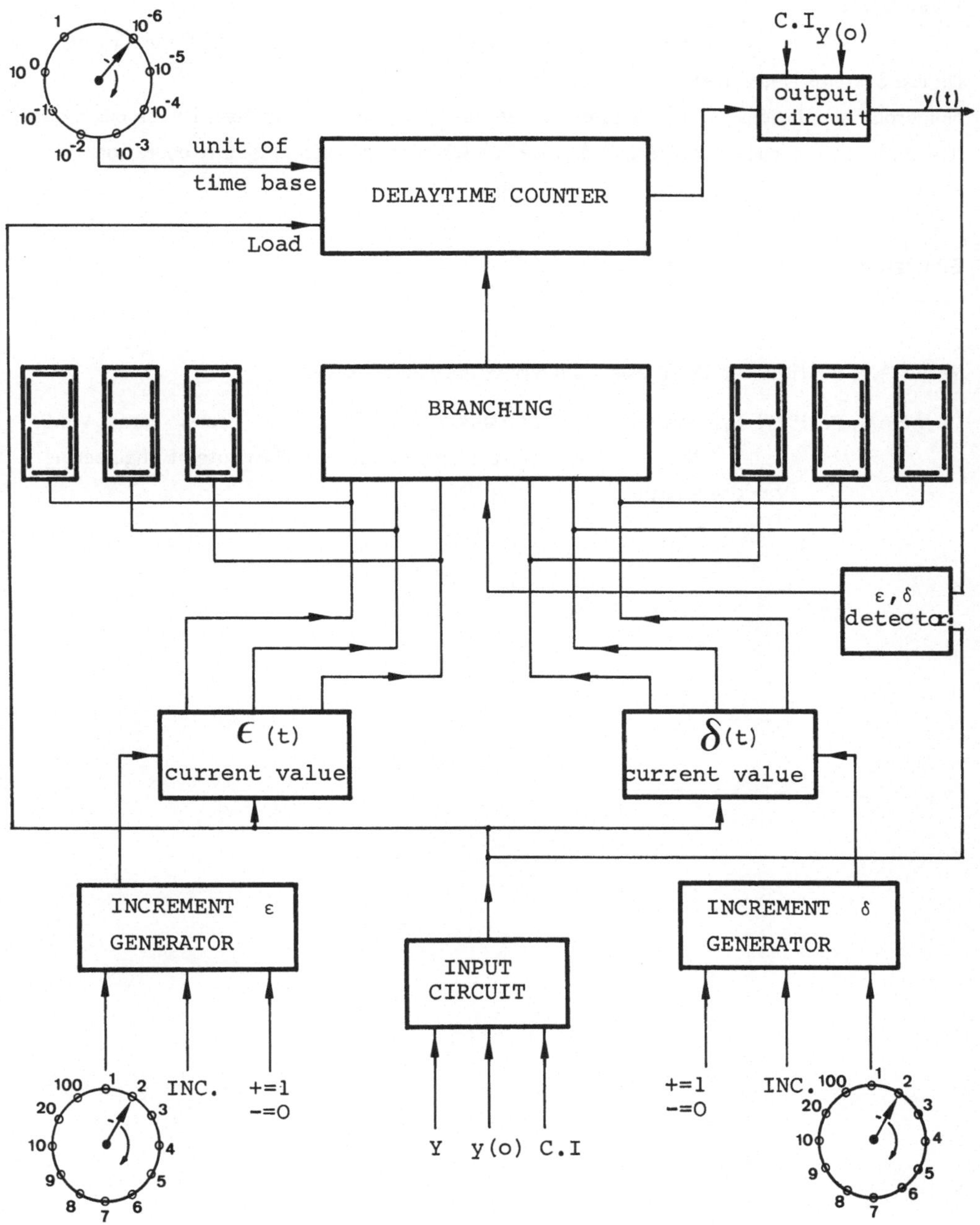

Fig.VIII/4.

The hardware detailed structure is built of integrated circuit technology.
It is not possible here to give a more detailed diagram of the whole machine than in figure VIII/4.

A set of 24 such delays have been constructed at the Logical and Digital Systems Department of Brussels University.
Numerous simulations up to 24 internal variables have tested the fiability of our system (Delphin) in many circumstances, so that we can say that the system is fully operational.

Bibliography

VAN HAM, P. (1975) Ph.D. Thesis - Université libre de Bruxelles.

VAN HAM, P. (1975) Symposium : Applied Aspects of the Automata Theory. Varna, Vol.2. , p. 728, Edit. Bulgarian Academy of Sciences, Institute of Engineering Cybernetics.

CHAPTER IX.

Computer Programs for simulation of logical models.

P. VAN VAM[+] & J.L. DEHOUCK[+]

IX.1. Introduction.

Systems involving a small number of variables have been analyzed in the preceeding chapters. As the number of variables, and especially, of memorization variables increases, there is an increasing need for computer programs to perform the various steps of the analysis :

- given a set of logical equations, compute the state table (programs TABET (BIN & DEC) see § IX.4.).
- given a set of logical equations and an arbitrarily chosen initial state, compute all the possible pathways (sequences of states). (program PRF03, see § IX.4.).
- given a set of logical equations

  a set of numerical values representing the turn-on and turn-off delays for each variable.

  an initial state

  compute the sequence of states and the duration of each transition (program PRFØ1, described in detail in § IX.2, see also Thomas et Van Ham, 1974).
- In the preceeding case, a well-defined value is associated to each time delay. In a population of (for instance) living cells, each delay will have a well-defined value for a given individual cell, but at the level of the population its value will be somehow distributed around an average value. In this case, each simulation (simulating an individual cell) will use a different set of time delays, resulting in a characteristic pathway and final state (Van Ham 1975, 1977). The frequencies of the various pathways, or of the various final states can be computed after a suitable (10^2, ... 10^5) number of simulations, thus giving an idea of the situation at the level of the population. Program PRAN2 (described in detail in § IX.3.) computes for each sample the final behaviour (a stable state or a cycle) of the system and a count of the number of occurences of each final behaviour computed. In the variant presented here one uses a simple rectangular distribution of the values of the time delays.

+ Service des Systèmes Logiques et Numériques - Université libre de Bruxelles.

- Once one knows which pathways are possible, one would like to know which conditions (linear inequalities) those paths imply on the values of the delays. This and the fact that those conditions are (or not) compatible can be computed with program PRF02 (see § IX.4.).

IX.2.1. Deterministic sequences simulation : PRFØ1.

The data of the problem are :

- A set of logical equations representing the logical structure of the model.
- A set of numerical values representing the turn-on and turn-off delays for each variable.
- An initial boolean state from which the system will proceed through some sequence of states.

In this section we keep each delay constant during the whole sequence of states.

Let us suppose that we have a set of eight equations :

$$Y_1 = \bar{y}_5 \quad (1)$$
$$Y_2 = (y_1 + y_2 + y_3)\bar{y}_5 \quad (2)$$
$$Y_3 = y_4 \quad (3)$$
$$Y_4 = (\bar{y}_2\bar{y}_3 + y_7)(\bar{y}_5 + y_6) \quad (4)$$
$$Y_5 = (\bar{y}_2\bar{y}_3 + y_7)\bar{y}_5 \quad (5)$$
$$Y_6 = (\bar{y}_2\bar{y}_3 + y_7)\bar{y}_5 \quad (6)$$
$$Y_7 = \bar{y}_2\bar{y}_3 y_8 + y_7 \quad (7)$$
$$Y_8 = (\bar{y}_2\bar{y}_3 + y_7) y_6 \bar{y}_5 \quad (8)$$

Note that functions Y_5 and Y_6 look identical ; in fact they are not because the delays are different.

An equation such as (4) for example, is written in FORTRAN IV :

```
YF (4) = (.NØT.YV (2) .AND ..NØT.YV (3).ØR.YV (7) ).AND. (.NØT.YV (5) . ØR.YV (6) ) . AND . 1
```

The program computes the next state from the set of equations and prints the time intervals (RI) between the present state and the next transition for each variable. Than it takes the minimum of those non zero intervals (MIN) ; this tells which transition will occur first and as a consequence the actual next state, which is then printed. Then the program computes next step in the same manner.

All the non-zero (RI) are decremented by the value of (MIN). If a variable is excited in this particular current state, the value of the corresponding delay is automatically used as (RI). If a variable is no longer (or not) excited in the current state, than (RI) is made equal to zero.

When the number of states in the sequence is greater than a program variable called EXT (which is chosen less than or equal to 999) a test stops the computing with that number of intermediary states. Otherwise, the sequence stops when a stable state is encountered, that is, when all the (RI) are zero. We give below a listing of PRF01 and a short example corresponding to the equations shown above. After this example we will give "directions for use" of this program.

```
PROGRAM PRF01        74/74    OPT=1                                   FTN 4.6+452

          PROGRAM PRF01(INPUT,OUTPUT)
          COMMON/Y/YF(30),YV(30),XV(15),PROB
          DIMENSION RI(30),DE(30),DD(30)
          INTEGER COUNT,EXT,YF,YV,PROB,DONN,XV
          REAL RI,DE,DD,RM
      100 COUNT=1
          READ 1,PROB,DONN
          IF(EOF(5LINPUT).NE.0) STOP
          READ 1,M,N,EXT
          READ 2,(DD(I),I=1,N)
          READ 2,(DE(I),I=1,N)
          PRINT 1000,PROB,DONN
          PRINT 1001,(DD(I),I=1,N)
          PRINT 1002,(DE(I),I=1,N)
          IF(M.EQ.0) GOTO 101
          READ 3,(XV(I),I=1,M)
          PRINT 1003,(XV(I),I=1,M)
      101 READ 3,(YV(I),I=1,N)
          PRINT 1004,(YV(I),I=1,N)
          DO 10 I=1,N
       10 RI(I)=0
      102 CALL EVAL
          PRINT 1005,(YV(I),I=1,N)
          DO 20 I=1,N
          IF(YV(I).NE.YF(I)) GOTO 21
          RI(I)=0
          GOTO 20
       21 IF(RI(I).NE.0) GOTO 20
          IF(YF(I).EQ.1) GOTO 22
          RI(I)=DD(I)
          GOTO 20
       22 RI(I)=DE(I)
       20 CONTINUE
          PRINT 1006,(RI(I),I=1,N)
          SOM=0
          DO 30 I=1,N
       30 SOM=SOM+RI(I)
          IF(SOM.EQ.0) GOTO 105
          RM=1E10
          DO 40 I=1,N
          IF(RI(I).EQ.0) GOTO 40
          RM=AMIN1(RM,RI(I))
       40 CONTINUE
          PRINT 1007,RM
          DO 50 I=1,N
          IF(RI(I).NE.RM) GOTO 52
          YV(I)=.NOT.YV(I).AND.1
          GOTO 51
       52 IF(RI(I).EQ.0.)GOTO 50
       51 RI(I)=RI(I)-RM
       50 CONTINUE
          COUNT=COUNT+1
          IF(COUNT.LE.EXT) GOTO 102
          PRINT 1009,EXT
          GOTO 100
      105 PRINT 1008,(YV(I),I=1,N)
          GOTO 100

PROGRAM PRF01        74/74    OPT=1                                   FTN 4.6+452
```

```
          1 FORMAT(3I3)
          2 FORMAT(12F6.2)
          3 FORMAT(30I2)
       1000 FORMAT(*1STRUCTURE NO. :*I2,5X,*DONNEES NO. :*I2)
       1001 FORMAT(*0DECL :*15F7.2/7X,15F7.2)
       1002 FORMAT(* ENCL :*15F7.2/7X,15F7.2)
       1003 FORMAT(*0ETAT EXTERNE :*15I3)
       1004 FORMAT(*0ETAT INITIAL :*30I3)
       1005 FORMAT(1H0,6X,15I7/7X,15I7)
       1006 FORMAT(* RI =  *15F7.2/7X,15F7.2)
       1007 FORMAT(* MIN   *F7.2)
       1008 FORMAT(///* ETAT FINAL :*30I3)
       1009 FORMAT(///* PAS D ETAT FINAL APRES *I3* ETATS INTERMEDIAIRES*)
            END
 SUBROUTINE EVAL        74/74    OPT=1                         FTN 4.6+452

            SUBROUTINE EVAL
            COMMON/Y/YF(30),YV(30),XV(15),PROB
            INTEGER YF,YV,XV,PROB
            GOTO (1,2,3,4,5) PROB
          1 CONTINUE
            YF(1)=.NOT.YV(5).AND.1
            YF(2)=(YV(1).OR.YV(2).OR.YV(3)).AND..NOT.YV(5).AND.1
            YF(3)=YV(4).AND.1
            YF(4)=(.NOT.YV(2).AND..NOT.YV(3).OR.YV(7)).AND.(.NOT.YV(5).OR.YV(6
           1)).AND.1
            YF(5)=(.NOT.YV(2).AND..NOT.YV(3).OR.YV(7)).AND..NOT.YV(5).AND.1
            YF(6)=(.NOT.YV(2).AND..NOT.YV(3).OR.YV(7)).AND..NOT.YV(5).AND.1
            YF(7)=((.NOT.YV(2).AND..NOT.YV(3).AND.YV(8)).OR.YV(7)).AND.1
            YF(8)=(.NOT.YV(2).AND..NOT.YV(3).OR.YV(7)).AND.YV(6).AND..NOT.YV(5
           1).AND.1
            RETURN
          2 CONTINUE
            RETURN
          3 CONTINUE
            RETURN
          4 CONTINUE
            RETURN
          5 CONTINUE
            RETURN
            END
```

```
STRUCTURE NO. : 1     DONNEES NO. : 1
DECL :  99.00  99.00  99.00  15.00   8.00   1.80  99.00   5.00
ENCL :  10.00  30.00   7.00   6.00   5.00   4.00  15.00   3.00
ETAT INITIAL :  0  0  0  0  0  0  0  0
             0      0      0      0      0      0      0      0
RI =     10.00   0.00   0.00   6.00   5.00   4.00   0.00   0.00
MIN       4.00
             0      0      0      0      0      1      0      0
RI =      6.00   0.00   0.00   2.00   1.00   0.00   0.00   3.00
MIN       1.00
             0      0      0      0      1      1      0      0
RI =      0.00   0.00   0.00   1.00   8.00   1.80   0.00   0.00
MIN       1.00
             0      0      0      1      1      1      0      0
RI =      0.00   0.00   7.00   0.00   7.00    .80   0.00   0.00
MIN        .80
             0      0      0      1      1      0      0      0
RI =      0.00   0.00   6.20  15.00   6.20   0.00   0.00   0.00
MIN       6.20
             0      0      1      1      0      0      0      0
RI =     10.00  30.00   0.00   8.80   0.00   0.00   0.00   0.00
MIN       8.80
             0      0      1      0      0      0      0      0
RI =      1.20  21.20  99.00   0.00   0.00   0.00   0.00   0.00
MIN       1.20
             1      0      1      0      0      0      0      0
RI =      0.00  20.00  97.80   0.00   0.00   0.00   0.00   0.00
MIN      20.00
             1      1      1      0      0      0      0      0
RI =      0.00   0.00  77.80   0.00   0.00   0.00   0.00   0.00
MIN      77.80
             1      1      0      0      0      0      0      0
RI =      0.00   0.00   0.00   0.00   0.00   0.00   0.00   0.00


ETAT FINAL :  1  1  0  0  0  0  0  0


14.20.59.UCLP, 24,      0.192KLNS.
```

IX.2.2. Directions for use of PRF01.

DATA :

(- number of the set of equations (PROB:I3) and number of the set of data (DON:I3) .

(- number of the external variables (M:I3) , number of the internal variables (N:I3), and EXT:I3.

ONE block of data

(- turn-off delays (DD(I) : 12F6.2) and

(=) eventually one or two "continuation cards".

(- turn-on delays (DE (I) : 12F6.2) and

(=) eventually one or two "continuation cards"

(- external variables values (:XV(I) : 15 (blank.I1))

(- internal variables initial values (YV (I) : 30 (blank.I1))

In this version of the program the maximum values for the different numbers are :

30 internal variables

15 external variables

5 different sets of equations

999 sets of data per set of equations

999 states per sequence

999.99 unit of time base for a delay.

The kth set of equations must be inserterted between the card "k CONTINUE" and the card "RETURN" in the SUBROUTINE "EVAL".

IX.3.1. Sequence simulations with random distributed delays : PRAN2.

We now want to make a large number of simulations with different delays. For each simulation from the same initial internal state, we will choose the delay values at random within a previously fixed range around a mean value :

$$\varepsilon_i = \langle \varepsilon_i \rangle \pm \Delta\varepsilon_i \qquad \text{MDE (I)} \pm \text{EDE(I)}$$

$$\delta_i = \langle \delta_i \rangle \pm \Delta\delta_i \qquad \text{MDD (I)} \pm \text{EDD(I)}$$

We suppose here that the successive values of a delay are uniformly chosen at random in the range MDE (I) - EDE (I) , MDE (I) + EDE (I)

As the number of samples may be as great as 100000, the program results is not of course, a list of all the computed sequences.

Program PRAN2 computes for each sample the final behaviour, which may be either a final stable state or a final stable cycle. Then a count of those final computed behaviours is made, in order to have a relative frequency of occurance for each of them.

Naturally, if the number of steps of some sequences exceeds the value of a fixed variable EXT (see PRF01), this sequence is counted in a special class of presumed cycles : "UNDETERMINED CYCLES".

IX.3.2. Cycle comparisons

Program PRAN2 detects a cycle when it finds in a sequence a previously encountered state with the same RI for each variable. (see PRFOl for RI definition).

As a consequence, it is possible to find a cycle of states which is not elementary, that is, with some states occuring more than once in the cycle. When program PRAN2 has detected a cycle, it must compare that cycle with the others found previously, but then there is a phase problem : for example the cycles A B C D and C D A B are the same with a phase shift.

To avoid spending too much time in trying all the possible comparisons (ABCD, BCDA, CDAB, DABC) with all the cycle classes found, the program calculates a unique presentation of a cycle.

All cycles are shifted so that the state described by the smallest number is the first. If this number is not unique, then the first state of the cycle will correspond to the minimal sum sequence of length 2 beginning at each minimum state. If the minimum is not unique, one tries again with sequences of length 3.a.s.o.

Example :

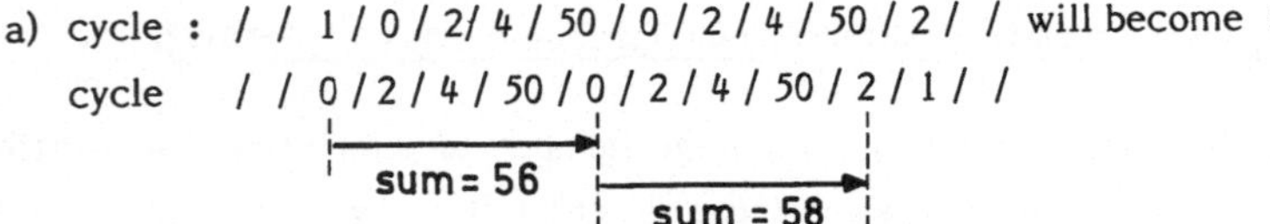

a) cycle : / / 1 / 0 / 2/ 4 / 50 / 0 / 2 / 4 / 50 / 2 / / will become :
cycle / / 0 / 2 / 4 / 50 / 0 / 2 / 4 / 50 / 2 / 1 / /
sum = 56
sum = 58

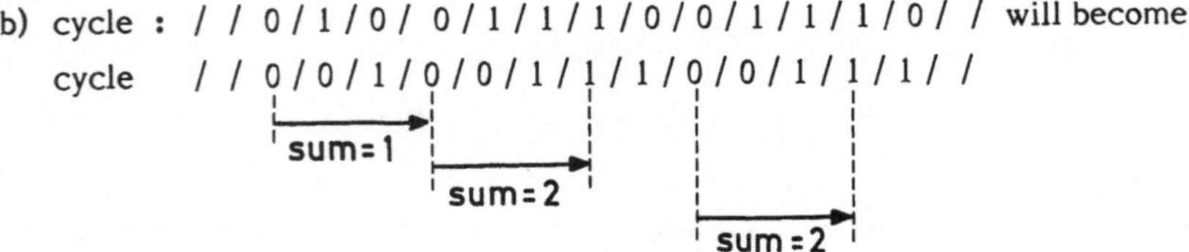

One sees that when the preceeding procedure does not have a unique solution, the reason is a periodicity or a symmetry in the cycle itself. (e.g. //0101010101// which has 5 equivalent solutions).

IX.3.3. Listing of PRAN2.

We give below a listing of PRAN2 and a short example.

One may see that PRF01 is a SUBroutine of PRAN2.

```
      PROGRAM PRAN2(INPUT,OUTPUT,TAPE1)
      INTEGER NETDIFF,NCYDIFF,NB,NETAT,NCYCLE,NINDCY,PROB,DONN,N,M,NECH,
     + EXT,XV,YV,EI,EQD,ETAT,CYLONG,CYCLE,LCYCLE,MAT,KBIN,LC
      REAL SAFE,MDD,MDE,EDD,EDE,RI,INC,AL,DD,DE
      DIMENSION EI(30),ETAT(50),NETAT(50),MDD(30),MDE(30),EDD(30),EDE(30
     +),CYCLE(50,100),NCYCLE(50),LCYCLE(50)
      COMMON/V123/YV(30)/V12/RI(30),N,EXT,EQD,MAT(100),CYLONG,DD(30),DE(
     +30)/V13/XV(15),PROB
      READ (1) SAFE
      CALL RANSET(SAFE)
      NETDIFF=50   $   NCYDIFF=20
  103 NB=1
      DO 10 I=1,NETDIFF
      NETAT(I)=0
   10 CONTINUE
      DO 15 I=1,NCYDIFF
      NCYCLE(I)=0
   15 CONTINUE
      NETDIFF=NCYDIFF=NINDCY=0
      READ 1,PROB,DONN
      IF(EOF(5LINPUT).NE.0) GOTO 200
      READ 1,N,M,NECH,EXT
      READ 3,(MDD(I),I=1,N)
      READ 3,(EDD(I),I=1,N)
      READ 3,(MDE(I),I=1,N)
      READ 3,(EDE(I),I=1,N)
      PRINT 1000,PROB,DONN
      PRINT1001,(MDD(I),I=1,N)
      PRINT1002,(EDD(I),I=1,N)
      PRINT 1003,(MDE(I),I=1,N)
      PRINT 1002,(EDE(I),I=1,N)
      IF(M.EQ.0) GOTO 120
      READ 2,(XV(I),I=1,M)
      PRINT 1004,(XV(I),I=1,M)
  120  READ 2,(EI(I),I=1,N)
      PRINT 1005,(EI(I),I=1,N)
       PRINT 1006,NECH
  102 DO 20 I=1,N
      YV(I)=EI(I)
      RI(I)=0.
      AL=RANF(INC)
      DD(I)=(2*AL-1)*EDD(I)+MDD(I)
      AL=RANF(INC)
      DE(I)=(2*AL-1)*EDE(I)+MDE(I)
   20 CONTINUE
      CALL PRF01 , RETURNS (100,106)
      IF(NETDIFF.EQ.0) GOTO 41
```

```
      DO 40 I=1,NETDIFF
      IF(EQD.NE.ETAT(I)) GOTO 40
      J=I   $   GOTO 104
   40 CONTINUE
   41 NETDIFF=NETDIFF+1
      IF(NETDIFF.GE.51)GOTO101
      J=NETDIFF
      ETAT(J)=EQD
  104 NETAT(J)=NETAT(J)+1
      GOTO 101
  100 IF(NCYDIFF.EQ.0) GOTO 51
      DO 50 I=1,NCYDIFF
      DO 55 J=1,CYLONG
      IF(MAT(J).NE.CYCLE(I,J)) GOTO 50
   55 CONTINUE
      M=I    $    `GOTO 105
   50 CONTINUE
   51 NCYDIFF=NCYDIFF+1
      IF (NCYDIFF.GE.51)GOTO 101
      M=NCYDIFF   $   LCYCLE(M)=CYLONG
      DO 60 I=1,CYLONG
      CYCLE(M,I)=MAT(I)
   60 CONTINUE
  105 NCYCLE(M)=NCYCLE(M)+1
      GOTO 101
  106 NINDCY=NINDCY+1
  101 NB=NB+1
      IF(NB.LE.NECH) GOTO 102
      IF(NETDIFF.EQ.0) GOTO 107
      PRINT 1020
      IF(NETDIFF.LE.50) GOTO 998
      PRINT 1013,NETDIFF
 1013 FORMAT(* NOMBRE D ETATS STABLES PLUS GRAND QUE 50 *,I5)
      NETDIFF=50
998   DO 70 I=1,NETDIFF
      KBIN=ETAT(I)
      DO 75 J=1,N
      M=N+1-J
      YV(M)=KBIN.AND.1
      KBIN=SHIFT(KBIN,-1)
   75 CONTINUE
      PRINT 1007,(YV(J),J=1,N)
      PRINT 1008,ETAT(I),NETAT(I)
   70 CONTINUE
  107 IF(NCYDIFF.EQ.0) GOTO 108
      PRINT 1021
```

```
      IF(NCYDIFF.LE.50)GOTO 999
      PRINT 1012,NCYDIFF
 1012 FORMAT(* NOMBRE DE CYCLES DIFFERENTS PLUS GRAND QUE 50 *,I5)
      NCYDIFF=50
  999 DO 80 I=1,NCYDIFF
      LC=LCYCLE(I)
      PRINT 1009,(CYCLE(I,L),L=1,LC)
      PRINT 1010,NCYCLE(I)
   80 CONTINUE
  108 IF(NINDCY.EQ.0) GOTO 103
      PRINT 1011,NINDCY
      GOTO 103
  200 CALL RANGET(SAFE)
      WRITE(1) SAFE
      STOP
    1 FORMAT(2I2,I5,I2)
    2 FORMAT(30I2)
    3 FORMAT(12F6.2)
 1000 FORMAT(*1STRUCTURE NO. *,I2,5X,*DONNEES NO. *I3)
 1001 FORMAT(*0DECL. MOYEN : *,15F7.2/15X,15F7.2)
 1002 FORMAT(*       ECART : *,15F7.2/15X,15F7.2)
 1003 FORMAT(* ENCL. MOYEN : *,15F7.2/15X,15F7.2)
 1004 FORMAT(* ETAT  D ENTREE    : *,15I2)
 1005 FORMAT(*0ETAT INITIAL INTERNE : *,30I2)
 1006 FORMAT(*0NOMBRE D ECHANTILLONS : *I5)
 1007 FORMAT(*0ETAT BINAIRE : *30I1)
 1008 FORMAT(1H+,46X,*EQUIV. DECIMAL :*I10/*NOMBRE D OCCURENCES : *I5)
 1009 FORMAT(*0SEQUENCE DECIMALE :*,10I10/9(21X,10I10/))
 1020 FORMAT(45X,*ETATS FINAUX.*/45X,13(1H*)/)
 1021 FORMAT(//45X,*CYCLES.*/45X,7(1H*)/)
 1010 FORMAT(* NOMBRE D OCCURRENCES :*,I5)
 1011 FORMAT(*0NOMBRE DE CYCLES INDETERMINES :*I5)
      END
      SUBROUTINE PRF01,RETURNS (A,B)
      COMMON/V123/YV(30)/V12/RI(30),N,EXT,EQD,MAT(100),CYLONG,DD(30),DE(
     +30)/V23/YF(30)
      INTEGER NB,EQD,YV,MATET,YF,POS,MIN,POSMIN,NEQ,EXCUR,COUNT,LN,LM,I,
     +J,K,CYLONG,POSMI,MAT,EXT,INTER
      REAL RI,MATRI,DD,DE,SOM,RM
      DIMENSION MATRI(100,30),MATET(100),POSMIN(50) ,INTER(100)
      NB=1
      EQD=0
      DO 5 I=1,N
      MATRI(1,I)=0.
      EQD=2*EQD+YV(I)
    5 CONTINUE
      MATET(1)=EQD
```

```
100 CALL EVAL
    DO 10 I=1,N
    IF(YV(I).NE.YF(I)) GOTO 11
    RI(I)=0.
    GOTO 10
 11 IF(RI(I).NE.0.) GOTO 10
    IF(YF(I).EQ.1) GOTO 12
    RI(I)=DD(I)
    GOTO 10
 12 RI(I)=DE(I)
 10 CONTINUE
    DO 45 I=1,N
    MATRI(NB,I)=RI(I)
 45 CONTINUE
    IF(NB.EQ.1) GOTO 104
    JNB=NB-1
    DO 35 I=1,JNB
    IF(EQD.NE.MATET(I)) GOTO 35
    DO 40 J=1,N
    IF(RI(J).NE.MATRI(I,J))GOTO 35
 40 CONTINUE
    POS=I
    GOTO 101
 35 CONTINUE
104 SOM=0.
    DO 15 I=1,N
    SOM=SOM + RI(I)
 15 CONTINUE
    IF(SOM.EQ.0.) RETURN
    RM=1E10
    DO 20 I=1,N
    IF(RI(I).EQ.0.) GOTO 20
    RM=AMIN1(RM,RI(I))
 20 CONTINUE
    DO 25 I=1,N
    IF(RI(I).NE.RM) GOTO 27
    YV(I)=.NOT.YV(I).AND.1
    GOTO 26
 27 IF(RI(I).EQ.0.) GOTO 25
 26 RI(I)=RI(I)-RM
 25 CONTINUE
    EQD=0
    DO 30 I=1,N
    EQD=2*EQD+YV(I)
 30 CONTINUE
```

```
      NB=NB+1
      MATET(NB)=EQD
      IF(NB.LE.EXT) GOTO 100
      RETURN B
  101 MIN=1E15
      DO 50 I=POS,JNB
      IF(MATET(I).LT.MIN) GOTO 51
      IF(MATET(I).EQ.MIN) GOTO 52
      GOTO 50
   51 MIN=MATET(I)
      POSMIN(1)=I
      NEQ=1
      GOTO 50
   52 NEQ=NEQ+1
      POSMIN(NEQ)=I
   50 CONTINUE
      CYLONG=NB -POS
      IF(NEQ.EQ.1) GOTO 102
      IF(POS.EQ.POSMIN(1)) GOTO 103
      LN=POSMIN(1) $ J=0
      DO 55 I=LN,JNB
      J=J+1
      INTER(J)=MATET(I)
   55 CONTINUE
      LN=POSMIN(1)-1
      DO 60 I=POS,LN
      J=J+1
      INTER(J)=MATET(I)
   60 CONTINUE
      EXCUR=POSMIN(1)-1
      DO 65 I=1,NEQ
      POSMIN(I)=POSMIN(I)-EXCUR
   65 CONTINUE
      DO 70 I=1,CYLONG
      MATET(I)=INTER(I)
   70 CONTINUE
  103 EXCUR=POSMIN(2)-POSMIN(1)-1
      LN=NEQ-1    $   COUNT=0
      DO 75 I=1,EXCUR
      DO 75 J=1,LN
      LM=NEQ-J
      DO 75 K=1,LM
      IF((POSMIN(J).EQ.0).OR.(POSMIN(J+K).EQ.0)) GOTO 75
      IF(MATET(POSMIN(J)+I).GT.MATET(POSMIN(J+K)+I)) GOTO 76
      IF(MATET(POSMIN(J)+I).LT.MATET(POSMIN(J+K)+I)) GOTO 77
      GOTO 75
   76 POSMIN(J)=0
      COUNT=COUNT+1
      GOTO 75
   77 POSMIN(J+K)=0
      COUNT=COUNT+1
   75 CONTINUE
      IF(COUNT.EQ.0) GOTO 102
      IF(COUNT.NE.(NEQ-1)) RETURN B
      DO 80 I=1,NEQ
      IF(POSMIN(I).EQ.0) GOTO 80
      POSMIN(1)=POSMIN(I)
      GOTO 102
   80 CONTINUE
```

```
  102 POSMI=POSMIN(1)
      J=0
      DO 90 I=POSMI,JNB
      J=J+1
      MAT(J)=MATET(I)
   90 CONTINUE
      IF(J.EQ.CYLONG) RETURN A
      LM=POSMI-1
      DO 95 I=POS,LM
      J=J+1
      MAT(J)=MATET(I)
   95 CONTINUE
      RETURN A
      END
      SUBROUTINE EVAL
      COMMON/V123/YV(30)/V13/XV(15),PROB/V23/YF(30)
      INTEGER YF,YV,XV,PROB
      GOTO(1,2,3,4,5,6,7,8,9,10,11,12,13,14,15)PROB
    1 CONTINUE
      RETURN
    2 CONTINUE
      RETURN
    3 CONTINUE
      RETURN
    4 CONTINUE
      RETURN
    5 CONTINUE
      RETURN
    6 CONTINUE
      RETURN
    7 CONTINUE
      RETURN
    8 CONTINUE
      RETURN
    9 CONTINUE
      RETURN
   10 CONTINUE
      RETURN
   11 CONTINUE
      RETURN
   12 CONTINUE
      RETURN
   13 CONTINUE
      RETURN
   14 CONTINUE
      RETURN
   15 CONTINUE
      RETURN
      END
```

```
STRUCTURE NO.  1      DONNEES NO.   3
DECL. MOYEN :    99.00   99.00   99.00    8.00   14.00    2.00   99.00   10.00
      ECART :     9.90    9.90    9.90     .80    1.40     .20    9.90    1.00
ENCL. MOYEN :    10.00   30.00    5.00    7.00    5.00    5.00    8.00    5.00
      ECART :     1.00    3.00     .50     .70     .50     .50     .80     .50
ETAT INITIAL INTERNE :  0 0 0 0 0 0 0 0
NOMBRE D ECHANTILLONS :    100
```

```
                                                  CYCLES.
                                                  *******

SEQUENCE DECIMALE :          2         6         14         10
NOMBRE D OCCURRENCES :   19
SEQUENCE DECIMALE :          2         10
NOMBRE D OCCURRENCES :   52
SEQUENCE DECIMALE :          34        38        46         62         58         42
NOMBRE D OCCURRENCES :   29
STRUCTURE NO.  1       DONNEES NO.   4
DECL. MOYEN :    99.00    99.00    99.00    8.00   32.00    2.00   99.00   10.00
       ECART :    9.90     9.90     9.90     .80    3.20     .20    9.90    1.00
ENCL. MOYEN :    10.00    30.00     5.00    7.00    5.00    5.00    8.00    5.00
       ECART :    1.00     3.00      .50     .70     .50     .50     .80     .50
ETAT INITIAL INTERNE :  0 0 0 0 0 0 0 0
NOMBRE D ECHANTILLONS :   100
                                                  ETATS FINAUX.
                                                  *************

ETAT BINAIRE : 11000000                             EQUIV. DECIMAL :        192
OMBRE D OCCURENCES :     23

                                                  CYCLES.
                                                  *******

SEQUENCE DECIMALE :          0         4         12         8
NOMBRE D OCCURRENCES :   18
SEQUENCE DECIMALE :          0         8
NOMBRE D OCCURRENCES :   59
STRUCTURE NO.  2       DONNEES NO.   1
DECL. MOYEN :    5.00   1.00
       ECART :    0.00   0.00
ENCL. MOYEN :    5.00   1.00
       ECART :    0.00   0.00
ETAT INITIAL INTERNE :  0 0
NOMBRE D ECHANTILLONS :    5
NOMBRE DE CYCLES INDETERMINES :    5
STRUCTURE NO.  2       DONNEES NO.   2
DECL. MOYEN :    3.00   1.00
       ECART :    0.00   0.00
ENCL. MOYEN :    3.00   1.00
       ECART :    0.00   '0.00
ETAT INITIAL INTERNE :  0 0
NOMBRE D ECHANTILLONS :    5

                                                  CYCLES.
                                                  *******

SEQUENCE DECIMALE :          0         1         0          3          2          3
NOMBRE D OCCURRENCES :   5
```

IX.3.4. <u>Directions for use of PRAN2.</u>

DATA	(	-	number of the set of equations (PROB : I2) and the number of the set of data (DON : I2)
	(	-	number of internal variables (N : I2), number of external variables (M : I2) ; number or samples for one logical structure and for one initial state (NECH : I5) and (EXT : 12)
ONE	(	-	Turn-off delays mean values (MDD (I)
blok	(	(=)	12F6.2) and eventually one or two "continuation
of DATA	(		cards".
	(	-	Range around the mean for each turn-off delay (EDD (I) : 12F6.2) and eventually one or two "continuation cards".
	(	-	Turn-on delays mean values (MDE (I) : 12F6.2)
	((	(=)	and eventually one or two "continuation cards".
	(	-	Range around the mean for each turn-on delay (EDE (I) :
	(	(=)	12F6.2) and eventually one or two "continuation cards".
	(		External variables values (XV(I) : 15 (blank I1))
	(	-	Internal variables initial values (YV(I) : 30 (blank I1))

In this program version, the maximum values for the different numbers ares :

30	internal variables
15	external variables
5	different sets of logical equations
99	blocks of data for one set of equations
99	states for one sequence (EXT)
99,999	samples for one initial state
999.99	units of time base for a delay

The kth set of equations must be inserted between the card "k CONTINUE" and the card "RETURN"in the SUBroutine "EVAL".

IX.4. Other software programs available

We cannot describe here in detail all the software tools which are available at present for logical model study.

Let us enumerate only the most useful :

- PRFO2 computes the linear system of inequalities between the delays (Max 9 turn-on and 9 turn-off delays) according to a given set of observed internal states sequences and a given set of logical equations. Finally, PRFO2 tests this system of linear inequalities for compatibility . (If there exists at least one non-negative set of values of the delays for which all the inequalities are fulfilled).

- PRFO3 draws a graph of all the possible states sequences according to one initial state and one set of logical equations (Maximum 9 internal variables). Simultaneous variable transitions are ignored, conserving only those where one internal variable turn-on or off at a time.

- TABET (BIN & DEC) are two similar programs which compute the state table corresponding to a set of logical equations.
  Two options BIN or DEC make it possible to choose a table printed in a binary form or in a decimal form.
  The stable states are marked to find them easily. Of course, it is also possible to have a printing of only those rows of the state table containing at least one stable state.

Bibliography.

THOMAS, R. (1973) J. Theor. Biol. 42, 563.

THOMAS, R. & VAN HAM, P. (1974) Biochimie 56, 1529.

VAN HAM, P. (1975). Ph.D. Thesis, Université libre de Bruxelles.

VAN HAM, P. (1977). IFAC Symposium : Discrete Systems, Vol. 5, p. 27. Dresden : KDT & WGMA.

Chapter X

Net structures for sequential logic.

J. FLORINE *

This chapter uses one of the examples of chapter V to outline new trends in the field of sequential problems. In the classical assignment of a state table, the main emphasis was put on the <u>reduction of the number</u> of internal functions. Only later did one try to simplify their structure, as well as that of the output functions.

The present tendency in the design of industrial automata adopts an inverse strategy;one tries to <u>simplify</u> as much as possible <u>the expressions</u> of the internal and output functions, even though this results in a considerable increase in the number of internal functions. In addition, one tries now to choose internal functions which have a physical meaning in the operation of the system, like the output functions.

In the former tendancy, the structure of these functions resulted from an arbitrary assignment of values to the internal variables. This usually leads to a number of internal functions equal to the number of distinct steps through which the system has to proceed.

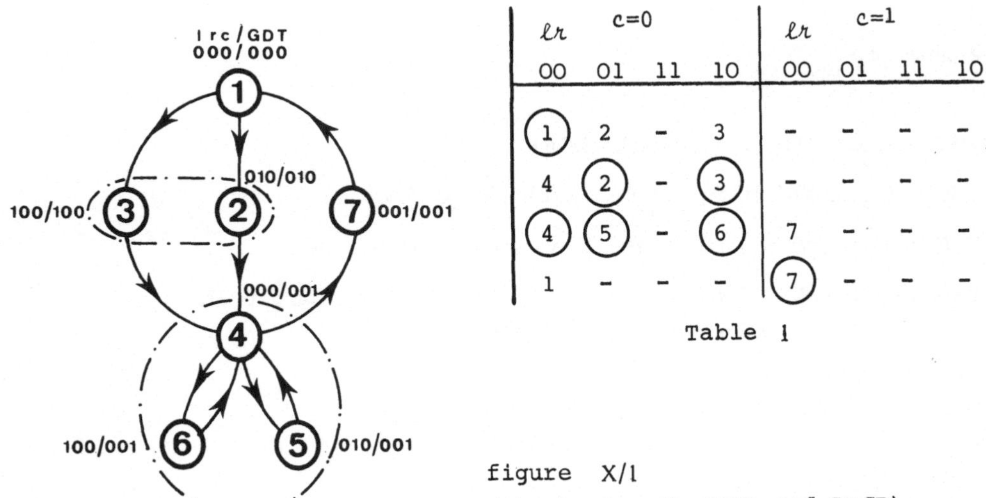

lr (c=0) 00	01	11	10	lr (c=1) 00	01	11	10
(1)	2	-	3	-	-	-	-
4	(2)	-	(3)	-	-	-	-
(4)	(5)	-	(6)	7	-	-	-
1	-	-	-	(7)	-	-	-

Table 1

figure X/1
(in chapter V, G≡GL and D≡GR)

* Directeur du Laboratoire des Systèmes Logiques et Numériques.
Faculté des Sciences Appliquées. Université de Bruxelles.

These steps can be obtained by grouping internal states into sets from which one can easily get the output function values.
Graph 1 reproduces fig. 5 of chapter V (the "simple minded cat").
In table 1, states 2 and 3 have been grouped; they don't correspond to equal values of the output functions, but states 2 and 3 both lead to state 4, and they can be easily distinguished from each other by their input state.Similarily, states 4, 5 and 6 have been grouped because they have in common the fact that the cat is trailing the mouse whatever the stimuli. See also the graph of fig. 2 which only contains four steps.

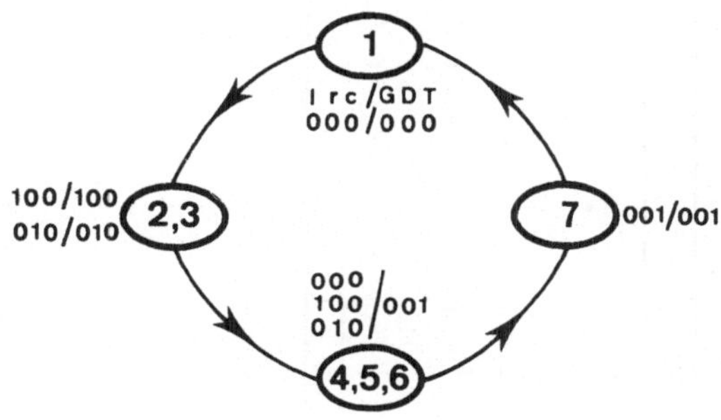

figure X/2

Note that table 1 derives from table 5 of chapter V; the grouping happened to be the same, although for different reasons (more technical above, more linked to the physical meaning here). In table 2, one uses <u>four</u> internal variables y_1 to y_4 and the four lines of table 1 have been assigned, respectively with the "single one" combinations of values 1000, 0100, 0010 and 0001 of these variables.
The transient state between 1000 and 0100 is necessarily 1100, between 0100 and 0010 it is 0110 in order to have only one change at a time. This avoids all the possible races. From table 2, one can derive table 3, in which the values of Y_1, Y_2, Y_3 and Y_4 are given for each stable and transient state.
Tables 4, 5, 6 and 7 give the value of Y_1, Y_2, Y_3 and Y_4 separately.

Table 2

$y_1y_2y_3y_4$	c=0 ℓr 00	01	11	10	c=1 ℓr 00	01	11	10
0000	-	-	-	-	-	-	-	-
<u>0001</u>	1	-	-	-	(7)	-	-	-
0011	-	-	-	-	7	-	-	-
<u>0010</u>	(4)	(5)	-	(6)	7	-	-	-
<u>0100</u>	4	(2)	-	(3)	-	-	-	-
0101	-	-	-	-	-	-	-	-
0111	-	-	-	-	-	-	-	-
0110	4	-	-	-	-	-	-	-
1100	-	2	-	3	-	-	-	-
1101	-	-	-	-	-	-	-	-
1111	-	-	-	-	-	-	-	-
1110	-	-	-	-	-	-	-	-
<u>1000</u>	(1)	2	-	3	-	-	-	-
1001	1	-	-	-	-	-	-	-
1011	-	-	-	-	-	-	-	-
1010	-	-	-	-	-	-	-	-

Table 3

$y_1y_2y_3y_4$	c=0 ℓr 00	01	11	10	c=1 ℓr 00	01	11	10
0000	----	----	----	----	----	----	----	----
0001	1001	----	----	----	0001	----	----	----
0011	----	----	----	----	0001	----	----	----
0010	0010	0010	----	0010	0011	----	----	----
0100	0110	0100	----	0100	----	----	----	----
0101	----	----	----	----	----	----	----	----
0111	----	----	----	----	----	----	----	----
0110	0010	----	----	----	----	----	----	----
1100	----	0100	----	0100	----	----	----	----
1101	----	----	----	----	----	----	----	----
1111	----	----	----	----	----	----	----	----
1110	----	----	----	----	----	----	----	----
1000	1000	1100	----	1100	----	----	----	----
1001	1000	----	----	----	----	----	----	----
1011	----	----	----	----	----	----	----	----
1010	----	----	----	----	----	----	----	----

Table 4.

Y_1	c=0 ℓr 00	01	11	10	c=1 ℓr 00	01	11	10
0000	-	-	-	-	-	-	-	-
0001	1	-	-	-	0	-	-	-
0011	-	-	-	-	0	-	-	-
0010	0	0	-	0	0	-	-	-
0100	0	0	-	0	-	-	-	-
0101	-	-	-	-	-	-	-	-
0111	-	-	-	-	-	-	-	-
0110	0	-	-	-	-	-	-	-
1100	-	0	-	0	-	-	-	-
1101	-	-	-	-	-	-	-	-
1111	-	-	-	-	-	-	-	-
1110	-	-	-	-	-	-	-	-
1000	1	1	-	1	-	-	-	-
1001	1	-	-	-	-	-	-	-
1011	-	-	-	-	-	-	-	-
1010	-	-	-	-	-	-	-	-

Table 5.

Y_2	c=0 ℓr 00	01	11	10	c=1 ℓr 00	01	11	10
0000	-	-	-	-	-	-	-	-
0001	0	-	-	-	0	-	-	-
0011	-	-	-	-	0	-	-	-
0010	0	0	-	0	0	-	-	-
0100	1	1	-	1	-	-	-	-
0101	-	-	-	-	-	-	-	-
0111	-	-	-	-	-	-	-	-
0110	0	-	-	-	-	-	-	-
1100	-	1	-	1	-	-	-	-
1101	-	-	-	-	-	-	-	-
1111	-	-	-	-	-	-	-	-
1110	-	-	-	-	-	-	-	-
1000	0	1	-	1	-	-	-	-
1001	0	-	-	-	-	-	-	-
1011	-	-	-	-	-	-	-	-
1010	-	-	-	-	-	-	-	-

Table 6.

Y_3	c=0 ℓr 00	01	11	10	c=1 ℓr 00	01	11	10
0000	-	-	-	-	-	-	-	-
0001	0	-	-	-	0	-	-	-
0011	-	-	-	-	0	-	-	-
0010	1	1	-	1	1	-	-	-
0100	1	0	-	0	-	-	-	-
0101	-	-	-	-	-	-	-	-
0111	-	-	-	-	-	-	-	-
0110	1	-	-	-	-	-	-	-
1100	-	0	-	0	-	-	-	-
1101	-	-	-	-	-	-	-	-
1111	-	-	-	-	-	-	-	-
1110	-	-	-	-	-	-	-	-
1000	0	0	-	0	-	-	-	-
1001	0	-	-	-	-	-	-	-
1011	-	-	-	-	-	-	-	-
1010	-	-	-	-	-	-	-	-

$y_1y_2y_3y_4$

Table 7.

Y_4	c=0 ℓr 00	01	11	10	c=1 ℓr 00	01	11	10
0000	-	-	-	-	-	-	-	-
0001	1	-	-	-	1	-	-	-
0011	-	-	-	-	1	-	-	-
0010	0	0	-	0	1	-	-	-
0100	0	0	-	0	-	-	-	-
0101	-	-	-	-	-	-	-	-
0111	-	-	-	-	-	-	-	-
0110	0	-	-	-	-	-	-	-
1100	-	0	-	0	-	-	-	-
1101	-	-	-	-	-	-	-	-
1111	-	-	-	-	-	-	-	-
1110	-	-	-	-	-	-	-	-
1000	0	0	-	0	-	-	-	-
1001	0	-	-	-	-	-	-	-
1011	-	-	-	-	-	-	-	-
1010	-	-	-	-	-	-	-	-

$y_1y_2y_3y_4$

A synthesis of Y_1, Y_2, Y_3 and Y_4 gives :

$$Y_1 = \bar{c} \cdot y_4 + \bar{y}_2 \cdot y_1 \quad *$$

$$Y_2 = (\ell + r) \cdot y_1 + \bar{y}_3 \cdot y_2$$

$$Y_3 = \bar{\ell} \cdot \bar{r} \cdot y_2 + \bar{y}_4 \cdot y_3$$

$$Y_4 = c \cdot y_3 + \bar{y}_1 \cdot y_4 \quad **$$

The effects of the second step on L and on R are distinguished by their respective entries.
The two steps which give the same value to T are grouped.

$L = \ell . y_2$
$R = r . y_2$
$T = y_3 + y_4$

First, it is seen that the number of logical equations used to represent the system is higher that in the initial resolution of chapter V (there are four internal functions instead of two).
However, the internal and output functions are simpler, and they can be derived easier.
The procedure with the "single one assignment" used above is already easier than the classical one, because it avoids the occurence of races; but we will see that it can become even much easier.
One can see that functions Y_1 to Y_4 all have the structure :

$$Y_s = E_s \cdot y_{s-1} + \overline{y_{s+1}} \cdot y_s$$

An internal function and variable having been associated to each step, the s index is related to the considered step. The (s-1) and (s+1) index concern respectively the preceeding and the following steps in the graph.

---

* or $\bar{y}_2\ \bar{y}_3\ \bar{c}$ The slightly more complex function in the text has been chosen for the reasons given below.

** or $c + \bar{y}_1\ y_4$ Same remark as in note *.

If the system is in the (s-1) step, thus if $y_{s-1} = 1$, the Y_s function will set to value 1 when the corresponding excitation order E_s will become 1.

The Y_s function will maintain this value 1 (second term of the expression) owing to value 1 of y_s AND as long as $y_{s+1} = 1$ hasn't happenned, that is until the next step arises (when $E_{s+1} = 1$).

In order to simplify the reasoning, we shall call

$$y_{s-1} = V_s \quad \text{and} \quad y_{s+1} = D_s$$

All the different symbols are now related to the same step : the excitation order E, the validation V of this excitation order by the preceeding step and the deletion D of Y = 1 memorization by the next step.

We shall than write : $Y = E \cdot V + \bar{D} \cdot y$

The logical circuit can thus be built with four modules of identical structure, one per step, connected in such a way that only one at a time has a value 1 at its output.

The output of a module is switched on only when the preceeding module output was on (validation) AND when its excitation input is activated.

When switched on, it switches off the preceeding module output, which looses its value 1 (deletion).

For the chosen example, the structure of the circuit is given in Fig.X/3.

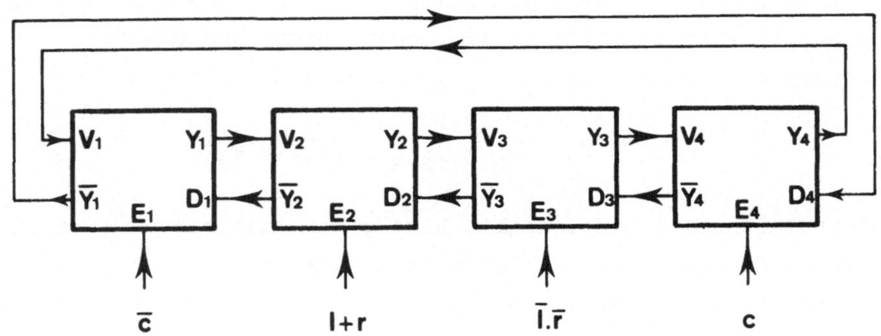

figure X/3.

If one comes back to the system of equations one shall see that, for such a modular structure, the synthesis procedure may be reduced to the identification of the excitation expressions E_s.
In our example :

$$E_1 = \bar{c}$$

$$E_2 = \ell + r$$

$$E_3 = \bar{\ell} \cdot \bar{r}$$

$$E_4 = c$$

In fact, these expressions can be found without going through tables 2 to 7. Table 1 can be directly converted into tables 8 and 9.

Table 8

	c=0 ℓr 00	01	11	10	c=1 ℓr 00	01	11	10
1000	(1)	2	*	2	-	-	-	-
0100	3	(2)	-	(2)	**	-	-	-
0010	(3)	(3)	-	(3)	4	-	-	-
0001	1	-	-	-	(4)	-	-	-
$y_1y_2y_3y_4$								

Table 9

$Y_1Y_2Y_3Y_4$	c=0 ℓr 00	01	11	10	c=1 ℓr 00	01	11	10
1000	1000	0100	----	0100	----	----	----	----
0100	0010	0100	----	0100	----	----	----	----
0010	0010	0010	----	0010	0001	----	----	----
0001	1000	----	----	----	0001	----	----	----
$y_1y_2y_3y_4$								

The excitation expression E_1 can be found by looking in wich row(s) there is a transient (not circled) state 1. Since this is so for row 4, one can tell that E_1 is validated by y_4. If one focuses on the values of Y_1 in this fourth line of table 9, it comes table 10.

Table 10

E_1	c=0 ℓr 00	01	11	10	c=1 ℓr 00	01	11	10
0 0 0 1 ($y_1y_2y_3y_4$)	1	-	-	-	0	-	-	-

Thus $E_1 = \bar{c}$

Similary, E_2 is validated by y_1, E_3 by y_2, E_4 by y_3, and one can construct tables 11, 12 and 13, in which one focuses, respectively, on the values of Y_2 in row 1, of Y_3 in row 2 and of Y_4 in row 3 :

Table 11

E_2	c=0 ℓr 00	01	11	10	c=1 ℓr 00	01	11	10
1 0 0 0 ($y_1y_2y_3y_4$)	0	1	* -	1	-	-	-	-

Table 12

E_3	c=0 ℓr 00	01	11	10	c=1 ℓr 00	01	11	10
0 1 0 0 ($y_1y_2y_3y_4$)	1	0	-	0	** -	-	-	-

Table 13

E_4	c=0 ℓr 00	01	11	10	c=1 ℓr 00	01	11	10
0 0 1 0 ($y_1y_2y_3y_4$)	0	0	-	0	1	-	-	-

This gives $E_2 = \ell + r$

$E_3 = \bar{\ell} \cdot \bar{r}$

$E_4 = c$

When several transitions leading to the same row appear in different rows, these rows can be treated independently; each partial solution is validated by the variable corresponding to the row in question. On the other hand, in table 8 one leaves row 1 only to reach row 2, row 2 only to row 3, 3 to 4 and 4 to 1.
But usually, several destinations may exist, when leaving a row. For instance, let us suppose that an additional 3 transient (not circled) is written in column ℓrc=110 , row 1 of table 8 (mark *). Instead of having $D_1 = y_2$ in expression $Y_1 = \bar{c} \cdot y_4 + \bar{y}_2 \cdot y_1$, we would then have :

$$D_1 = y_2 + y_3 \quad \text{and} \quad \overline{D_1} = \overline{y_2 + y_3} = \overline{y_2} \cdot \overline{y_3}$$

The expression of Y_1 would then become :

$$Y_1 = \bar{c} \cdot y_4 + \bar{y}_2 \cdot \bar{y}_3 \cdot y_1$$

DESIGN PROCEDURE

1) Compose table 1 from the desiderata.
2) Write table 8 directly from table 1.
   Be sure a stable state is provided everywhere in a row, in the columns from which a departure is unacceptable. "Don't care" marks may remain wherever an excitation order never appears in the considered step (row).
3) Without going through table 9, write tables 10 to 13 according to the procedure below, in order to find Y expressions (for instance Y_1)
   - Look for columns where transient 1 appears.
   - Consider separately each row containing such a transient. (only row 4, here). The related y variable determines the validation to be used ($V_1 = y_4$, here).
   - To find the excitation order E_1, build up table 10 directly from row 4 of table 8 : so, put a 1 value in place of the considered transient (transient 1, here). Put a 0 value where other diffe-

rent transients would be, and also for all stable states of the row. Synthesis of table 10 gives the excitation expression ($E_1 = \bar{c}$).

4) Still from table 8, determine the deletion function for each row by looking for several transients of departure. The $\bar{D}$ factor in Y expression is composed by NORing the related destination variables.

NOTE 1 - Improved simplification of the excitation orders

In the above procedure, the rows may always be treated separately. Accordingly, the complexity of the resolution, as well as the complexity of the obtained expressions, only depends on the number of columns, thus on the number of input variables. Sequential problems can therefore be formally treated as simply as purely combinational ones.
The absence of combinational interaction between rows results from the chosen assignment where, excepting for transients, the internal variables never have simultaneously a 1 value.
In a similar way, a reduction or an elimination of interaction between the columns may yet simplify a great deal the expressions of the excitation orders.
This situation naturally occurs in the particular problems where the input variables never have simultaneously a 1 value.
In a general way, resorting to edge-detection (variosensors) of the input variables (ref. 2 and 5), to edge sequences-detection (sequosensors) of them (ref.6) or to a mixed form (ref. 7), trends to reduce the combinational interaction between columns and lead to much simpler expressions, more intelligible for the designer and the user. These advantages are to be joined to those resulting from the non-interaction of rows described in this chapter.

NOTE 2 - Risks resulting from an euristic synthesis of a net structure

When a state table is completely specified, the described procedure leads automatically to logical expressions such that the excitation orders of two consecutive modules never have simultaneously a 1 value, even if the input variables don't fulfill this condition.
When the state table is uncompletely specified, two excitation expressions relative to consecutive steps may have value 1 simultaneously

for a specific input state that never occurs in the first of these two steps. So, in table 8, input state ℓrc = 001 never occurs when in step 2 (mark **). But, this input state must lead to row 4, when in step 3. The two excitation expressions $E_3 = \bar{\ell} \cdot \bar{r}$ and $E_4 = c$ coming from rows 2 and 3 (tables 12 and 13) have a common cover : both have value 1 for ℓrc = 001.

On the other hand, if an adress 1 (go to step 1), (2) (stay in step 2) or 4 (go to step 4) would be present at place ** in table 8, a 0 value should have been put at the same place in table 12. One would have then found $E_3 = \bar{\ell} \cdot \bar{r} \cdot \bar{c}$ which has no more cover with $E_4 = c$ for input state 001.

The systematic synthesis thus avoid overlappings between excitation orders which could lead, in practice, to jump undeliberately steps due to spontaneous validations in cascade.

When one tries to find the expressions of the excitation orders in an euristic way, for instance directly on a graph which is not a complete representation by nature, it is not easy to foresee and avoid the overlappings of these excitation orders.

During the design, when the graph is not yet completed, that is, when all the desiderata are not yet fulfilled, the simple example of an uncompletely specified problem given above shows how such an euristic synthesis would be still more risky.

GENERALIZATION AND CONCLUSIONS

The present example comprises only four steps linked by the single cycle of graph 2. But the same method can be applied to more linked networks. In fact, one can now solve in a systematic way control networks for which one had formerly to use the formalism of Petri (ref. 1) or its derivatives. In these more linked structures, each module can be validated by several others, and can validate several others. Several modules can then have value 1 at the same time. Such structures can be represented by state tables similar to table 8, but in which more than one y variable can have value 1 in a same row.

It is well know that in contrast with the classical state table, the graph formalism doesn't efficiently help the conceptor because it doesn't show all the imaginable evolutions of the system.
It only gives a provisional support for the reasoning without giving evidence of uncompatibilities with the whole set of working desiderata. However the graph is a convenient representation after complete design, for instance in the aim of maintenance.

Petri nets and their derivatives have the same disadvantages.
In contrast, extensions of tables such as table 8 indicate all the imaginable evolutions that have not yet been considered and provide a certain amount of parallelism in the information treatment by the logical net. This parallelism is interesting in some special industrial problems and in formalizing logical models. In fact, it seems to be possible to represent with peculiar advantages, concurrent phenomena(critical races) by means of this parallelism.

References

1) BLANCHARD,M. Rapport de lacommission de normalisation AFSET:Représentation du cahier des charges d'un automatisme logique. Revue Automatisme, Paris. Mars-avril 1978.

2) FLORINE, J. Automatismes à séquences et commandes numériques Ch. 4 et 5. Dunod-Paris, 1969.

3) VERLINDEN,M. Méthode directe de synthèse d'un circuit séquentiel à partir de la table des états primitive. Proceedings International Symposium - Design and Applications of logical systems p. 889, Editions AICA Bruxelles 1969.

4) DAVID, R. Réalisation de systèmes séquentiels asynchrones par interconnexion simple de cellules séquentielles identiques - Thèse de Doctorat en Sciences, Université de Grenoble 1969.

5) LEUSSLER,A. Synthèse impulsionnelle de systèmes séquentiels. Thèse de Doctorat en Sciences Appliquées, Université de Bruxelles 1975.

6) FLORINE, J. Complex module trends in design IFAC Symposium : Discrete Systems Proceedings vol. 3 p. 1 , Dresden 1977.

7) YENERSOY,O. Systèmes logiques cellulaires à excitations mixtes. Thèse de Doctorat Sc.A.- U.L.B.à paraitre en 1979.

P A R T III

More refined methods

CHAPTER XI

MATHEMATICAL ASPECTS OF BIOLOGICAL REGULATORY PROCESSES

G. Nicolis,
Faculté des Sciences,
Université Libre de Bruxelles, Belgium

I. INTRODUCTION - STATEMENT OF THE PROBLEM

It has already been stressed by René Thomas and Stuart Kauffman that one of the very characteristic features of regulation is to endow living organisms with a variety of distinct regimes, or states. It suffices to recall the occurrence of uninduced and of induced cells in a population of E. Coli , the phenomenon of immunity in phage λ infected bacteria or the committment of undifferentiated cells of drosophila to become wing, leg or antenna cells.

It is important to realize that such regulatory processes in living organisms imply two largerly separated scales of phenomena :

(i) An ensemble of processes occurring at the molecular level such as the conformational transformation of a macromolecule or the action of a repressor molecule on a gene.

(ii) A series of mechanisms which permit the molecular phenomenon to manifest itself coherently at a macroscopic scale, for instance at the level of a population of single cells, or even at the supercellular level of dimensions comparable to those of a morphogenetic field in a developing organism.

This passage from an isolated molecular event to a collective phenomenon involving the coupling between a great number of variables is responsible for the maintenance of biological order. Without it the various microscopic events would add up with completely random phases resulting in noiselike behavior, instead of the observed coherent behavior of living organisms.

The purpose of these lectures is to show you the type of insight one gets on these problems by the methods of physics, physical chemistry and continuous mathematics. Indeed, both physics and physical chemistry deal with this continual interplay between macroscopic level and properties at the atomic scale, to such an extent in fact, that the explanation of the observable properties of macroscopic bodies (heat capacity, thermal conductivity, etc...) in terms of the properties of atoms and their interactions is the object of statistical mechanics, a special branch of physics. For the purposes of our discussion, one of the most important conclusions of this type of analysis of macroscopic systems is that the macroscopic description introduces some new qualitative aspects. Think, for a moment, of the conduction of heat across a fluid submitted to a temperature difference ΔT Locally, each particle undergoes a disordered thermal motion. And yet, on the whole, a macroscopic quantity of energy is transmitted through the fluid, which is largerly independent of the details of thermal motion and is simply proportional to ΔT. Naturally, such macroscopic properties are perfectly compatible with the microscopic behavior of the individual particles. The only point is that the microscopic language is simply not adequate once one exceeds a certain scale of phenomena.

Following this type of philosophy, we will look at molecular biology to obtain information on what we called above the "microscopic level". We will then be interested in large scale phenomena accessible by macroscopic measurements, and in constructing models that couple among themselves several variables, each one referring to a particular molecular event.

The very fact that we are interested in the modelling of macroscopic phenomena permits us to adopt a contracted description in terms of a limited number of observables. In a problem of regulation these are obviously the numbers of particles $X_1, \ldots, X_n$ of the cellular or intercellular chemical constituents participating in the process, and the temperature T. In fact, hereafter we assume that T is fixed and focus on the composition variables only.

The basic problem is to derive the laws of evolution of X_i. By definition, the X_i change only by jumps : 1, 2, etc... They do so because of regulation, metabolism, etc... (that is, because of various <u>chemical reactions</u>), because of diffusion etc... (that is, became of <u>transport phenomena</u>), because of direct cell to cell interactions, and because of cell division. Obviously, all cells or all parts of a cell cannot be in exactly the same state. Hence, the evolution of the X's will be of a random, or <u>statistical nature</u>. Postponing a discussion of the foundations of statistical methods to a later part of these lectures, we only point out here that the central quantity to evaluate is the probability function $P(X_1, \ldots, X_n, t)$ whose evolution will be associated to the series of discontinuous transitions (see Fig. 1)

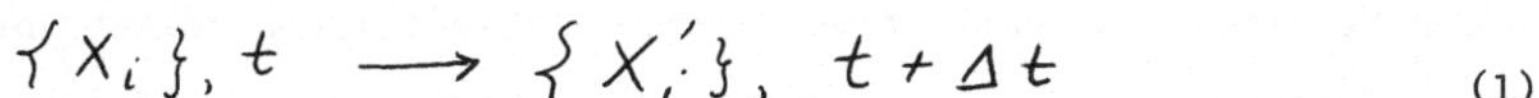

$$\{X_i\}, t \longrightarrow \{X_i'\}, t+\Delta t \qquad (1)$$

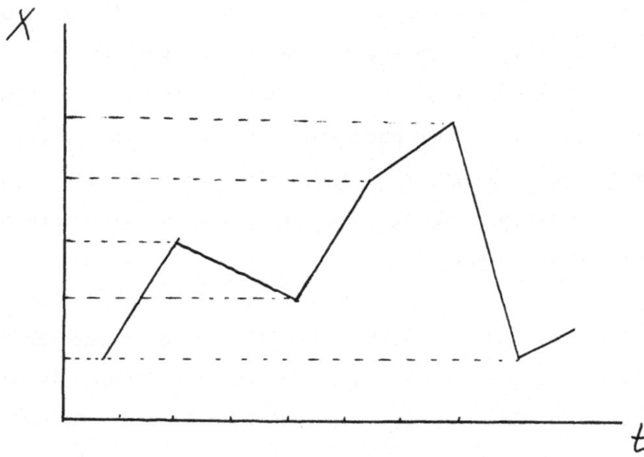

<u>Figure 1</u> : Discontinuous transitions of the state variable X.

Although this description sounds like a Boolean one, its relation to the latter will depend on the structure of a quantity known as <u>transition probability</u> which will be defined later on in detail and will be used more extensively by David Rigney in his lectures.

Now, to each realization of the random process (1) one can associate a smooth variable by taking some kind of envelope or, to be more precise, a mean value over the values of $X_i(t)$ during an appreciable time inter-

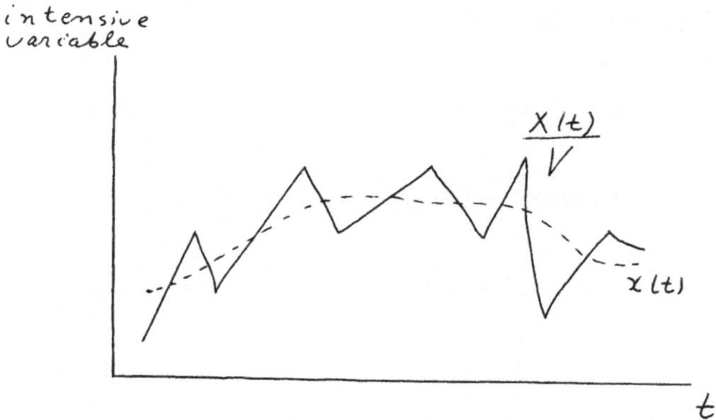

Figure 2 : Statistical average $x(t)$ associated to the intensive variable $X(t)/V$

val or over a great number of replicas of the system (see Fig. 2). We also prefer to work with intensive variables by dividing X with a reference volume V. This leads to :

Extensive variable $X_i(t) \longrightarrow$ Intensive variable $\frac{X_i(t)}{V} \rightarrow$ mean value $x_i(t)$

The latter is in fact the usual chemist's concentration variable and is, by definition, continuous.

Assuming now one can construct an equation for P, there is in principle no reason why the operation of taking mean values should lead to a closed form equation for $x_i(t)$. Whenever this is possible we say that we have a continuous macroscopic description of the system. The basic features of this description one outlined in Section 2.

2. CONTINUOUS DESCRIPTION

We first consider the effect of chemical interactions alone. The rate of change of $x_i(t)$ is obviously the sum of partial changes arising from the chemical reactions present :

$$\frac{dx_i}{dt} = \sum_{\rho} (\text{rate of reaction } \rho) \times (\text{number of molecules of i created or destroyed in reaction } \rho) \equiv \sum_{\rho} \nu_{i\rho} w_{\rho} \qquad (2)$$

where $\nu_{i\rho}$ is the stoichiometric coefficient of constituent i in reaction ρ and w_ρ is the velocity of reaction ρ.

In many cases of interest, w_ρ can be expressed in terms of the instantaneous values of the composition variables $x_i(t)$. This highly nontrivial simplification becomes possible if the state of the system is near a <u>local equilibrium</u> regime : In each small volume far from the boundaries the system is characterized by the same state variables satisfying the same formal relations as in the state of thermodynamic equilibrium (Glansdorff and Prigogine, 1971). We express this by the <u>phenomenological relations</u>

$$w_\rho = w_\rho(x_1, \dots, x_n) \qquad (3)$$

Examples of (3) are the law of mass action of classical chemical kinetics, the Langmuir adsorption characteristic, or the Michaelis-Menten law of transformation of a substrate S into a product P mediated by an enzyme E :

$$E + S \underset{k_{-1}}{\overset{k_1}{\rightleftarrows}} ES \xrightarrow{k_2} E + P \qquad (4a)$$

$$\frac{dx_s}{dt} = -\frac{k_2 E_0 x_s}{K_s + x_s} \qquad (4b)$$

where $$K_s = \frac{k_{-1}+k_2}{k_1}$$

$$E_0 = x_E + x_{ES}$$ = total enzyme concentration.

As is clear from the above example, relationship (3) is system-dependent. In particular, in a regulation problem involving feedbacks w_ρ will be nonlinear, almost by definition of what a feedback is. We thus arrive at the general form of the regulation equations :

$$\frac{dx_i}{dt} = f_i(\{x_j\}) \qquad (5)$$

$$x_i \geq 0$$

f_i nonlinear

As a rule, some of the terms in the right hand side of eq. (5) will reflect the influence of constraints acting on the system like, for instance, inputs of energy or matter. A simple illustration is provided by the scheme

$$A \xrightarrow{k_1} X \xrightarrow{k_2} X^* \qquad (6)$$

where X is an active intermediate, X^* an inactive conformation of X and A a control variable in the sense that its concentration is maintained fixed by some external action.

The rate equation

$$\frac{dx}{dt} = k_1 a - k_2 x \qquad (7)$$

exhibits then an input term $k_1 a$ acting as a constraint on the evolution of X.

The presence of such terms reflects the fact that in biology one deals, as a rule, with open systems. Moreover, the exchanges of matter, energy and information with the external world occur under nonequilibrium conditions and give rise to systematic fluxes traversing the system. In irreversible thermodynamics one shows that both elements are necessary prerequistes for biological order (Glansdorff and Prigogine, 1971). I will not insist on this however, as I would like to develop mainly the kinetic aspects of the problem.

The solution of the nonlinear differential equations (5) confronts us with an arduous task, and is still a subject of active mathematical research. To tackle biological problems, one can adopt one the following attitudes :

(i) Analyze specific forms of eq. (5) motivated by concrete problems. It would be nice if this could be done completely analytically, but usually one needs sooner or later some numerical simulations.

(ii) Analyze the qualitative behavior of families of eq. (5) and try to classify it. This is essentially a search of archetypes, and is usually limited to representative, if simple, situations not directly related to specific data.

Let us illustrate briefly the two types of approach described above.

A. An example of the control equations : β - galactosidase induction in E.coli

The regulation of β -galactosidase synthesis has been already described in René Thomas' lectures. Fig. 3 gives a global view of the process.

E, M stand respectively, for β-galactosidase and permease i_e is the external sugar concentration, i_i the intracellular sugar concentration.

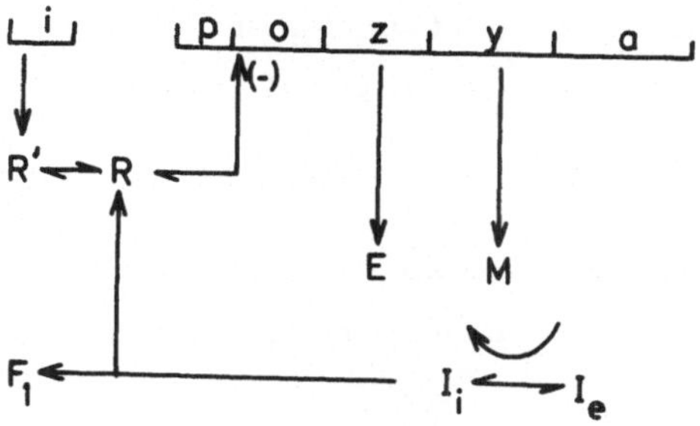

Figure 3 : Regulation of lac operon circuit in E.coli.

The gene i synthesizes a monomer [R'] which is assumed to be in equilibrium with the oligomeric form [R] of the repressor. The latter can be blocked by n_I molecules of I_i ; othermise, it binds to the genome and blocks, through the operator O, the expression of genes z, y, a. In fact, R comprises two allosteric forms in conformational equilibrium, of which one (active) binds the operator and the other (inactive) is stabilized by the inducer.

$$
\begin{aligned}
R' &\underset{k'_1}{\overset{k_1}{\rightleftharpoons}} R \\
R + O^+ &\underset{k'_2}{\overset{k_2}{\rightleftharpoons}} O^- \\
R + n_I I_i &\underset{k'_3}{\overset{k_3}{\rightleftharpoons}} F_1 \\
\eta + O^+ &\xrightarrow{k_4} O^+ + E + M \\
M + I_e &\underset{k'_5}{\overset{k_5}{\rightleftharpoons}} M + I_i \\
M &\xrightarrow{k_6} F_2 \\
E &\xrightarrow{k_7} F_3
\end{aligned}
\qquad (8)
$$

The last two steps describe the dilution of M and E as the cells grow and divide, whereas the fourth step gives a global representation of the biosynthesis of E and M.

Assuming that the second step is rapid with respect to the others one can eliminate the operator variables by taking also into account the conservation condition

$$O^{+} + O^{-} = c = \text{Constant}$$

One obtains

$$O^{+} = \frac{k_2' c}{k_2 r + k_2'}$$

$$O^{-} = \frac{k_2 c r}{k_2 r + k_2'}$$

Substituting into the rate equations for R, I_i, E and M and treating R', I_e as control variables, one obtains :

$$\frac{dr}{dt} = k_1 r' - k_1' r - k_3 r i_i^{n_I} + k_3' f_1$$

$$\frac{di_i}{dt} = k_5 i_e m - k_5' i_i m - n_I k_3 r i_i^{n_I} + n_I k_3' f_1$$

$$\frac{de}{dt} = \eta k_4 \frac{k_2' c}{k_2 r + k_2'} - k_7 e$$

$$\frac{dm}{dt} = \eta k_4 \frac{k_2' c}{k_2 r + k_2'} - k_6 m \qquad (9)$$

Because of the non linearities (terms in $i_i^{n_I}$, $i_i m$, $(k_2 r + k_2')^{-1}$), these equations can admit multiple solutions. Fig. 4 describes the steady-

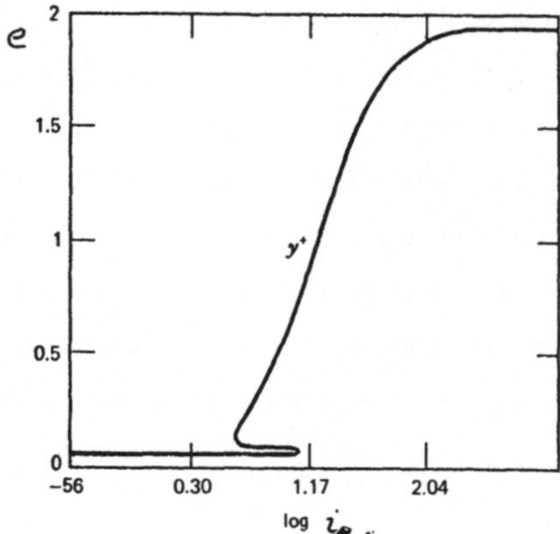

Figure 4 : Theoretical curves pf stationary-state solutions for e in lac operon model obtained by using experimental constants and $k_1 = 0.08$, $k_1' = 10$. y^+ denotes the wild type strain.

state level of e in the case of $n_I = 2$. For an interval of values of the control variable l'_e the system exhibits steady states with widely different concentrations, together with the phenomenon of hysteresis. Only some of these states exhibit stability, that is, the ability to damp the perturbations acting on the system. Others, like the intermediate branch of states in Fig. 4, are unstable and hence impossible to maintain against perturbations. These features endow the system with excitability and all-or-none behavior, which are among the most characteristic features of biological regulation.

Numerical analyses using appropriate values of the parameters reproduce correctly the 10^3- fold increase of the level of e associated with induction, as well as the time-dependent behavior observed experimentally.

B. Qualitative aspects of the control equations

The preceding example, as well as those treated by René Thomas and Stuart Kauffman suggest that a characteristic feature of the control equations (5) should be (i) multiplicity of solutions, and (ii) a concomitant exchange of stability between various branches. Our purpose now is to understand the mechanism of onset of these phenomena. In fact, I shall only illustrate this deep question on a simple case where all computations can be performed in detail.

Consider a regulatory circuit of the form

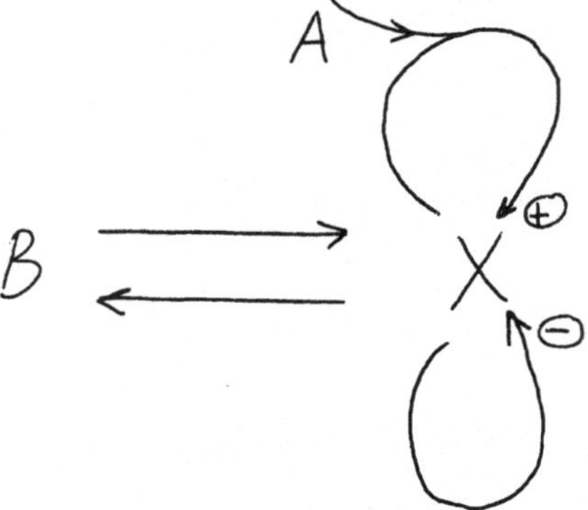

Figure 5 : Schematic representation of model (10a).

The active species X is being pumped into the system thanks to some precursor B. It can then participate to a positive feedback pathway, e.g. a direct autocatalytic reaction or an enzyme catalyzed process. On the other hand, if its concentration tends to become high a negative feedback process becomes dominant which tends to consume X. These are very common phenomena in such diverses fields as biochemistry, free tadical chemistry, or even population dynamics.

For a single dependent variable, the lowest nonlinearity needed to observe nontrivial phenomena is cubic. This leads to the following chemical model (Schlögl, 1971, 1972) :

$$A + 2X \underset{k_2}{\overset{k_1}{\rightleftarrows}} 3X$$

$$X \underset{k_4}{\overset{k_3}{\rightleftarrows}} B \tag{10a}$$

The rate equation for the concentration of X is :

$$\frac{dx}{dt} = -k_2 x^3 + k_1 a x^2 - k_3 x + k_4 b \tag{10b}$$

We introduce the following scaled variables and parameters :

$$\begin{aligned} \tau &= k_2 t \\ \frac{k_1 a}{k_2} &= 3\alpha \\ \frac{k_4 b}{k_2} &= \alpha^3 (1+\delta') \\ \frac{k_3}{k_2} &= \alpha^2 (3+\delta) \\ x &= \alpha (1+u) \end{aligned} \tag{11}$$

Eq. (10b) becomes :

$$\frac{du}{d\tau} = -u^3 - \delta u + \delta' - \delta \qquad (12)$$

We see that we achieved a considerable reduction of the number of parameters thanks to the scaling described by eq. (11). Catastrophe theory - a special branch of modern mathematics - provides systematic information on this reduction procedure for special forms of the control equations (5), namely those deriving from a potential (Thom, 1972). Obviously any system described by a single variable, like scheme (10a), belongs to this class.

Let us focus on the steady-state solutions of eq. (2) :

$$\delta' = u^3 + \delta(u+1) \qquad (13)$$

For $\delta = \delta' = 0$ it admits the triple root

$$u = 0 \qquad \text{or} \qquad x = \alpha \qquad (14)$$

We want to explore the behavior of u as δ', δ approach to the value $\delta' = \delta = 0$. The theory of cubics gives us the following picture (Fig. 6) of the behavior of the roots of eq. (13) in the parameter space (δ, δ').

The region of multiple roots is enclosed by a curve displaying a singular point at the origin $\delta = \delta' = 0$, the <u>cusp</u>. On crossing this curve one switches between the regime of a single real solution and that of three solutions.

Let us now follow this passage in the "physical space" where u is plotted against the parameters. As a matter of fact, we shall fix the relation $\delta = \delta'$ among δ, δ' in which case eq. (13) yields :

$$u^3 + \delta u = 0 \qquad (15)$$

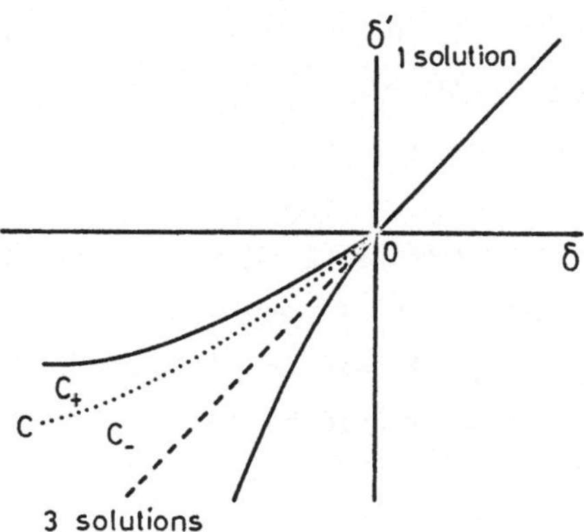

Figure 6 : Stability diagram for eq. (12) and (13). 0 = cusp singularity. C = coexistence line of stable steady states.

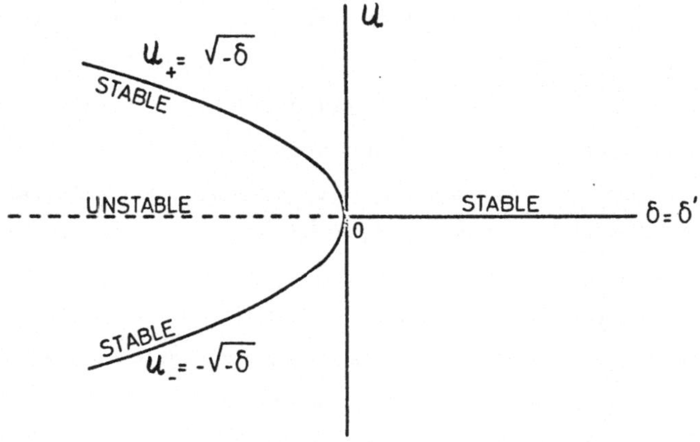

Figure 7 : Bifurcation phenomena associated with the passage through the cusp singularity along the line $\delta = \delta'$.

Fig. 7 describes the behavior of u with respect to δ. At $\delta = 0$ the solution $u = 0$ which exists for all δ, branches with the solutions $u_{\pm} = \pm\sqrt{-\delta}$ which emerge in the region of $\delta < 0$. This is the phenomenon of bifurcation, which implies a non-analytic dependence of the solution on the system's parameters.

We see that smooth-looking equations can develop transition phenomena ! This fact is nowadays widely recognized. Two special branches of mathematics, bifurcation theory (Nicolis and Prigogine, 1977) and catastrophe theory (Thom, 1972) develop methods for tackling these problems in a systematic way.

The stability of the various branches depicted in Fig. 7 can be determined by testing the response of the system to various disturbances tending to remove it from one of the "reference" steady states. One finds in this way that the intermediate branch u = 0 is unstable, whereas branches $u_{\pm}$ are stable. It is amazing to see that a system as simple as (10a) can develop regulatory properties by becoming capable of switching between different concentration levels of the substances involved !

One might be tempted to react as follows to the foregoing analysis : all this is very nice, but it sounds a bit pathological. After all, $\delta = \delta' = 0$ or $\delta = \delta'$ are exceptional situation in parameter space and there is little chance that they correspond exactly to an actual physiological situation (in a sense, this is the analog of "critical race" of Boolean circuits which would operate following a synchronous mode)?.The answer to this is that bifurcation loci merely serve as landmarks separating qualitatively different types of behavior. If for some reason one of the parameters varies enough (for instance, the lactose or glucose concentration in the induction problem of E.coli), then the system will be able to cross a bifurcation point and undergo a radical change in behavior. In other situations the system is already stabilized in one of the regions of parameter space and as a result bifurcation is not perceived as such, although ultimately is responsible for the observed behavior.

Needless to say, we have been concerned here with the simplest possible bifurcation. Higher order nonlinearities, two or more variables, or spatially distributed systems give rise to more complex phenomena (Nicolis and Prigogine, 1977), although the basic principles remain the same. For instance in systems involving two variables and nonlinearities of third degree or more, or in systems of three variables with quadratic or higher order nonlinearities, one can get sustained oscillations of the chemical concentrations of the type shown in Fig. 8. Their period and amplitude are stable and

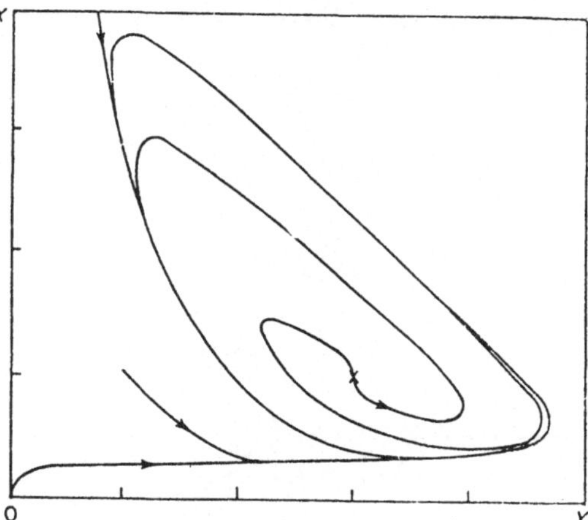

Figure 8 : Limit-cycle trajectory in a two-variable model system

independent of initial conditions. Mathematically, they correspond to limit cycle solutions of the rate equations. Their importance lies in the fact that they constitute models of rhythmic phenomena observed in nature, like biological or chemical clocks. Another interesting class of phenomena becomes possible in spatially distributed systems, say systems undergoing chemical reactions and diffusion. In particular, one may have (see Fig. 9) symmetry-breaking transitions leading to the formation of spatial compartments in an initially uniform medium.

The bifurcation diagrams of these more complex systems differ from Fig. 7 in the following points (see Fig. 10) :

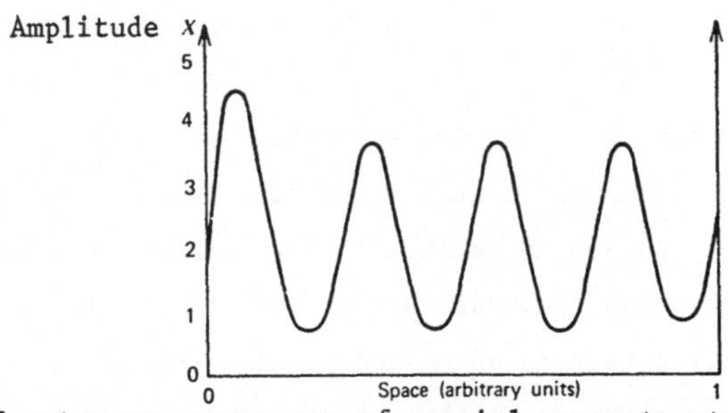

Figure 9 : Spontaneous emergence of special compartments in a two-variable model system.

(i) More than one bifurcation points from the reference solution u = 0 are possible. The branches emerging at these points are referred to as primary branches.

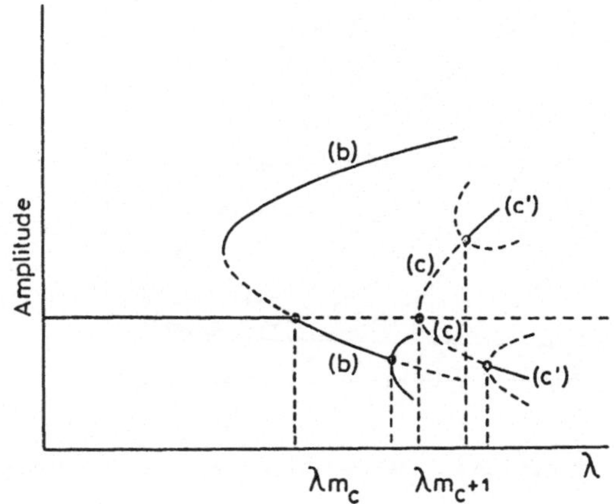

Figure 10 : Illustration of the concept of cascading bifurcations (b) = stable primary branch. (c) = unstable primary branch. (c') = branch(c) stabilized thanks to a secondary bifurcation.

(ii) A primary branch can in turn have its stability properties changed via <u>secondary</u> (or higher order) <u>bifurcations</u>.

(iii) Inverted, or <u>subcritical bifurcations</u> become possible.

Examples of systems showing this behavior can be found in the recent monograph by Nicolis and Prigogine.

C. Discussion

From this analysis we can already begin to realize the shortcomings of the macroscopic description, bearing also in mind the remarks made in the Introduction. First, we have the fact that near bifurcation points the system has to choose between several possible branches of solutions of the rate equations. But nothing in these equations justifies preference for any particular choice. Therefore, statistical or stochastic elements will presumably enter into account. We can think of them as self-generated perturbations around $x_i(t)$ (see Fig. 2), which modify continuously the initial conditions. Hereafter we refer to them as <u>fluctuations</u>.

Second, we have the fact that in the bistability region a system is expected split into two subpopulations, depending on the initial conditions. This will compromise the very meaning of the sampling that gave us $x_i(t)$ in Fig. 8. In a sense, $x_i(t)$ will no longer be representative of either of the subpopulations. Again, we need a finer description involving fluctuations.

Finally, in some of the systems of biological interest the number of molecules of certain types is small. This will further enhance the randomness, as it will increase the sensitivity of the overall behavior with respect to the details of the behavior of the individual molecules.

3. STOCHASTIC DESCRIPTION : FOUNDATIONS

Let us forget about cellular regulation for a few minutes. Consider a coin which is fair, in the sense that it has heads or tails on each of its twosides and that it is symmetrically constructed. The coin is tossed several times, and the frequency of H or T is measured. It will be found equal. We call $\text{Prob}(H) = \frac{\text{number of H outcomes}}{\text{total number of tosses}}$ Similarly for T. We have :

$$p = \text{Prob}(H) = q = \text{Prob}(T) = \frac{1}{2}$$

Consider now two successive tosses of the coin. We describe the outcomes by (HH, HT, TH, TT). As before, we assign equal probabilities to all of them. Consider the statements : "A H turns up on the first toss", "A H turns up on the second toss", and "H turns up on the first and the second toss". Then :

$$\text{Prob}(H_1 H_2) = \frac{1}{4}, \quad \text{Prob}(H_1) = \text{Prob}(H_2) = \frac{1}{2} \text{ that is,}$$

$$\text{Prob}(H_1 H_2) = \text{Prob}(H_1)\,\text{Prob}(H_2)$$

We say that the two events H_1, H_2 are statistically independent.

Repeated independent events (or trials) with only two possible outcomes characterized by the probabilities p and q for each trial $(p+q=1)$ and having the same probability throughout the trials are called Bernoulli trials.

A. Binomial distribution

Consider now an arbitrary number of successive tosses of a coin. The set of the outcomes of these trials (that is $HTHH----$, or $TTHTHHH\cdots$) is a stochastic process. We call H

"success", T "failure", $Prob(H) = p$, $Prob(T) = q$.
If the coin is fair $p = q$, but here we consider a more general case.

We are interested in the total number of successes after n successive trials, independently of thei: order. Thus, we want to compute $Prob$ (k successes in n trials). Now, the event 11 trials result in k successes and $n-k$ failures" can happen in as many ways as k elements H can be distributed among n places. Obviously this number is $\binom{n}{k}$ and, by definition, each particular realization has the probability $p^k q^{n-k}$. This yields the well-known binomial distribution (Feller, 1957) :

$$P_b(k; n, p) = \binom{n}{k} p^k q^{n-k} = \frac{n!}{k!(n-k)!} p^k q^{n-k} \tag{16}$$

Note that $\sum_{k=0}^{n} P_b(k) = 1$. This is known as normalization condition.

Let me convince you that this distribution is useful elsewhere than gambling (Feller, 1957). Suppose that the normal rate of infection of a certain discase in cattle is 25 % . To test a newly discovered serum, healthy animals are injected with it. How are we to evaluate the result of the experiment ? If the serum is absolutely worthless, the probability that exactly k of the n test animals remain free from in ection may be equated to $P_b(k; n, 0.75)$. For $k = n = 10$ this is about 0.056, and for $k = n = 12$ only 0.032. Thus, if out or 10 or 12 test animals none catches infection, this may be taken as an indication that the serum has had an effect. Obviously, the experiment is more conclusive if it is carried out with a great number of animals.

One can easily plot (16) versus k . One finds a bell-shaped curve having a maximum at $k = m$, where

$$(n+1)p - 1 < m \leq (n+1)p$$

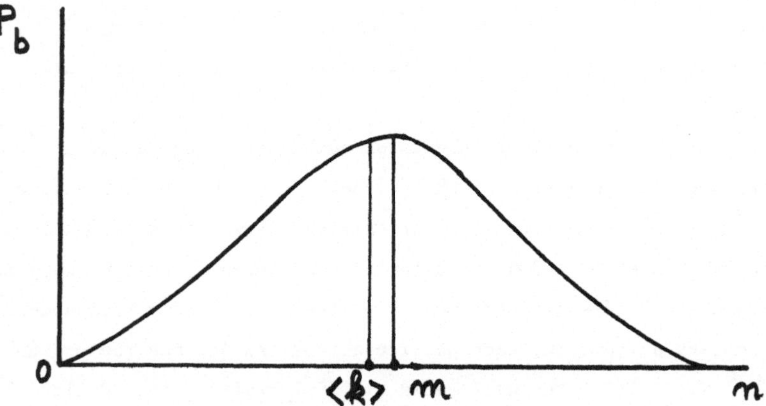

Figure 11 : Schematic representation of the binomial distribution.
m = most probable value. $\langle k \rangle$ = average value

Let us now imagine the following two operations.

(i) For n trials, we count the number of successes H_n and compute the "average" number of successes H_n/n.

(ii) We make repeated sequences of n trials and compute the statistical average

$$\langle k \rangle = \sum_{k=0}^{n} k P_b = \sum_{k=1}^{n} \frac{n!}{(k-1)!\,(n-k)!} p^k q^{n-k} =$$

$$= \sum_{k'=0}^{n-1} \frac{n!}{k'!\,(n-1-k')!} p^{k'+1} q^{n-1-k'} =$$

$$= np$$

that is

$$\frac{\langle k \rangle}{n} = p$$

How are p and H_n/n related ?

The following theorem, which I am stating without proof, answers this question (Feller, 1957) :

$$\mathrm{Prob}\left\{ \left| \frac{H_n}{n} - p \right| < \epsilon \right\} \longrightarrow 1 \quad \text{as } n \to \infty \tag{17}$$

This is the celebrated <u>law of large numbers</u>, which turns out to be true for a far greater variety of problems than those covered by the binomial distribution.

Essentially, the law of large number underlines the importance of the notion of <u>statistical average</u>, which is shown to be close to the outcome of an experiment under certain conditions.

A still finer measure of the closeness of statistical and arithmetic average is the variance. For the binomial distribution :

$$\langle (k - \langle k \rangle)^2 \rangle \equiv \langle (\delta k)^2 \rangle = n p (1-p) \tag{18}$$

It expresses the width of P_b around its most probable value *) $m \simeq np$. In eq. (18) we see that despite its quadratic character, $\langle (\delta k)^2 \rangle$ is proportional to n , just like $\langle k \rangle$. This is a consequence of an important law known as the <u>central limit theorem</u>, which will be enunciated later on.

B. <u>Passage to Poissonian</u>

In many applications we deal with Bernoulli trials where, comparatively speaking, n is large and p is small, whereas the product $\lambda = np$ is of moderate magnitude. In such cases it is convenient to use an approximate representation for P_b which is due to Poisson.

Skipping all details, we find the Poisson distribution (Feller, 1957) :

---

*) Note that, for n large, average and most probable value are essentially identical

$$P_P = e^{-\lambda} \frac{\lambda^k}{k!} \tag{19}$$

In fact, this distribution arises in a great number of problems in its own right, and not only as an approximation to the binomial as it was done in our particular mode of presentation. An example will be analysed in Section 5.

The number of observations fitting the Poissonian is surprisingly large. Let us consider a biological illustration. We are given a Petri plate with bacterial colonies, which are visible under the microscope as dark spots. The plate is divided into small squares. The number of squares having exactly k dark spots fits the Poissonian perfectly.

C. Passage to Gaussian

We start again with the binomial distribution but, contrary to the preceding case, we take now p finite and $n \rightarrow \infty$. One can then evaluate eq. (16) using a well-known approximation for the factorials known as Stirling's formula :

$$n! \simeq \sqrt{2\pi}\, n^{n+\frac{1}{2}}\, e^{-n} \tag{20}$$

Skipping again details one has a celebrated result due to Laplace and De Moivre (Feller, 1957) :

$$P_b \simeq P_G = \frac{1}{\sqrt{2\pi n p q}}\, e^{-\frac{(k-np)^2}{2npq}} \quad , \text{ n large} \tag{21}$$

This is the Gaussian distribution and is to be described in terms of the <u>continuous</u> variable

$$\xi = k - np \quad , \quad -\infty < \xi < \infty$$

It is easily found that

$$\langle \xi \rangle = 0 \qquad \text{(reflecting the fact that } \langle k \rangle = np \text{)}$$

and
$$\langle \xi^2 \rangle = npq \tag{22}$$

As in the case of the Poissonian, the Gaussian is an important distribution in its own right and not just as an approximation to the binomial. The reason is the central limit theorem, which I am stating without proof (Feller, 1957) .

Let $\{x_i\}$ be a sequence of mutually independent random variables with a common distribution, and suppose that both $\langle x_i \rangle = m$ and $\langle (\delta x_k)^2 \rangle = \sigma^2$ exist. Then, for every fixed β and for $n \to \infty$:

$$\text{Prob}\left\{ \frac{\frac{1}{n}(x_1 + \cdots + x_n) - m}{(\sigma / n^{1/2})} < \beta \right\} \longrightarrow \Phi(\beta)$$

where
$$\Phi(\beta) = \frac{1}{\sqrt{2\pi}} \int_{-\infty}^{\beta} dy \, e^{-\frac{1}{2} y^2} \tag{23}$$

The number of problems solved thanks to Gaussian is again impressive. For instance, suppose we want to determine the fraction of smokers of a large population (Feller, 1957). This is done by random sampling. Usually it is desired to find p with an error not exceeding a certain small value, say 0.005. How large should the sample size be ? The Gaussian gives the answer,but I leave this as an exercise !

In the light of the central limit theorem, we now begin to understand the reason why the variance of fluctuations is of order n rather than

n^2 (see eq. (18) and (22)). The consequences of this property are for reaching. Indeed, consider the following quantity, which measures the relative importance of the fluctuations :

$$\mu = \left(\frac{\langle(\delta k)^2\rangle}{\langle k\rangle^2}\right)^{1/2}$$

From eq. (22) or (18) one has :

$$\mu = \left(\frac{npq}{n^2p^2}\right)^{1/2} = \left(\frac{q}{p}\right)^{1/2}\frac{1}{n^{1/2}} \qquad (24)$$

As $n \longrightarrow \infty$, $\mu \longrightarrow 0$. In other words, fluctuations become unimportant for macroscopic systems. As long as this is valid, we have a justification of the macroscopic description outlined in Section 2. Things are not always so simple, however. We discuss this point further on in Section 5.

4. MARKOV CHAINS

So far we have been concerned with independent trials such that the joint probability of a sequence of events satisfies the multiplicative property. We now consider a generalization provided by the theory of Markov processes (Feller, 1957). We permit the outcome of any trial to depend on that of preceding trials. In fact, for simplicity, we only deal with situations where this influence extends only over the trial preceding immediately a given trial.

In addition to the probabilities of single events introduced before, we now must introduce conditional probabilities W_2, ...

$$W_2(k, t_1 \mid \ell, t_2) \geq 0$$

Obviously, $$\sum_{\ell} W_2(k, t_1 \mid \ell, t_2) = 1 \qquad (25)$$

If k is discrete (we then speak of Markov chains) we can arrange k and ℓ along the rows and columns of a matrix. We then obtain a metrix whose row sums are equal to one. Such matrices are kown as stochastic matrices. W_2 is then called the transition matrix.

As an example, consider a population whose size is kept constant by selection of N individuals in each successive generation (Feller, 1957). Suppose that a particular gene assumes the forms A and a and hence is present in $2N$ replicals in the population. We say that the system is at state j at time n, if in the nth generation A occurs j times and a occurs $2N-j$ times. Assuming random mating, the composition of the $(n+1)$st generation will be determined by $2N$ trials in which the probability of "sucess" is $j/2N$. We thus obtain a Markov chain with

$$W_2(j|k) = \binom{2N}{k}\left(\frac{j}{2N}\right)^k\left(1-\frac{j}{2N}\right)^{2N-k} \tag{26}$$

Here k denotes the number of "successes", that is, the number of times the gene appears in form A within the population.

We now derive equations of evolution for $P(k,t)$. Suppose $W_2(k,t_1|\ell,t_2)$ can be determined in some way. Then, if $\Delta t = |t_1 - t_2|$ is the duration of a transition and if one assumes that $W_2(k,t_1|\ell,t_2) = W(k|\ell,\Delta t)$ one has by definition of a Markov process :

$$P(k,t+\Delta t) = \sum_{\ell} W(\ell|k,\Delta t)\, P(\ell,t) \tag{27}$$

where we sum over all transitions from or to state k. This is known as the master equation. It can be transformed into a differential equation if t can be regarded as continuous variable :

$$\Delta t \frac{dP}{dt} + O(\Delta t^2) = \sum_{\ell} \left[W(\ell|k,\Delta t) - \delta_{\ell k}^{kr} \right] P(\ell,t)$$

We define

$$w_{k\ell} = \lim_{\Delta t \to 0} \frac{W(k|\ell,\Delta t) - \delta_{k\ell}^{kr}}{\Delta t} = \text{transition probability/unit time}$$

with $\sum_{\ell} w_{k\ell} = 0$

or

$$w_{kk} = - \sum_{\ell} w_{k\ell} < 0$$

Eq. (27) then yields

$$\frac{dP(k,t)}{dt} = \sum_{\ell} w_{\ell k} \; P(\ell,t) =$$

$$= \sum_{\ell \neq k} \left[w_{\ell k} P(\ell,t) - w_{k\ell} P(k,t) \right] \qquad (28)$$

To have a well-posed problem we must be able to assign to the system well-defined transition frequencies $w_{k\ell}$. These are system-dependent. Their construction and the corresponding solution of the master equations are illustrated in the next section. See also David Rigney's lecture for further biological illustrations of Markov processes.

5. SIMPLE EXAMPLES

A. Unimolecular Reactions

The simplest chemical chain operating as an open system far from equilibrium is

$$A \xrightarrow{k_1} X \xrightarrow{k_2} E \qquad (29)$$

Let X be the number of particles of constituent X $(X = 0, 1, 2, \ldots)$. Successive reactions will yield a random sequence of values of X ranging from 0 to ∞ . This defines an infinite chain, and we want to compute the probability $P(X,t)$. By construction, reaction k_1 gives rise to one additional molecule of X, and reaction k_2 destroys one molecule of X. We call this a birth and death process (Nicolis and Prigogine, 1977) and write

$$\frac{dP(X,t)}{dt} = w_1(X-1|X)\,P(X-1,t) - w_1(X)\,P(X,t) + \\ + w_2(X+1|X)\,P(X+1,t) - w_2(X)\,P(X,t)$$

To construct W_1, W_2 we recall that the speed of a reaction is proportional to the number of molecules present (or to the product of the numbers of molecules, if we have several different reacting species). We express this as follows :

$$w_1 = k_1 A$$
$$w_2(X) = k_2 X$$

We thus obtain :

$$\frac{dP}{dt} = k_1 A\,[P(X-1,t) - P(X,t)] + \\ + k_2(X+1)\,P(X+1,t) - k_2 X\,P(X,t) \qquad (30)$$

Let us construct the stationary solution for $P(X,t)$. We expect this solution to exist since, by (29), there is a steady state in the macroscopic descritpion given by :

$$x = \frac{k_1 a}{k_2} \tag{31}$$

We have :

$$k_1 A P_s(X-1) - k_2 X P_s(X) =$$

$$= k_1 A P_s(X) - k_2(X+1) P_s(X+1)$$

This relation is of the form

$$G(X) = G(X+1)$$

where $$G(X) = k_1 A P_s(X-1) - k_2 X P_s(X)$$

The only solution is $G(X) =$ constant, and we choose the constant to be zero. This is known as the detailed balance solution. It yields :

$$P_s(X) = \frac{k_1 A}{k_2 X} P_s(X-1) \tag{32}$$

It is easy to see that relation (32) is satisfied by the Poisson distribution :

$$P_s(X) = e^{-\frac{k_1 A}{k_2}} \frac{(k_1 A/k_2)^X}{X!} \tag{33}$$

From eq. (33) we may compute

$$\langle X \rangle = \frac{k_1 A}{k_2}$$

in agreement with eq. (31) as well as

$$\langle (\delta X)^2 \rangle = \frac{k_1 A}{k_2} = \langle X \rangle$$

From these relations one deduces (see also eq. (24)) :

$$\left(\frac{\langle (\delta X)^2 \rangle}{\langle X \rangle^2} \right)^{1/2} = \frac{1}{\langle X \rangle^{1/2}} \longrightarrow 0 \quad \text{as } \langle X \rangle \to \infty$$

There is therefore a clear cut distinction between macroscopic description and fluctuations. The former is completely justified as long as one deals with systems involving a macroscopically large number of particles within the reaction volume.

B. Breakdown of macroscopic description

We have already stated briefly at the very end of Section 2, that bistability often means inadequacy of the macroscopic approach. To see this clearly from the probabilistic treatment one should solve the master equation for a bistable system like for instance model (10a). This turns out to be rather difficult analytically (Nicolis and Turner, 1977 ; Nicolis and Lefever, 1977). David Rigney is going to report the results of numerical simulations in a bistable problem of biological interest. Generally speaking, bistability implies under certain conditions a bimodal distribution with peaks of equal height (see Fig. 12). Because of this, the fluctuations of an extensive variable X turn out to be as important as the mean itself which is no longer representative of the state of the system. For instance

$$\langle (\delta X)^2 \rangle \sim \langle X \rangle^2$$

$$\langle (\delta X)^3 \rangle \sim \langle X \rangle^3$$

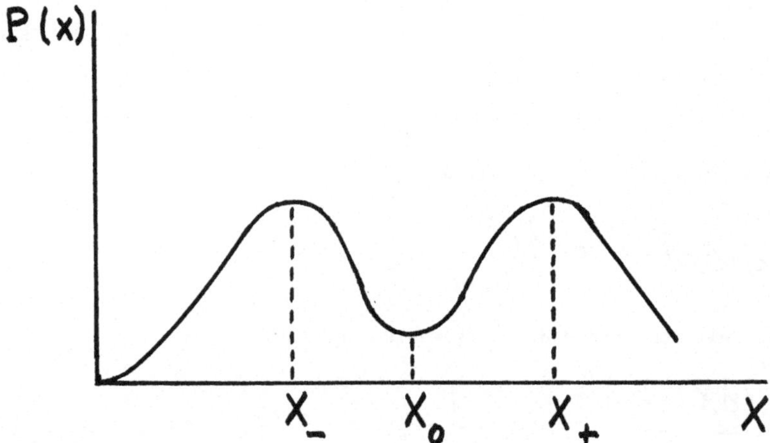

Figure 12 : Bimodal probability distribution corresponding to a system with simultaneously stable steady states.

The consequence is that the evolution of $\langle X \rangle$ no longer obeys to a closed equation but is coupled, instead to the evolution of $\langle (\delta X)^2 \rangle$ etc.... In other words, the state of the system is determined by a large number of degrees of freedom.

I would like to finish with a different example of breakdown of the macroscopic description which has the merit to be amenable to a complete analytic study. It refers to a model of population dynamics described by the rate equation (May, 1973) :

$$\frac{dx}{dt} = k x (N - x) - \rho x \tag{34 a}$$

The first term takes into account growth in a medium of limited resources, whereas the second term describes the death of the individuals.
The deterministic approach yields the following two steady-state solutions :

$$\begin{aligned} x_{01} &= 0 \\ x_{02} &= N - \frac{\rho}{k} \end{aligned} \tag{34 b}$$

If $N - \frac{\rho}{k} > 0$, x_{02} is stable and x_{01} is unstable. (see Fig. 13)

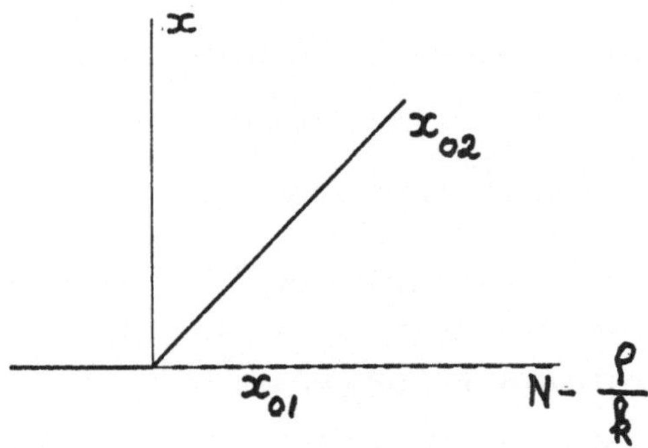

Figure 13 : Bifurcation diagram for eq. (34). x_{01} = unstable trivial branch. x_{02} = non trivial stable branch of solutions.

Consider now the stochastic analog of this system. The master equation reads :

$$\frac{dP}{dt} = kN(X-1)\,P(X-1,t) - kNX\,P(X,t) + \rho(X+1)\,P(X+1,t) - \rho X\,P(X,t) + \frac{kX(X+1)}{2}\,P(X+1,t) - \frac{kX(X-1)}{2}\,P(X,t) \tag{35}$$

The detailed balance condition at the steady state yields :

$$P_s(X)\left[\rho X + \frac{kX(X-1)}{2}\right] = kN(X-1)\,P_s(X-1)$$

We obtain, successively :

for $X = 1$: $P_s(1) = 0$

for $X = 2$: $P_s(2) = 0$

and so on.

Thus, the only normalized solution of the master equation is :

$$P_s(X) = \delta^{kr}_{X,0} \tag{36}$$

It describes the phenomenon of extinction (Malek-Mansour and Nicolis, 1975), whereby the population collapses to zero as a result of the fluctuations. This situation is well-known in biology, and is to be contrasted with the predictions of the macroscopic analysis summarized in eq. (34b) and Fig. 12. Note, however, that if initially the system had an appreaciable size, extinction would involve an exceedingly long time scale. Thus, the state of the system would remain close to that predicted by the macroscopic analysis for an extended period of time.

REFERENCES

1. Babloyantz A. and Sanglier, M., FEBS Letters 23, 364 (1972)

2. Feller, W., An Introduction to Probability Theory and its Applications, Wiley :, New York (1957).

3. Glansdorff P. and Prigogine, I., Thermodynamics of Structure, Stability and Fluctuations, Wiley, New York (1971).

4. See e.g. May, R., Model Ecosystems , Princeton Univ. Press, Princeton (1973).

5. Malek-Mansour, M. and Nicolis, G., J. Stat. Physics 13, 197 (1975).

6. Nicolis, G. and Lefever, R., Phys. Letters 62A, 469 (1977).

7. For an account of applications of bifurcation theory and stochastic theory in nonlinear kinetics,see the monograph by Nicolis, G. and Prigogine, I. Self-organization in nonequilibrium Systems, Wiley, New York (1977).

8. Nicolis, G. and Turner, J.W., Physica, 89A, 326 (1977).

9. Schlögl, F., Z. Physik 248, 446 (1971) ; ibid. 253, 446 (1972).

10. Thom, R., Stabilité Structurelle et Morphogénèse, Benjamin, New York (1972).

CHAPTER XII

ON THE RELATION BETWEEN BOOLEAN METHODS AND THE THEORY OF FINITE MARKOV CHAINS(*)

David R. RIGNEY
Faculté des Sciences, Service de Chimie Physique II.
Université libre de Bruxelles, Bruxelles 1050 (Belgium) and
Center for Statistical Mechanics and Thermodynamics
University of Texas at Austin, Texas 78712 (U.S.A.)

§1. INTRODUCTION

One of the objectives of these lectures is to compare the different theoretical methods available to persons wishing to make models of genetic regulatory processes. This might be done in two ways. On the one hand, we may select some concrete system, analyze it by the various theoretical techniques, and then compare the results. The trouble with this approach is that in constructing a model for analysis by any technique, there is a certain freedom allowed to the model builder in translating the qualitative aspects of a concrete system into the language of the theoretical method. For example, in a Boolean representation a system may sometimes be described using either logical AND or logical OR, depending on whether or not some substance is absolutely necessary for the occurence of an effect ; the selection of either logical function is at the discretion of the model builder. And in constructing a deterministic, kinetic model of the same system, we may decide to use quadratic kinetics rather than, say, cubic kinetics. Since the conclusions

(*) This lecture follows closely the recent article
Rigney, D. Bull. Cl. Sc. Acad. Roy. Belg. 63, 516-533 (1977)
Markov Chain Analysis of operon - type genetic control
Networks and its Relation to the Boolean Formalism of Thomas.

drawn depend on the particular theoretical representation, it is not always clear whether different predictions made by different approaches are the result of something inherent in the methods or whether they result from the particulars of each model.

Another way of comparing different methods is to look for mathematical isomorphisms which are model independent. In this lecture I would like to follow this second approach in comparing Boolean methods with the theory of finite Markov chains.

§2. Finite Markov Chains

At the risk of repeating some of the things discussed by Dr. Nicolis, I will first outline the relevant features of Markov chains.

Random processes are usually classified on the basis of their types of variables. Suppose that Z represents some random variable. It may be either discrete or continuous. If Z represents, say, the number of persons born in some village next year, then Z is discrete since this number can only be an integer. If Z represents the temperature of the room we are in, there is no such restriction on the values it can take; it is a continuous variable.

Depending on the way in which a random variable may change the associated random process is further classified as either continuous or discrete. For example, the number of persons born in successive years in a village is inherently discrete, since this random process is specified by writing down the sequence of numbers of birth $Z_0, Z_1, Z_2, \ldots$, where a subscript is added to the variable in order to distinguish successive years. A random process with discrete values is said to be a chain (the successive values of the random variable being the "links" in the chain). A continuous process, on the other hand, would have to be specified by drawing a graph of Z.

Notice that there may be a random chain embedded in a time-continuous process, as shown in figure 1. The discrete variable is seen to be changing value in continuous time, but we may identify the sequence of values $Z_0=3, Z_1=7, Z_2=6, Z_3=2, \ldots$ as an embedded random chain, the discrete index of which has nothing to do with the actual length of time that the system remains in one state before passing to another.

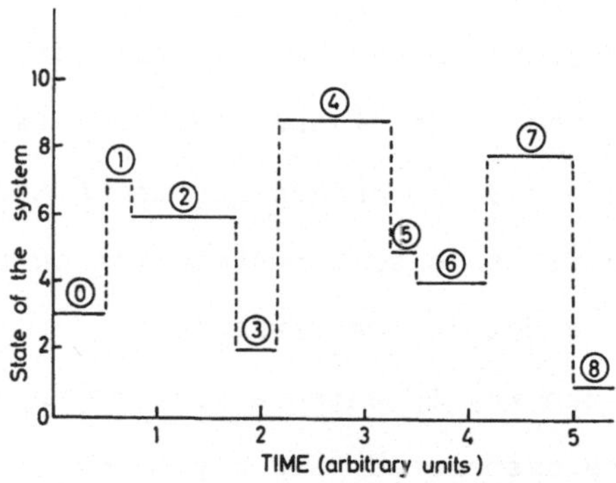

FIG. 1. — The state of a system changing randomly as a function of time. The possible states of the system are indexed by integers. The circled numbers label the sequence of states that the system passes through.

We will be concerned with a special type of random chain, the finite Markov chain. By the term "finite" we mean that the number of possible states that the random variable may occupy is not infinite. Then, the random variable is necessarily discrete, and the possible states which the system may occupy can be labelled (in some arbitrary fashion) by the integers 0,1,2,...,N.

The Markovian property of the finite Markov chain requires a more lengthy explanation. Take the example mentioned previously, where Z_i represents the number of persons born in a village in year number i. Now let $Y_i = Z_0+Z_1+Z_2+\ldots+Z_i$, the total number of persons born over the period of i+1 years. Obviously, the value of Z_i (and therefore Y_i) is not independent of the value of Y_j for $i > j$, since persons born in year j may eventually have children born in year i. Now, in order to make any predictions about the random variable Y_i, it is necessary to make some assumptions about the way in which its value in one year influences the value in a future year. As an idealization we might say that a person whose age is greater than r years never has children; so, the value of Y_j directly influences the values of Y_i only for $i-j < r$. The random chain is then said to be a (infinite) Markov chain of order r: the properties of the chain at any given point in time are influenced only by its previous r values. This is what is meant by the Markov property.

For simplicity, let us restrict our attention to

finite Markov chains of order 1. The extension to chains of arbitrary order is straightforward, if somewhat tedious. Now, for such a process, suppose that the system is known to be in state $Z_i=a$. There is a certain probability that in the next time step, it will be in some state $Z_{i+1}=b$. A stochastic model consists of specifying the various probabilities of making particular transitions from a to b, for all possible values of b. The set of all possible states is called the <u>state space</u>, and the probabilities of going from a to b are called the matrix of <u>transition probabilities</u>. They will be denoted by P_{ab}.

It may be that $P_{ab}=0$ for some values of b. For example, for the birth-death processes discussed by Dr. Nicolis, $P_{ab}=0$ unless $b=a+1$ (birth) or $b=a-1$ (death). Thus, for $P_{ab}=0$, direct transition from a to b is forbidden. Denote by S_{ab} the <u>selection rule matrix</u>: $S_{ab}=1$ if $P_{ab}\neq 0$ and $S_{ab}=0$ if $P_{ab}=0$. If S_{ab} equals zero for all values of b except for b=a, and if the system is in state a, it is trapped there forever. Then, state a is called an <u>absorbing state</u>.

Absorbing states are special cases of what are known as *recurrent* states (sometimes also called *persistent* states), which we define as follows. Let us denote by ρ_{ab} the probability that a system starting in state a will at some later step be in state b; in particular ρ_{aa} is the probability that the system starting in a will be in state a at some later step. If $\rho_{aa} = 1$ the state is recurrent. If $\rho_{aa} \neq 1$, the state is said to be *transient*. In order to determine which states are recurrent and which are transient, we need to inspect the selection rules S_{ab} in order to first determine which states *lead to* which other states. That is to say, we first find the states for which $\rho_{ab} \neq 0$. Obviously, a leads to b if $S_{ab} = 1$. And if $S_{bc} = 1$, then $\rho_{ac} \neq 0$. Thus, it is possible by direct enumeration to determine which states lead to which others.

A set of states is said to be *closed and irreducible* if every state in the set leads to every other state in the set, but no state in the set leads to any state outside of the set. Such closed and irreducible sets may contain one or more members. We are now in a position to explain what is meant by the *decomposition of state space*. For the (finite) systems which we are considering, a state is recurrent if and only if it is the member of a closed and irreducible set; otherwise it is transient. The set of recurrent states may be grouped into disjoint sets, each of which is closed and irreducible. That is to say, the recurrent states may be decomposed into separate sets according to whether or not they lead to one another. If the system enters one of these sets, it will remain there indefinitely; and if the system is initially in a transient state, it will eventually enter (be *absorbed into*) one of the closed and irreducible sets.

Each of the closed and irreducible sets exhibited by the system may be classified as either *periodic* or *aperiodic*; and if it is periodic, it is characterized by a certain integer *period*. These terms are defined as follows. Let us select any member of a closed and irreducible set and determine whether or not it is possible for the system to be in that state after w steps, given that it was initially in that state. Then the period of the set is defined to be the largest integer which is an integer divisor of any number of steps w for which it is possible to be in the state under consideration. If the period is one, the set is aperiodic; otherwise it is periodic.

As examples of closed and irreducible sets with different periods, consider those of figure 2. Each consists of three members. We remark that in general, closed and irreducible sets may contain any number of members, less than or equal to the number of states in state space. Restrictions inherent in the way in which selection rules are constructed may, however, disallow sets of certain sizes; for example, the rules which we will construct will disallow odd-numbered closed and irreducible sets (except for absorbing states).

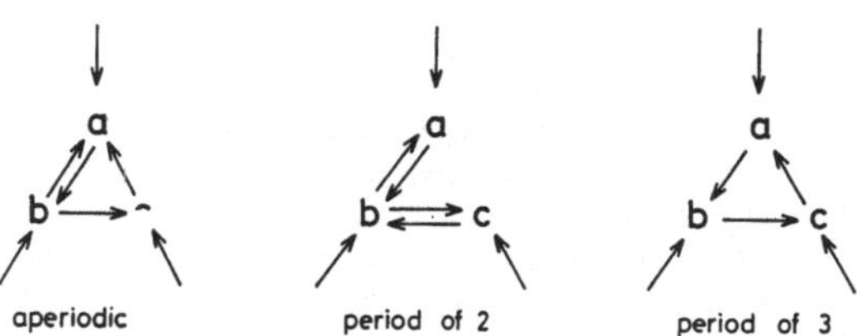

FIG. 2. — Examples of periodicities. See the text for an explanation.

§3. Operon-Type Genetic Control Network

The relation between the Boolean description and the theory of finite Markov chains is easiest to see if we treat a general type of system. The example which I have selected is an operon-type genetic control network.

Consider a set of N operon-like genetic systems with the following features. Each contains one or more genes, the transcription of which begin in a regulatory region. The regulatory region contains a promoter, to which the DNA-dependent-RNA polymerase binds; it also contains one or more sites to which regulatory protein molecules may bind. These sites may be positive or negative. By positive site we mean that transcription may not occur unless the site is occupied. An example of such a site the region to which the catabolite repressor protein binds in the lactose operon. By negative site we mean that transcription may not occur if it is occupied. An example of such a site is the operator of the lactose operon. For simplicity, we assume that there are either one or two regulatory sites per regulatory region.

At any instant the various sites will be either occupied or not occupied. Suppose that there is a total of M sites among the N operons. Then clearly, there will be a total of 2^M possible configurations. Denote by X_i whether or not the ith site is occupied:

$$X_i = 1 \quad \text{if site } i \text{ is occupied}$$

$$X_i = 0 \quad \text{if site } i \text{ is not occupied.}$$

By the *state* of the system we mean the values of *all* of the X_i.

The state space will consist of 2^M possible states.

For convenience, let us index each of the possible states by an integer. The way in which this indexing is done is arbitrary. We may, for example, choose to label the states by the octal integer corresponding to the string of binary digits which represent the particular values of the set $\{X_i\}$. Alternatively, we may use some decimal scheme.

At any instant, a single cell containing the regulatory network will be in one the states of state space. We allow for the possibility that eventually, the state will change. This will be due to the binding of a regulatory molecule to a previously unoccupied site or to the dissociation of a regulatory molecule from a previously occupied site. Both the times at which state changes occur and the eventual state resulting from a transition are assumed to be random, so, let us add a subscript j to $\{X_i\}$ to indicate the sequence of states through which the system passes: $\{X_i\}_j$.

The restrictions on the states to which the system may change, given that it is presently in a particular state, are of two types. First there is the restriction that only one of the X_i among the set $\{X_i\}$ may change at a time — if the probability that any X_i change over a small time interval Δt is proportional to Δt, then the probability that two of the X_i change over the same Δt is proportional to $(\Delta t)^2$; in the limit that $\Delta t \to 0$, the possibility of two simultaneous changes in X_i may be ignored. Second, there are restrictions which define the mechanism of the genetic regulation. We will have more to say about them in a moment.

Now, if we select some matrix of transition probabilities for this regulatory network, then we have a well-defined finite Markov chain. From the P_{ab} we could then determine the selection rules S_{ab} and then undertake a classification of states, as described previously.

But if we are interested mainly in the algebraic structure of the network, we might just as well construct a model for the S_{ab} directly, rather than obtain them from some P_{ab}. This may be done with the aid of Boolean functions, as is shown in the next section.

§4. Boolean Determination of the Selection Rules

a. Mechanism of the regulatory network .

Let the total number of proteins coded for by the N operons be denoted by Q. Then denote by E_{ij} whether or not the ith regulatory region regulates the production of the jth protein:

$E_{ij} = 1$ if regulatory region i regulates production of protein j $(i = 1,N; j = 1,Q)$

$E_{ij} = 0$ otherwise.

Let us denote by F_{ij} whether or not the ith protein binds to the jth regulatory site:

$F_{ij} = 1$ if protein i binds to site j $(i = 1,Q; j = 1,M)$

$F_{ij} = 0$ otherwise.

Since we assume that there are either one or two regulatory sites per regulatory region, each regulatory site may be classified as one of six types:

Type 1: it is in a regulatory region containing a single negative site.

Type 2: it is in a regulatory region containing a single positive site.

Type 3: it is the positive site of a region containing a positive and a negative site.

Type 4: it is the negative site of a region containing a positive and a negative site.

Type 5: it is in a region containing two positive sites.

Type 6: it is in a region containing two negative sites.

Denote by G_{ijk} whether or not site i is of type j and is in regulatory region k:

$G_{ijk} = 1$ if site i is of type j and is in regulatory region k $(i = 1,M; j = 1,6; k = 1,N)$

$G_{ijk} = 0$ otherwise.

The *mechanism* of the regulatory network is contained in the Boolean matrices E, F, and G. We remark that if the transcription of a particular gene may originate at more than one regulatory region, it may be convenient to consider that the gene product of this gene is actually of several types (corresponding to the several regulatory regions), each of which has the same F_{ij}.

b. PRODUCTION OF THE PROTEINS

Given that the system is in the state $\{X_i\} = \{x_i\}$, we now require to know which of the Q proteins may be produced. That is, we require the logical function $R_i(\{x_i\})$ where

$R_i = 1$ if protein i is being produced if the systems is in the state $\{X_i\} = \{x_i\}$

$R_i = 0$ otherwise.

As suggested by the mechanism of the system,

$$R_i = \sum_{j=1}^{M}\sum_{k=1}^{N}\Bigg[G_{j1k}\cdot E_{ki}\cdot \bar{x}_j + G_{j2k}\cdot E_{ki}\cdot x_j + \sum_{l=1}^{M}(G_{j3k}\cdot G_{l4k}\cdot E_{ki}\cdot x_j\cdot \bar{x}_l + G_{j5k}\cdot G_{l5k}\cdot E_{ki}\cdot x_j\cdot x_l + G_{j6k}\cdot G_{l6k}\cdot E_{ki}\cdot \bar{x}_j\cdot \bar{x}_l)\Bigg]$$

where summation indicates logical OR, multiplication indicates logical AND, the bar above the x indicates logical NOT, and where i ranges from 1 to Q.

c. DETERMINATION OF THE S_{ab}

Given that our system is in some state a, we are now in a position to determine S_{ab}. Evidently, the regulatory site i may change state if either i is occupied and no protein is being produced which may bind to this site, or i is not occupied and some protein is being produced which may bind to it. Thus, denote by H_i whether or not the ith site may change, given that the present state of the system is $\{X_i\} = \{x_i\}$:

$H_i = 1$ if site i may change

$H_i = 0$ if site i may not change.

Evidently,

$$H_i = x_i\cdot \overline{\sum_{l=1}^{Q} R_l\cdot F_{li}} + \bar{x}_i\cdot \sum_{l=1}^{Q} R_l\cdot F_{li}$$

Now, the set $\{H_i\}$ may be expressed as the logical sum of the vectors $\{H'_i\}$, where each $\{H'_i\}$ has the property that exactly one of its elements equals 1. If no such $\{H'_i\}$ may be constructed, then we define $\{H'_i\}$ to be identically zero, for each of its members.

Then, the selection rules S_{ab} may be determined as follows. For each state $\{X_i\} = a$, we construct the set of states $\{c\}_a$, each member of which has the property that it may be expressed in the form $\{Y_i\} = c$, where $Y_i = \bar{x}_i$ if $H'_i = 1$ and $Y_i = x_i$ if $H'_i = 0$, for some $\{H'_i\}$ constructed for the state $a = \{x_i\}$.

Then, $S_{ab} = 1$ if b is a member of the set $\{c\}_a$

$S_{ab} = 0$ if b is not a member of the set $\{c\}_a$.

Examples

Consider the systems shown in figures 3 and 4. Since there are four regulatory sites for each of the systems, there are 16 possible combinations of the $\{X_i\}$. These are tabulated in Table 1. By using the equation for R_i we infer which states produce which proteins (Table 1). For example, $G_{424} = 1$ and $E_{44} = 1$, so that any state for which $X_4 = 1$ (such as state 16) will produce protein number 4 (i.e., $R_4 = 1$).

We then apply the equation for the H_i to determine the S_{ab} (Table 2 for the system of Figure 3; Table 3 for the system of Figure 4). For example, if the system of Figure 3 is in state 1 (i.e., $X_1 = 0$; $X_2 = 0$; $X_3 = 0$; $X_4 = 0$) it will be producing proteins 1 and 2 (i.e., $R_1 = 1$; $R_2 = 1$; $R_3 = 0$; $R_4 = 0$). Now, since protein 1 may bind to site 3 (i.e., $F_{13} = 1$) and site 3 is unoccupied ($\bar{X}_3 = 1$), then $\bar{X}_3 \cdot R_1 \cdot F_{13} = 1$. This is a term in the equation for H_3, so $H_3 = 1$. Similarly, $H_4 = 1$, but H_1 and H_2 equal zero. Now, the set $\{H_i'\}$ may be expressed as the

TABLE 1. — Proteins produced by the various states of the systems shown in Figures 3 and 4.

State Index	State X_i X_1 X_2 X_3 X_4	Proteins Produced by the state X_i
1	0 0 0 0	1 2
2	1 0 0 0	2
3	0 1 0 0	1
4	1 1 0 0	
5	0 0 1 0	1 2 3
6	1 0 1 0	2 3
7	0 1 1 0	1 3
8	1 1 1 0	3
9	0 0 0 1	1 2 4
10	1 0 0 1	2 4
11	0 1 0 1	1 4
12	1 1 0 1	4
13	0 0 1 1	1 2 3 4
14	1 0 1 1	2 3 4
15	0 1 1 1	1 3 4
16	1 1 1 1	3 4

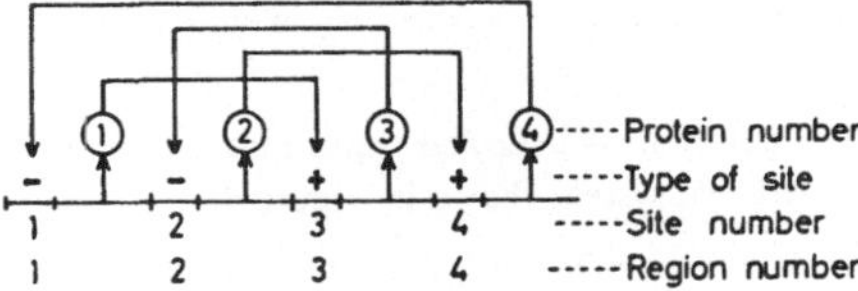

FIG. 3. — Genetic regulatory system. The possible states are shown in Table 1. Its allowed transitions and algebraic structure may be found in Table 2.

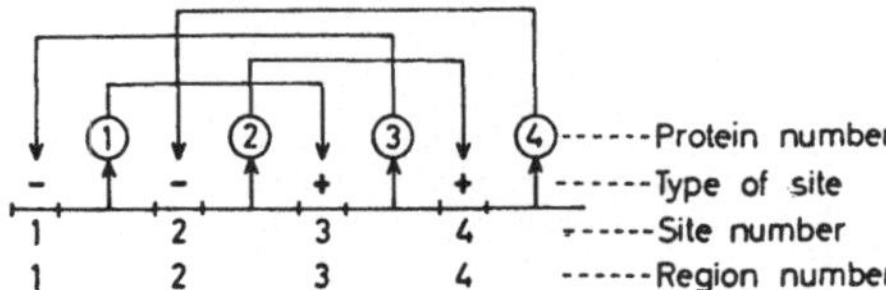

FIG. 4. — Genetic regulatory system consisting of two independent regulatory loops. The possible states are shown in Table 1. Its allowed transitions and algebraic structure may be found in Table 3. A representation of the allowed transitions may be found in Figure 5.

logical sum {0,0,1,1} = {0,0,1,0} + {0,0,0,1}. That is, we may decompose $\{H_i\}$ into the sum of two $\{H_i'\}$, each of which indicates that a site may change (i.e., sites 3 and 4). When these sites change, the result is either state 5 or state 9 of Table 1. Thus, $S_{15} = S_{19} = 1$, but S_{1b} equals zero for other values of b.

The search for closed and irreducible sets is then undertaken, and for those which are found, their periodicities are determined.

The system shown in figure 3 exhibits two closed and irreducible sets. They are both absorbing states: state 7 is one, and state 10 is the other. These two sets are necessarily aperiodic. To see how the analysis goes, consider for example that since $S_{15} = 1$ and $S_{57} = 1$, then state 1 leads to state 7. But since $S_{7b} = 1$ only if $b = 7$, then if the system starts in state 1, it may eventually be trapped in state 7. This illustrates the fact that state 1 is transient and state 7 is absorbing.

TABLE 2. — Allowed transitions for the system shown in Figure 3.

State Index a	$H_1\ H_2\ H_3\ H_4$	S_{ab} is nonzero for these values of b
1	0 0 1 1	5 9
2	1 0 0 1	1 10
3	0 1 1 0	1 7
4	1 1 0 0	2 3
5	0 1 0 1	7 13
6	1 1 1 1	2 5 8 14
7	0 0 0 0	7
8	1 0 1 0	4 7
9	1 0 1 0	10 13
10	0 0 0 0	10
11	1 1 1 1	3 9 12 15
12	0 1 0 1	4 10
13	1 1 0 0	14 15
14	0 1 1 0	10 16
15	1 0 0 1	7 16
16	0 0 1 1	14 15

States 7 and 10 are absorbing. All other states are transient and lead to all 16 states.

TABLE 3. — Allowed transitions for the system shown in Figure 4.

State Index a	$H_1\ H_2\ H_3\ H_4$	S_{ab} is nonzero for these values of b
1	0 0 1 1	5 9
2	1 0 0 1	1 10
3	0 1 1 0	1 7
4	1 1 0 0	2 3
5	1 0 0 1	6 13
6	0 0 1 1	2 14
7	1 1 0 0	5 8
8	0 1 1 0	4 6
9	0 1 1 0	11 13
10	1 1 0 0	9 12
11	0 0 1 1	3 15
12	1 0 0 1	4 11
13	1 1 0 0	14 15
14	0 1 1 0	10 16
15	1 0 0 1	7 16
16	0 0 1 1	8 12

All states lead to all other states. The periodicity of the closed and irreductible set consisting of all 16 states is 4. See Figure 5.

The system shown in Figure 4 exhibits a closed and irreducible set consisting of all 16 members of state space. This means that if the system starts in any state, it may eventually pass through any other state. The set has a periodicity of 4, as may be verified by examining Figure 5, which indicates possible transitions of the system. The period of 4 is the same as that of the two separate, independent loops if treated individually.

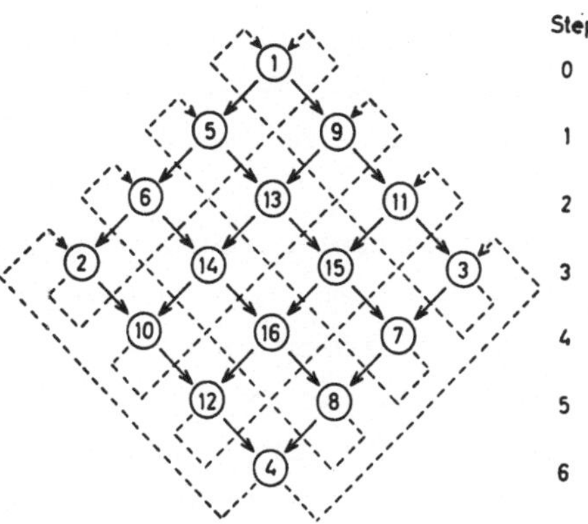

FIG. 5. — Possible sequences of states of the system shown in Figure 4. The circled numbers represent the states found in Table 1. The periodicity of this systems is four.

Examples of other systems and their algebraic structure as determined by the foregoing method may be found in Figures 6 and 7. The one in Figure 6 bears a resemblance to the immunity region of phage λ, provided that we identify protein 1 as N, protein 2 as cro, protein 3 as the repressor *c*I, and protein 4 as either *c*II or *c*III. We remark that the properties of such systems are not obvious until the algebraic analysis is performed; for example, it is not obvious (to the author, anyway) that the system shown in figure 7 has the property that no matter which state the system starts in, it will eventually pass through any other state — with the sole exception of the state $\{1,1,1,1,1,1\}$, which is transient; and that if the system starts in any of the other states, it is possible to reenter that state only after an even number of transitions (i.e., the periodicity of the single, closed and irreducible set consisting of 63 members is two).

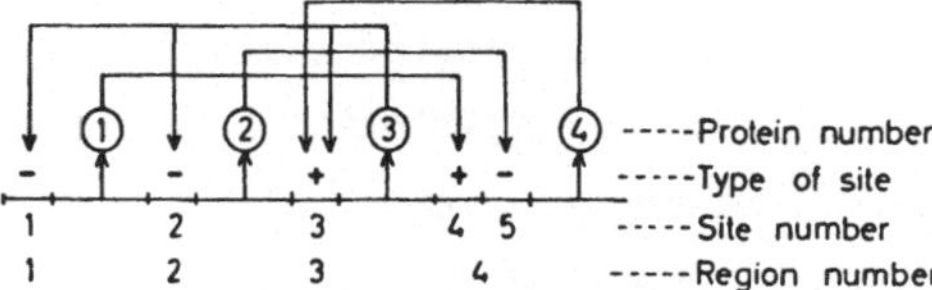

FIG. 6. — Genetic regulatory system similar to the immunity region of phage λ. The system exhibits two absorbing states. Either protein 3 alone is produced or only proteins 1 and 2 are produced.

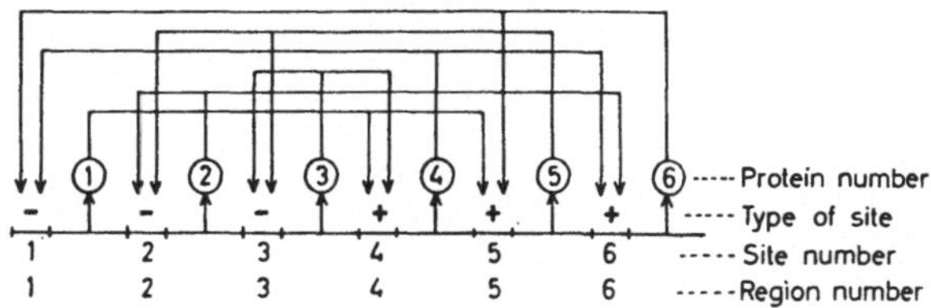

FIG. 7. — Genetic regulatory system. The network has a single closed and irreducible set containing all possible states except for the one for which all sites are bound. This latter state is transient. The periodicity of the set is two.

&6. Discussion.

In general terms, the analysis which I have described is a restatement of the Boolean analyses which were developed in previous lectures by Profs. Kauffman and Thomas. Prof. Kauffman describes Boolean networks in which randomness is involved in the changing of states ; Prof. Thomas describes "critical races" in which the "winner of the race" is random. What I hope to have done is to show how their approach to developmental problems may be viewed in terms of classical stochastic theory. The benefit is that additional elements of the theory of Markov chains may be used to enrich the Boolean theory.

It is possible to write a computer program which performs the algebraic analysis which I have decribed. You tell the program the mechanism of the system (i.e., the input is the matrices E,F,and G), and the output is the number of closed and irreducible sets and their members, as well as the periodicity of each set. The algorithm for the program is the decomposition of state space indicated in section 2. For a complicated system such as the one shown in fig.7, analysis by hand may take several minutes ; it takes the computer about one second to do the same task.

It should be apparent that the terms "stable steady state" and "oscillation" mentioned in connection with Boolean analysis are identical to the "absorbing state" and "periodicity" of the theory of Markov chains. A purist would insist, therefore, that the same name be given to these mathematically identical objects. We are not purists, however ; the important point is simply that we recognize that the same ideas are present in both the Boolean and stochastic approaches.

The algebraic analysis discussed here is only the first stage in the standard approach to problems involving Markov chains. Let me indicate the type of problem which is treated in later stages. For genetic regulatory systems, the biological interpretation of a closed and irreducible set is a phenotype. We may be interested in the probability that a system, beginning in some state a, eventually exhibit one of several possible phenotypes, where it is understood that state a is one which is uncommitted (transient). Let the various phenotypes (closed and irreducible sets) be indexed by the letter d, and let the probability that the system beginning in state a eventually be trapped in phenotype d be denoted by A_{ad}. Then, a standard theorem states that to find the A_{ad}, we need to solve the set of linear algebraic equations:

$$A_{ad} = \sum_{\substack{\text{recurrent} \\ b}} P_{ab} + \sum_{\substack{\text{transient} \\ b}} P_{ab} A_{bd}$$

To use this theorem, it is , of course, necessary to introduce some model for the P_{ab}, the transition probabilities.

In conclusion, the Boolean and stochastic methods are closely related when examined algebraically. To make contact with other approaches, we might say that the Markov chain described here is embedded in a more general, time-continuous random process. That is to say, it is part of a hierarchy of more complicated stochastic descriptions. Deterministic approaches, such as classical chemical kinetics, might then be obtained by averaging the probability distributions present in the higher levels of the hierarchy.

Legends

Figure 1: The state of a system changing randomly as a function of time. The possible states of the system are indexed by integers. The circled numbers label the sequence of states that the system passes through.

Figure 2: Examples of periodicities. See the text for an explanation.

Figure 3: Genetic regulatory system. The possible states are shown in Table 1. Its allowed transitions and algebraic structure may be found in Table 2.

Figure 4: Genetic regulatory system consisting of two independent regulatory loops. The possible states are shown in Table 1. Its allowed transitions and algebraic structure may be found in Table 3. A representation of the allowed transitions may be found in Figure 5.

Figure 5: Possible sequences of states of the system shown in Figure 4. The circled numbers represent the states found in Table 1. The periodicity of this system is four.

Figure 6: Genetic regulatory system similar to the immunity region of phage λ. The system exhibits two absorbingstates either protein 3 akone is produced or only protein 1 and 2 are produced.

Figure 7: Genetic regulatory system. The network has a single closed and irreducible set containing all possible states except for the one for which all sites are bound. This latter state is transient. The periodicity of the set is two.

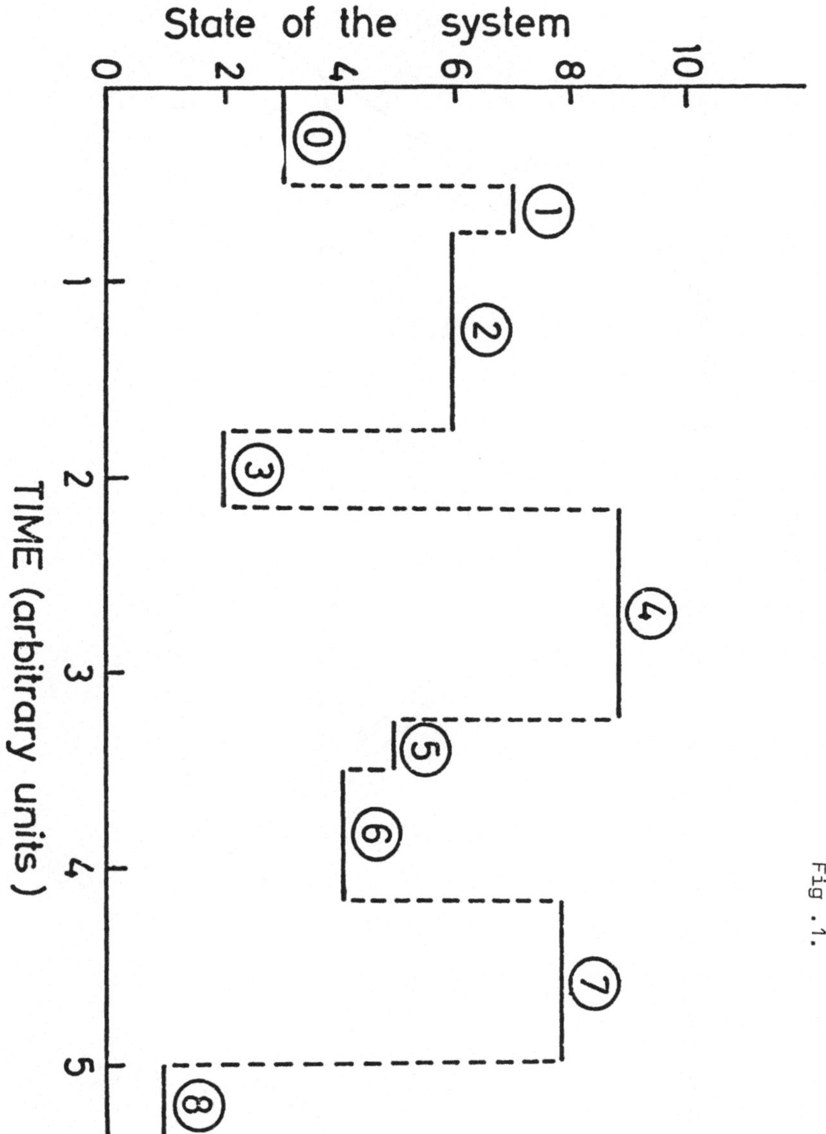
State of the system
0
2
4
6
8
10
TIME (arbitrary units)
1
2
3
4
5
0
1
2
3
4
5
6
7
8

Fig .1.

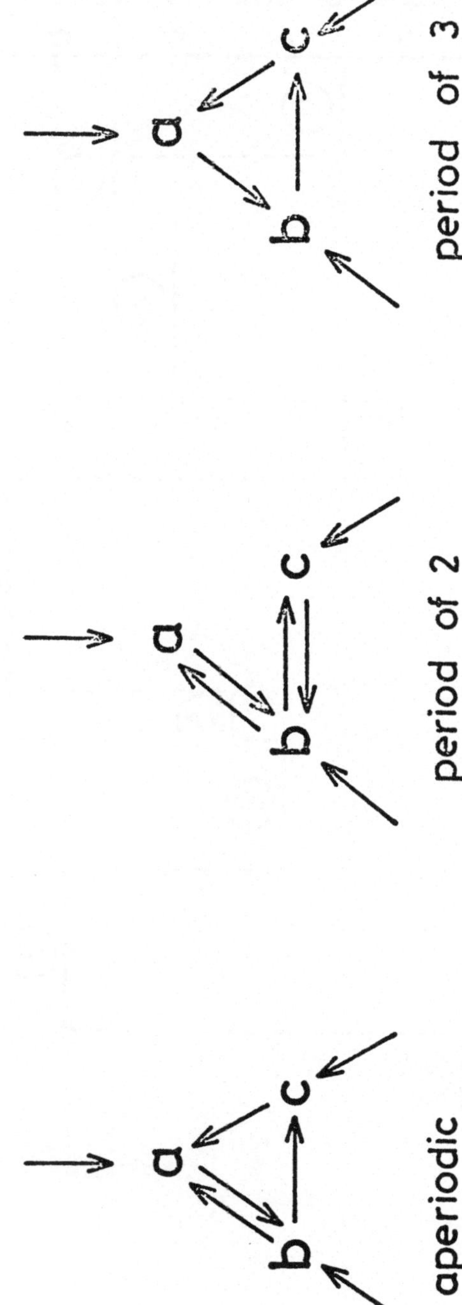

Fig.2.

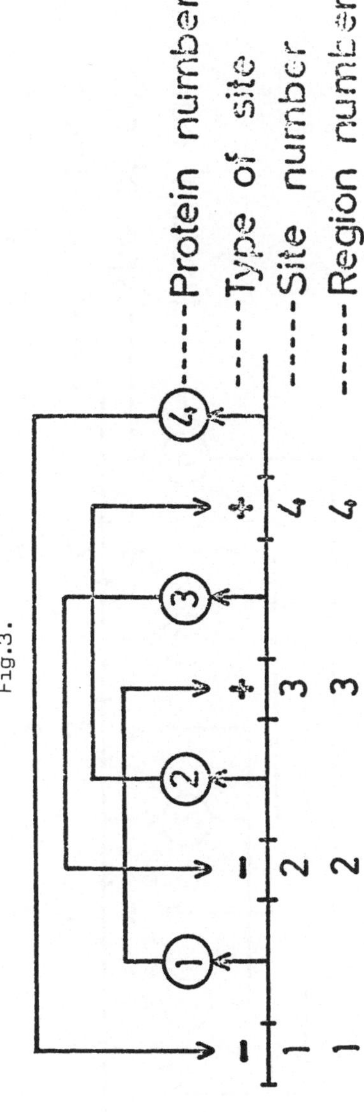

1
2
3
4
1
2
3
4
Protein number
Type of site
Site number
Region number

Fig.3.

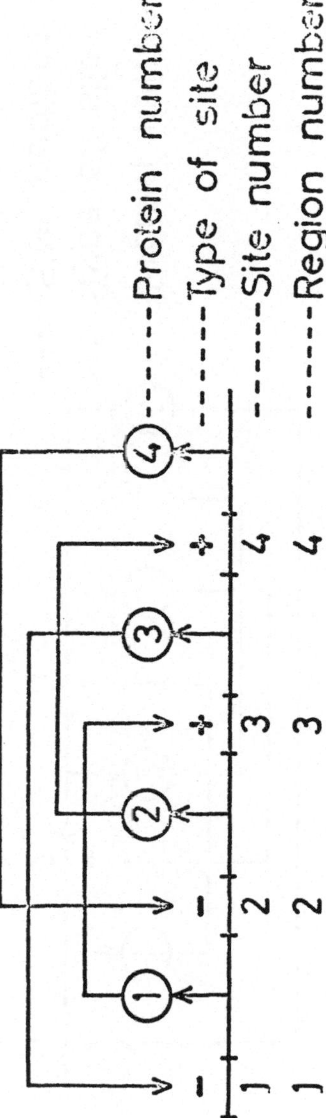

Protein number
Type of site
Site number
Region number
1
2
3
4
-
-
+
+
1
2
3
4
1
2
3
4

Fig.4.

Fig.5.

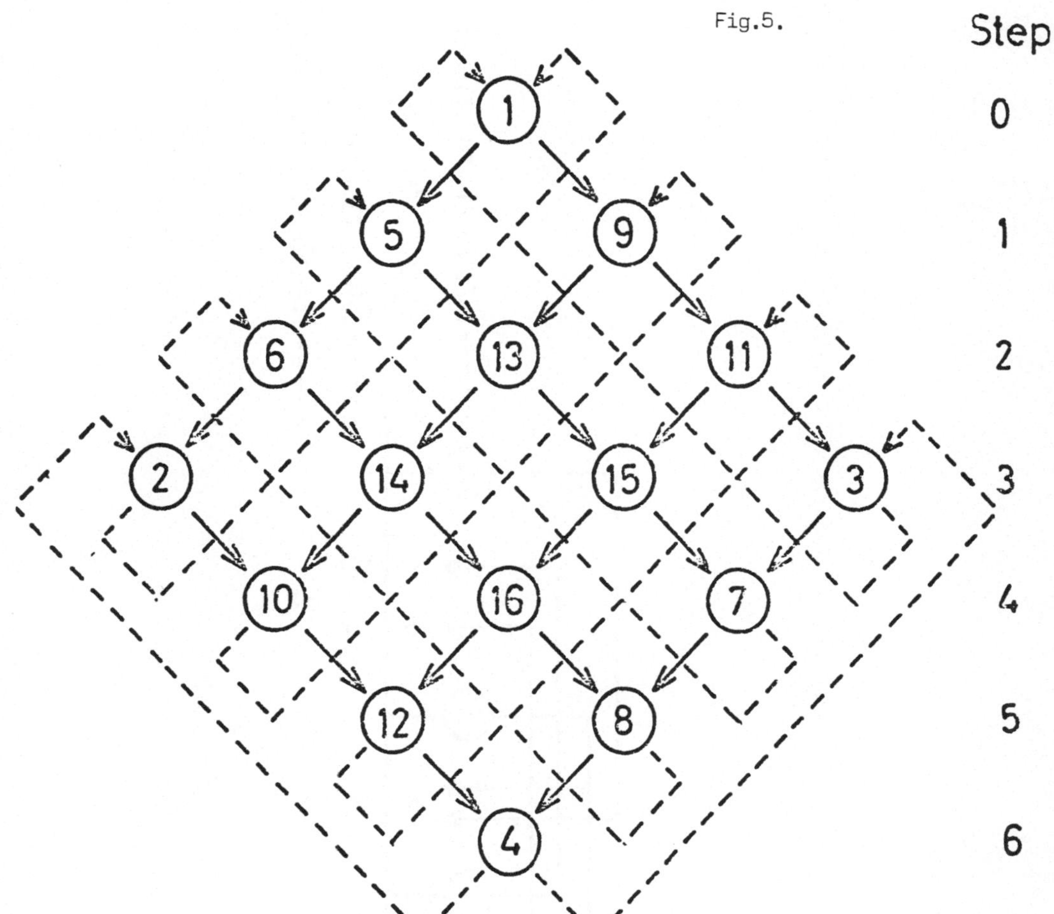

Fig.6.

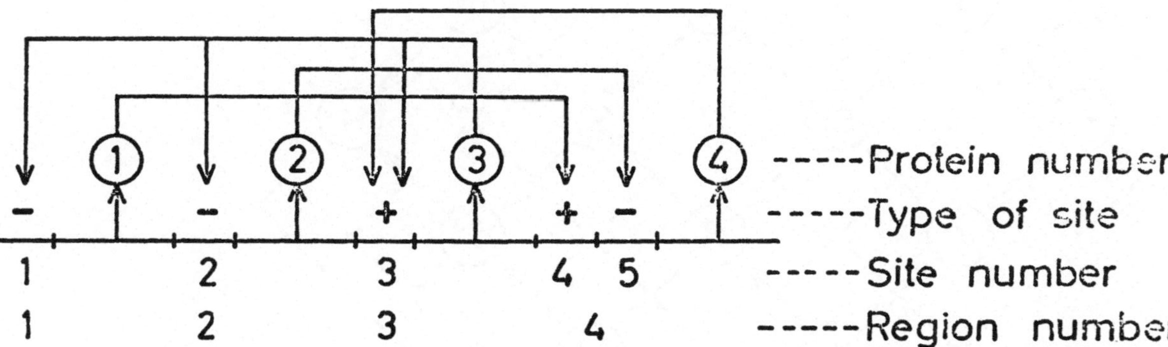

1 2 3 4
- - + + -
1 2 3 4 5
1 2 3 4
-----Protein number
-----Type of site
-----Site number
-----Region number

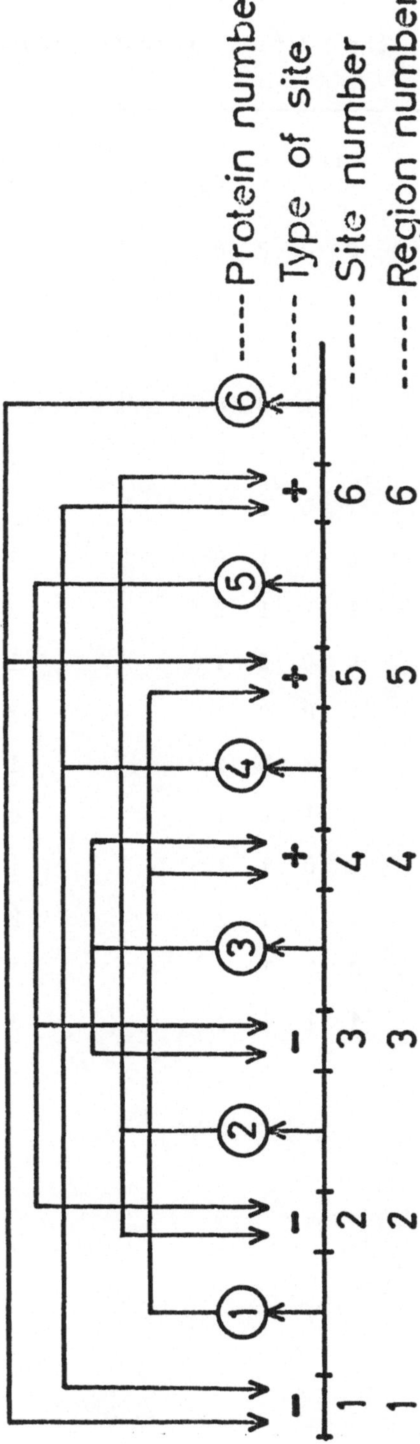
Protein number
Type of site
Site number
Region number
1
2
3
4
5
6
− − − + + +
1 2 3 4 5 6
1 2 3 4 5 6

Fig.7.

TABLE 1: Proteins produced by the various states of the systems shown in figures 3 and 4.

State Index	State X_i X_1	X_2	X_3	X_4	Proteins Produced by the state X_i
1	0	0	0	0	1 2
2	1	0	0	0	2
3	0	1	0	0	1
4	1	1	0	0	
5	0	0	1	0	1 2 3
6	1	0	1	0	2 3
7	0	1	1	0	1 3
8	1	1	1	0	3
9	0	0	0	1	1 2 4
10	1	0	0	1	2 4
11	0	1	0	1	1 4
12	1	1	0	1	4
13	0	0	1	1	1 2 3 4
14	1	0	1	1	2 3 4
15	0	1	1	1	1 3 4
16	1	1	1	1	3 4

TABLE 2: Allowed transitions for the system shown in Figure 3

State Index a	H_1	H_2	H_3	H_4	S_{ab} is nonzero for these values of b
1	0	0	1	1	5 9
2	1	0	0	1	1 10
3	0	1	1	0	1 7
4	1	1	0	0	2 3
5	0	1	0	1	7 13
6	1	1	1	1	2 5 8 14
7	0	0	0	0	7
8	1	0	1	0	4 7
9	1	0	1	0	10 13
10	0	0	0	0	10
11	1	1	1	1	3 9 12 15
12	0	1	0	1	4 10
13	1	1	0	0	14 15
14	0	1	1	0	10 16
15	1	0	0	1	7 16
16	0	0	1	1	14 15

States 7 and 10 are absorbing. All other states are transient and lead to all 16 states.

TABLE 3: Allowed transitions for the system shown in Figure 4

State Index a	H_1	H_2	H_3	H_4	S_{ab} is nonzero for these values of b
1	0	0	1	1	5 9
2	1	0	0	1	1 10
3	0	1	1	0	1 7
4	1	1	0	0	2 3
5	1	0	0	1	6 13
6	0	0	1	1	2 14
7	1	1	0	0	5 8
8	0	1	1	0	4 6
9	0	1	1	0	11 13
10	1	1	0	0	9 12
11	0	0	1	1	3 15
12	1	0	0	1	4 11
13	1	1	0	0	14 15
14	0	1	1	0	10 16
15	1	0	0	1	7 16
16	0	0	1	1	8 12

All states lead to all other states. The periodicity of the closed and irreducible set consisting of all 16 states is 4. See Figure 5.

CHAPTER XIII

STOCHASTIC MODELS OF CELLULAR VARIABILITY

David R. RIGNEY

Faculté des Sciences, Service de Chimie Physique II

Université libre de Bruxelles, Bruxelles 1050 (Belgium)

and

Center for Statistical Mechanics and Thermodynamics

University of Texas at Austin, Austin Texas 78712 (U.S.A.)

§1. INTRODUCTION

Most of the systems treated in these lectures involve cells having identical genomes but which may exhibit one of several phenotypes. Thus, an E.coli cell infected by phage λ may develop either a lytic or lysogenic response, and developmental processes in general involve the emergence of histologically novel cell types among genetically homogeneous cell lines. In this lecture I propose to describe certain stochastic techniques which are extensions of classical kinetic methods and which are useful for analyzing this type of cellular variability. I think it appropriate to preface the discussion with some remarks of an historical nature.

Prior to the dramatic advances in molecular biology of recent decades, theoretical models broadly similar to those which I will describe in this lecture were prominent in the debates over the physical nature of heredity and over the way in which novel phenotypes (mutations) arise. The school of thought which advanced these models held that "the persistence of chromosomes is not a material continuity but rather the persistence of the integrated conditions of dynamic order" (L.Bertalanffy, quoted by Kühn, (10). The occurence of different cell types (e.g., the existence of rare cells which are able to adapt to such environmental insults as the absence of some essential nutrient)was then explained as the shift of occasional cell's chemical equilibria to new, favorable equilibria which could be perpetuated cytoplasmically for many generations (Hinshelwood,[9]). The opposing school of thought insisted on a less aetherial template basis of heredity, and the controversy came to be regarded as a debate between Lamarckism and Darwinism, especially among microbial geneticists. At the time, the debate was settled in favor of the latter even before a clear picture of the material structure of the gene emerged, and the experiments that decided the issue (e.g.,the fluctuation test and replica plating) are now the classics taught in all genetics courses. From a modern perspective, of course, we may identify circumstances in which differences between cells are attributable not to different genotypes, but to regulatory mechanisms involving more global,cooperative effects. But they remain some features of the old controversy that warrant reexamination.

According to Hayes ([8],page 180), the debate between these two schools of thought remained inconclusive for so long because "it was conducted not by geneticists, but by bacteriologists who were accustomed to thinking in terms of the activity of <u>cultures</u> containing thousands of millions of cells, rather than the behavior of individual bacteria and their descendents, that is, in terms of <u>clones</u>. Thus, an analytical way of thinking about bacterial populations, as well as the techniques necessary for decisive experiments, were slow to evolve" (original emphasis).

I mention this bit of history in order to underscore a feature of chemical-theoretical models which has historically tended to make geneticists suspicious of conclusions drawn from them: although these models describe cellular activity, there is nothing cellular about them---nowhere in the kinetic equations is there anything to suggest that the chemical reactions being modelled are occurring in a living cell which has a growth and division cycle. And in practice, the models' kinetic variables are compared not with chemical concentrations in individual cells, but rather with the concentration of chemical species which have been measured in a sample containing millions of cells.

As a consequence, some objections may be raised. Clearly, molecules in one cell cannot react with molecules in another cell; so, kinetic models based on mass action principles are conceptually single cell rather than cell population models. Comparison of a model's variables with data obtained by sampling an entire population is not obviously legitimate. To give a trivial example, if half of the cells in a culture contain chemical species A but not species B, while the other half of the cells in the culture contain chemical species B but not species A, the "reaction" $A+B \rightarrow C$ will deviate from mass action kinetics if viewed at the population level.

But viewed at the level of the single cell, the conventional chemical kinetic model does not seem to be entirely appropriate. In the first place, many of the macromolecules participating in reactions are present in extremely small amounts. As discussed by Dr. Nicolis, a birth-death, discrete stochastic formulation would appear to be a better description than the continuous formulation of chemical kinetics. Second, problems arise in any model which does not include a cell's growth and division cycle. This is true independently of any questions having to do with whether or not events of the cell cycle directly influence metabolic reactions:

a cell in culture will be growing, so that the rates of the various chemical reactions should be explicit functions of that cell's volume, which in turn depends on its age. Cell growth as such is not included in conventional kinetic models but is introduced indirectly through the artifice of a "loss by dilution" term in the equation for each chemical species.

The purpose of including a dilution term (especially for the chemical species which are perfectly stable) is to allow for the existence of steady states, which play such a prominent role in classical kinetic thinking. But from the point of view of a single cell, the notion of a steady state is not entirely meaningful. The only time independent state a cell in culture will ever achieve is to disappear from the culture by giving rise to independent daughter cells.

In brief, to many biologists, chemical-theoretical models sometimes seem less than totally convincing for the simple reason that they ignore the most primitive of biological facts--- namely, that life is compartmentalized into cells which arise only from other cells. This is especially so for the systems with which we are concerned since we wish to understand not just the way in which phenotypic differences can arise in genetically homogeneous cell lines, but also how these differences are propagated clonally.

Actually, it is possible to adapt kinetic methods in such a way as to accommodate cell growth and division. In this lecture, I would like to describe one such stochastic method which, for pedagogical reasons, is easiest to present by treating the relatively simple problem of constitutive protein synthesis. Then I will use this technique to consider a system that was mentioned in earlier lectures, the sub-optimally induced lac operon; the feature that I would like to emphasize is the one which Novick and Weiner [14] recognized in their experiments: that the presence or absence of lac proteins is a heritable property.

§2. Constitutive Protein Levels in Growing and Dividing Cells

a. The Model

Suppose that we are interested in the kinetics of some stable, constitutive protein which is present in bacteria growing in culture. To make things interesting, let us assume that the protein is present in only trace amounts (a few molecules per cell on the average). For example, it might be a repressor molecule or some coenzyme. If we were able to measure the amount of this protein in individual cells, we would undoubtedly find that the number of molecules would vary from cell to cell. In other words, it is expected that there is some statistical distribution in the number of molecules of protein per cell. An assay of millions of cells for the total protein content would reveal the average number of molecules per cell but would tell us nothing about the way in which the total is distributed among the individuals of the sample.

It is not difficult to identify the factors which determine the exact number of molecules in any given cell. If the cell is newly divided, then the number of molecules in the cell is simply the number which it inherited from its mother.

As the cell grows older, the gene coding for the protein may be transcribed, and subsequent translations of the resulting messenger will increase the cell's content of the protein. Evidently, the older the cell, the more likely it is to have an elevated protein content.

This being the case, it is sensible to distinguish statistical distributions corresponding to cell subpopulations of different ages: it is better to compare newly-divided cells with other newly-divided cells and cells about to divide with other cells about to divide. Let $\pi_A(x)$ denote the probability that a cell selected at random from the subpopulation of cells having age A contain exactly x molecules. We desire a model for calculating these probabilities.

Four things must be decided on before we can proceed. First, it is necessary to be specific about the way in which molecules are partitioned to daughter cells at the time of cell division. Second, we must decide on a model for the transcription of the protein's gene. Third, we must say something about the translation of the mRNA. And fourth, the timing of two cell cycle events must be specified: the duplication of the protein's gene and the septation and division of the cell. In each case, we will adopt the simplest possible model since our concern now is with method rather than with realistic predictions.

1. Partition of Molecules: Suppose that a cell about to divide is known to contain exactly x_0 molecules. What is

the probability that either of the daughter cells receive x molecules? The simplest assumption is that each of the x_o molecules has undergone a random walk within the mother cell so that the probability of eventually being trapped in either of the daughter cells is 1/2. Thus, the probability that either daughter cell receive x molecules is binomial:

$$\frac{x_o!}{(x_o-x)!\,x!}\left(\frac{1}{2}\right)^{x_o}$$

2. <u>Transcription</u>: If we examine electron micrographs of genetic activity, such as those prepared by Miller, Hamkalo, and Thomas [12], we observe that there is a randomness in the spacing of the DNA-dependent-RNA polymerase along a given gene. This randomness is presumably due to two factors, random collision times of the polymerase molecules with a gene's promoter (resulting in random inter-initiation times) and irregular rates of mRNA elongation by the individual polymerase. If we ignore any variation in chain elongation rates, the observed inter-polymerase spacings divided by the assumed constant rate of polymerase movement along the DNA will have some statistical distribution, a model for which we must now specify. The one which is simplest is the exponential distribution.

A related random variable is the number of times that transcription has been initiated over a given time period. If the inter-initiation times are exponentially distributed, then, the number of initiations over some time interval t will have a Poisson distribution with an average value proportional to t. Stated another way, we are assuming that the probability per unit time per gene copy of a mRNA initiation is some constant, λ.

3. Translation: After a messenger RNA is initiated by an RNA polymerase, it will eventually be translated a certain number of times before being degraded. Let this number be denoted by S. There is always a time delay between the mRNA initiation and the time at which the first protein molecule appears, and the S successive polypeptides appear thereafter in temporal succession. For constitutive protein synthesis, the sequential nature of transcription and translation is of small importance, and we will ignore it. J. Richelle will discuss a regulatory model in which the time delay is included, since in that case the delay may be responsible for oscillatory behavior. For present purposes, assume that once an mRNA has been initiated, S protein molecules are immediately produced (i.e., S is to be treated simply as a constant), and any questions as to assembly of protein from subunits will be ignored.

4. Cell Cycle: The events of the cell cycle enter this model in two ways. First, when the gene coding for the protein is duplicated, the potential rate of protein synthesis doubles due to a gene dosage effect. Denote the age at which a cell's replication apparatus passes the gene by T_1. How to go about specifying T_1 need not concern us here. Suffice it to say that auxiliary models similar to that of Cooper and Helmstetter [2] may be introduced to predict T_1 as functions

of the gene's position on the conventional gene map and of the generation time of the cell, which we denote by T_2. The generation time is the second time parameter which must be given, and we treat both T_1 and T_2 as constants.

b. <u>Mathematical Formulation of the Problem</u>

To obtain the desired probabilities for arbitrary age A, $\pi_A(x)$, it is easiest to construct them from the distribution among cells having age T_2. This may be accomplished by use of the expression

$$\pi_A(x) = \sum_{x_0=0}^{\infty} \pi_{T_2}(x_0) \cdot P_A[x_0 \rightarrow x] \qquad (1)$$

where $P_A[x_0 \rightarrow x]$ is the probability that a cell at age T_2, possessing x_0 protein molecules, give rise to a daughter cell which possesses x molecules at age A. This expression states the triviality that in order for a cell of age A to contain x molecules, it must first have arisen from a mother cell which had some number of molecules (x_0) at the time of its division; to find the probability that a cell contain x molecules at age A, we need to add the probabilities corresponding to each possible way that this could happen. If A equals to T_2, then the equation gives us a means of calculating the distribution among cells of age T_2, $\pi_{T_2}(x)$, from the distribution among cells of age T_2 during the previous generation, $\pi_{T_2}(x_0)$. Evidently, the $\pi_{T_2}(x)$ so calculated may then be used to obtain the distribution for the following generation, and the resulting distribution may in turn be used to predict the distribution after yet another generation time has elapsed. If equation 1 is used iteratively an infinite number of times, then for $A=T_2$, we obtain the steady state distribution for cells of age T_2.

The problem now is to find this steady state distribution and then to use it to calculate the steady state distribution for cells of an arbitrary age.

Equation 1 is a type of master equation which may be solved by the methods described by Dr. Nicolis; for this particular problem, the simplest method is to use generating functions. Since you may not be familiar with them, a brief mathematical digression is in order.

c. Generating Functions

Let $\pi(x)$ represent some arbitrary probability distribution, where x may take on any non-negative integer value. Then the distribution's generating function is defined by the equation

$$\gamma(z) = \sum_{x=0}^{\infty} z^x \cdot \pi(x) \tag{2}$$

For example, if $\pi(x)$ is the binomial distribution

$$\pi(x) = \begin{cases} \frac{N!}{(N-x)!\, x!}\left(\frac{1}{2}\right)^N & 0 \leq x \leq N \\ 0 & \text{otherwise} \end{cases} \tag{3}$$

then the generating function is

$$\gamma(z) = \sum_{x=0}^{N} \frac{N!}{(N-x)!\, x!}\left(\frac{1}{2}\right)^N z^x = \left[1+\tfrac{1}{2}(z-1)\right]^N \tag{4}$$

If $\Pi(x)$ is the Poisson distribution

$$\Pi(x) = \frac{e^{-\lambda} \lambda^{x}}{x!} \tag{5}$$

the generating function is

$$\gamma(z) = \sum_{x=0}^{\infty} \frac{e^{-\lambda} \lambda^{x} z^{x}}{x!} = e^{\lambda(z-1)} \tag{6}$$

A generating function completely specifies its distribution through the following relation:

$$\Pi(x) = \frac{d^{x}}{dz^{x}} \gamma(z) \Big|_{z=0} \Big/ x! \tag{7}$$

where the superscripts indicate that the xth derivative of $\gamma(z)$ is to be taken before z is set equal to zero.

The generating function may also be used to calculate averages. If we define the rth factorial moment to be

$$\alpha_{[r]} = \sum_{x=0}^{\infty} \Pi(x) \cdot [x \cdot (x-1) \cdots (x-r+1)] \tag{8}$$

then, it is easy to see that the factorial moments may be obtained from the expression

$$\alpha_{[r]} = \frac{d^{r}}{dz^{r}} \gamma(z) \Big|_{z=1} \tag{9}$$

It is also easy to show that the first factorial moment, $\alpha_{[1]}$, is simply the average of the distribution function and that the variance, σ^2, is related to the first and second factorial moments by the equation

$$\sigma^2 = \alpha_{[2]} - \alpha_{[1]}(\alpha_{[1]} - 1) \qquad (10)$$

The usefulness of generating functions derives from two extremely powerful theorems; they are proved in Feller's text ([6], chapters 11 and 12).

Theorem 1: if $X = X_1+X_2+\ldots+X_k$ is the sum of k independent (but not necessarily identically distributed) random variables, then the generating function for the distribution of X is the product of the generating functions of the contributing k random variables X_i.

Theorem 2: If $X = X_1+X_2+\ldots+X_k$ is the sum of k identically distributed random variables, each having the generating function g(z), and if k is itself random and independent of the X_i, then the generating function for X equals h(g(z)), where h(z) is the generating function for the distribution of the random variable k.

d. The Transition Function $P_{T_2}[x_0 \rightarrow x]$

Returning now to the problem of solving the master equation , we must first use the model described in section a to calculate the transition function $P_{T_2}[x_0 \rightarrow x]$. This function, you will recall, is the probability that a cell having age T_2 give rise to a daughter cell which contains x molecules when it has itself reached the age of T_2, given that the mother cell contained x_0 molecules.

Rather than attempt to directly construct an expression for the transition probabilities, we will use theorem 1 to obtain its generating function. Note that x may be expressed as the sum of three independent random variables:

$x = x_1 + x_2 + x_3$

where x_1 = the number of molecules inherited from the mother cell ($x_1 \leq x_0$)

x_2 = the number of molecules produced by the daughter cell when it had one gene copy (i.e., had an age between 0 and T_1).

x_3 = the number of molecules produced by the daughter cell when it had two gene copies (i.e., had an age between T_1 and T_2).

Now, we have already decided that the partitioning of molecules at the time of cell division is governed by the binomial distribution. Hence, the (binomial) generating function for x_1 is

$$\left[1 + \tfrac{1}{2}(z-1)\right]^{x_0}$$

The random variable x_2 is S multiplied by the number of messenger RNA initiations over the time period zero to T_1. That is to say,

$$X_2 = \xi_1 + \xi_2 + \xi_3 + \dots + \xi_K$$

where k is the number of messenger RNA initiations, and ξ_i is a "random" variable having the property that $\xi_i = S$ with a probability of 1. The generating function for ξ_i is therefore z^S. Now, the model which was selected for the random variable k was a Poisson distribution having a mean value proportional to the time of allowed transcription, λ t. Thus, the generating function for k is $\exp(\lambda T_1(z-1))$. Then by theorem 2, the generating function for x_2 is $\exp(\lambda T_1(z^S-1))$.

Similarly, the generating function for x_3 is $\exp(2\lambda(T_2-T_1)(z^S-1))$. The two enters this expression since there are two genes upon which transcription may occur.

Hence, theorem 1 gives us the generating function for $P_{T_2}[x_0 \to x]$:

$$\gamma(z) = \left[1+\tfrac{1}{2}(z-1)\right]^{x_0} \exp\left\{\lambda(2T_2-T_1)(z^S-1)\right\} \tag{11}$$

As mentioned in the previous section , the actual probabilities may be obtained by taking derivatives of the generating function. Figure 1 shows these probabilities for selected values of x_0, λ , T_1, T_2, and S.

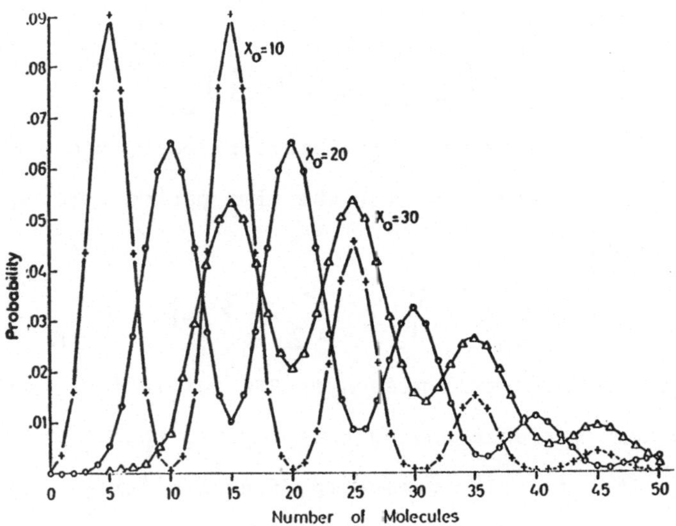

Figure 1 Transition probabilities $P(x_0 \to x)$ associated with three values of x_0. The parameters used are τ_2 =3000 seconds, τ_1 =1000 seconds, λ =1/5000 per second, μ =0 and s=10.

e. Steady State Generating Functions

Suppose that we begin a clone with a single cell containing, say, 15 molecules at the time of cell division. Then, an initial distribution function may be defined by

$$\pi_{T_2,0}(x) = \begin{cases} 1 & \text{if } x=15 \\ 0 & \text{otherwise} \end{cases} \qquad (12)$$

The first equation tells us that after one generation, the probability that either daughter cell contain x molecules is

$$\pi_{T_2,1}(x) = \sum_{x_0=0}^{\infty} \pi_{T_2,0}(x_0)\cdot P_{T_2}[x_0 \to x]$$

$$= P[x_0=15 \to x] \qquad (13)$$

and after j generations,

$$\pi_{T_2,j}(x) = \sum_{x_0=0}^{\infty} \pi_{T_2,j-1}(x)\cdot P_{T_2}[x_0 \to x] \qquad (14)$$

where you will note that a subscript has been added to $\pi_{T_2}(x)$ to indicate the generation number. As shown in Fig.2, the $\pi_{T_2,j}$ converge rapidly to a steady state.

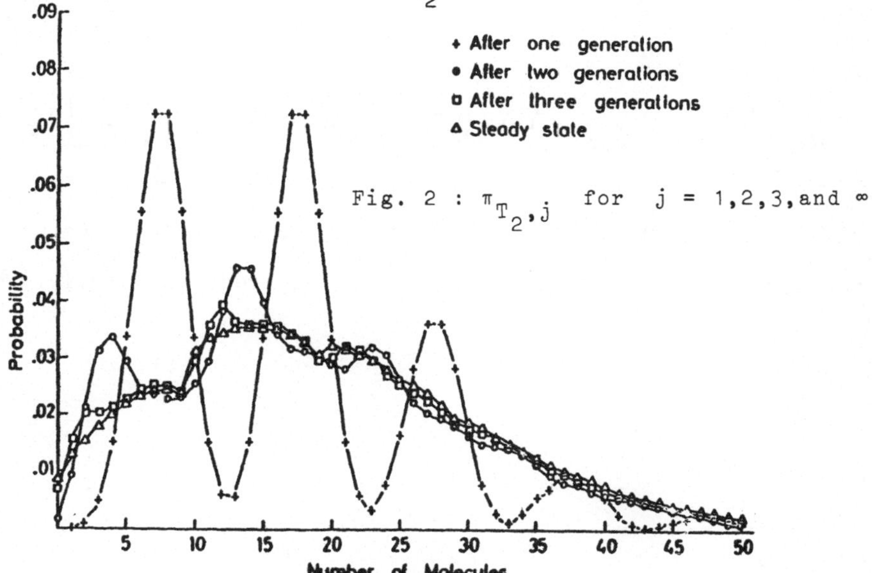

Fig.2. Approach to the steady state. In the zeroeth generation a cell (age T_2) contains exactly 15 molecules. The probability that daughter cells contain x molecules (at age T_2) after successive generations is calculated. By the third generation, the probability distribution approximates the steady state. The parameters used are the same as in Fig.1.

The unique property of the steady state distribution is that it reproduces itself when substituted into the master equation:

$$\Pi_{T_2,\,\substack{\text{steady}\\ \text{state}}}(x) = \sum_{x_0=0}^{\infty} \Pi_{T_2,\,\substack{\text{steady}\\ \text{state}}}(x_0)\cdot P_{T_2}[x_0 \to x] \qquad (15)$$

Let $\phi_{T_2}(z)$ denote the generating function for the steady state. Then multiplying the master equation by z^x and summing over x, we obtain

$$\phi_{T_2}(z) = \sum_{x_0=0}^{\infty} \Pi_{T_2,\,\substack{\text{steady}\\ \text{state}}}(x_0)\cdot \gamma(z;x_0) \qquad (16)$$

Now, substituting the previously derived expression for $\gamma(z)$, the master equation becomes

$$\phi_{T_2}(z) = \sum_{x_0=0}^{\infty} \Pi_{T_2,\,\substack{\text{steady}\\ \text{state}}}^{(x_0)} \left[1+\tfrac{1}{2}(z-1)\right]^{x_0} e^{\lambda(2T_2-T_1)(z^S-1)} \qquad (17)$$

or

$$\phi_{T_2}(z) = \exp\left\{\lambda(2T_2-T_1)(z^S-1)\right\}\cdot \phi_{T_2}\left(1+\tfrac{1}{2}(z-1)\right) \qquad (18)$$

which is a functional equation for the steady state generating function in terms of itself. It may be solved iteratively. (Appendix)

The result is

$$\phi_{T_2}(z) = \exp\left\{\lambda(2T_2-T_1)\left[z^s-1+\sum_{\upsilon=1}^{s}\frac{s!}{(s-\upsilon)!\,\upsilon!}\,\frac{(z-1)^\upsilon}{2^\upsilon-1}\right]\right\} \tag{23}$$

The generating function for cells of arbitrary age A, $\phi_A(z)$, may now be calculated using $\phi_{T_2}(z)$. For example, for newly divided cells, $P_0[x_0 \to x]$ is simply the binomial distribution, and equation 1 may be rewritten in terms of generating functions as

$$\phi_0(z) = \sum_{x_0=0}^{\infty} \Pi_{T_2,\,\text{steady state}}(x_0)\cdot\left[1+\tfrac{1}{2}(z-1)\right]^{x_0} = \phi_{T_2}\left(1+\tfrac{1}{2}(z-1)\right) \tag{24}$$

But from equation 18 this is simply

$$\phi_0(z) = \exp\left\{-\lambda(2T_2-T_1)(z^s-1)\right\}\cdot\phi_{T_2}(z) \tag{25}$$

$$= \exp\left\{\lambda(2T_2-T_1)\sum_{\upsilon=1}^{s}\frac{s!}{(s-\upsilon)!\,\upsilon!}\,\frac{(z-1)^\upsilon}{2^\upsilon-1}\right\}$$

Similarly, for cells of arbitrary age A, we obtain

$$\phi_A(z) = \exp\left\{\mu(A)\cdot(z^s-1)\right\}\cdot\phi_0(z) \tag{26}$$

where

$$\mu(A) = \begin{cases} \lambda A & \text{for } A \le T_1 \\ \lambda(T_1+2[A-T_1]) & \text{for } T_1 < A \le T_2 \end{cases}$$

By taking derivatives of the generating functions, we may recover the actual probability distributions as well as the averages, as described in section c. Fig.3. shows distributions corresponding to three different cell ages, and Fig.4. shows the mean and variance as a function of cell age :

$$\text{Average} = s\left(\mu(A)+\mu(T_2)\right) \tag{27}$$

$$\text{Variance} = s(s-1)\left[\mu(A)+\frac{\mu(T_2)}{3}\right] + s\left[\mu(A)+\mu(T_2)\right] \tag{28}$$

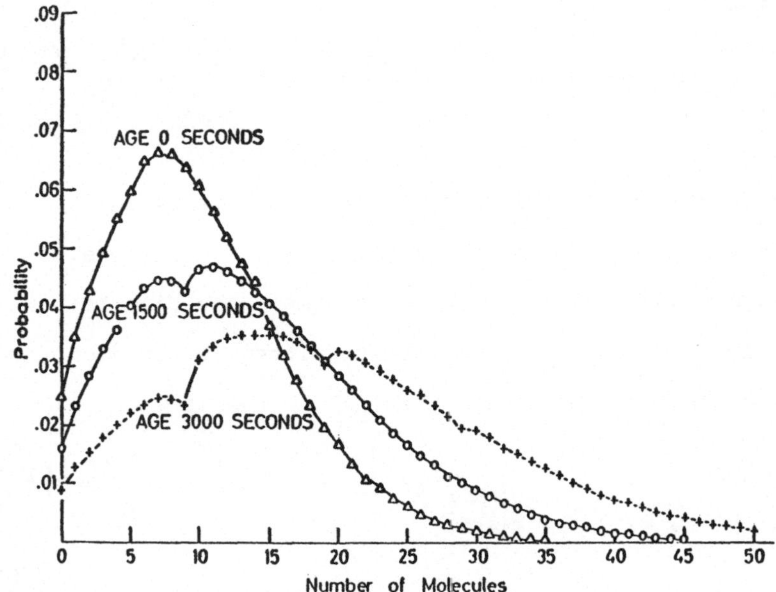

Fig.3. Steady state distributions corresponding to cells of different ages. The parameters are the same as in Fig.1.

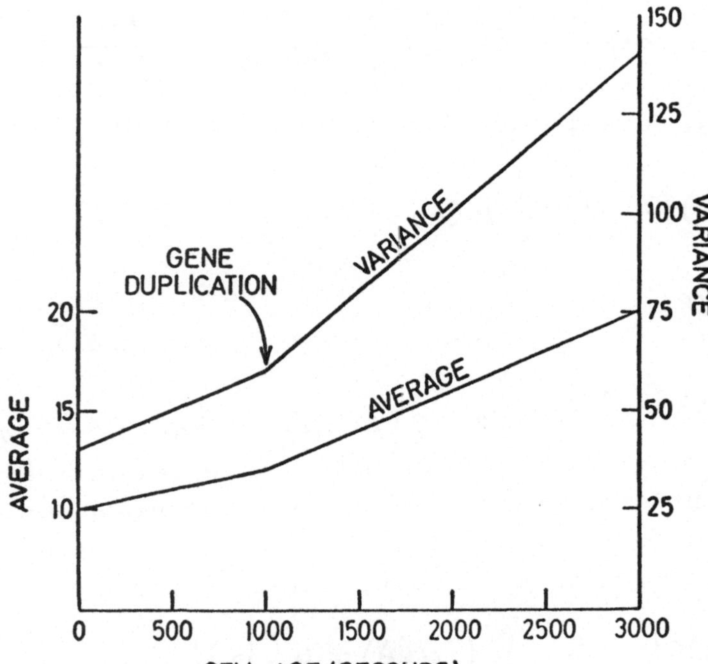

Fig.4.The mean and variance of steady state distributions as a function of cell age.The parameters used are the same as in Fig.1.

§3. Stochastic Model of the Lactose Operon Maintenence Effect

Turning now to a system which is relevant to the theme of this course consider the experiments performed by Novick and Weiner [14] involving the *lac* operon of *E. coli*. They found that if an uninduced cell population and a fully induced cell population are mixed and cultivated in a medium containing gratuitous inducer at certain low concentrations, the amount of *lac* protein per bacterium remains at its initial value, when viewed at the population level. Their interpretation of this phenomenon was that an uninduced cell, due to the presence of permease in only basal amounts, is unable to accumulate enough inducer to set the autocatalytic cycle in motion; but a previously induced cell possesses enough permease to maintain its internal inducer concentration at a level which assures the production of enough permease to maintain this level.

The experiments were performed over time periods much longer than a single generation time, so the phenomenon is appropriately interpreted at the clonal level. It is conceivable that some cells in a clone derived from an initially uninduced cell will eventually become induced and that some cells in a clone derived from an initially induced cell will eventually become uninduced. If this is the case, and if these two changes of state occur at roughly the same rate, then the measurements performed at the population level would be identical with those predicted by a model which proposed that neither change of state could occur.

In other words, given that induction and noninduction are heritable properties, how stable are they?

Experimentally, the question may be decided only by performing measurements on single cells and their clones. Evidence in favor of almost perfect stability (at some concentrations of gratuitous inducer) comes from the observation by Novick and Weiner that when single cells are cultivated, the clones derived from them are found to contain either basal protein levels or fully induced levels, but no intermediate levels. More direct evidence is obtained by performing measurements on individual cells (Rotman [16], Maloney and Rotman [11], Fangman et. al [15]).

The situation may be analyzed by the method presented in the previous section. Things are slightly more complicated here since we are dealing with a two-variable problem, rather than with a single-variable problem. So, it is necessary to begin with a catalog of the relevant properties of two-variable generating functions, a straightforward extension of those properties outlined in section 2c.

a. Two-Variable Generating Functions

Let $\Pi(X,Y)$ be the probability that both X and Y occur, where X and Y are non-negative integers. That is, $\Pi(X,Y)$ is a two-dimensional joint probability distribution. The distributions for X and Y individually are given by

$$\Pi_X(X) = \sum_{Y=0}^{\infty} \Pi(X,Y) \quad \text{and} \quad \Pi_Y(Y) = \sum_{X=0}^{\infty} \Pi(X,Y) \tag{29}$$

The generating function for $\Pi(X,Y)$ is defined to be

$$\gamma(z_x, z_y) = \sum_{X=0}^{\infty}\sum_{Y=0}^{\infty} z_x^X z_x^Y \, \Pi(X,Y) \qquad (30)$$

Then, $\gamma(z_x,1)$ is the generating function for $\Pi_x(X)$ and $\gamma(1,z_y)$ is the generating function for $\Pi_y(Y)$. $\gamma(1,1)=1$ since the sum of all probabilities must equal 1.

The generating function $\gamma(z_x,z_y)$ may be used to recover the original distribution through the relation

$$\Pi(X,Y) = \left. \frac{\frac{\partial^X}{\partial z_x{}^X} \frac{\partial^Y}{\partial z_y{}^y} \gamma(z_x, z_y)}{X!\,Y!} \right|_{\substack{z_x=0\\ z_Y=0}} \qquad (31)$$

where the superscripts indicate that X and Y partial derivatives must be taken before setting z_x and z_y equal to zero.

The $r_x r_y$ factorial moment is defined to be

$$\alpha_{[r_x, r_Y]} = \sum_{X=0}^{\infty}\sum_{Y=0}^{\infty} \Pi(X,Y)\,[X(X-1)\cdots(X-r_x+1)][Y(Y-1)\cdots(Y-r_Y+1)] \qquad (32)$$

For example, $\alpha_{[1,0]}$ is the average value of X, and $\alpha_{[0,1]}$ is the average value of Y. For the variances we have

$$\text{Variance}_X = \alpha_{[2,0]} - \alpha_{[1,0]}(\alpha_{[1,0]} - 1)$$

$$\text{variance}_Y = \alpha_{[0,2]} - \alpha_{[0,1]}(\alpha_{[0,1]} - 1)$$

$$\text{covariance}_{X,Y} = \alpha_{[1,1]} - \alpha_{[1,0]}\alpha_{[0,1]} \qquad (33)$$

The $\alpha_{[r_x, r_y]}$ may be obtained from the generating function by:

$$\alpha_{[r_x, r_Y]} = \left. \frac{\partial^{r_x}}{\partial z_x{}^{r_x}} \frac{\partial^{r_Y}}{\partial z_Y{}^{r_Y}} \gamma(z_x, z_Y) \right|_{\substack{z_x=1\\ z_Y=1}} \qquad (34)$$

The two previously mentioned theorems become

<u>Theorem</u> 1: if $(X,Y)=(U_1,V_1)+(U_2,V_2)+\ldots+(U_k,V_k)$ is the sum of k statistically independent vectors (U_i,V_i), then the generating function for (X,Y) is the product of the generating functions for the (U_i,V_i). Note that U_i and V_i need not be independent, in contrast to U_i and U_j, V_i and V_j, and U_i and V_j for $i \neq j$.

<u>Theorem</u> 2: if in addition the (U_i,V_i) are identically and independently distributed, each having generating function $b(z_x,z_y)$, and if k is itself randomly distributed, having generating function $a(z)$, then the generating function for the random vector (X,Y) is $a(b(z_x,z_y))$.

b. <u>The Linear Problem</u>

The autocatalytic mechanism of the <u>lac</u> operon suggests that two extreme situations will arise. In the case of the optimally induced population, the fraction of cells in which repressor is bound to the operator is negligible; the rate of transcription of the operon then becomes promoter-limited, and the production of permease is uncoupled from the internal inducer concentration. At the other extreme of very low external inducer concentrations, the small number of inducer molecules present in the cell will be insufficient to increase transcription of the operon above the basal rate.

Both cases may be treated in the following way. Let (x,y) denote the state of a cell, where x is the number of permease present,and y is the number of internal inducer molecules.

The transitions from (x,y) to some other state may occur by two mechanisms. The production of permease, which is assumed to occur in batches of S after each mRNA initiation, has a constant probability per unit time of taking place. Assume also that the entry of a single inducer molecule into the cell has a constant probability per unit time per permease of occurring. Then,

$x \rightarrow x+S$ occurs with probability per time of $G\lambda$

$y \rightarrow y+1$ occurs with probability per time of kx

where k and λ are simple parameters, and where G is the number of operon genes in the cell: G=1 if the cell's age is less than T_1 and G=2 if the cell's age is between T_1 and T_2. The transport parameter k is some function of the external inducer concentration: linearly proportional to the concentration at low concentrations and saturating at high external inducer concentration.

To first approximation, we are ignoring the degradation of permease and the entry or exit of inducer via non-specific membrane routes.

At the age of cell division, T_2, we will assume that the permease and inducer are distributed among daughter cells according to the binomial distribution. This is consistent with the experiments of Aussiter,Jaffe,and Kepes [1] .

We now desire the probability that a cell selected at random from a population of cells contain x permease and y internal inducer molecules. As with the constitutive protein model, we will decompose (x,y) for a cell of age T_2

into independent random variables, in order to calculate the generating function for $P_{T_2}\left[(x_0,y_0)\rightarrow(x,y)\right]$, the probability that a mother cell containing x_0 and y_0 permease and inducer, respectively, give rise to a daughter cell with state (x,y) at age T_2.

Let $(x,y)=(0,y')+(x_1,y_1)+(x_2,y_2)+(x_3,y_3)$ where

$(0,y')$ = the number of inducer molecules inherited from the mother.

(x_1,y_1) = the number of permease inherited from the mother and the number of inducer molecules transported by them over the life of the daughter cell.

(x_2,y_2) = the number of permease produced while the cell has one gene copy and the number of inducer molecules transported by them until the time of division.

(x_3,y_3) = the number of permease produced while the cell has two gene copies and the number of inducer molecules transported by them until the time of cell division.

The generating function for $(0,y')$ is that of the binomial: $[1+(1/2)(z_y-1)]^y$, where z_y is the variable associated with the random variable y.

Let $(x_1,y_1)=(1,\xi_1)+(1,\xi_2)+\ldots+(1,\xi_{x_1})$,where ξ_i is the number of inducer molecules transported by one of the permease partitioned into the daughter cell. Since transport per permease occurs at a constant rate, ξ_i will be governed by a Poisson distribution. So, the generating function for

$(1,\xi_i)$ will be $z_x \exp(kT_2(z_y-1))$. The number of terms contributing to (x_1,y_1) has a binomial distribution, so, the application of theorem 2 yields its generating function:

$$\left[1+\tfrac{1}{2}(z_x\theta-1)\right]^{X_0} \tag{35}$$

$$\text{where} \quad \theta = \exp\left\{kT_2(z_Y-1)\right\}$$

Let $(x_2,y_2)=(S,\xi_1)+(S,\xi_2)+\ldots+(S,\xi_j)$ where ξ_i is the number of inducer molecules present at time T_2 due to the S permease produced at a time selected at random on the time interval from zero to T_1. If the S permease are produced at a known time t, the number of inducer molecules present at time T_2 will be distributed according to a oisson having mean value $Sk(T_2-t)$. Thus, the distribution for ξ_i should be the integral of the Poisson from 0 to T_1 over the integration variable t, divided by T_1. The resulting expression for the probability that $\xi_i=y$, where y is a non-negative integer, is

$$\frac{1}{kST_1}\cdot\frac{1}{Y!}\left(\Gamma_{kST_2}(Y+1)-\Gamma_{kS[T_2-T_1]}(Y+1)\right) \tag{36}$$

where Γ is an incomplete gamma function. The generating function for (S,ξ_i) is then obtained by multiplying the above expression by $(z_x)^S\,(z_y)^y$ and summing over y from zero to infinity (Appendix). Then using the fact that j is distributed according to the Poisson with mean value λT_1, theorem 2 implies the following generating function for (x_2,y_2):

$$\exp\left\{\lambda T_1\left[\frac{z_x^S}{kST_1}\left(\frac{e^{kST_2(z_Y-1)}-e^{kS[T_2-T_1](z_Y-1)}}{(z_Y-1)}\right)-1\right]\right\} \tag{37}$$

The generating function for (x_3, y_3) is found by exactly the same argument as that for (x_2, y_2); then using theorem 1, the generating function for (x,y) is the product of the above four generating functions:

$$\gamma(z_x, z_y) = \left[1+\tfrac{1}{2}(z_x\theta-1)\right]^{X_0}\left[1+\tfrac{1}{2}(z_y-1)\right]^{Y_0}$$

$$\cdot \exp\left\{\frac{\lambda z_x^s}{kS}\left[\frac{\theta^s+\theta^{s\left[\frac{T_2-T_1}{T_2}\right]}-2}{(\ln\theta)/kT_2}\right]-\lambda(2T_2-T_1)\right\}$$

$$\text{where } \theta = \exp\{kT_2(z_y-1)\} \qquad (40)$$

We now desire the stationary probability distribution, which will satisfy the equation

$$\Pi_{T_2,\text{Steady State}}(x,y) = \sum_{x_0=0}^{\infty}\sum_{y_0=0}^{\infty}\Pi_{T_2,\text{steady state}}(x_0,y_0)\,P_{T_2}[(x_0,y_0)\to(x,y)] \qquad (41)$$

As with equation 18, this equation may be solved iteratively using generating functions (appendix) :

$$\phi_{T_2}(z_x, z_y) = \exp\{\beta(z_1, \theta(z_2))\} \quad \text{where}$$

$$\beta = \sum_{j=0}^{\infty}\left\{\frac{\lambda T_2}{S}\left[1+\frac{1}{2^j}\left(z_1\theta^{2-(\frac{1}{2})^{j-1}}-a_j\right)\right]^s\left[\frac{\theta^{s(\frac{1}{2})^j}+\theta^{s\left[\frac{T_2-T_1}{T_2}\right](\frac{1}{2})^j}-2}{(\frac{1}{2})^j\ln\theta}\right]-\lambda(2T_2-T_1)\right\}$$

where $$a_{j+1}(\theta) = a_j(\theta)\cdot\theta^{(\frac{1}{2})^j} + 2\left(1-\theta^{(\frac{1}{2})^j}\right) \qquad j=0,1,2,\ldots$$

and $$a_0 = 1\,; \qquad (44)$$

and where $\phi_{T2}(Z_x, Z_y)$ is the steady state generating function.

The steady state generating function for cells whose age is not T_2 may now be obtained, as was done for the constitutive protein problem. For example, we may obtain the generating function for cells which are newly divided by replacing z_x by $1+(1/2)(z_x-1)$ and z_y by $1+(1/2)(z_y-1)$ in equation 44.

The averages and variances of x and y may be obtained by taking derivatives of $\phi_{T_2}(z_x,z_y)$ and evaluating at $z_x=z_y=1$; the resulting values are (appendix):

$$\text{average}_x = 2s\lambda(2T_2-T_1) \tag{46}$$

$$\text{average}_y = \lambda k T_2^2 s\left\{\left(1+\left[\frac{T_2-T_1}{T_2}\right]^2\right)+2\left(1+\left[\frac{T_2-T_1}{T_2}\right]\right)\right\} \tag{47}$$

$$\text{variance}_x = 4\lambda s(s-1)(2T_2-T_1)/3 + 2s\lambda(2T_2-T_1) \tag{48}$$

$$\begin{aligned}\text{variance}_y = {} & \lambda k^2 T_2^3 s^2\left\{\frac{4}{9}\left(1+\left[\frac{T_2-T_1}{T_2}\right]^3\right)\right. \\ & \left.+\frac{4}{9}\left(1+\left[\frac{T_2-T_1}{T_2}\right]^2\right)+\frac{20}{27}\left(1+\frac{T_2-T_1}{T_2}\right)\right\}+\lambda k^2 T_2^3 s\cdot\frac{40}{27}\left(1+\frac{T_2-T_1}{T_2}\right) \\ & +\lambda k T_2^2 s\left\{\left(1+\left[\frac{T_2-T_1}{T_2}\right]^2\right)+2\left(1+\frac{T_2-T_1}{T_2}\right)\right\}\end{aligned} \tag{49}$$

C. Continuation of the Linear Solution for the Non-linear Problem

The problem just solved should represent the stationary distribution of permease and inducer (among cells of age T_2) only under the extreme situations of very low or very high external inducer concentrations. The question then arises as to the possibility of approximating the distribution under intermediate circumstances by the linear model.

We must first adapt the transition probabilities of the linear model so as to accommodate autocatalysis. To that end, we may replace the (constant) initiation parameter λ by a function of y:

$x \rightarrow x+S$ occurs with probability per unit time G f(y) where

$$f(Y) = \lambda + \frac{K\,Y^2}{[K_m V(a)]^2 + Y^2} \tag{53}$$

In the absence of inducer, production of permease occurs at the basal level λ . At very high concentrations of inducer, it occurs at the maximum rate $\lambda + K \cong K$. At intermediate levels (determined by the parameter K_m) the production rate depends strongly on the internal concentration of inducer, $y/V(A)$, where V(A) is the volume of the cell at age A. As a first approximation, we might let the volume of the cell as a function of cell age be $V_0(1+A/T_2)$ with V_0 equal to the volume of a newly-divided cell. The quadratic dependence of permease production on inducer concentration is suggested by experiment (see Yagil [19] for a review) and has as its origin the fact that two or more inducer molecules must bind to a repressor molecule in order that it lose its affinity for the operator. Other transition functions may be motivated by appeal to experimental results, but this is perhaps the simplest to give the appropriate qualitative behavior.

By continuation of the linear solution we mean the replacement of y in the above expression for f by the average value of y given in the previous section. Then,

$$\langle Y \rangle = f(\langle Y \rangle) \cdot \mu_Y \qquad (54)$$

$$\langle X \rangle = f(\langle Y \rangle) \cdot \mu_X \qquad (55)$$

where the parameters

$$\mu_X = 2S(2T_2 - T_1) \qquad (56)$$

$$\mu_Y = k T_2^2 S \left\{ 1 + \left[\frac{T_2 - T_1}{T_2} \right]^2 + 2\left(1 + \frac{T_2 - T_1}{T_2}\right) \right\}$$

are obtained from equations 46 and 47. (57)

Equation 54 implies a cubic equation for y :

$$\langle Y \rangle^3 - \mu_Y (\lambda + K) \langle Y \rangle^2 + [K_m V(T_2)]^2 \langle Y \rangle - \mu_Y \lambda [K_m V(T_2)]^2 = 0 \qquad (58)$$

which has either one or three real solutions, depending on the values of the parameters. The parameter μ_Y is variable in that it is a function of k which is in turn a function of the experimentally variable external inducer concentration.

For some particular values of the parameters, the roots of equation 58 were calculated and substituted into equation 53 in order to obtain the values of $\langle x \rangle$ given by equation 55. Figure 5 shows the values of $\langle x \rangle$ as a function of the variable k (external inducer concentration).

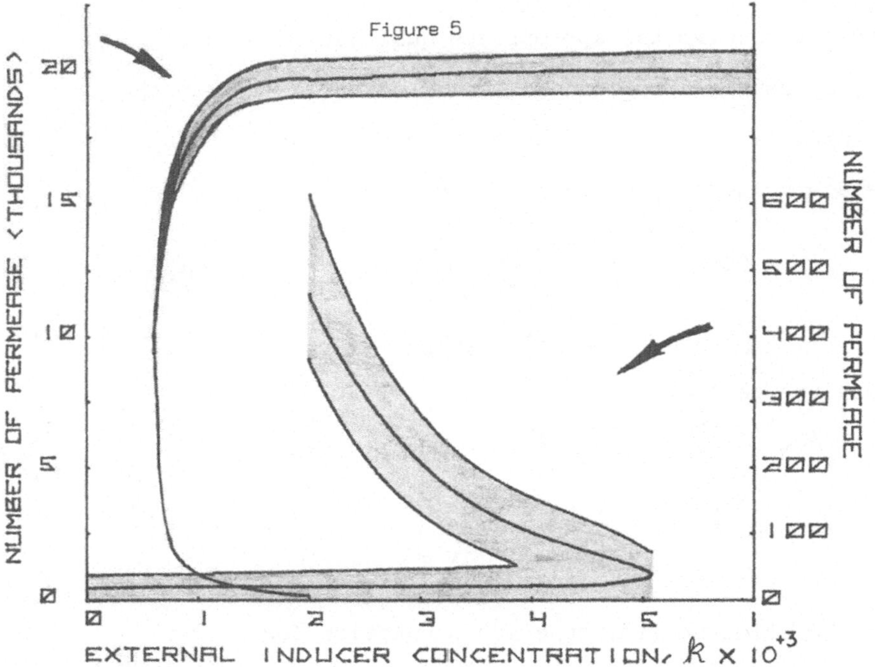

The shaded region surrounding the average indicates the location of the probability distribution to within two standard deviations, calculated with the aid of equation 48. The analagous curve for the inducer (not shown) has the same features as figure 5.

The validity of the approximation which we are using requires for self-consistency that the value of f be approximately constant within the region of the most likely values of inducer. The extent to which this is true may serve as an indicator of the stability of the distribution. Let us calculate the value of f at plus and minus two standard deviations from the average value of y to construct the diagnostics :

$$B = \frac{f(\langle Y\rangle) - f(\langle Y\rangle - 2\sigma)}{f(\langle Y\rangle)} \times 100 \qquad (60)$$

$$A = \frac{f(\langle Y\rangle + 2\sigma) - f(\langle Y\rangle)}{f(\langle Y\rangle)} \times 100 \qquad (59)$$

Values of A and B along the three branches of figure 5 are listed in Table 1, for representative values of k.

TABLE 1

$k (\times 10^{-3})$	A_{Lower}	B_{upper}	A_{middle}	B_{middle}
1.0	2	7×10^{-1}	15	14
1.5	6	3×10^{-1}	25	23
2.0	10	1×10^{-1}	36	30
2.5	16	9×10^{-2}	46	37
3.0	23	6×10^{-2}	56	42
3.5	32	5×10^{-2}	66	47
4.0	43	4×10^{-2}	75	50
4.5	56	3×10^{-2}	82	52
5.0	75	2×10^{-2}	86	47

From these values we conclude that the approximation is good for the upper branch and for the lower branch at small values of k, but it is poor for the middle branch.

It is suggested, therefore, that the maintenance effect will surely be observed for the steady states of figures 5 for k between 1 and 2×10^{-3}

For k between 4 and 5×10^{-3}, the uninduced cell will probably become uninduced. For the intermediate values of k, it is difficult to form an opinion on the basis of the approximation we are pursuing.

d. Monte Carlo Simulation of the Maintenance Effect

Although the linear approximation of the previous section suggests the existence of external inducer concentrations at which the uninduced population remains uninduced indefinitely due to the absence of fluctuations sufficiently large to be amplified, it is desirable to verify this by Monte Carlo simulation. The simulation techniques are described by Tocher [18] and Hammersley and Handscomb [7]. Using the non-linear, time dependent transition probabilities mentioned previously, figure 6 shows the evolution of a lineage of newly-divided cells for which the founder cell contained 50 permease and no inducer. It is observed that if the initial number of permease is of the same order of magnitude as expected of an uninduced cell, the fluctuations in permease and inducer remain in a fairly narrow band, even if the simulation continues for hundreds of generations.

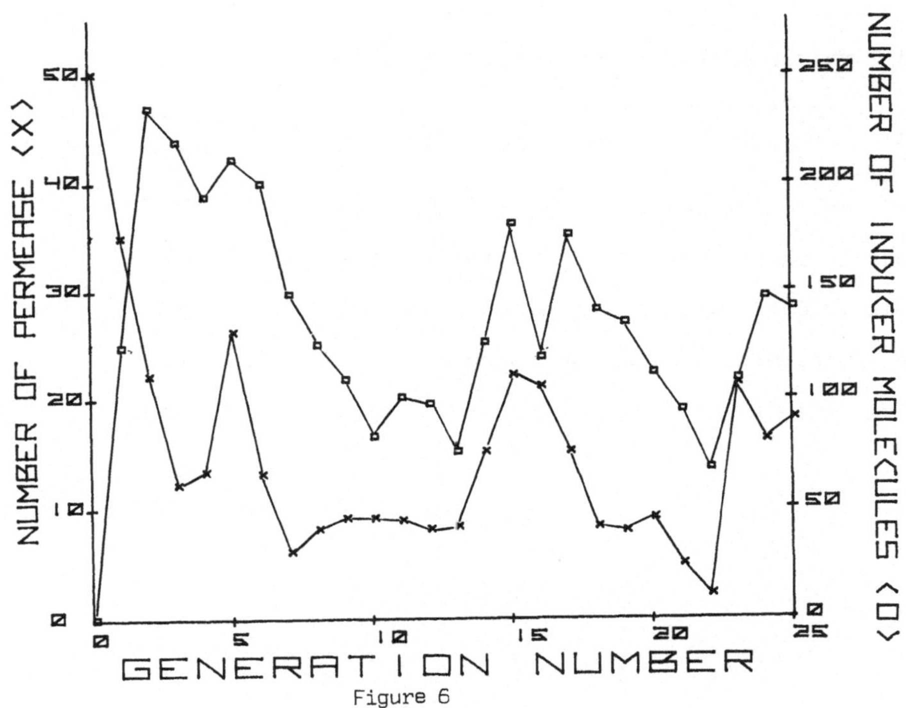

Figure 6

However, if the initial permease content is an order of magnitude greater than that of uninduced cells (but still much less than the values expected of initially induced cells), within one generation the autocatalytic nature of the induction process becomes apparent. And if the initial cell of a lineage contains inducer as well as permease (as we would expect of preinduced cells) the cell will, within half a generation, double its initial permease content and eventually reach a high level of inducer (Figure 7).

The simulations then demonstrate the following point. If a cell ever achieves a permease content which is over an order of magnitude greater than that which characterizes basal synthesis, autocatalysis will drive the cell to an even higher permease content. But cells which initially contain basal levels of permease, and the descendents of these cells, do not exhibit fluctuations sufficiently large for this to occur (at least within the time scale which is typical of experiments demonstrating the maintenance effect).

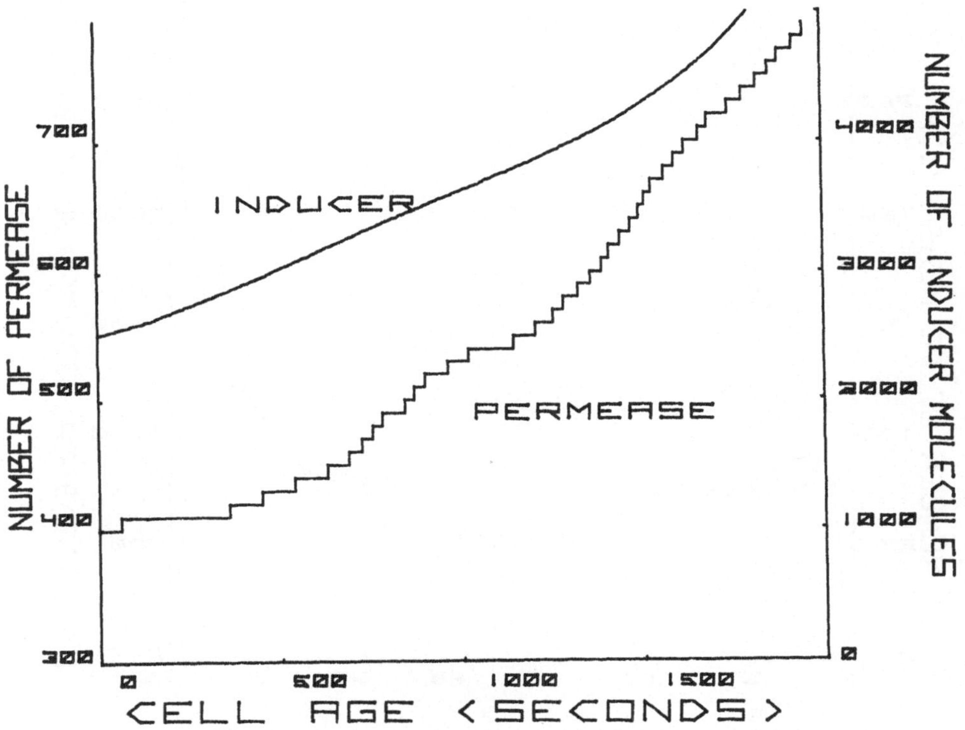

Figure 7.

The model does not suggest, however, that basal synthesis is an asymptotically stable state, for it we are prepared to wait long enough, at least one of the descendents of an initially uninduced cell will exhibit a fluctuation which is sufficiently large to be amplified by the autocatalytic mechanism. Figure 8, for example, shows a lineage of a cell which initially contained 200 permease and no inducer. For over 20 generations, descendents of this cell fluctuated at levels slightly greater than basal levels. After about 30 generations, a fluctuation was sufficiently large to be amplified, and the lineage more or less abruptly increased in permease content, as was the case with figure 7. This demonstrates the "all-or-none" character of the autocatalytic process.

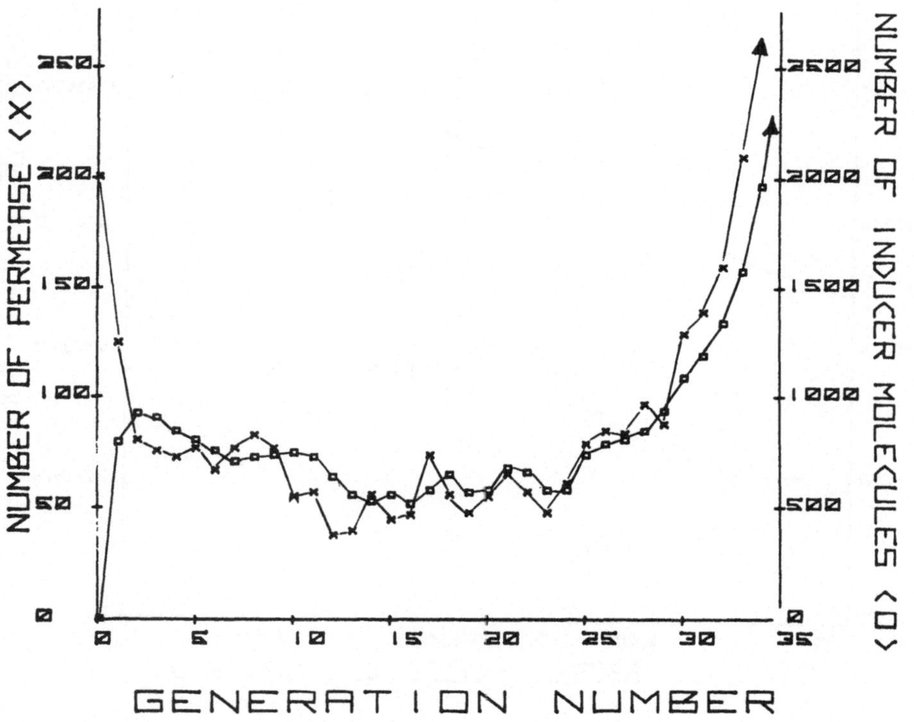

figure 8.

Thus, stability of the basal state must always be qualified by a statement of the length of observation. For, while a lineage may exhibit occasional large fluctuations when observed for periods of weeks, they will almost never be observed in an experiment extending over a day or so.

Let us define a cell as induced if it contains over 600 permease and greater than 4000 inducer molecules. Then , if we begin a lineage with a cell containing a fixed number of permease and inducer, it is possible by repeated simulation to estimate the fraction of lineages which will be induced at the end of 10 generations. When this is done, it is found that the initial values resulting in a fixed fraction of inductions lie approximately along a straight line (figure 9). The levels characterizing basal synthesis are seen to lie in the trough from which essentially no cells escape to become induced.

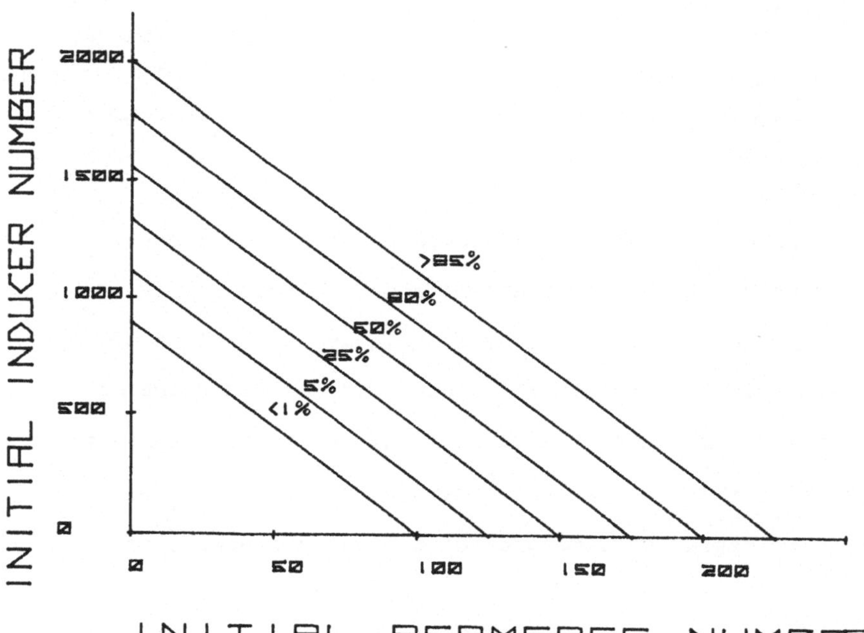

Figure 9

§4. Discussion

In view of the added difficulties which arise when we construct models that explicity include cell growth and division, it is useful to identify those situations when it is appropriate to include them. Three instances come to mind. These are

1. if the measurements with which theoretical predictions are to be compared have been performed on individual cells. Owing to the small size of cells this is technically difficult and is therefore not ordinarily done. In fact, the lac operon system is one of the few systems for which a single cell enzyme assay has been developed (Rotman [16], Maloney and Rotman [11]). However, the ease with which single-cell measurements may be accomplished has increased considerably over the last few years due to the availability of cell sorting and counting devices (reviewed by Crissman, Mullaney, and Steinkamp [3]).

Thus, the types of things which may be predicted by the models presented here (and by other models derived using similar argument) may be commonplace measureables in the near future : steady state cellular distributions of macromolecules as functions of cell age or volume (e.g. Figure 3) and the approach to steady state (e.g. Figure 2). Extensions of the model of section 3 might also be used to predict the random distribution of threshold times (first passage times in the jargon of Markov chains) such as that observed in the Monte-Carlo simulation of Figure 8.

Other types of single-cell measurements have also appeared in the recent literature (Spudich and Koshland [17]).

2. if the events of the cell cycle are themselves of interest or if they directly influence the system which is of interest. For example, if a cell culture is synchronized according of size or age, the kinetics of enzyme synthesis is found to depend on (among other things) the number of genes coding for the enzyme. Thus, the discontinuity in average cellular enzyme content shown in figure 4 is experimentally observable (Mitchison [13], Donachie and

Masters [5]). It is likely that cell cycle events play an even greater role in general metabolic processes than we now realize. For example, to what extent does the physical conformation of the replicating chromosome determine cell properties?(see Donachie ,Jones, and Teather [4] for a review of this and related matters).

3. if variability among individual cells is expected to be pronounced, so that a cell population description does not adequately describe the events occurring in individual cells. In addition to threshold phenomena, as we have seen with the suboptimally induced <u>lac</u> operon, this situation will arise if oscillatory behavior is present. In other words, if oscillations occur in individual cells, they will not be observed at the population level if the various cells are out of phase. J. Richelle will have more to say about this.

The suboptimally induced <u>lac</u> operon is treated elsewhere in these lectures-- by Dr. Thomas using Boolean methods and by Dr. Nicolis using classical kinetic methods. I would like to conclude by commenting on how the method which I described complements these other approaches. Obviously, the three methods resemble one another in not violating the experimentally obtained fact that under identical conditions, the <u>lac</u> operon may exist in two states--induced and non-induced. The methods differ not in this regard, but in the types of things that may be predicted.

A Boolean description's strength lies in the simplicity of its representation of a biological state. It does not require the user to specify exactly how many molecules a cell must have before it qualifies to be considered as induced: the notions of induction and non-induction are rendered plausible only by appeal to intuition and experiment. The method which I described complements the Boolean description by being somewhat more specific about the meaning of induction in molecular terms.

A much closer comparison can be made with the model described by Dr. Nicolis. This is not surprising, of course, since the model presented here is a stochastic extension of classical kinetic methods. Apart from the fact that we are able to make predictions concerning cellular variability, the difference between the two approaches concerns the notion of stability. Whereas the deterministic model considers only infinitesimal stability, a stochastic approach treats stability in a more global way. For example, from the deterministic point of view, the lower branch of figure 5 would be considered stable throughout its extent, but from the stochastic viewpoint, the system may leave the lower branch even before the far right portion of the lower branch is reached. This is because large scale fluctuations are allowed by the description.

LEGENDS TO FIGURES 5 - 9

FIGURE 5 :

Steady state permease levels as a function of external inducer concentration (k). The parameters used are $\lambda = 1/5400 \ sec^{-1}$, $K = 1000\ \lambda$, $V_{\cdot}=2$, $T_1 = 600$ sec, $T_2 = 3000$ sec, $K_m = 7000$, and $S = 10$.

FIGURE 6 :

Simulation of a lineage of newly-divided cells. The parameters are as in figure 5 with $k = 2.5 \times 10^{-3}$

Figure 7 :

Simulation of induction in a single cell.
The parameters are as in Figure 6

Figure 8 :

Simulation of a lineage of newly-divided cells. The parameters are as in Figure 6.

Figure 9 :

The probability that a lineage becomes induced by ten generations, as functions of permease and inducer levels in the newly-divided founder cell.
The parameters are as in Figure 6.

APPENDIX : Intermediate Calculations

1. Derivation of equation 23

Equation 18 may be substituted into itself to obtain a new expression for $\phi_{T_2}(z)$.

For example

$$\phi_{T_2}\left(1+\tfrac{1}{2}(z-1)\right) = \exp\left\{\lambda(2T_2-T_1)\left(\left[1+\tfrac{1}{2}(z-1)\right]^{s}-1\right)\right\} \cdot \phi_{T_2}\left(1+\tfrac{1}{2}\left(\left[1+\tfrac{1}{2}(z-1)\right]-1\right)\right) \quad (19)$$

So,

$$\phi_{T_2}(z) = \exp\left\{\lambda(2T_2-T_1)\left(z^{s}+\left[1+\tfrac{1}{2}(z-1)\right]^{s}-2\right)\right\} \cdot \phi_{T_2}\left(1+\tfrac{1}{4}(z-1)\right) \quad (20)$$

After n iterations,

$$\phi_{T_2}(z) = \exp\left\{\lambda(2T_2-T_1)\left[(z^{s}-1)+\sum_{\upsilon=1}^{s}\frac{s!}{(s-\upsilon)!\,\upsilon!}(z-1)^{\upsilon}\sum_{j=1}^{n}\left(\frac{1}{2^{\upsilon}}\right)^{j}\right]\right\} \cdot \phi_{T_2}\left(1+\left[\tfrac{1}{2}\right]^{n+1}(z-1)\right) \quad (21)$$

But in the limit of large n ,

$$\lim_{n\to\infty} \phi_{T_2}\left(1+\left[\tfrac{1}{2}\right]^{n+1}(z-1)\right) = \phi_{T_2}(1) = 1 \quad (22)$$

$$\sum_{j=1}^{\infty}\left(\frac{1}{2^{\upsilon}}\right)^{j} = 1/(2^{\upsilon}-1)$$

So, the final result is

$$\phi_{T_2}(z) = \exp\left\{\lambda(2T_2 - T_1)\left[z^s - 1 + \sum_{\nu=1}^{s} \frac{s!}{(s-\nu)!\,\nu!}\,\frac{(z-1)^\nu}{2^\nu - 1}\right]\right\} \tag{23}$$

2. Properties of gamma functions used in deriving equation 37.

$$\Gamma_c(p+1) = p\,\Gamma_c(p) - c^p e^{-c} \tag{38}$$

$$\sum_{p=0}^{\infty} \frac{\alpha^p}{p!}\,\Gamma_c(p+1) = \frac{e^{c(\alpha-1)} - 1}{\alpha - 1} \tag{39}$$

3. Derivation of equation 44.

The generating function for $P_{T_2}[(x_0, y_0) \to (x, y)]$ is given by equation 40, so we are able to write equation 41 in terms of generating functions :

$$\phi_{T_2,\,\text{steady state}}\left(z_x, 1 + \frac{\ln\theta}{kT_2}\right) = \exp\left\{\frac{\lambda T_2}{s} z_x^s \left[\frac{\theta^s + \theta^{s\left[\frac{T_2 - T_1}{T_2}\right]} - 2}{\ln\theta}\right] - \lambda(2T_2 - T_1)\right\}$$

$$\cdot\ \phi_{T_2}\left(1 + \tfrac{1}{2}(z_x\theta - 1),\ 1 + \frac{\ln\theta^{1/2}}{kT_2}\right) \tag{42}$$

where ϕ_{T_2} is the generating function for the steady state probability distribution. Equation 42 implies that

$$\phi_{T_2}\left(1+\frac{1}{2^j}\left(z_x\theta^{2-(\frac{1}{2})^{j-1}}-a_j(\theta)\right),\ 1+\frac{\ln\theta^{(\frac{1}{2})^j}}{kT_2}\right)=$$

$$\exp\left\{\frac{\lambda T_2}{S}\left[1+\frac{1}{2^j}\left(z_x\theta^{2-(\frac{1}{2})^{j-1}}-a_j(\theta)\right)\right]^S\left[\frac{\theta^{S(\frac{1}{2})^j}+\theta^{S\left[\frac{T_2-T_1}{T_2}\right](\frac{1}{2})^j}-2}{(\frac{1}{2})^j\ln\theta}\right]-\lambda(2T_2-T_1)\right\}$$

$$\cdot\ \phi_{T_2}\left(1+\frac{1}{2^{j+1}}\left(z_x\theta^{2-(\frac{1}{2})^{j}}-a_{j+1}(\theta)\right),\ 1+\frac{\ln\theta^{(\frac{1}{2})^{j+1}}}{kT_2}\right)$$

$$\text{where}\quad a_{j+1}(\theta)=a_j(\theta)\cdot\theta^{(\frac{1}{2})^j}+2\left(1-\theta^{(\frac{1}{2})^j}\right)\quad,\ j=0,1,2,\ldots$$

$$\text{and}\quad a_0=1 \tag{43}$$

which may be used interatively to give

$$\phi_{T_2}(z_x,z_y)=\exp\left\{\beta(z_x,\theta(z_y))\right\}\quad\text{where}$$

$$\beta=\sum_{j=0}^{\infty}\left\{\frac{\lambda T_2}{S}\left[1+\frac{1}{2^j}\left(z_x\theta^{2-(\frac{1}{2})^{j-1}}-a_j\right)\right]^S\left[\frac{\theta^{S(\frac{1}{2})^j}+\theta^{S\left[\frac{T_2-T_1}{T_2}\right](\frac{1}{2})^j}-2}{(\frac{1}{2})^j\ln\theta}\right]-\lambda(2T_2-T_1)\right\} \tag{44}$$

We have used the fact that

$$\lim_{j\to\infty}\phi_{T_2}\left(1+\frac{1}{2^j}\left(z_x\theta^{2-(\frac{1}{2})^{j-1}}-a_j(\theta)\right),\ 1+\frac{\ln\theta^{(\frac{1}{2})^j}}{kT_2}\right)$$

$$=\phi_{T_2}(1,1)=1 \tag{45}$$

4. Calculations used in obtaining equations 46 - 49.

$$A_j \equiv \lim_{\substack{Z_X \to 1 \\ Z_Y \to 1}} \frac{d}{d\theta}\left[\frac{Z_X \theta^{2-(\frac{1}{2})^{j-1}}}{2^j} - \frac{a_j(\theta)}{2^j} \right] = \frac{1}{2} A_{j-1} + \frac{1}{2^j} = \frac{j}{2^j} \qquad (50)$$

$$C_j \equiv \lim_{\substack{Z_X \to 1 \\ Z_Y \to 1}} \frac{d^2}{d\theta^2}\left[\frac{Z_X \theta^{2-(\frac{1}{2})^{j-1}}}{2^j} - \frac{a_j}{2^j} \right] = \frac{1}{2} C_{j-1} + \left(\frac{1}{4}\right)^{j-1}(2j-1) \qquad (51)$$

$$\left[\frac{\theta^{s(\frac{1}{2})^j} + \theta^{s\left[\frac{T_2-T_1}{T_2}\right](\frac{1}{2})^j} - 2}{(\frac{1}{2})^j \ln \theta} \right] = \sum_{i=1}^{\infty} \frac{\left[(\frac{1}{2})^j k T_2 (Z_Y - 1)\right]^{i-1} \left[s^i \left(1 + \left[\frac{T_2-T_1}{T_2}\right]^i\right)\right]}{i!} \qquad (52)$$

REFERENCES

[1] Aussiter,F.,Jaffe,A.&Kepes,A. Molecular and General Genetics 112,275.(1971).

[2] Cooper,S.& Helmstetter,C.E. J. Mol. Biol. 31,519. (1968).

[3] Crissman,H.A., Mullaney,P.F. & Steinkamp,J.A. Methods in Cell Biology 9,179. (1975).

[4] Donachie,W.D.,Jones,N.C., & Teather,R. Symp. Soc. Gen. Microbiol. 23,9. (1973).

[5] Donachie,W.D. & Masters,M. in The Cell Cycle: Gene Enzyme Interactions (G.M. Padilla,G.L. Whitson, & I.L. Cameron,eds.) New York:Academic Press. (1968).

[6] Feller,W. An Introduction to Probability and its Applications.Volume 1. Third edn. New York:Wiley. (1968).

[7] Hammersley,T.M. & Handscomb,D.C. Monte Carlo Methods. London: Methuen. (1964).

[8] Hayes,W. The Genetics of Bacteria and their Viruses. Oxford: Blackwell Scientific Publ. (1968).

[9] Hinshelwood,C.N. The Chemical Kinetics of the Bacterial Cell. London: Oxford Univ. Press. (1946).

[10] Kühn,A. Lectures in Developmental Physiology. New York: Springer-Verlag. (1971).

[11] Maloney,P.C. & Rotman,B. J. Mol. Biol. 73,77.(1973).

[12] Miller,O.L.,Hamkalo,B.A., & Thomas,C.A. Science 169,394. (1970).

[13] Mitchison,J.M. The Biology of the Cell Cycle.Cambridge: University Press. (1971).

[14] Novick,A. & Weiner,M. Proc.Natl.Acad.Sci.USA 43,553.(1957).

[15] Fangman,W.L.,C. Gross, and A. Novick. J. Mol. 29, 317 (1967)

[16] Rotman,B. Proc.Natl.Acad.Sci.USA.47,1981.(1961).

[17] Spudich,J.L. & Koshland,D.E. Nature 262,467. (1976).

[18] Tocher,K.D. The Art of Simulation. London:Unibooks,Hoddler, & Stoughton. (1963).

[19] Yagil,G. Curr. Top. Cell. Regulation 9,183.(1975).

CHAPTER XIV

J. RICHELLE (1)

Comparative analysis of negative loops by continuous, boolean and stochastic approaches.

"Ce qui nous rend ces solutions périodiques si précieuses, c'est qu'elles sont, pour ainsi dire, la seule brèche par où nous puissions pénétrer dans une place jusqu'ici réputée inabordable".

POINCARE

14.1. Introduction.

Since the work of B.G. Goodwin (1965) who pointed out the biological relevance of negative feedback loops, the dynamics of the corresponding differential equations has been extensively analysed (2). Most of these studies deal with regulatory chains in which the last element, Y_m, exerts a negative control on the production of the first one, Y_1 (Figure 1).

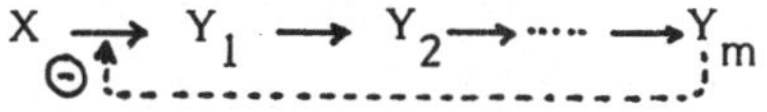

$$X \xrightarrow[\ominus]{} Y_1 \longrightarrow Y_2 \longrightarrow \cdots \longrightarrow Y_m$$

Figure 1.

This system is usually described by the following set of differential equations (Othmer, 1976)

$$\begin{cases} \dfrac{d[Y_1]}{dt} = \dfrac{k_1}{1+\alpha\cdot[Y_m]^n} - k_{-1}\cdot[Y_1] \\ \\ \dfrac{d[Y_i]}{dt} = k_i\cdot[Y_{i-1}] - k_{-i}\cdot[Y_i] \quad i = 2,\ldots,m \end{cases} \tag{I}$$

(X is supposed to come out of a pool and there is no need to consider explicitly its concentration which is constant ; $[Y_i]$ describes the concentration of product Y_i).
This model was first used by B.C. Goodwin to describe "phenomena of feedback inhibition and feedback repression whereby enzymatic activities are controlled at the level of the enzyme and the gene ..." (Goodwin, 1963) ; however the meaning of the equations (I) is different in the two cases. This important distinction will be dealt with in the following section (see also the last section of Chapter VII).

The first term of $d[Y_1]/dt$ expresses the negatively regulated production of Y_1 : when there is no Y_m, the production is maximum and it decreases as the concentration of Y_m increases (figure 2).

(1) "Aspirant" of the Fonds National de la Recherche Scientifique (Belgium).
(2) For purpose of clarity later on this approach will be referred to as "continuous description" or "continuous approach".

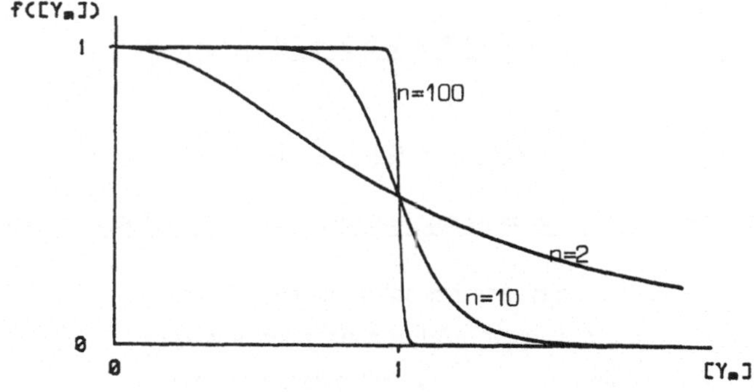

Figure 2. Plotting of the function $f([Y_m]) = \dfrac{k_1}{1 + \alpha . [Y_m]^n}$;

$k_1 = 1$, $\alpha = 1$, and $n = 2, 10, 100$

The sigmoid shape of the curves arises from the cooperativity of the regulatory interaction ; n is an index of this cooperativity and the higher its value the steeper the sigmoid.

The analysis of the system of differential equations (I) (Othmer, 1976) shows that sustained oscillations are possible if, and only if, there are at least three elements in the loop ($m \geqslant 3$) and the cooperativity (n) is sufficiently high. More specifically, if $m = 3$, n must be $\geqslant 8$; if m is higher, sustained oscillations are possible for lower values of n ($n \to 1$ when $m \to \infty$).

An alternative description of systems of regulation is the boolean approach. This description has mainly been developed for the analysis of nets of genetic regulations. The set of boolean equations corresponding to the negative loop showed in figure 1 is

$$\begin{cases} y_1 = \bar{\upsilon}_m \\ y_i = \upsilon_{i-1} \quad (i = 2,\ldots,m) \end{cases} \tag{II}$$

(y_j describes the state - on or off - of expression of gene j , and υ_j described the presence - or absence - of the product of gene j, $j = 1,\ldots,m$).
From equations (II) one expects, for any number of elements in the loop (even if m = 1), permanent oscillations.

One is therefore faced with two different types of predictions for the same regulatory structure. Why is it that the boolean description predicts permanent oscillations in all cases while the continuous description predicts sustained oscillations only in well defined conditions ?

One element is that in the boolean description each regulatory interaction is treated as if it were infinitely non - linear : there exists, implicitly, a threshold of concentration below which a substance is considered to be present at an inefficient level. Figure 3 shows the value of the boolean function describing the state of expression (on : 1 ; off : 0) of a negatively regulated gene as a function of the concentration of the regulatory substance.

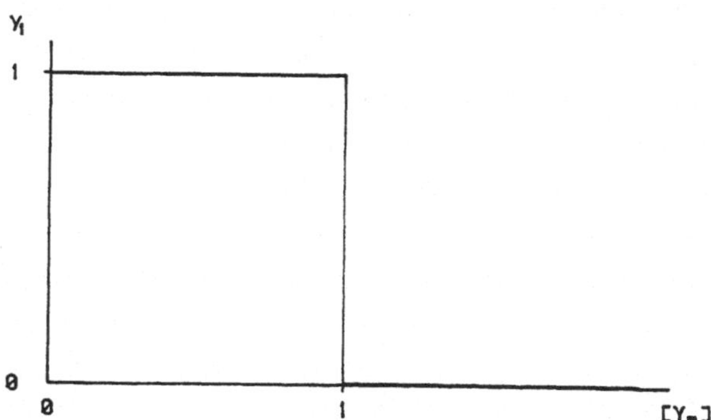

Figure 3. Plotting of y_1 , the boolean function describing the state of expression of the gene involved in thesynthesis of substance Y_1 , as a function of the concentration, $[Y_m]$, of the regulatory substance. The boolean variable describing the presence of substance Y_m is υ_m ; the boolean equation corresponding to this negative regulation is $y_1 = \bar{\upsilon}_m$. The threshold of concentration, mentioned in the text, equals 1.

The curve obtained corresponds to the limit, for n (the index of the cooperativity) $\rightarrow \infty$, of the function $k_1 / (1 + \alpha \cdot [Y_m]^n)$ (figure 2) used in the differential equations (I) for the rate of synthesis of substance Y_1 (°). Therefore, in the boolean model of the negative feedback loop, all regulations are described as infinitely cooperative ; this is evidently not the case in the continuous model. However this explanation cannot account for the fact that the continuous approach predicts also asymptotically stable behaviour for some values of the parameters and that sustained oscillations are excluded for one-element loops. In this context, it would be interesting to consider continuous models in which not only the negative, but also the positive steps (described by the linear terms $k_{-1} \cdot [Y_{i-1}]$ in (I)) involve non-linear (cooperative) interactions, and see for which values of the parameters (number of steps, m , in the loop, cooperativity, n) there are sustained oscillations. This is under study.

In this paper, I shall propose three elements of response to the question arising from the discrepancy between the boolean and the continuous approaches.

The genetic systems taken as models involve the synthesis of macromolecules. This implies <u>absolute time delays</u> of a significant duration. As will be seen

---

(°) The boolean threshold corresponds to $(1/\alpha)^n$.

in section 14.3, even negative loops comprising a single element can be shown by the continuous approach to undergo sustained oscillations, provided the regulatory interaction is slightly cooperative and there is a sufficient time delay.

Another point is that the two approaches apply to different situations. The simple boolean approach gives an appropriate description of an individual independent cell ; it can also be used to simulate the behaviour of a population of non-interacting cells by assuming a stochastic point of view (section 14.4). On the other hand, the continuous approach is usually not suitable for the description of individual cells as some of the regulatory elements are present in very small numbers so that the laws of large numbers cannot be applied. In addition the continuous description implicitly assumes that the cells are fully interacting. The rôle played by intercellular communication is studied in section 14.5 thanks to a stochastic approach in which one allows or not intercellular diffusion.

14.2. Enzymatic vs. genetic regulation.

I pointed out in the preceding section that the negative feedback loop of figure 1 may correspond to either an enzymatic or a genetic regulation. In fact, one is dealing with very different types of ssystems :

a) $\rightarrow Y_i \longrightarrow Y_{i+1} \longrightarrow$ is a chain of reaction such that Y_i is transformed into Y_{i+1} . In such cases, the only properly regulatory step of the loop is the negative control exerted by Y_m on the production of Y_1 . This situation is found in metabolic chains but may be generalized to other chemical reactions.

b) $\rightarrow Y_i \longrightarrow Y_{i+1} \longrightarrow$ means that Y_i is required for the production of Y_{i+1} without being used up in the process. In other words, Y_i has a catalytic effect or exerts a positive control on the synthesis of Y_{i+1} . This is the case when a gene is involved in the synthesis of a messenger RNA (transcription) itself involved in the synthesis of a protein (translation). It is also the case when, Y_1 , Y_2 ,... being the products of gene 1, 2,... respectively, the product of gene i is required for the expression of gene i+1 , that is the synthesis of Y_{i+1} .

We have to be aware that the biological systems taken as models in various works are in fact "hybrids" between both types ; Goodwin writes, for instance : " ... a genetic locus ... produces messenger ribonucleic acid (mRNA) ... . This mRNA then combines with ribosomes to form active protein-synthesizing aggregates (polysomes) ... This protein, assumed to be an enzyme then directs a metabolic transformation giving rise to a metabolic species ... which passes through a cellular pool ... A fraction of the metabolites in the pool feeds back to the genetic locus where it serves to repress the activity of the gene, presumably in association with a macromolecule, the aporepressor". (Goodwin, 1963 ; see also Walter, 1970).

Whether one deals with a situation of type (a) or of type (b) the dynamics of a negative feedback loop is usually described by the same set of differential equations (I) but the meaning of various terms is different. $k_i . \left[Y_{i-1}\right]$, the first term of the second equation of (I) describes the production of Y_i . In the case (a), and in particular in the case of a cascade of chemical reactions, this term means that the rate of transformation of Y_{i-1} into Y_i is proportional to the concentration of Y_{i-1}. In case (b), and in particular in a loop comprising successive genetic regulations, $k_i . \left[Y_{i-1}\right]$ means that one assumes a positive regulation that is non-cooperative ; it is in fact not at all obvious that it should be so. The second term in each of the equations also has a different meaning depending on the type of the system. If it belongs to the "type b", the term simply describes the loss (by degradation, denaturation, ...) of substance Y_i (its form implies that this process follows an exponential kinetics, an assumption which is probably correct in many cases). In the "type a" systems, the second term comprises in fact, in addition to the loss of Y_i by any kind of degradation, as considered previsouly, the loss due to its transformation into Y_{i+1} . If the rate of the first process were negligible, k_{-i} would equal k_{i+1} .

Equations like (I) can be applied to both types of systems. However, if one deals with systems (type b) in which each element plays a catalytic role in the appearance of the next one, these equations represent a rather particular case in which each of the positive regulatory interaction would be non-cooperative.

Relative to the boolean equations (II) proposed in the preceding section to describe a negative loop, one notices that

1° The rate of decay of each substance is not explicitly expressed : it is taken into account by the lag, usually noted $t_{\bar{v}_i}$, between the extinction of the gene i and the disappearance of its product, Y_i.

2° These equations can be used to describe systems involving transcription and translation ("type b" systems). For "type a" systems one may have to use a different set of boolean equations (discussed by Thomas in Chapter VII of this book).

14.3. Time delays.

In the preceding section, two extreme types of systems have been considered in which $Y_{i-1} \longrightarrow Y_i$ meant a transformation of Y_{i-1} into Y_i or a catalytic requirement of Y_{i-1} for the synthesis of Y_i .

These two types of loops are exemplified respectively by metabolic chains, and by "genetic systems" involving the synthesis of macromolecules. Here I want to stress another fundamental difference between these two systems, namely that the synthesis of macromolecules involves absolute delays, contrary to the simple chemical transformation involved in the metabolic chains.

When a signal switches on the expression of a gene, it takes a significant time before the very first molecule of messenger RNA has been built by the sequential condensation of hundreds or thousands of nucleotides and the first molecule of the protein coded by the gene has been built by sequential condensation of hundreds of aminoacids. In addition to this irreducible time it takes time before the protein has reached an efficient concentration at its site of action (accumulation and transport times).

Such irreducible delay cannot be neglected and it must be taken into consideration in any model involving genetic regulation. In the continuous approach, this can be achieved by the introduction of time delays in the differential equations (1) (Landahl - 1969 (2), Mac Donald - 1977, and Richelle - 1977). In this section the use of time delays is exemplified by a model of a simple genetic regulation.

Let us consider the simplest possible negative loop, in which the synthesis of Y is inhibited by Y itself. For this system the set of equations (I) given previously is reduced to :

$$\frac{d[Y]}{dt} = \frac{k_1}{1+\alpha\cdot[Y]^n} - k_{-1}\cdot[Y] \qquad \text{(III)}$$

To take into account the time of synthesis of macromolecule Y, we use the following equation

$$\frac{d\,Y_{(t)}}{dt} = \frac{k_1}{1+\alpha\cdot[Y]^n\,(t-\tau)} - k_{-1}\cdot[Y]_{(t)} \qquad \text{(IV)}$$

($[Y]_{(t)}$ means the concentration of Y at time t).

The first term of (IV) expresses the rate of production of substance Y : it is at the stage of its initiation that the synthesis of new macromolecules Y is (negatively) controlled by the concentration of Y already present at that moment ; thus the rate of production of Y at time t depends on the concentration of Y at time $(t-\tau)$, τ being the lag between the initiation and completion of the macromolecule synthesis. The second term expresses that the rate of degradation of Y at time t is proportional to its concentration at the same time.

Whereas the steady state of equation (III) is, in any case, globally asymptotically stable, equation (IV) yields sustained oscillations for appropriate values of the parameters, provided $n > 1$ (Landáhl - 1969 ; for an analytical proof, see appendix B). Figure 4 gives a numerical integration of equations (III) and (IV) for the same values of the parameters, k_1 , k_{-1} , α and n (see legend).

---

(1) Notice that these delays (in differential equations) do not mean exactly the same as the boolean delays used elsewhere in this book ; the distinction will become clearer in the next section. To avoid confusion, in the rest of this chapter, "transition times" will be used in place of boolean "delays".

(2) Notice that Landahl (1969) introduces time delays to take into account the transport process but, as Glass and Kauffman (1972) have done, this can be achieved by considering spatial diffusion.

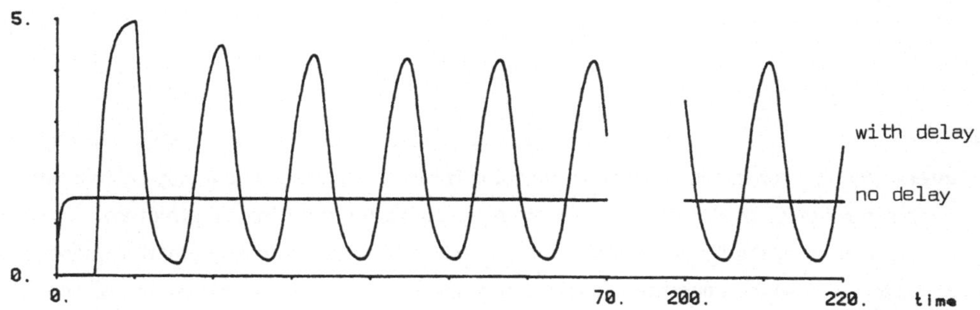

Figure 4 . Numerical integration of equations (III) and (IV) ; plot of $[Y]_{(t)}$; $k_1 = 5$, $k_{-1} = 1$, $\alpha = 1$, $n = 2$. Equation (III) yields an asymptotically stable solution ; equation (IV) yields a sustained oscillation, the delay τ equals 5 units of time.

In summary : 1. In systems such as those involving macromolecular synthesis there are absolute time lags and these lags can and should be introduced into the description by differential equations.

2. When this is done, the only general condition for the existence of sustained oscillations is that n , the index the cooperativity of the negative step, be greater than 1 : provided the regulatory interaction is slightly cooperative, there exists a value of the delay such that sustained oscillations are possible.

3. However, the behaviour of the system, following the continuous model comprising delays, is not restricted to sustained oscillation ; depending on the value of the parameters, it may also reach a stable steady state via globally asymptotic motion or damped oscillation.

4. Even one-element negative loops can display three characteristic behaviours, contrary to the results of the simple continuous approach (without delay) - in which such loops are committed to a stable steady state - and contrary to the results of the simple boolean approach - in which they are committed to a permanent oscillation. This last remark means that time delays are not the unique reason for the discrepancy between the results of the boolean and the continuous approaches.

14.4. Stochastic approach of boolean models.

When one considers a population of living cells which display little interaction (as, for instance, some bacterial cultures) the global system may be considered as a great number of small independent systems involving the same logical structure. Given this logical structure, the "plain" boolean analysis tells which are the possible pathways (sequence of states) and which conditions (initial state and values of the transition times) determine each pathway. As already noticed by Thomas (1973, 1978) if each transition time had a well-defined value all the cells would follow simultaneously the same pathway. However it is reasonable to assume that even in a "homogeneous" cell culture, the value of each transition time is not identical among the individual cells but is subject to random fluctuations arising from the molecular nature of the processes involved. Given the distribution of the value of each transition time, one can thus evaluate the relative frequencies of the various pathways in the cells of the population (see Chapter IX by Van Ham and Chapter XVII, section 3 , by Thomas).

Here follows a more theoretical analysis of the relation between the behaviours of individual cells and that of the population in the case of the negative loops. To achieve this study, one simulates the behaviour of individual cells by applying the classical boolean description and assuming random transition times. For simple cases, the analytical methods of the stochastic theory can give exact solutions describing the behaviour of the population ; for the other cases, this behaviour is obtained by summing a large number of individual simulations (°).

In this section boolean negative loops are examined from a stochastic point of view. In the first two models the transition are assumed to be Poisson processes, i.e. processes for which the chance of a transition from one state to another is not affected by the time elapsed since this transition became possible ; the transition may thus occur as soon as the state of the system allows it. In the last model, the switchon transition involved in a one-element negative loop cannot occur during a lag period after the moment the system enters a state in which this transition is allowed (such transitions are no longer Poisson processes).

(°) Contrary to Van Ham, we have chosen to use different random realization of the transition times each time a given transition has to occur in a cell ; see appendix C for a comparison of these different points of views.

14.4.1. One-element loop with transition at random instant.

Let us first consider the one element negative loop. The set of boolean equations (II) given previously is reduced to :

$$y = \bar{\upsilon} \qquad \text{(V)}$$

If the transition times, t_{υ} for the switch-on transition and $t_{\bar{\upsilon}}$ for the switch-off transition, were constant, all the cells of a population would exhibit the same behaviour (figure 5).

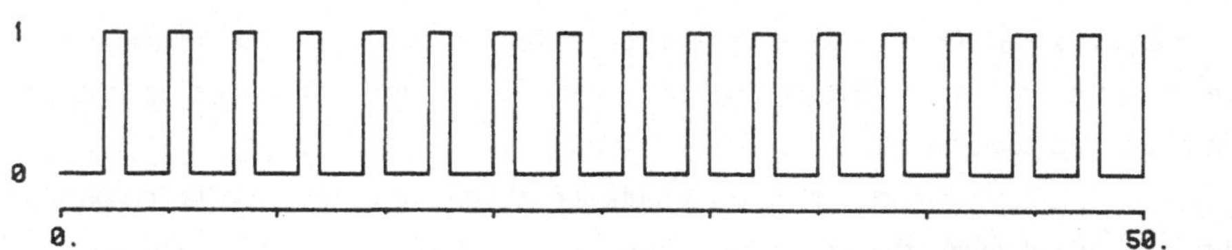

Figure 5. State of the variable in a cell involving a one-element negative loop $y = \bar{\upsilon}$ in which t_{υ} , the switch-on transition time, and $t_{\bar{\upsilon}}$ the switch off transition, are constant and equal 2 and 1 units of time respectively.

As outlined in the beginning of this section, the assumption of constant transition times is quite unrealistic in the case of biological systems. Indeed such systems involve processes which are subject to random fluctuations. One way to take these fluctuations into account is by assuming that the transitions are Poisson processes. Figure 6a shows the probability that the transition $0 \rightarrow 1$, considered as a Poisson process, occurs before time t if the system is initially in state 0 at time 0 (figure 6b, for the transition $1 \rightarrow 0$ and state 1 occupied at $t = 0$).

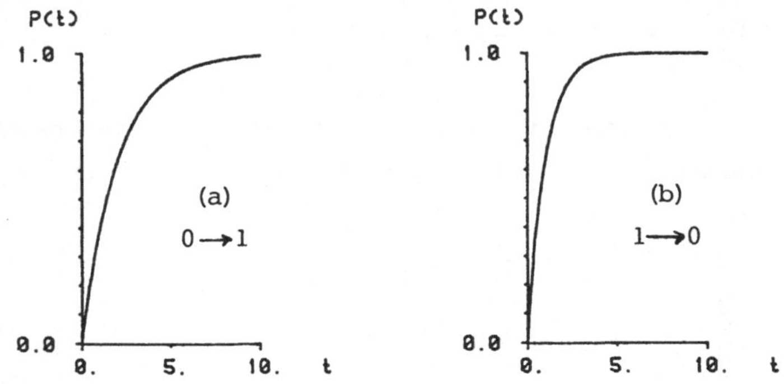

Figure 6. Plot of the probabilities that transitions $0 \rightarrow 1$ (a) and $1 \rightarrow 0$ (b) occur before time t when state 0 (a) and state 1 (b) respectively are occupied at time 0 ; the probability rate of the transition equal (a) $f_0 = 1 / t_{\upsilon} = 1.0$, (b) $f_1 = 1 / t_{\bar{\upsilon}} = 0.5$

Let us denote f_0 the probability rate of transition $0 \rightarrow 1$ (f_1 for $1 \rightarrow 0$) ; such that the probability that the transition occurs in a small interval Δt equals $f_0 \cdot \Delta t$. In the case of a Poisson process, the stochastic theory establishes that the mean transition time, i.e. the average time the system remains in a given state, is equal to the inverse of the probability rate. If one chooses $f_0 = 1/t_{\upsilon}$ and $f_1 = 1/t_{\bar{\upsilon}}$, the average time between entering and leaving state 0 (or 1) is t_{ν} (or $t_{\bar{\nu}}$).

This model exemplifies a cellular system where a metabolite inhibits the enzyme responsible for its synthesis. In this case the variable ν describes the presence of the metabolite in a concentration greater ($\nu = 1$) of lesser ($\nu = 0$) than a threshold ; this threshold corresponds to the concentration of metabolite required to inhibit the enzyme activity. The metabolite can be synthesized as soon as the enzyme is no more inhibited, but, depending on the rate of synthesis, it may take some (random) time before the first molecules are produced again. Notice that, choosing Poisson transition, this boolean model can describe an enzymatic system. One understands that, depending on the system considered, one has to choose different types of distribution of transition of times in order to respect the characteristics of the system.

Figure 7 shows a simulation of the permanent oscillation obtained by this boolean model when the transition times are random. The probability rate of the transitions are such that the mean transition times equal the (constant) transition times of figure 5. The period of this oscillation is now fluctuating.

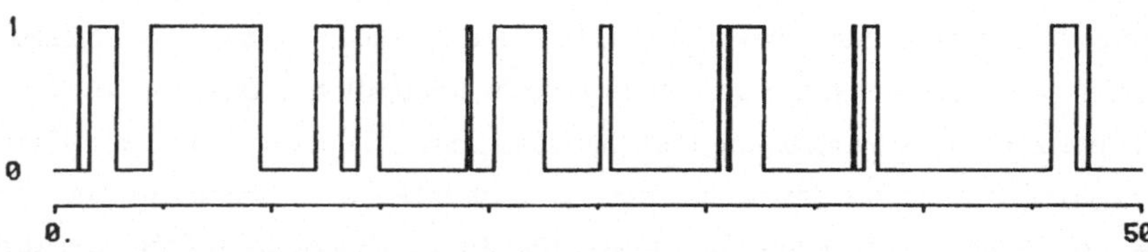

Figure 7. State of the variable ν in a cell involving a one-element negative loop $y = \bar{\nu}$, in which the transition times are randomly distributed ; the probability rate of the switch-on transition is $f_0 = 1/t_{\nu} = 1.0$, the probability rate of the switch-off transition is $f_1 = 1/t_{\bar{\nu}} = 0.5$.

In the stochastic theory this two-state process (see Cox and Miller - p.172 - for a complete analysis) is described by the following differential equations.

$$\begin{cases} \dfrac{d\,p_0}{dt} = -\,f_0 \cdot p_0 + f_1 \cdot p_1 \\[2ex] \dfrac{d\,p_1}{dt} = f_0 \cdot p_0 - f_1 \cdot p_1 \end{cases} \qquad \text{(VI)}$$

the p_i's are time functions and express the probability that a cell is in state i ($p_0+p_1 = 1$) ; they also represent the fraction of a population of cells in state i ; the f_i's, as stated previously, are the probability rates of transition. The first equation of (VI) means that p_0, the fraction of the population in state 0 , is modified by (1) the individuals leaving state 0 at a rate f_0 : their number is proportional to the population in state 0 ; this process gives the term $-f_0.p_0$; (2) the individuals in state 1 which leave it and enter state 0 at a rate f_1 : their number is proportional to the population in state 1 ; this process gives the term $f_1.p_1$. One can describe in the same way the terms of the second equation of (VI).

If initially all the cells of the population are in state 0, the evolution of the fraction of the population in state 1 is given by

$$p_1(t) = \frac{f_0}{f_0 + f_1} \cdot (1-e^{-(f_0+f_1) \cdot t}) \qquad \text{(VII)}$$

As $t \to \infty$, p_1 tends to the state $f_0/(f_0+f_1)$. In no case can damped oscillations occur. Figure 8 shows a plot of equation (VII) with the same values of f_0 and f_1 as in figures 6 and 7 corresponding to mean transition times equal to t_{υ} and $t_{\bar{\upsilon}}$ of figure 5.

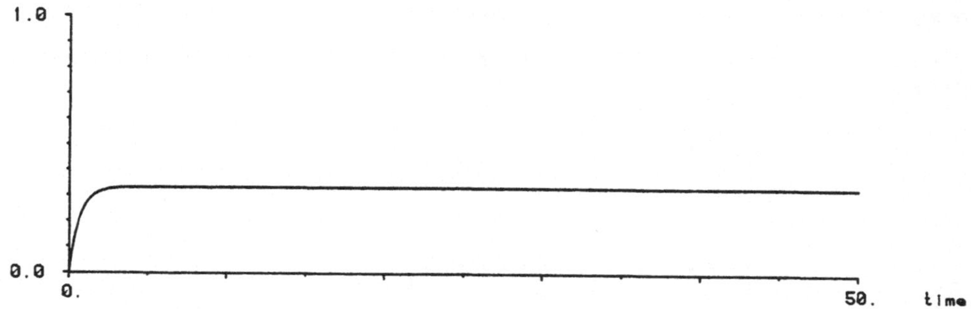

Figure 8. Behaviour of a population of cells, each one involving a one-element negative loop $y = \bar{\upsilon}$. Plot of the equation (VII) giving the fraction of the population where $\upsilon = 1$. cf. legend of figure 7 for the value of f_0 and f_1.

14.4.2.. <u>Two-element loop.</u>

Let us now consider a two-element negative loop for which the boolean equations, derived from (II), are

$$\begin{cases} y_1 = \bar{\upsilon}_2 \\ y_2 = \upsilon_1 \end{cases}$$

The graph obtained from these equations is

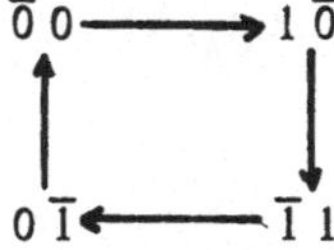

The methods to write out this graph have been described by Thomas (1973 ; Chapter VI of this book). In short, $\bar{1}$ 0 means that $\upsilon_1 = 1$ and $\upsilon_2 = 0$; the $^-$ above 1 means that the value of the function y_2 is different from the value of the associated memorization variable υ_2, thus $y_2 = 1$; the absence of $^-$ above the value of a variable means that both the variable and the function have the same value. For the simplicity of the notations in the following development, the labels 1, 2, 3 and 4 (1) are arbitrarily attributed to the states $\bar{0}\,0$, $1\,\bar{0}$, $\bar{1}\,1$ and $0\,\bar{1}$ respectively ; the graph of the model becomes

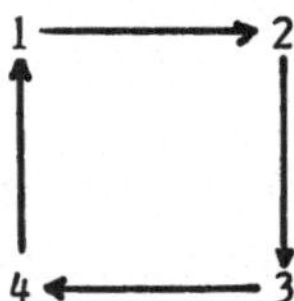

The f_i's are defined as the probability rate of transition from state i to the next one in the graph and the p_i's, as the fractions of the population in state i .

The behaviour of this system may be described by the following set of differential equations :

$$\begin{cases} \dfrac{d\,p_1}{dt} = -\,f_1 \cdot p_1 + f_4 \cdot p_4 \\ \dfrac{d\,p_2}{dt} = -\,f_2 \cdot p_2 + f_1 \cdot p_1 \\ \dfrac{d\,p_3}{dt} = -\,f_3 \cdot p_3 + f_2 \cdot p_2 \\ \dfrac{d\,p_4}{dt} = -\,f_4 \cdot p_4 + f_3 \cdot p_3 \end{cases}$$

These differential equations are linear and the general solution of the system (IX) is of the form

$$p_i(t) = c_{i1} \cdot e^{\lambda_1 t} + c_{i2} \cdot e^{\lambda_2 t} + c_{i3} \cdot e^{\lambda_3 t} + c_{i4} \cdot e^{\lambda_4 t} \quad (i = 1, 2, 3, 4,) \qquad \text{(X)}$$

(the c_{ij}'s are real constants). The λ_j's are the roots of the characteristic equation (XI) (2) of the system (IX)

$$(f_1 + \lambda)(f_2 + \lambda)(f_3 + \lambda)(f_4 + \lambda) - f_1 f_2 f_3 f_4 = 0 \qquad \text{(XI)}$$

that may be rewritten

$$\lambda^4 + (f_1 + f_2 + f_3 + f_4)\lambda^3 + (f_1 = F_2 + f_1 + f_3 + f_1 + f_4 + f_3 f_4)\lambda^2 + (f_1 f_2 f_4 + f_1 f_3 f_4 + f_1 f_3 f_4 + f_2 f_3 f_4)\lambda = 0 \qquad \text{(XI')}$$

---

(1) The octal (or decimal) equivalents of the values of the boolean states would give rise to the succession of states 0 - 1 - 3 - 2 - , that may be confusing because different from the natural order.

(2) This characteristic or secular equation is obtained thanks to the following determinant derived from system (IX).

$$\begin{vmatrix} -f_1 - \lambda & 0 & 0 & f_4 \\ f_1 & -f_2 - \lambda & 0 & 0 \\ 0 & f_2 & -f_3 - \lambda & 0 \\ 0 & 0 & f_3 & -f_4 - \lambda \end{vmatrix}$$

Equation (XI') has always one null root ; we have found by solving numerically this polynome that, in the case where all f_i's are equal, there is always a pair of complex conjugate roots with negative real parts and one negative real root. This means that the system will tend to a stable state (all real parts of the roots are negative) but owing to the complex conjugate roots it will exhibit damped oscillations.

14.4.3. One-element loop with transition times involving irreducible delays.

Let us go back to the one-element negative loop. As seen in section 14.4.1. this model involves two transitions : $0 \to 1$ and $1 \to 0$. It as been considered till now that, once state 0 was reached, the transition $0 \to 1$ had the possibility to occur at any (random) time. Let us now assume that the transition is forbidden during some irreducible delay after state 0 is entered. This delay is introduced to take into account the lag between initiation and completion of macromolecule synthesis, as discussed previously in section 14.3.

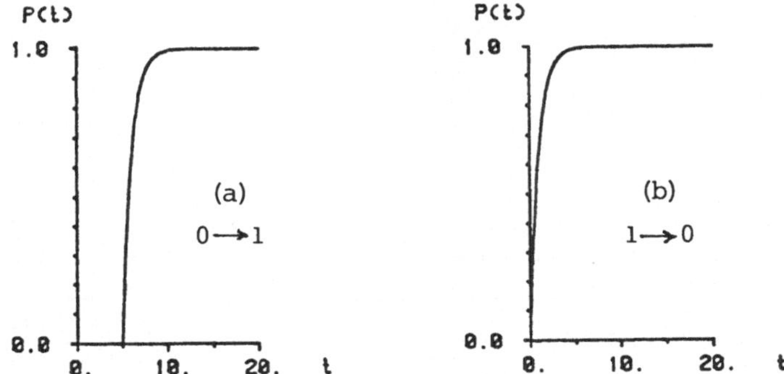

Figure 9. Plot of the probabilities that transition $0 \to 1$ (a) and $1 \to 0$ (b) occur before time t when state 0 (a) and 1 (b) are entered at time 0 ; transition $0 \to 1$ involves an irreducible delay (equal to 5 units of time) during which the transition is forbidden ; the probability of transition are (a) - after the delay - $f_0 = 1.0$, (b) $f_1 = 1.0$.

With the rules of transition of figure 9, one gets individual simulation like that of figure 10.

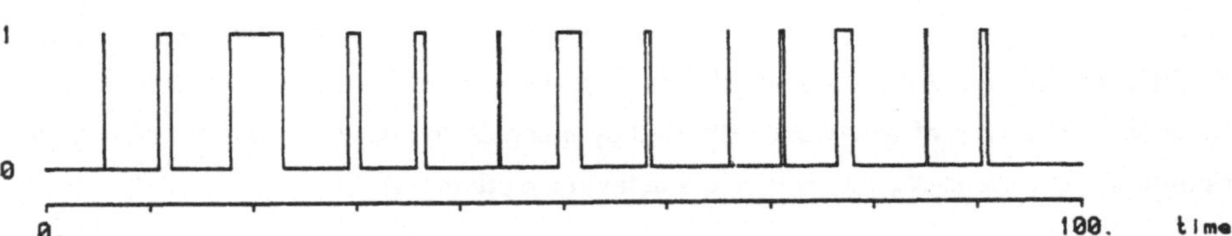

Figure 10. Simulation of a cell involving a one-element loop $y = \bar{\upsilon}$ in which the transition times are distributed as in figure 9 ; (see legend for the values of the probability rates and of the delay). Plot of the state of the variable υ .

The behaviour of a population is simulated by summing the results of a great number of individual simulations. The populations show either a globally asymptotically stable behaviour (figure 11) or a damped oscillation (figure 12) depending upon the length of the irreducible delay (see legend of figures 11 and 12).

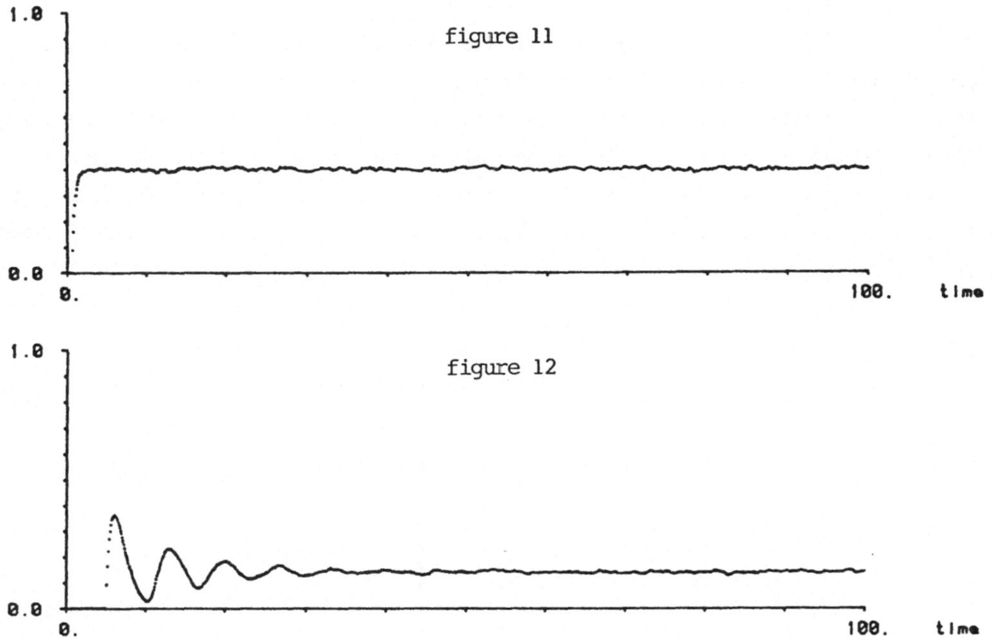

Figures 11 and 12.

Behaviour of a population of cells, each one involving a one-element negative loop $y = \bar{\upsilon}$; the switch-on transition involves an irreducible delay τ = 0.5 (11), τ = 5.0 (12). Plots giving the fractions of the population where $\upsilon = 1$. See legend of figure 10 for the values of the probability rates of transition.

To summarize : 1° The boolean permanent oscillations are meaningful at the level of the individual cell but not at the level of the population.

2° These permanent oscillations are compatible with different behaviours of the population : globally asymptotical behaviours or damped oscillations. The type of behaviour exhibited by a population depends upon the values of the parameters (for instance τ, the irreducible delay, in section 14.4.3).

3° These individual oscillations cannot produce a structured behaviour like sustained oscillation at the level of a population of individuals. This is because the individual boolean oscillators undergo transitions at random moments, and, even if initially in the same state, they will become progressively out of phase since there is no possibility for the individual systems to be synchronized.

4° A population of cells involving a one-element boolean negative loop, whose transitions are simply Poisson processes, will never exhibit any oscillation. However damped oscillations can take place if the model is modified in one of two different ways : either by introducing an irreducible delay in one of the transitions or by introducing one more element into the loop. The situation is exactly the same for the equivalent continuous system (see Appendix B) : the classical one-element negative feedback loop never exhibits oscillation, although one-element negative feedback loop with delay or two-element negative feedback loop may exhibit damped oscillation.

14.5. Stochastic approach of systems with many discrete states.

In the preceding section I pointed out that, in the stochastic theory one does not expect sustained oscillation of a population of random boolean oscillators controlled by a negative loop. However this regulatory mechanism is responsible for sustained oscillation in the continuous analysis. It seemed reasonable to think that this discrepancy could arise from the fact that, contrary to the continuous approach, our stochastic boolean approach deals with a population of non interacting cells. Effectively in the continuous approach one assumes the system as ideal and consisting of a continuous medium ; this would mean, in the case of a population of cells, that the system is composed of the content of all the cells of the population without their walls. If true, this explanation implies that the abolition of the cell walls in the stochastic approach should induce the system to display the sustained oscillation found using the continuous approach. In this section the effect of intercellular communication on the behaviour of a population of cells is examined.

In order to allow a quantitative comparison with the results of the continuous approach, one will consider a stochastic model with many discrete states describing explicitly the number of molecules per cell. Except for this aspect, the system is again a negative loop : a gene product (repressor), by binding to the operator of its coding gene, prevents its own synthesis.

14.5.1. Molecular description of a system of negative feedback.

Let us assume that the system is composed of cells, each one containing one copy of a gene coding for a regulatory macromolecule (repressor). A lag time, τ, is assumed to elapse between the initiation and the completion of the synthesis of each molecule of repressor. Let us call production of a molecule of repressor, the appearance in a cell of a new complete molecule. Let us suppose that the synthesis of repressor is uniquely controlled at the stage of initiation and that the rate of production of molecules of repressor, at time t, equals the rate of the initiation of syntheses at time $(t - \tau)$. It is through the binding of repressors to a genetic site, the operator, that the initiation process is controlled.

One will only consider the simultaneous binding of two molecules of repressor. This process is a rather good model of a situation where repressor dimers, that bind to the operator, are in very rapid equilibrium with repressor monomers (that do not bind to the operator). This binding precludes completely the initiation of synthesis, but not the completion of previously initiated syntheses. In other words, when repressors are bound to the operator, the rate of initiation of synthesis is zero ; when the operator is free of repressor, syntheses are initiated at a constant rate. The binding of repressors to the operator is assumed to be reversible : this reverse process of freeing occurs at a constant rate and results in the simultaneous release of both molecules of repressor. Finally the unbound molecules of repressor are subject to degradation. The rate of degradation per molecule is constant so that the chance that one degradation occurs is proportional to the number of molecules of repressor in a cell. The rate constants of the various processes will be referred to as k_1 for the binding, k_2 for the freeing, k_3 for the initiation and k_4 for the degradation.

14.5.2. The continuous model.

The system is described by the following set of differential equations :

$$\begin{cases} \dfrac{d\,O_{(t)}}{dt} = k_1 \cdot M^2_{(t)} \cdot O_{(t)} + k_2 \cdot (1 - O_{(t)}) \\[2ex] \dfrac{d\,M_{(t)}}{dt} = -2.k_1.M^2_{(t)} \cdot O_{(t)} + 2.k_2.(1 - O_{(t)}) + k_3.O_{(t-\tau)} + k_4 \cdot M_{(t)} \end{cases} \qquad \text{(XII)}$$

(For convenience, the unit of volume used to express the concentrations is the average volume of a cell ; thus the concentration of repressor gene and of operator are 1, since there is on the average one gene and one operator per cell volume) $M_{(t)}$ and $O_{(t)}$ are the concentrations, at time t, of repressor and of free operator respectively ; $O_{(t)}$ gives the fraction of the cells of the population where the operator is free, $(1 - O_{(t)})$, where it is bound ; $M_{(t)}$ gives the mean number of molecules of repressor per cell.

The term $-k_1.M^2_{(t)} \cdot O_{(t)}$ of the first equation of (XII) corresponds to the binding of two molecules of repressor to the operator ; the velocity of this process is proportional to the concentration of free operator and to the square of the concentration of repressor (two molecules of repressor are involved in the process). The term $k_2 \cdot (1 - O_{(t)})$ corresponds to the freeing of the operator and its velocity is proportional to the concentration of bound operator $(1 - O_{(t)})$. As the first equation describes the variation of concentration of free operator $(d\,O_{(t)} / dt)$, the stoichiometric coefficient of these factors equals 1 . In the second equation of (XII), describing the variation of concentration of repressor $(dM_{(t)} / dt)$, one finds these terms with a stoichiometric coefficient equal to 2 since two molecules of repressor are used up in the binding process or liberated in the freeing process. $k_3.O_{(t-\tau)}$, the third term of the second equation means that the concentration of repressor is increased by the production process whose velocity, at time t , is proportionnal to the concentration of free operator at the time $(t-\tau)$ of the initiation of the syntheses that are completed at time t. Finally, $k_4 \cdot M_{(t)}$, the fourth term corresponds to the decrease of the repressor concentration by the degradation process.

14.5.3. The stochastic model.

Due to the complexity of the system considered (which involves time delays), it cannot be analytically resolved and one has to sum the simulations of various individual cells to obtain the behaviour of a population of cells (in Appendix A one can find details on the methods of simulation).

The state of a cell is defined, at time t , (i) by a set of three variables, $\{M, N, O\}_t$: M_t is the number of repressor in the cell, N_t , the number of syntheses in progress (M_t and N_t can only have integer values) and O_t, the state (0 : free, 1 : bound) of the operator ; (ii) by a set of times, $\{t_i\}$ $(i = 1, N_t)$, at which the initiations of the N_t syntheses in progress at time t have occurred ; the t_i's are arranged so that t_1 corresponds to the oldest initiation (it follows that $t_1 + \tau$ is the nearest time at which a synthesis will be completed).

Let us uxamine the various processes modifying the state $\{M, N, O\}$ (°) of a cell.

1° The binding process : binding can only occur when the operator is free (O = 0) and the cell contains more than one molecule of repressor. This process may be represented by $\{M, N, 0\} \longrightarrow \{M - 2, N, 1\}$. This means that at a given moment the free operator (O = 0) becomes bound (O = 1) and that two molecules of repressor are used up ($M \longrightarrow M - 2$). The probability rate of such a transition equals $k_1 . M . (M - 1)$, where $M . (M - 1)$ is the number of different pairs of molecules of repressor that can bind to the operator (one molecule out of M molecules with one out of the (M - 1) remaining molecules).

2° The freeing process : $\{M, N, 1\} \longrightarrow \{M + 2, N, 0\}$

when this process can occur (i.e. when the operator is bound) its rate is not affected by the state of the other variables ; the probability of freeing is constant and equals k_2. By the act of freeing, two molecules of repressor are liberated ($M \longrightarrow M + 2$).

3° The initiation process :

$$\{M, N, 0\}_t \longrightarrow \{M, N + 1, 0\}_t$$

$$\{t_i\}_t \longrightarrow \{t_i, t\}_t \qquad (i = 1, N_t)$$

when the operator is free, this process can occur and results in the synthesis of one more molecule ($N \longrightarrow N + 1$). Therefore one has to add the actual time t to the set of times of initiation occurrence $\{t_i\} \longrightarrow \{t_i\, ,\, t\}$. The probability rate of initiation is constant (when O = 0) and equals k_3.

4° The production process : $\{M, N, O\}_{t=t_1+\tau} \longrightarrow \{M + 1, N - 1, O\}_t$

This process is independent of the state of the operator. The completion of a synthesis in progress ($N \longrightarrow N - 1$) results in the appearance of one more molecule of repressor ($M \longrightarrow M + 1$). This transition occurs at time t as a result of the completion of a synthesis, the initiation of which took place at time $t - \tau$. In other words, the transition occurs if t_1, the first time in the set of initiation times $\{t_i\}$, is such that $t_1 = t - \tau$ (or $t = t_1 + \tau$). The set $\{t_i\}$ must be updated by suppressing t_1, the initiation time of the synthesis just completed :

$$\{t_1\, ,\, t_j\}_t \longrightarrow \{t_j\}_t \equiv \{t_i\}_t \qquad \begin{cases} j = 2, N_t \\ i = 1, (N_t - 1) \end{cases}$$

5° The degradation process $\{M, N, O\} \longrightarrow \{M - 1, N, O\}$

By this process there is one molecule less in the cell ($M \longrightarrow M - 1$). It occurs with a probability rate equal to $k_4 . M$. This is because the chance that one molecule is degraded increases with the number of molecules.

(°) In the following, when no confusion is possible, M (N or O) will be used to mean M_t (N_t or O_t).

The evolution of the state of a cell is simulated thanks to a computer program. Without going into the details of the conception and the organization of the program, one can say that it computes random times of transition occurrences and, in fact, keeps in memory, for each process, the nearest time of transition occurence, among which it searches for the first transition and simulates the transition by modifying the state of the cell according to the rules outlined in the preceding paragraph. The program then updates the transition times of the processes whose probability rates are affected by the modification of the cell state (°) and computes a new time of occurrence relative to the process responsible of the transition just realized. As a set of observations, the program memorizes the state of the cell at regular intervals of time. Large numbers of such sets of observations on different individual simulations are used to simulate a population of cells.

The free diffusion of the molecules of repressor among the different cells of the population is simulated by the random redistribution of all the molecules of repressor among the cells at regular and short intervals of time (the transition times are obviously adapted at each redistribution).

14.5.4. Results

Figures 13, 14 and 15 show, for different values of the rates of binding and freeing, the behaviour of a population of cells simulated by (A) the differential equations (system XII) in the frame of the continuous approach, (B) the stochastic approach and (C) the stochastic approach incorporating the intercellular diffusion. In each figure, plots A, B and C have been combined in D to facilitate the comparison. Figures 13 and 14 present cases where the qualitative behaviours obtained by the three approaches are identical : either globally asymptotical evolution (figure 13) or damped oscillation (figure 14). In addition in both cases the results of the stochastic approach with diffusion (part C of the figures) are quantitatively equivalent to those of the continuous approach (part A of the figures). In figure 15 the qualitative behaviours are no longer all identical : the "simple" stochastic approach (part B) gives a damped oscillation while both the continuous approach (part A) and the stochastic approach incorporating diffusion (part C) give sustained oscillation.

(°) Following the modification of the cell state, some transition times may become longer (eventually infinitely long if the transition is forbidden until another modification) or shorter. For instance, consider a cell containing one molecule, if a degradation occurs, no more degradation will occur before the next production event. On the other hand, if a production event occurs in a cell containing one molecule of repressor the probability of a degradation will double and the corresponding transition time will become shorter.

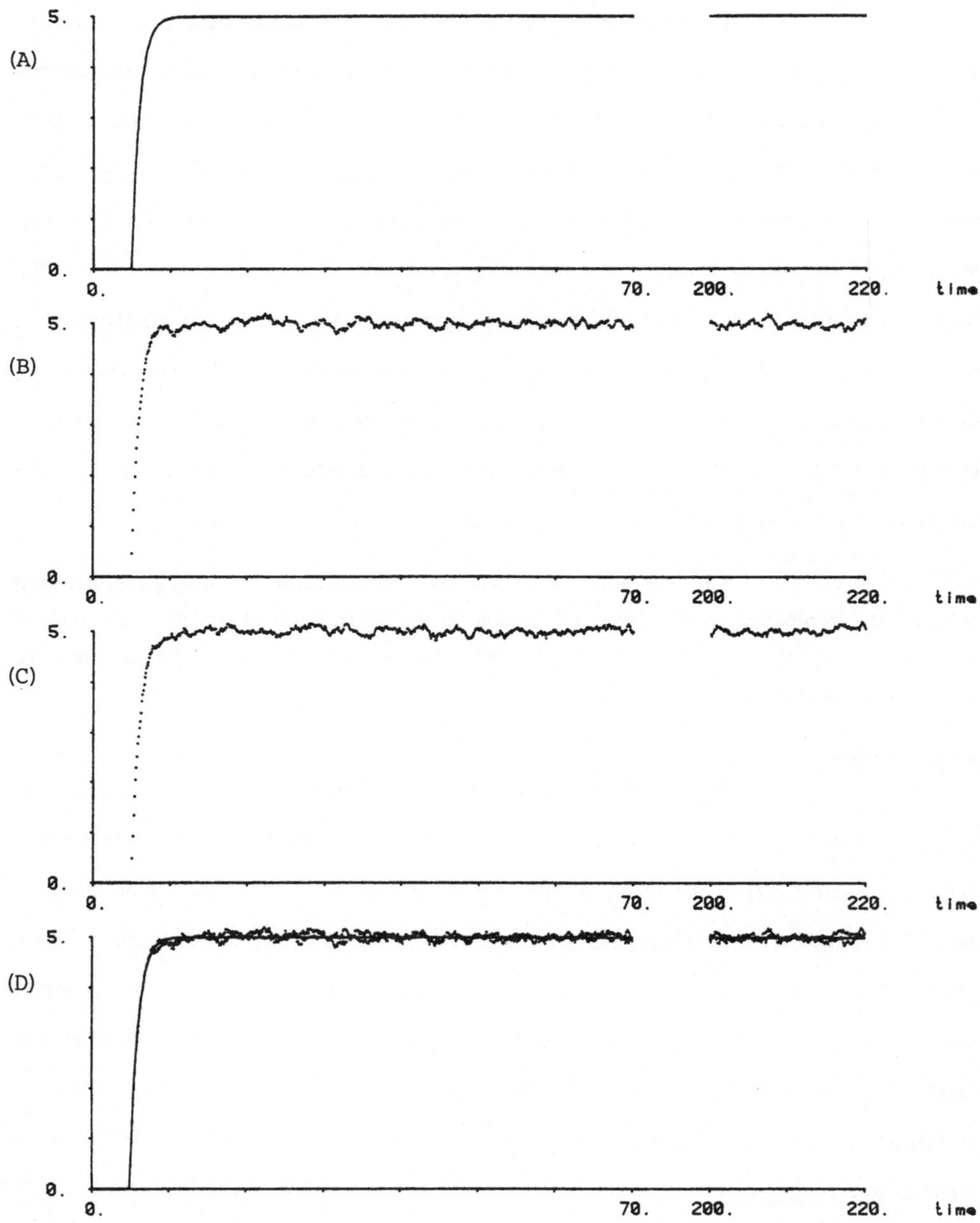

Figure 13. Globally asymptotical evolution, simulation of the mean number of molecules of repressor per cell in a population of cells involving a negative feedback regulation. Simulation obtained (A) by the integration of the system of differential equations. (XII), (B) by the stochastic approach and (C) by the stochastic approach incorporating intercellular diffusion ; (D) superposition of (A), (B) and (C) ; the stochastic simulation have been done over 1000 cells. For (C) the redistribution of the molecules simulating the diffusion is realized 100 times by unit of time.

$k_1 = 0.001$, $k_2 = 10.0$, $k_3 = 5.0$, $k_4 = 1.0$ and $\tau = 5.0$

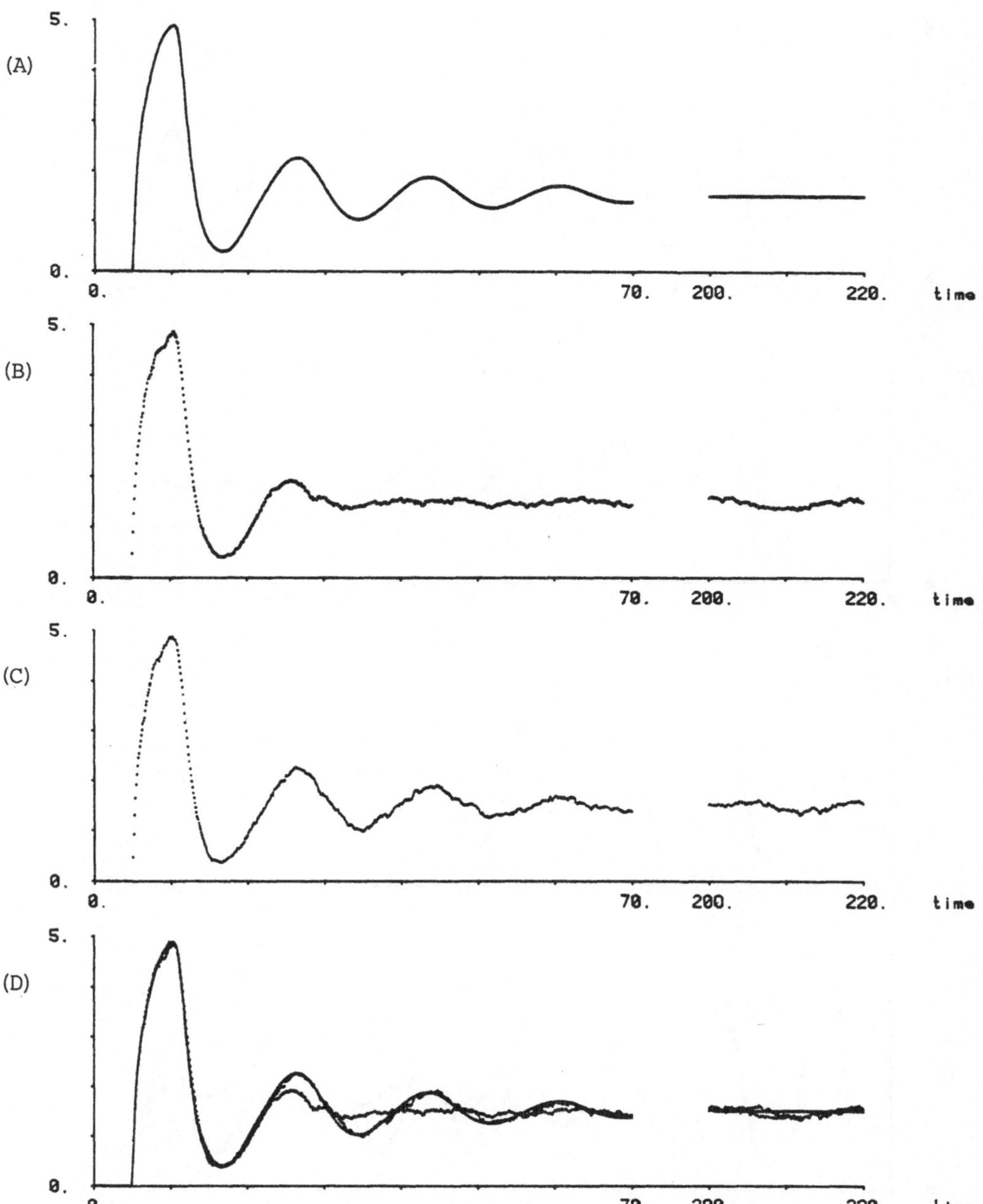

Figure 14. Damped oscillation (see legend of figures 13).

$k_1 = 0.1$, $k_2 = 0.1$, $k_3 = 5.0$, $k_4 = 1.0$ and $\tau = 5.0$

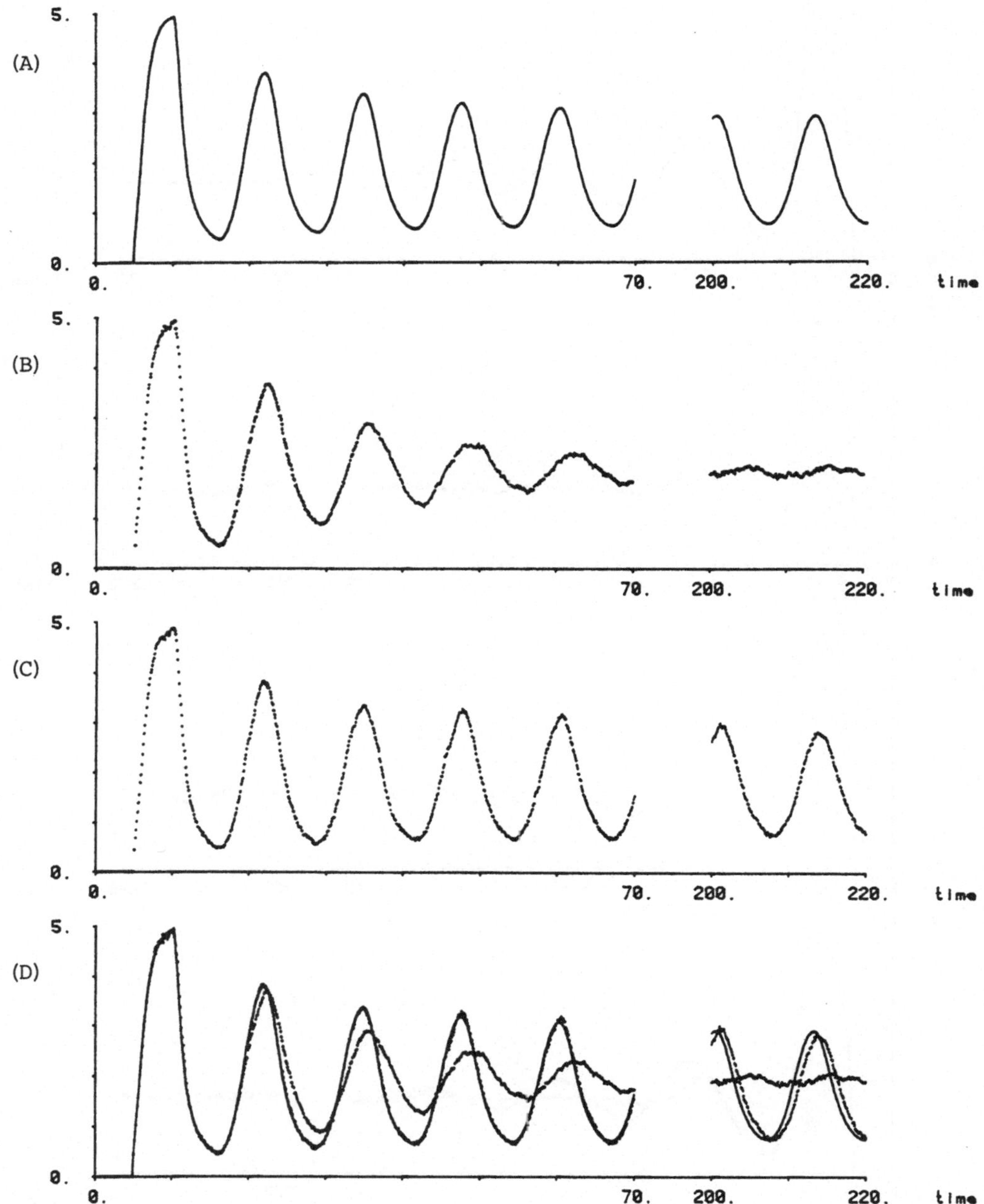

Figure 15. Sustained oscillation when intercellular diffusion exists (see legend figure 13). $k_1 = 5.0$, $k_2 = 5.0$, $k_3 = 5.0$, $k_4 = 1.0$ and $\tau = 5.0$
One notices the slight difference between the periods of the sustained oscillations obtained by either the continuous approach (A) or the stochastic approach incorporating diffusion (C). We think that this discrepancy can be attributed to the rather "rough" way diffusion is simulated i.e. as a punctal process of redistribution of molecules at discrete time intervals although the molecules travel continuously over all the system.

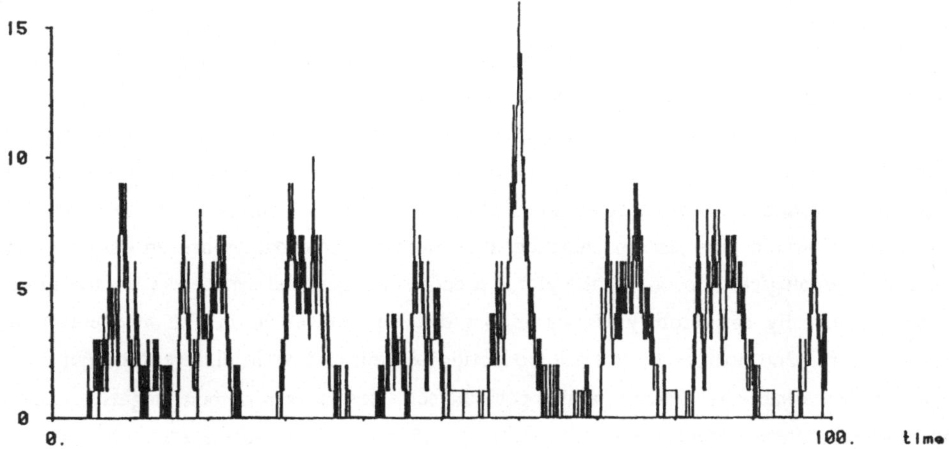

Figure 16. Individual random oscillation of the number of molecule of repressor in a "stochastic" cell obtain the method described in section 14.4.2. $k_1 = 5.0$, $k_2 = 5.0$, $k_3 = 5.0$, $k_4 = 1.0$ and $\tau = 5.0$

Figure 16 shows the simulation, by the "pure" stochastic approach, of the behaviours of a cell : the number of molecules of repressor fluctuates randomly but one notices an alternation of periods of variable length where the number of molecules is either low or high. One does not expect, as previously emphasized (see p.16), that a population of independent individuals undergoing such random oscillations should give rise to sustained oscillation at the level of the population : the individual random oscillations will become progressively out of phase (even if initially all the cells are in the same state).

To summarize one can point out that :

1° As far as individual and independent cells are concerned, the stochastic approach used here gives a description which corresponds well to the description obtained previously (section 14.4) with the boolean approach. The periods where the number of repressor molecules is low in figure 16 can be represented by the value 0 of a boolean variable describing the presence of repressor (see, for instance, figures 5, 7, 10 corresponding to the boolean equation (V)), the periods where the number of repressor molecules is high corresponding to the boolean value 1 ; the systems oscillates between these two types of states.

2° It is remarkable that the stochastic approach <u>when diffusion is simulated</u> gives rise to sustained oscillations, a behaviour excluded in the "pure" stochastic approach of a population of independent cells. Moreover the results obtained are not only qualitatively but also quantitatively equivalent to those of the continuous approach ; this quantitative agreement supports the idea that our two approaches are based upon the same molecular local assumptions.

14.6. Discussion.

Many points have already been discussed in the various sections of this chapter. Here I want to outline the most salient contributions of this study.

If one deals with systems involving syntheses of macromolecules, it is necessary to take into account the time lags between the initiation and the completion of the syntheses. An exact description of all the molecular processes involved in the synthesis of macromolecules would give rise to unmanageable models ; time delays are an easy tool to give a phenomenological account of the global process of sequential synthesis ; furthermore this method is fully satisfactory if one is not directly interested in the mechanism of synthesis itself. One notices that the introduction of time delays in differential equations gives rise to remarkable behaviours : stable periodic solutions for one-element negative loops (Landahl - 1969, Mackey and Glass - 1977, Richelle - 1977 and this chapter), stable periodic or chaotic solutions for one-element loops involving a complex regulation (Mackey and Glass - 1977). For systems involving negative feedback regulation, the classical continuous analysis predicts sustained oscillations in very restricted conditions: there must be at least three elements in the loop and a sufficiently cooperative regulatory interaction. These restrictive conditions nearly vanish provided one introduces a sufficient delay : a somewhat cooperative regulatory interaction is sufficient in order to allow sustained oscillations.

The behaviour of an individual, independent cell endowed with a negative feedback mechanism is characterized by a permanent oscillatory behaviour by the boolean and stochastic approaches ; this behaviour appears as a characteristic property of the fundamental and logical structure of negative loops.

At the global level, following the stochastic approach,a population of independent cells is not expected to exhibit sustained oscillations. Sustained oscillations can only arise if either (1) all the molecular species are present in large amounts (this is assumed in the continuous approach and can be perfectly described by the stochastic approach) but this situation is generally not realized in one bacterial cell, or (2) there exists intercellular communication : this is implicitly assumed in the continuous approach and can be introduced in the stochastic approach by the constant redistribution of the content of the cells of the population.

This study leads to the following conclusions about the applicability of three theoretical approaches to a biological system composed of living cells. The continuous approach is inadequate to describe these systems, because (i) a population of non-interacting cells cannot be assimilated to a continuous medium, (ii) the law of large numbers cannot be applied to systems where one or more molecular species are present in small numbers. On the contrary, the stochastic approach gives a well-adapted description of closed systems even if some of the molecular species considered in the model are present down to one or two exemplars. While the stochastic approach essentially deals with probabilities of occupation of states or mean states of the cells of a population, the boolean approach, for its part, gives an all or none description that appears appropriate at the level of the individual cell. The system realized by a population of bacterial cells is an extreme situation from the point of view of diffusion : bacteria are generally considered as closed and independent microscopic systems. For these kind of systems, the stochastic approach seems the privileged one. Intercellular communication has been shown to occur for some bacteria, for unicellular eucaryotes and in cell tissues. Even if this diffusion is limited to a few types of molecules it may play a critical rôle in the spatial and /or temporal organization of the global system. Moreover, the dynamics of the system may well be affected by the buffer rôle played by the intercellular medium where the diffusible substances are in transit. All these aspects of intercullular communication can be introduced in the stochastic treatment . The stochastic approach appears, therefore,as the most suitable to study a large variety of biological systems.

Appendix A. Random numbers and Monte Carlo simulations.

The Monte Carlo method consists essentially in the utilization of random numbers to simulate a given process. This method is of particular use when an exact (analytical) solution to a problem is not available Here I shall review briefly some simulation techniques and their application to problems of biological interest. For a complete account of Monte Carlo methods see Hammersley and Handscomb (1964); for the probabilistic and stochastic theory, see Feller (1971) and Cox and Miller (1965).

A1. Random number generator.

Various algorithms have been proposed to generate random numbers. One of them (Lehmer, 1948) is :

$$
\begin{aligned}
X_{n+1} &= (a X_n + C)_{\text{mod } m} \ (°) \\
U_{n+1} &= \frac{X_{n+1}}{m}
\end{aligned}
\qquad \text{(A1)}
$$

in which X_n are integers whose maximal value is m ; m, a and c are arbitrary parameters. The series $(U_1, ..., U_n)$, whose values are real and comprised between 0 and 1, is a sequence of numbers that behave exactly as random numbers from any statistical point of view (see Knuth - 1969 - and Devillers et al - 1973 - for extensive statistical analyses of this generator).

To generate a sequence, an initial value, X_0, has to be given. On the computer systems where this generator is implemented, such a value is usually provided by the software at the first call of the generator in the run of a program (each call of the generator furnishes the following number of the sequence). This means that different runs will produce the same sequence of numbers. Nevertheless the software of the computer systems usually allow the choice of the initial value, X_0. In particular, the X_n value at the end of a run may be used as initialising value for a new run. This permits various uses of the generator : repetition of a given sequence, generation of various independent sequences, creation of very long sequences.

The numbers $(U_1, ... U_n)$ are uniformly distributed between 0 and 1 : this means that the probability that a number (U_i) be comprised in any interval included in $[0, 1]$ is equal to the length of this interval and does not depend upon the position of the interval in $[0,1]$, obviously the probability that a number be comprised in $[0,1]$ is 1 . Figure A1 shows the distribution of 100,000 numbers in the 10 intervals of width 0.1 comprised in $[0,1]$: the probabilities relative to each interval are almost equal to 0.1 .

(°) $(Y)_{\text{mod } m}$ means the remainder ofthe integer division of Y by m .

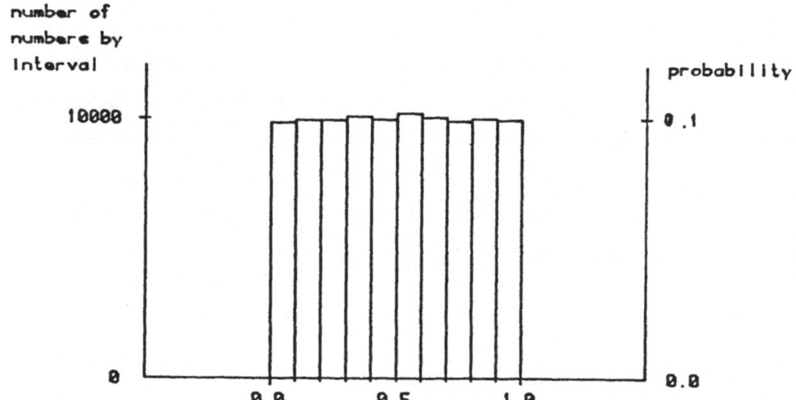

Figure A1. Utilization of a computer routine based on the algorithm of Lehmer : distribution of 100,000 random numbers in the ten intervals of length 0.1 comprised in [0,1].

It follows that the probability that one of these numbers lies between 0 and p ($0 < p < 1$) is p. In other words, the probability that one of these numbers be smaller than p is p. A concrete example is found in table 1, in which the probability is computed for five different values of p.

P	PROBABILITY	NUMBER OF NUMBERS THROWN	NUMBER OF NUMBERS LESS THAN P
.100	.100	100000	10015
.300	.301	100000	30062
.500	.499	100000	49901
.700	.701	100000	70106
.900	.900	100000	89961

Table 1 For various values of p ($0 < p < 1$), one draws 100,000 random numbers between 0 and 1, and count how many are $< p$. It is found that the fraction of numbers $< p$ is equal to p.

A2. Completion of protein chains as a Poisson process.

It seems reasonable to treat the successive acts of completion (or liberation) of new protein chains from a constitutive gene, for instance, as a sequence of random events occurring at constant rate, say ρ. This type of process is known as "Poisson process" (Cox and Miller, 1965). In mathematical terms, one writes that, if $N_{t,\,t+\Delta t}$ is the variable describing the number of events occurring in the time interval $(t, t+\Delta t)$, the conditions for a Poisson process of rate ρ are, for $\Delta t \to 0$,

$$\begin{aligned} \text{prob}\,(N_{t,t+\Delta t} = 0) &= 1 - \rho.\Delta t + 0_1(\Delta t) \\ \text{prob}\,(N_{t,t+\Delta t} = 1) &= \rho.\Delta t + 0_2(\Delta t) \qquad \text{(AII)} \\ \text{prob}\,(N_{t,t+\Delta t} > 1) &= 0_3(\Delta t) \end{aligned}$$

in which $0_1(\Delta t)$,$0_2(\Delta t)$ and $0_3(\Delta t)$ denotes quantities which tends to 0 more rapidly than Δt. Notice that $N_{t,t+\Delta t}$ does not depend upon what happens before time t. One can show (Cox and Miller, 1965) that, in these conditions, the distribution of the time intervals between successive events is exponential. If an event has occurred at time $t = 0$, the probability that the next event occurs in the time interval (0,t) is given by

$$P\,(t) = 1 - e^{-\rho t} \qquad \text{(AIII)}$$

for a Poisson process of rate ρ (figure A2).

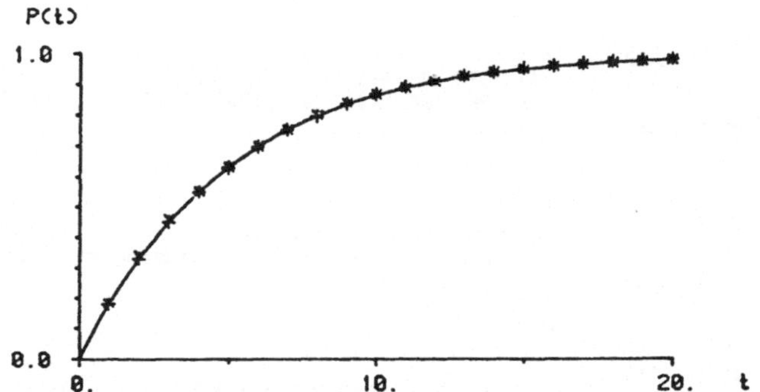

Figure A2. Plot of the probability that the first event (after $t = 0$) occurs in an interval of time of length t for a Poisson process of rate ρ = 0.1

A3. Simulation of a Poisson process : first method
(see Cox and Miller (1965) p. 6-7).

A computer program that simulates the realization of a Poisson process can be easily written. The time scale is divided into intervals Δt. For each interval a random number is generated and one considers that the event takes place if the number is smaller than $\rho . \Delta t$. Note that multiple occurrences of the event within an interval are excluded; the smaller the interval Δt the greater the accuracy of the simulation (the probability of multiple events tends to 0 more rapidly than Δt).

A4. Simulation of a Poisson process : second method
(see Knuth 1969, p. 113, 114).

In the preceding method, if one takes $\rho . \Delta t = 0.001$, about 100 random numbers will have to be drawn to stimulate an event; this may result in large computing times. It is fortunately possible to draw only one random number for each event. The method is based on the fact if the random variable U is uniformly distributed over $[0,1]$, then $(-\ln(1-U)/\rho)$, the inverse function of the probability distribution function (AIII), gives numbers exponentially distributed. This is used to directly obtain the time intervals between succesive events of a Poisson process.

A5. Protein degradation as a death process
(see Cox and Miller 1965, p. 142).

Let us now consider a system in which a protein species is degraded (or inactivated) at a constant rate μ per molecule. This means that a given molecule, present at time t, obeys the following rules

$$\text{prob (the molecule is not degraded in } (t,t+\Delta t)) = 1 - \mu . \Delta t$$

$$\text{prob (the molecule is degraded in } (t,t+\Delta t)) = \mu . \Delta t \qquad \text{(AIV)}$$

If there are k molecules of the protein in the system

$$\text{prob (no molecule is degraded in } (t,t+\Delta t)) = \prod_{i=1}^{k} (1-\mu . \Delta t) = (1-\mu . \Delta t)^k$$

$$= 1 - k.\mu.\Delta t + 0_1 (\Delta t) \qquad \text{(AV)}$$

$$\text{prob (only 1 molecule is degraded in } (t,t+\Delta t)) = \sum_{i=1}^{k} \mu . \Delta t (1-\mu . \Delta t)^{k-1} = k.\mu.\Delta t + 0_2 (\Delta t)$$

This is a Poisson process (section A2), of rate $k.\mu$. It can be simulated by the methods previously mentioned, if one notes that k must decrease one unit each time a molecule is degraded.

A6. Process with variable rates : a special case of the Poisson process.

In the preceding section, it was assumed that the number of molecules of protein was modified uniquely by the process of degradation : between two degradation events the number of molecules was constant. Let us now assume that another process may change the number of molecules independently of the degradation process. (For example, the number of molecules increases one unit if a molecule is synthesized, it decreases two units if two molecules bind to form a dimer). For simplicity one will assume that this process takes place once, at time d (>0) and results in the instantaneous transition of the number of molecules from k_1 to k_2. Before this transition, the degradation is a Poisson process of rate k_1 so that one can write

$$(\text{for } t \leqslant d) \quad P(t) = 1 - e^{-k_1 \cdot \mu \cdot t} \qquad \text{(AVI)}$$

describing the probability that an event of degradation occurs before time t. From (AVI) one deduces that the probability that the first event of degradation occurs after the transition is

$$\text{prob } (1^{st} \text{ event occurrence after time } d) = e^{-k_1 \cdot \mu \cdot d} \qquad \text{(AVII)}$$

After the transition the degradation is a Poisson process of rate k_2 . Hence the probability that the first event of degradation occurs after the time $t > d$ is

$$\text{prob } (1^{st} \text{ event occurrence after } t > d) = \text{prob } (1^{st} \text{ event occurrence after } d).\ \text{prob (no event in } [d,t]\text{)}$$

$$= e^{-k_1 \cdot \mu \cdot d} \cdot e^{-k_2 \cdot \mu \cdot (t-d)} \qquad \text{(AVIII)}$$

$$= e^{-k_1 \cdot \mu \cdot d - k_2 \cdot \mu \cdot t + k_2 \cdot \mu \cdot d}$$

From (AVIII), one determines the probability that the first degradation event occurs before time $t > d$:

$$(\text{for } t > d) \quad P(t) = 1 - e^{-k_1 \cdot \mu \cdot d - k_2 \cdot \mu \cdot t + k_2 \cdot \mu \cdot d} \ (°) \qquad \text{(AIX)}$$

One can simulate this situation with the method described in section A3 : for each interval of time Δt before the transition, the random numbers drawn are compared with $k_1 \mu \Delta t$, after the transition, the random numbers are compared with $k_2 \mu \Delta t$, and one considers that the event takes place when a random number is smaller than $k_i \cdot \mu \Delta t$. But one can also proceed as in the method described in section A4 by using the inverse functions of the probability functions describing the process of degradation :

(°) if $k_1 = k_2 = k$, $P(t) = 1 - e^{-k\mu d - k\mu t + k\mu d} = 1 - e^{-k\mu t}$: if there is no modification of the number of molecules at time d, one gets a probability function describing a Poisson process of constant rate $k\mu$.

inverse function of (AVI) (for $t \leqslant d$) : $$t = -\frac{\ln(1-U)}{k_1\mu} \quad \text{(AX)}$$

inverse function of (AIX) (for $t > d$) : $$t = d - \frac{k_1}{k_2} d - \frac{\ln(1-U)}{k_2\mu} \quad \text{(AXI)}$$

In practice, one replaces, in both equations (1-U) by one random number, R, pertaining to a uniform distribution between 0 and 1. If the time t obtained by (AX) is greater than d (the time of transition), one has to use the time t given by the second equation (AXI).

Figure A3 gives plots of the probability functions (AVI) and (AIX) together with the probabilities calculated by computer simulations with the two preceding methods.

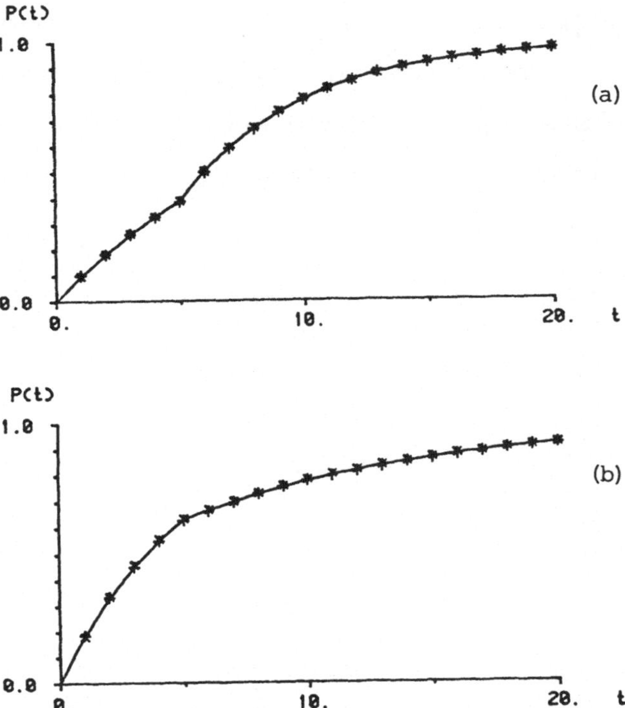

Figure A3. Probability of first degradation before time t ; (a) there are $k_1 = 1$ and $k_2 = 2$ molecules respectively before and after the transition at time $d = 5$, (b) there are $k_1 = 2$ and $k_2 = 1$ molecules respectively before and after the transition at time $d = 5$; the rate of degradation per molecule is $\mu = 0.1$; the line is a plot of equations (AVI) and (AXI), before and after transition respectively ; + : the probability is determined over 10000 simulations with the first method (section A3 and section A5), the time interval Δt equals 0.01 ; x : the probability is determined over 10000 simulations with the second method (section A4, equations (AX) and (AXI)).

A7. Simulation of a random distribution of molecules between the cells of a population.

In a population of cells, the number of molecules of a given species is not identical in each cell of the population : on the contrary, it is randomly distributed. In some situations it is necessary to simulate the state of the cells of a population and hence to determine, for each cell, a random number of molecules. This necessity occurs also when one wants to redistribute continuously among the cells of a population all their molecules (as in section 14.5).

How does one realize a random distribution of m molecules between c cells ? One draws m random numbers (pertaining to an homogeneous distribution in $[0,1]$ and one counts the numbers, N_i ($i=1, ..., c$) of random numbers falling in each of the c intervals dividing exactly the interval $[0,1]$. The set of numbers $\{N_i\}$ provides the distribution expected. Figure A1 gives an example in which the number of molecules, m , equals 100,000 and the number of cells, c , equals 10 .

Appendix B. Steady state and stability analysis of simple negative feedback loops.

In this appendix, the steady state and the stability of three simple models of negative feedback loop are analyzed. The first model describes the simplest negative loop which involves only one element ; the second involves two elements and the third is a one-element loop involving a time delay. These models are the "continuous" equivalent of the boolean models studied in section 14.4.

14.B.1. One-element feedback loop.

For this system, the set of differential equations (I), given in section 14.1., is reduced to

$$\frac{d\,Y_t}{dt} = \frac{k_1}{1 + \alpha \cdot Y_t^{\,n}} - k_{-1} \cdot Y_t \qquad \text{(B1)}$$

(the meaning of the parameters is given in section 14.1).

The steady state solution, $Y°$, is obtained by solving the equation

$$\frac{k_{-1}}{k_1} \cdot Y° = \frac{1}{1 + \alpha \cdot Y°^{n}} \quad , \qquad \text{(B2)}$$

derived from (B1) where one assumes $\frac{dY}{dt} = 0$.

Using the replacement $$y = \frac{k_{-1}}{k_1} \cdot Y° \quad , \qquad \text{(B3)}$$

(B2) is rewritten as $$y = \frac{1}{1 + \alpha \cdot \left(\frac{k_1}{k_{-1}}\right)^n \cdot y^n} \quad . \qquad \text{(B2')}$$

If $$a^n = \left(\frac{k_1}{k_{-1}}\right)^n \cdot \alpha \quad , \qquad \text{(B4)}$$

(B2') becomes now $$y = \frac{1}{1 + a^n \cdot y^n} \qquad \text{(B2'')}$$

or $$a^n \cdot y^{n+1} + y - 1 = 0 \quad . \qquad \text{(B5)}$$

By the Descartes rules, there exists only one real positive solution for (B5).

This solution is easily visualized graphically thanks to (B2"): it corresponds to the intersection of the functions $f_1(y) = y$ and $f_2(y) = \frac{1}{1 + a^n \cdot y^n}$: the circled point in figure B1.

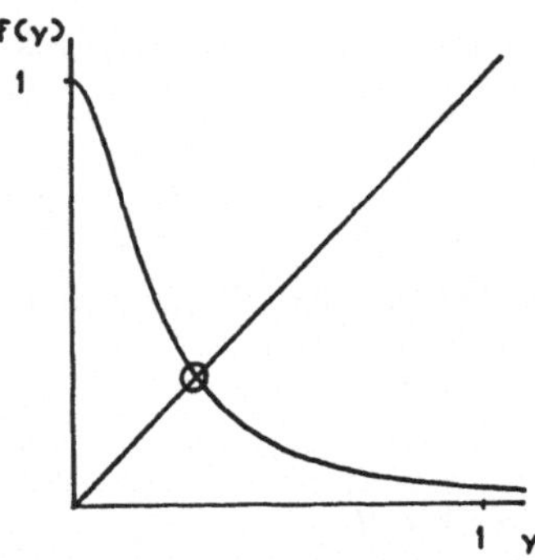

Figure B1. Graphical solution of equation (B2").

A simple analysis indicated that $\quad 0 < y < 1 \quad .$ (B6)

The stability of the steady state, Y°, is analyzed by studying its response to perturbations. If the perturbations are small, (B1) can be linearized and becomes

$$\frac{d\,\upsilon(t)}{dt} = - \frac{n \cdot k_1 \cdot \alpha \cdot Y^{\circ\,(n-1)}}{(1 + \alpha \cdot Y^{\circ n})^2} \cdot \upsilon(t) - k_{-1} \cdot \upsilon(t) \quad , \qquad \text{(B7)}$$

where $\upsilon(t) = Y(t) - Y^\circ << 1$.

From (B2) one gets

$$1 - \frac{k_{-1}}{k_1} \cdot Y^\circ = 1 - \frac{1}{1 + \alpha \cdot Y^{\circ n}} = \frac{\alpha \cdot Y^{\circ n}}{1 + \alpha \cdot Y^{\circ n}} \quad . \qquad \text{(B8)}$$

Using (B2), (B3) and (B8), one can write

$$\frac{k_1 \cdot \alpha \cdot Y^{\circ\,(n-1)}}{\left(1 + \alpha \cdot Y^{\circ n}\right)^2} = \frac{k_1}{Y^\circ} \cdot \frac{\alpha \cdot Y^{\circ n}}{1 + \alpha \cdot Y^{\circ n}} \cdot \frac{1}{1 + \alpha \cdot Y^{\circ n}} = \frac{k_{-1}}{y} \cdot (1 - y) \cdot y = k_{-1} \cdot (1 - y) \qquad \text{(B9)}$$

so that (B7) becomes

$$\frac{d\,\upsilon(t)}{dt} = - k_{-1} \cdot \left(n \cdot (1 - y) + 1\right) \cdot \upsilon(t) \quad . \qquad \text{(B7')}$$

Due to (B6) the real coefficient of $\upsilon(t)$ in (B7') is always negative; one concludes that the unique steady state of (B1) is globally asymptotically stable : the solution of (B1) will never exhibit oscillation, sustained or damped.

14.B.2. Two-element feedback loop.

For a loop involving two elements (m=2), the set of a differential equation (I) (section 14.1) becomes

$$\begin{cases} \dfrac{d\,Y_1}{dt} = \dfrac{k_1}{1 + \alpha \cdot Y_2^n} - k_{-1} \cdot Y_1 \\ \dfrac{d\,Y_2}{dt} = k_2 \cdot Y_1 - k_{-2} \cdot Y_2 \end{cases} \qquad (°)(B10)$$

Assuming $\frac{d\,Y_1}{dt} = \frac{d\,Y_2}{dt} = 0$ in (B10) one can calculate the steady state solutions, Y_1° and Y_2°, of (B10) :

$$\begin{cases} Y_1^{\circ} = \dfrac{k_{-2}}{k_2} \cdot Y_2^{\circ} & \text{(B.11.1)} \\ \dfrac{k_{-1}}{k_1} \cdot Y_1^{\circ} = \dfrac{1}{1 + \alpha \cdot Y_2^{\circ n}} \quad \text{or} \quad \dfrac{k_{-1} \cdot k_{-2}}{k_1 \cdot k_2} \cdot Y_2^{\circ} = \dfrac{1}{1 + \alpha \cdot Y_2^{\circ n}} & \text{(B.11.2)} \end{cases}$$

Using the replacements

$$y = \frac{k_{-1} \cdot k_{-2}}{k_1 \cdot k_2} \cdot Y_2^{\circ} \quad \text{and} \quad a^n = \alpha \cdot \left(\frac{k_1 \cdot k_2}{k_{-1} \cdot k_{-2}}\right)^n \qquad \text{(B12), (B13)}$$

one gets from (B.11.2)

$$y = \frac{1}{1 + \alpha \cdot \left(\dfrac{k_1 \cdot k_2}{k_{-1} \cdot k_{-2}}\right)^n \cdot y^n} = \frac{1}{1 + a^n \cdot y^n}$$

As for the preceding model, (B10) admits only one steady state , that can also be graphically determined as in the preceding section (figure B1).

The stability of (B10) is studied by analysing the stability of the linearized system derived from (B10)

$$\begin{cases} \dfrac{d\,\upsilon_1(t)}{dt} = -k_{-1} \cdot \upsilon_1(t) - \dfrac{n \cdot k_1 \cdot \alpha \cdot Y_2^{\circ\,(n-1)}}{\left(1 + \alpha \cdot Y_2^{\circ n}\right)^2} \cdot \upsilon_2(t) \\ \dfrac{d\,\upsilon_2(t)}{dt} = k_2 \cdot \upsilon_1(t) - k_{-2} \cdot \upsilon_2(t) \end{cases} \qquad \text{(B14)}$$

The stability features of (B14) depend upon the values of the solutions , λ_1 and λ_2 of the characteristic, or secular equation of the system (B14).

$$\begin{vmatrix} -k_{-1} - \lambda & -\dfrac{n \cdot k_1 \cdot k_2 \cdot \alpha \cdot Y_2^{\circ\,(n-1)}}{\left(1 + \alpha \cdot Y_2^{\circ n}\right)^2} \\ k_2 & -k_{-2} - \lambda \end{vmatrix} = 0 \qquad \text{(B15)}$$

(°) Y_1 and Y_2 are functions of the time, but for the sake of clarity Y_1 and Y_2 are used instead of $Y_1(t)$ and $Y_2(t)$.

or $$\lambda^2 + (k_{-1} + k_{-2})\cdot\lambda + k_{-1}\cdot k_{-2} + \frac{n\cdot k_1\cdot k_2\cdot\alpha\cdot Y_2^{o\,(n-1)}}{\left(1 + \alpha\cdot Y_2^{o\,n}\right)^2} = 0 \; . \qquad \text{(B15')}$$

In a similar way as in the preceding section (14.B.1, transformations B8 and B9), one can write (B15') as

$$\lambda^2 + (k_{-1} + k_{-2})\cdot\lambda + k_{-1}\cdot k_{-2} + n\cdot k_{-1}\cdot k_{-2}\cdot(1 - y) = 0 \; . \qquad \text{(B15'')}$$

The solutions of (B15") are

$$\lambda = \frac{1}{2}\cdot\left(- (k_{-1} + k_{-2}) \pm \sqrt{(k_{-1} + k_{-2})^2 - 4\cdot k_{-1}\cdot k_{-2} - 4\cdot n\cdot k_{-1}\cdot k_{-2}\cdot(1 - y)} \right) \qquad \text{(B16)}$$

As the term under the square root is less than the square of $(k_{-1} + k_{-2})$ - the term out of the square root - both solutions will be always negative. The condition to have damped oscillations is that the term under the square root is negative (its square root would be imaginary); after simplification, this condition is

$$(k_{-1} - k_{-2})^2 - 4.n.k_{-1}.k_{-2}.(1 - y) < 0 \qquad \text{(B17)}$$

If one chooses $k_{-1} = k_{-2} = 1$, for instance, (B17) becomes

$$- 4\cdot n\cdot(1 - y) < 0 \; , \qquad \text{(B17')}$$

which is always true. One sees that there exist values of the parameters allowing damped oscillatory behaviours of the two-element feedback loop. Sustained oscillations are not possible since (B16) cannot become purely imaginary, $(k_{-1} + k_{-2}) \neq 0$.

14.B.3. One-element feedback loop with delay.

In section 14.3., the following differential equation is proposed to describe a system of feedback loop comprising a unique element and involving a time delay in the term corresponding to the synthesis of a macromolecule.

$$\frac{d\,Y_t}{dt} = \frac{k_1}{1 + \alpha\cdot Y^n_{(t-\tau)}} - k_{-1}\cdot Y_t \; . \qquad \text{(B18)}$$

With respect to the steady state, the analysis is exactly the same as for the differential equation (B1) of the first section of this annex (14.B.1.).

By linearization one gets from (B18) the following differential equation

$$\frac{d\,\upsilon(t)}{dt} = - \frac{n\cdot k_1\cdot\alpha\cdot Y^{o\,(n-1)}}{\left(1 + \alpha\cdot Y^{o\,n}\right)^2}\cdot\upsilon(t-\tau) - k_{-1}\cdot\upsilon(t) \; . \qquad \text{(B19)}$$

This equation differs only from (B7) by the presence of the time delay τ in $\upsilon(t-\tau)$ of the first term of the right hand member. As in 14.B.1., one can rewrite (B19) as

$$\frac{d\,\upsilon(t)}{dt} = - n\cdot k_{-1}\cdot(1 - y)\cdot\upsilon(t-\tau) - k_{-1}\cdot\upsilon(t) \qquad \text{(B19')}$$

The general solution of (B19') is

$$\upsilon(t) = \upsilon_o\cdot e^{\Omega t} \qquad \text{(with } \Omega \text{ complex and } \upsilon_o \text{ real).} \qquad \text{(B20)}$$

After replacement with (B20) and simplification, (B19') becomes

$$\Omega = - n\cdot k_{-1}\cdot(1 - y)\cdot e^{-\Omega t} - k_{-1} \; . \qquad \text{(B21)}$$

If Ω crosses the imaginary axis when $\tau = \tau_o$ the steady state becomes unstable for $\tau > \tau_o$ and a time periodic solution may arise (Hopf bifurcation). At $\tau = \tau_o$, $\Omega = \pm i\omega_o$

and one gets from (B21) two relations between τ_o and ω_o, corresponding to the real and imaginary parts, that must be simultaneously verified

$$\begin{cases} -\omega_o = -n \cdot k_{-1} \cdot (1-y) \cdot \sin(\omega_o \cdot \tau_o) \\ k_1 = -n \cdot k_{-1} \cdot (1-y) \cdot \cos(\omega_o \cdot \tau_o) \end{cases} \tag{B22}$$

The trigonometric factors are eliminated by summing the square of each member of equations (B22) ; one gets a unique equation

$$\frac{\omega_o^2}{k_{-1}^2 \cdot n^2} = y^2 - 2 \cdot y + \frac{n^2 - 1}{n^2} . \tag{B23}$$

$\frac{\omega_o^2}{k_{-1}^2 \cdot n^2}$ being always positive, the condition necessary for a periodic solution becomes now

$$y^2 - 2 \cdot y + \frac{n^2 - 1}{n^2} > 0 . \tag{B24}$$

The function $f(y) = y^2 - 2y + \frac{n^2 - 1}{n^2}$ has two zeros corresponding to $y = 1 - \frac{1}{n}$ and $y = 1 + \frac{1}{n}$, and a negative minimum between them in y = 1. Thus the condition (B24) is verified for $y < 1 - \frac{1}{n}$ or $y > 1 + \frac{1}{n}$. This together with (B6) finally gives the condition

$$0 < y < 1 - \frac{1}{n} . \tag{B25}$$

The necessary condition on n in order to observe a bifurcation point at τ_o is $n > 1$ (B26) independently of the values of the other parameters.

On the other hand, (B5) may be rewritten as $1 = y + a^n \cdot y^{n+1}$ (B27)

If we denote y_o the upper limit of y, one has

$$1 = y + a^n \cdot y^{n+1} < y_o + a_o^n \cdot y_o^{n+1} \tag{B28}$$

Replacing y_o by its value, $1 - \frac{1}{n}$, one gets after a slight transformation, the condition on a which allows periodic solution

$$a^n > \frac{n^n}{(n-1)^{n+1}} . \tag{B29}$$

For example, if $n = 2$, $a = \sqrt{\alpha} \cdot \frac{k_1}{k_{-1}} > 2$.

The minimum critical delay τ_o (°) necessary to destabilize the steady state is calculated from either equations (B22)

$$\tau_o = \frac{\text{arc cos} \{-1/(n \cdot (1-y))\}}{\omega_o} \tag{B30}$$

where ω_o is obtained by the resolution of equation (B23).

14.B.4. Conclusion.

A simple one-element loop (without delay) can only exhibit a stable steady state. Oscillatory behaviours can appear if either (at least) one more element or a time delay is introduced in the loop. One notices that sustained oscillations are only allowed either if a time delay is present or if the loop comprises at least three elements (Othmer, 1976).

(°) In fact, there is a succession of critical values $\tau_i = \tau_o + \frac{2i\pi}{\omega_o}$, but by simple numerical analysis no additional instabilities have been noticed for delays greater than these successive values.

Appendix C. About the fluctuations of the transition times in boolean systems.

As mentioned by Thomas (1973, 1978) there is no reason to assume that each of the transition times involved in a biological boolean model has a rigourously constant value ; rather their length should be somehow distributed around an average value.

This is taken into account in a computer program developed by Van Ham & Dehouck (this book, Chapter IX). The purpose of this program is to simulate the behaviour of a population of living cells, each treated as an independent boolean system. Each simulation refers to an individual cell. Van Ham & Dehouck assume that if a transition time is used repeatedly, its successive values within one cell will remain essentially the same, and that the fluctuations take place essentially from cell to cell. Consequently, for each simulation (representing an individual cell) each transition time is ascribed a fixed value, and only from one simulation to the other does this value change.

We feel rather that the fluctuations in time within a given cell and the fluctuations between cells proceed from the same origin and should have the same order of magnitude. The aim of this appendix is to compare the predictions of these two points of view. In practice, as shown in figure C1, the first approach individualizes each cell by conferring upon it a distinct set of values for its delays, while the second approach does not allow individual cells to be distinguished from each other according to that criterion.

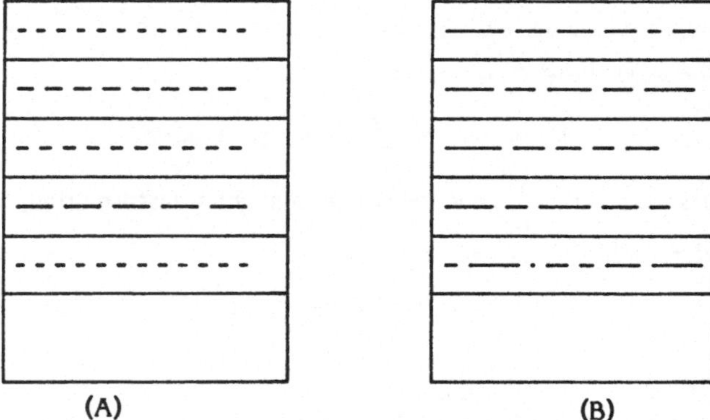

Figure C1. Successive realizations of the transition time of a given transition in a population of individuals following the two approaches described in the text ; each row of black dashes corresponds to a different individual, the dashes in the row represent the successive realizations of the transition time. (A) individualized cells,(B) indistinguishable cells.

As an example, let us consider the following two-variable system

$$\begin{cases} r_1 = \bar{\rho}_2 \\ r_2 = \bar{\rho}_1 \cdot \bar{\rho}_2 \end{cases}$$

whose graph of states is

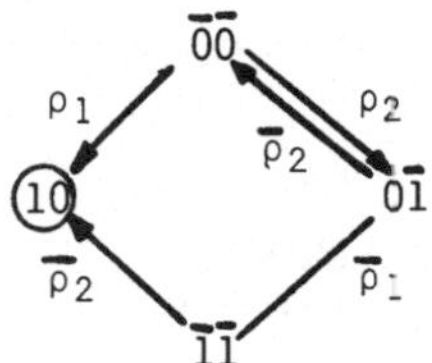

Such a logical structure offers (starting from $\bar{0}\,\bar{0}$) a choice between a stable state $1\,0$ (circled) and an oscillation between $\bar{0}\,\bar{0}$ and $0\,\bar{1}$.

Figures C3 to C10 present the behaviour of this system simulated in various conditions according to both points of view. In the first set (part A of figures C3-C10) a value is ascribed to each transition time at the start of each simulation, whereas in the second set (Part B of figures C3 - C10) a new value is drawn for each occurrence of the transition.

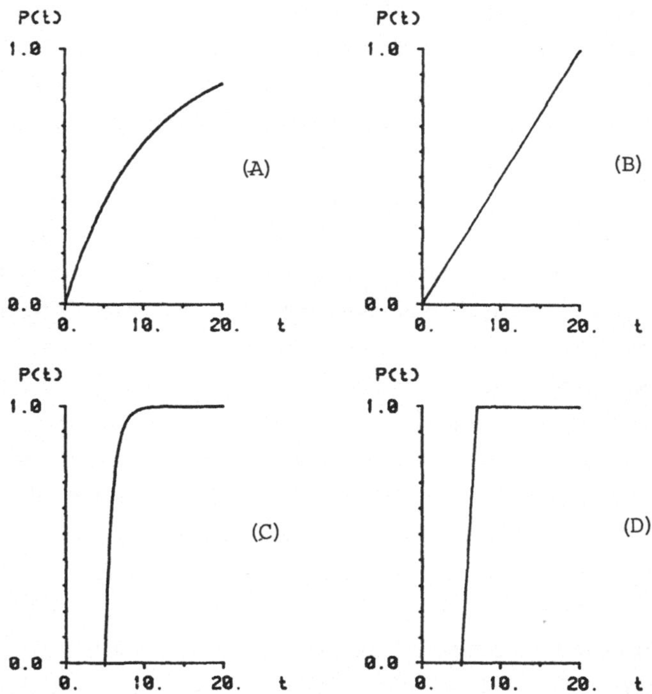

Figure C2. Various distributions of probability that a given transition occurs before time t. The time is measured relative to the moment the state of the system allow the transition.
(A) : the transition is a Poisson process of rate 0.1 (unit of time)$^{-1}$ or (u.t.)$^{-1}$; (B) : the transition occurs at random in the interval of time (0,20) ; in (C) and (D) the transition is forbidden in the initial interval of time (0,5), afterwards, in (C), it is a Poisson process of rate 1.0 (u.t.)$^{-1}$ and, in (D), it occurs at random in the interval of time (5,7).

Both conceptions have been examined with four different types of distributions of transition times. In the first case one assumes that the transitions are Poisson processes (see sections 14.4.1. and 14.A2.) (distribution : figure C2A, simulations : C3 and C4). In the second case, the transition occurs at random in an interval of time starting as soon as the state of the system allows the transition (distribution : figure C2B, simulations :

figures C5 and C6). For the third case, one uses a distribution corresponding to a delayed Poisson processes (see section 14.4.3.) (distribution : figure C2 C, simulations : figures C7 and C8). Finally in the fourth case the transition occurs at random, but within a specific time starting after the state of the system allows the transition. (distribution : figure C2 D, simulations : figures C9 and C10).

In figures C3, C5, C7 and C9, the average value of the transition time t_{ρ_1} (of transition $\bar{0}\bar{0} \longrightarrow$ ⑩) is appreciably greater than the average value of the transition time t_{ρ_2} (of transition $\bar{0}\bar{0} \longrightarrow 0\bar{1}$) . One sees that both approaches (exemplified respectively in part A or in part B of the figures) predict that almost all the individual cells will undergo the permanent oscillation $\bar{0}\bar{0} \rightleftarrows 0\bar{1}$: the mean state of variable ρ_1 is 0 and the variable ρ_2 tends to a mean state equal to 0.5 corresponding to the half of the population having $\rho_2 = 0$ and the other half having $\rho_1 = 1$. In view of the randomness of the values of the transition times the cells are more or less rapidly out of phase depending upon the type of distribution of the transition times.

In figures C4, C6, C8 and C10, the average values of the transition times, t_{ρ_1}, and t_{ρ_2} , are equal. According to the first approach (part A of the figures), roughly half of the population proceeds to state ⑩ , while the other half indefinitely oscillates between $\bar{0}\bar{0}$ and $0\bar{1}$: the mean state of variable ρ_1 is 0.5. Only half of the population may have $\rho_2 = 1$, it corresponds to the individuals having a value of t_{ρ_2} shorter than the value of t_{ρ_1} , thereby (t_{ρ_2} having the same distribution in each case) these individuals will also stay less time in state $\bar{0}\bar{0}$ than in state $0\bar{1}$; this is why the mean state of ρ_2 tends to ~ 0.3 . Following the second approach, the two possibilities ($\bar{0}\bar{0} \longrightarrow$ ⑩ and $\bar{0}\bar{0} \longrightarrow 0\bar{1}$) will occur, and for the cells which choose ($\bar{0}\bar{0} \longrightarrow 0\bar{1}$) this choice will be repeated : they maintain the capacity to enter ⑩ so that progressively the whole population will be in the stable state ⑩ . This situation is shown in part B of the figure C4, C6, C8 and C10.

Thus, the two approaches (one assuming that the values of the transition times fluctuate from cell to cell but not with time within a cell, the other assuming fluctuations both from cell to cell and with time) may lead to different results (figures C4, C6, C8 and C10). As the absence of fluctuations with time within a cell is very unikely in the case of bacterial systems, this point if view might lead to erroneous results and a very careful examination of the system under study is indispensable before choosing one or the other point of view.

General legend of figures C3 to C10.

Simulation of populations of 10000 cells.
In ordinate : mean state of the variables ρ_1 and ρ_2 (indicated by the arrows on the right of the curves). Part A of the figures : the values of the transition times fluctuate from cell to cell <u>but not</u> with time ; part B of the figures : the values of the transition times fluctuate from cell to cell <u>and</u> with time.

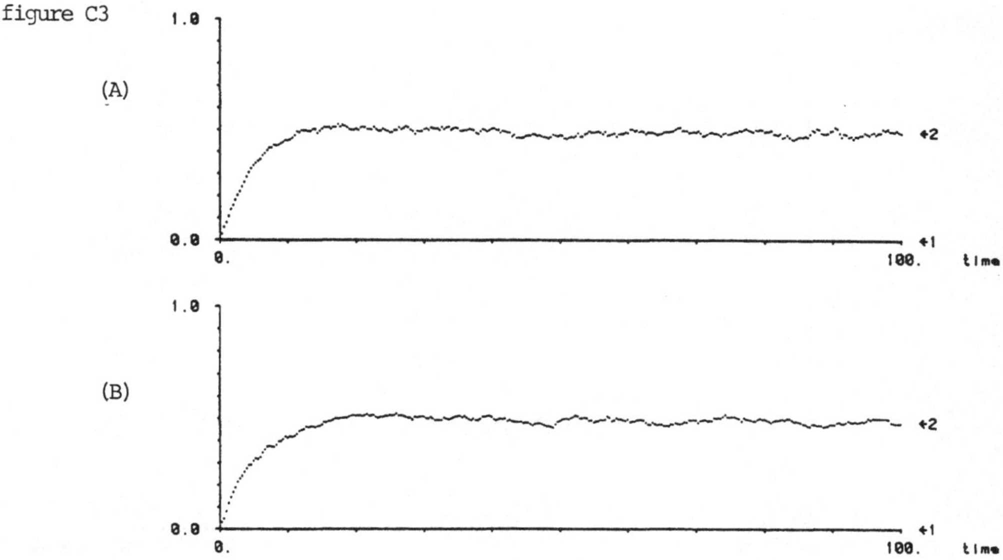

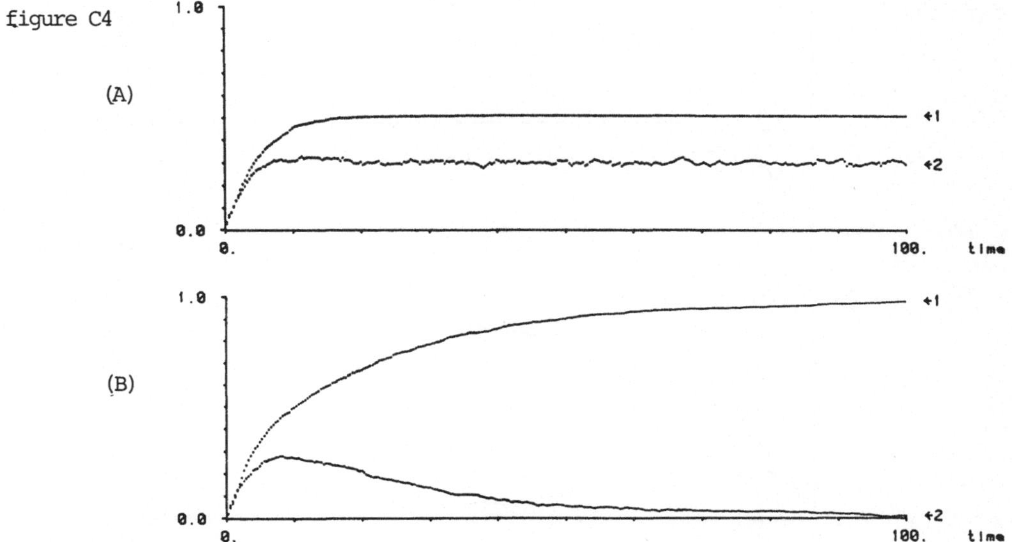

Figures C3 and C4.

Use of the exponential distribution of figure C2A.

Rate of transitions :

C3 ρ_1 : $10^{-10}(\text{u.t.})^{-1}$, ρ_2 and $\bar{\rho}_2$: $0.1\ (\text{u.t.})^{-1}$

C4 ρ_1, ρ_2 and $\bar{\rho}_2$: $0.1\ (\text{u.t.})^{-1}$

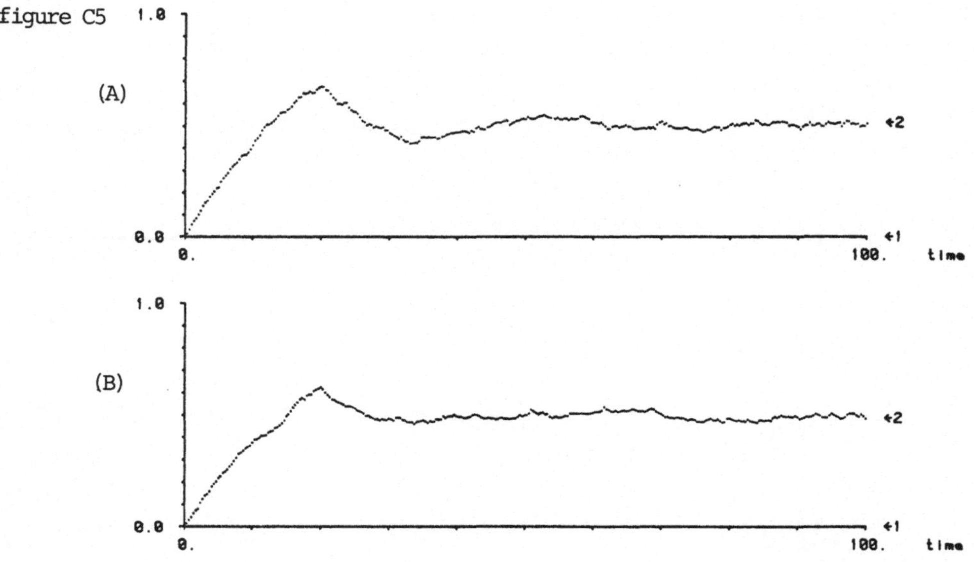

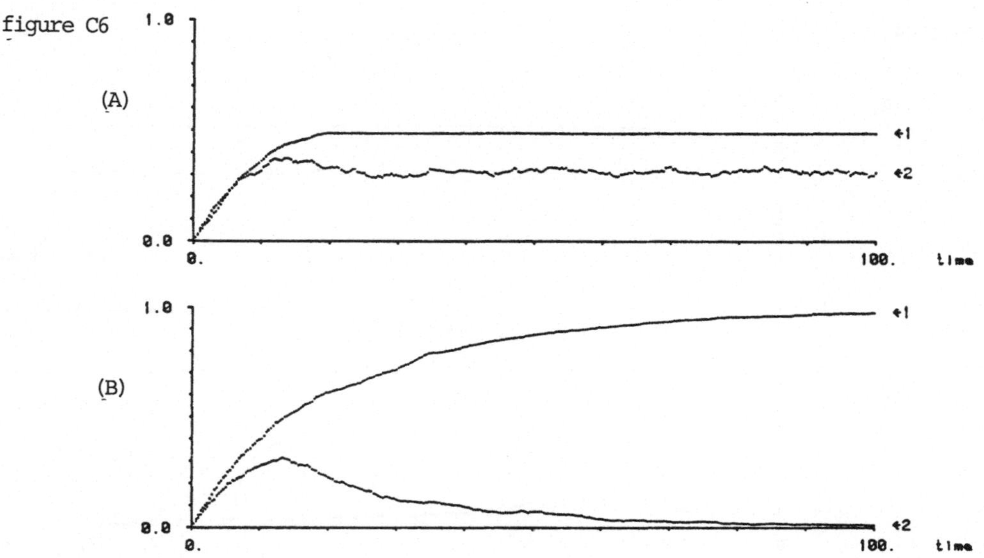

Figures C5 and C6.

Use of the uniform distribution of figure C2B.

Interval of time in which the transitions occur :

C5 ρ_1 : $(0,10^{10})$, ρ_2 and $\bar{\rho}_2$: $(0,20)$

C6 ρ_1, ρ_2 and $\bar{\rho}_2$: $(0,20)$

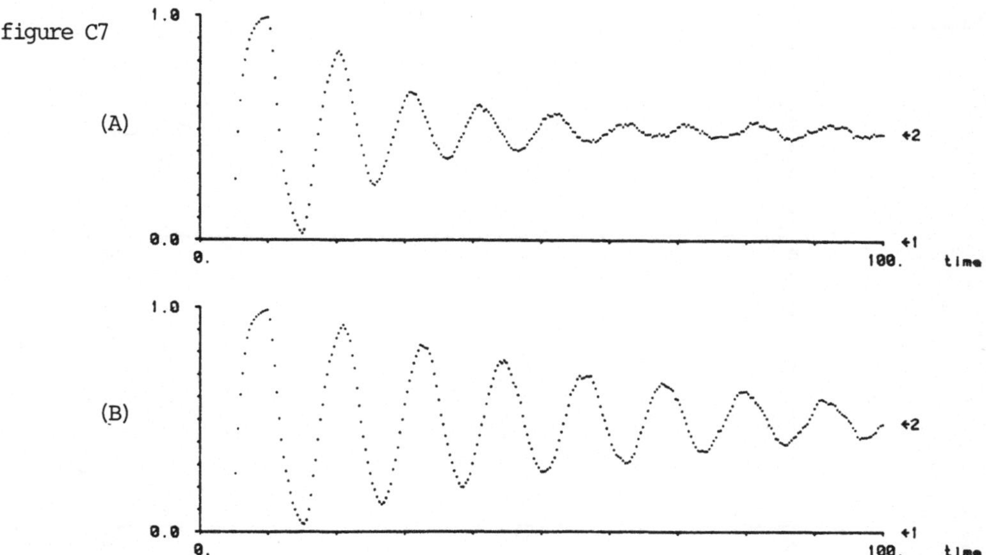

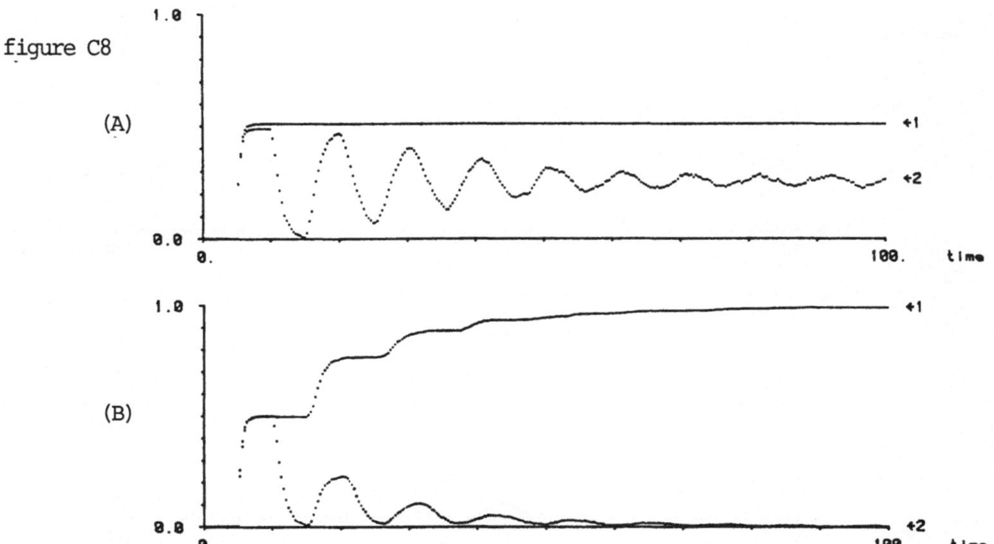

Figures C7 and C8.

Use of the delayed exponential distribution of figure C2 C.

Rate of the transitions after the delay of 5 units of time :

C7 ρ_1 : $10^{-10}(\text{u.t.})^{-1}$, ρ_2 and $\bar{\rho}_2$: $1.0\ (\text{u.t.})^{-1}$

C8 ρ_1, ρ_2 and $\bar{\rho}_2$: $1.0\ (\text{u.t.})^{-1}$

figure C9

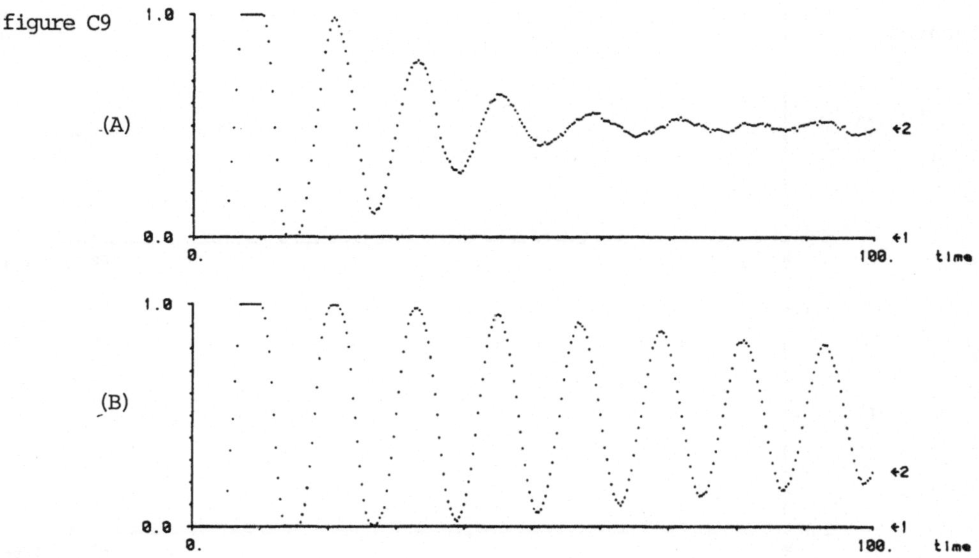

figure C10

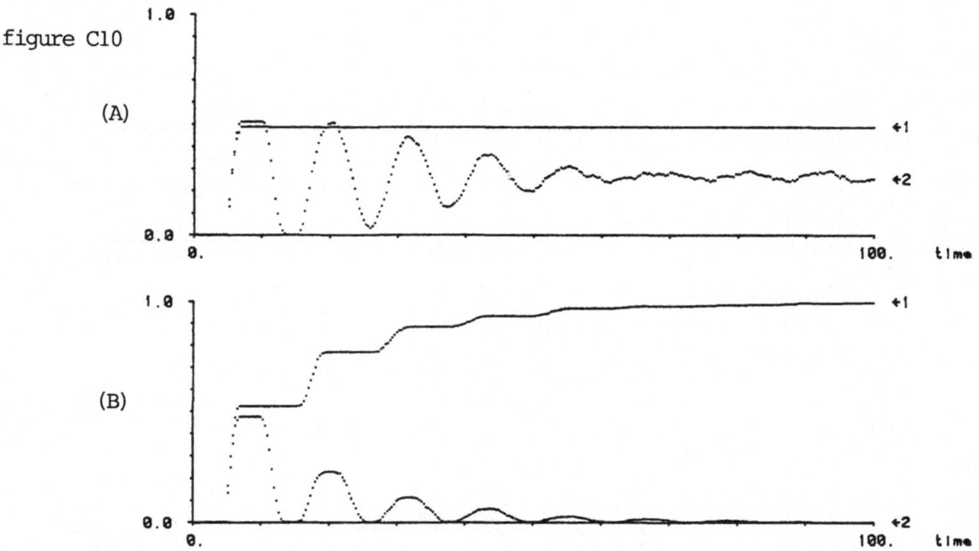

Figures C9 and C10.

Use of the delayed uniform distribution of figure C2 D.

Interval of time in which the transitions occur :

C9 ρ_1 : $(5,10^{10})$, ρ_2 and $\bar{\rho}_2$: (5,7)

C10 ρ_1, ρ_2 and $\bar{\rho}_2$: (5,7)

References.

Cox, D.R. and Miller, D.O. (1965). The theory of stochastic processes. Lond. Methuen.

Devillers, R., Dumont, J.J. and Latouche, G. (1973) Bull. Class. des Sciences, 5ème série, Tome LIX, 703-724.

Feller, W. (1971). An introduction to probability theory and its application. 3rd ed. Vol. 1. N.Y. Wiley.

Glass, L. and Kauffman, S.A. (1972) J. Theor. Biol., 34, 219-237.

Goodwin,B.C. (1963) . Temporal organization in cells. Lond. and N.Y. Academic Press.

Goodwin, B.C. (1965) . Adv. Enzyme Regulation, 3, 425-438.

Hammersley,J.M. and Handscomb, D.C. (1964) Monte Carlo methods. Lond. Methuen.

Knuth, D. (1969). The art of computer programming. Vol.2. Addison. Wesley.

Landahl, H.D. (1969). Bull. Math. Biophysics, 31, 775-787.

Lehmer, D.M. (1948) Proc. 2nd. Symp. on Large-scale digital computing machinery. Cambridge, Harvard University Press.

Mac Donald, N. (1977) J. Theor. Biol. 67, 549-556.

Mackey, M.C. and Glass, L. (1977) Science 197, 287-289.

Othmer, H.G. (1976) J. Math. Biol. 3, 53-78.

Richelle, J. (1977). Bull. Classe des Sciences, 5ème série, Tome LXIII, 534-546.

Thomas, R. (1973)., J. Theor. Biol. 42, 565-583.

Thomas., R. (1978) J. Theor Biol. 73, 631-656.

Walter, C.F. (1970) J. Theor. Biol. 27, 259-272.

CHAPTER XV

How to deal with variables with more than two levels

P. VAN HAM [+]

1. Introduction

The general philosophy of this book is that the qualitative behaviour of a system can often be treated in terms of two-valued variables. Obviously, this is not always so ; a variable can have a number of values which sometimes do not correspond to a quantitative scale (for instance, the variable "colour" can have the values "red", "yellow", "blue", "green", a.s.o.), or which can be ordered along a quantitative scale.
In the latter case, one might use a continuous treatment for the variable(s) which seem to require it, or approximate this continuous treatment by dividing the range of this variable into equal intervals. We propose here a radically different approach, which consists of ascribing to variables additional levels only to the extent that these levels have a physical (or biological) signifiance.

To achieve this goal we see at least three possibilities :

a) effectively using a n-ary logic, also called multivalued logic.
b) take advantage of the fact that associating p binary variables to a given quantity endows it with 2^p levels. This allows one the benefit of all the facilities of binary technology and it is economical from the viewpoint of the number of variables. If one follows this attitude, a level, say 9, whose binary equivalent is 1001, can be coded with this configuration of four variables a, b, c, d . Unfortunately, the individual values of these variables have no physical meaning by themselves.
c) for this reason, I choose another attitude, which may seem less economical from the viewpoint of the number of binary variables used. Instead of coding for 2^p levels, p binary variables are used here to specify $p+1$ levels only. However, as we will see, the values of these variables have a concrete meaning.

2. Continuity conditions

Suppose we have physical reasons to focus our attention on several intermediary levels of a variable between the levels labelled 0 and 1 . These intermediary levels are not necessarily equidistant.

+ Service des Systèmes Logiques et Numériques. Université libre de Bruxelles.

It is NOT a regular quantization of the interval between 0 and 1 The purpose is not to make more precise the numerical value of a variable, but to add some values of the variable which are physically significant with respect to the model (Figure 1).

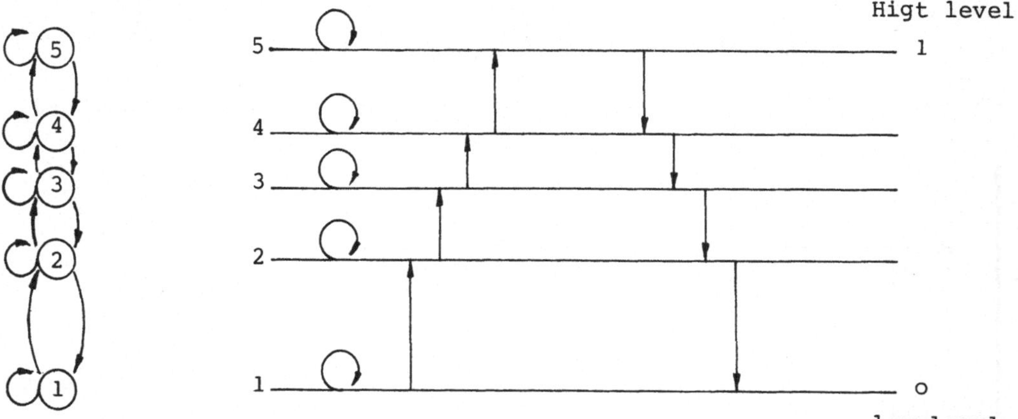

Figure 1.

In most cases a very simple condition must be satisfied : the variable cannot jump over a level but must pass through each intermediary level. It is a continuity condition in its simplest form : the only possible transitions are performed between adjacent levels. It is not trivial to impose such a continuity condition, because the levels do not necessarily have a numerical meaning and the general case of qualitative levels does not exclude jumps (e.g. it is not necessary to go from "salted" to "sugared" by passing through an intermediary "bitter" level).

A variable v with p + 1 levels will be represented as shown in Figure 2.

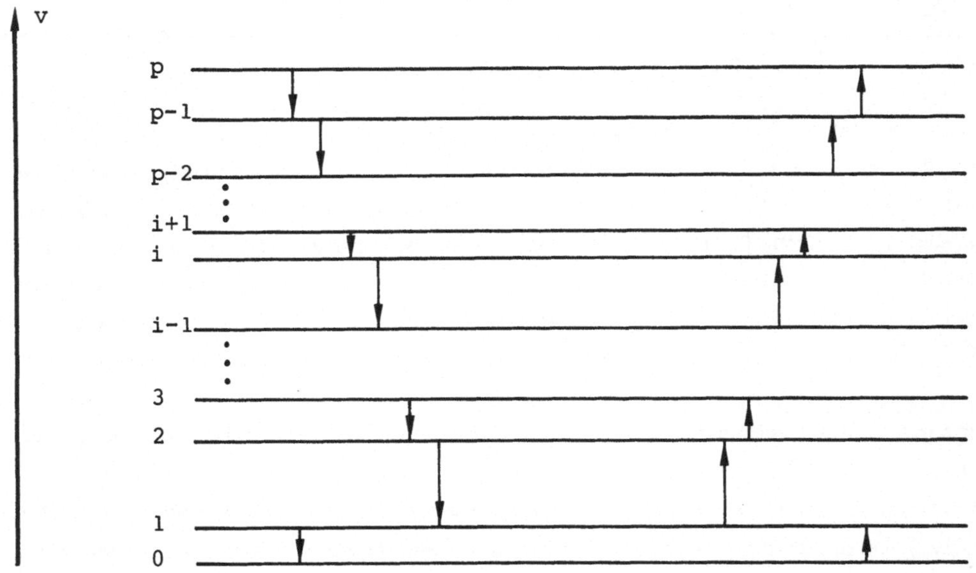

Figure 2.

The continuity condition requires that to go from level i to j the variable v must pass through levels i + 1, i + 2, ..., j - 2, j - 1.

3. What about the delays in multilevel variables ?

We shall suppose that when the variable v goes from level i to level i + 1, there is a time delay ε_{i+1}. This delay is the interval between the moment when v starts its transition from level i and the moment when it has reached the level i + 1 (or the range of numerical values belonging to this level).

Similarly, we shall call δ_{i+1} the time delay for a transition between the levels i + 1 and i.

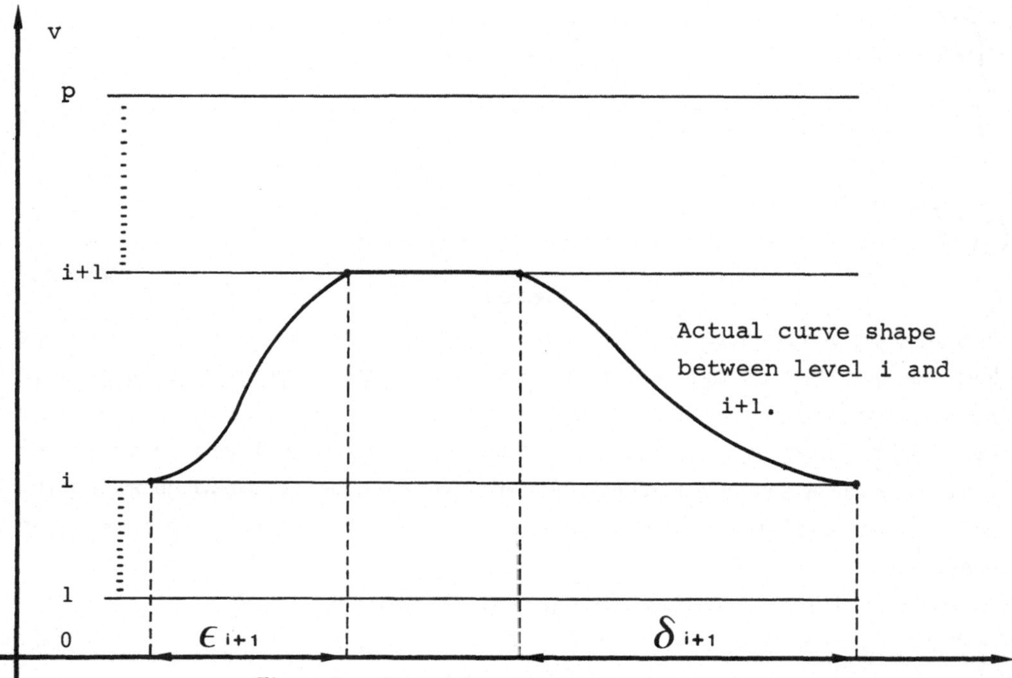

Figure 3. Time delays between levels.

We assume that $y_{i+1} = 1$ when the value of v is $\geqslant i+1$ (that is, $v = i + k$, with $k = 1, 2, \ldots p-i$), $y_{i+1} = 0$ when $v < i+1$. Thus, the transition of the multilevel variable v from the level i to the level i + 1 is formalized by the transition of the boolean variable y_{i+1} from 0 to 1 ; the delay ε_{i+1} of the transition from $v = i$ to $v = i + 1$ is also the turn-on delay of variable y_{i+1}. Similarily, when v decreases from the level i + 1 to the level i, the boolean variable y_{i+1} drops from 1 to 0 ; the delay δ_{i+1} of the transition of v from i + 1 to i can also be used as the turn-off delay of y_{i+1}.

For a (p + 1) levels variable (0,1,2,...,p), we have thus p turn-on delays $(\varepsilon_1, \ldots, \varepsilon_p)$ and p turn-off delays $(\delta_1, \ldots, \delta_p)$.

Each of these levels (except level 0) can be associated with a pair of boolean quantities (Y_i, y_i) similar to the pair (internal function, internal variable) of the classical sequential machine. When function Y_{i+1} is turned on, immediately variable v starts its transition from level i to level i + 1 (and the boolean variable y_{i+1}, from 0 to 1) with a time delay ε_{i+1}. Similarily, when function Y_{i+1} is turned off variable v immediately starts its transition from level i + 1 to level i (and y_{i+1}, from 1 to 0) with a time delay δ_{i+1}.

Each delay will be considered as inertial, that is, if two successive inverse transitions occur in a time interval less than or equal to the typical delay of the first, they annihilate each other.

The value of function Y_{i+1} depends on the value of the boolean variables considered in the model itself, including its associated variable y_{i+1}. This function (Y_{i+1}) is a part of the model and expresses in logical formalism the reason why the variable v starts its transitions to or from the level i + 1 . We will see more precisely the form of the Y_i function in the next section.

Fig.4 summarizes the dynamic behaviour of a multilevel variable with (p + 1) levels.

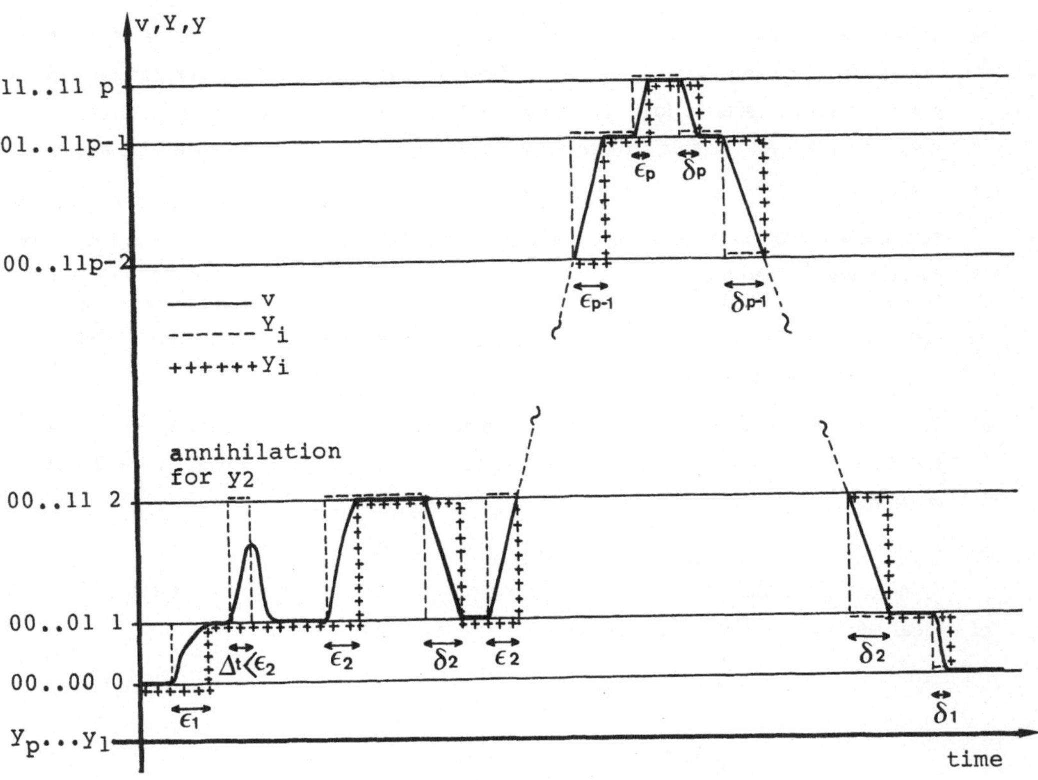

Figure 4. Dynamic behaviour of a multilevel variable.

Note : We have chosen to represent any temporal diagram of a pair (Y_i, y_i) such as :

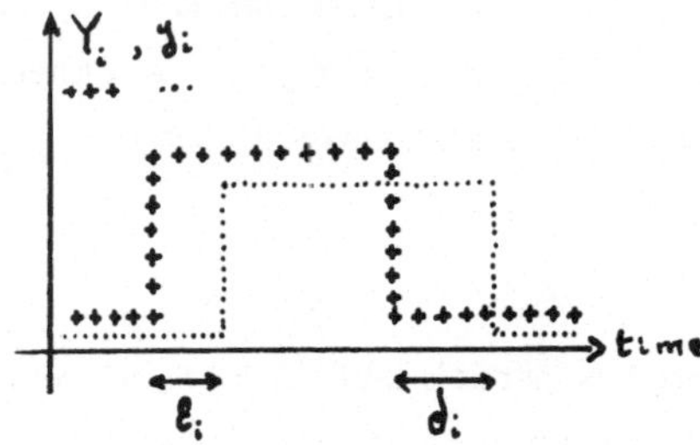

by a linear representation as one can see below :

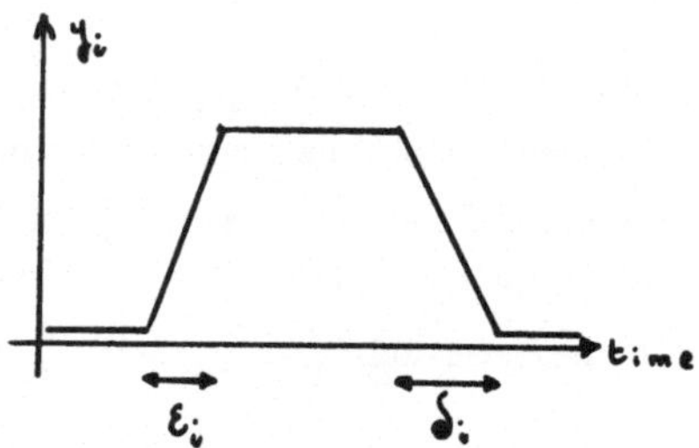

4. Remarks on the boolean coding of the levels

a) Two adjacent levels differ only by the value of one variable.

b) There are p boolean variables to encode only (p + 1) levels. It is possible of course to encode (p + 1) levels with a number of boolean variables given by the lowest integer greater than or equal to $\log_2 (p + 1)$, but in this case we lose all the advantages of the delayed multilevel variable structure. (see the introduction of this chapter, § b and c).

c) If variable $y_i = 1$, than $y_j = 1$ for $j \leq i$. This property is a consequence of the continuity condition : a variable y_i can turn-on if the other variables y_j with $j < i$ have successively turned on.

d) If variable $y_i = 0$, than $y_j = 0$ for $j \geq i$. This property is also a consequence of the continuity condition : a variable y_i may turn off if the other variables y_j with $j > i$ have successively turned off.

e) The modular structure of a multilevel variable suggests that the equations describing the functions y_i have a normalized general form. This form is given in the following section.

5. Logical structure of the multilevel variable

A multilevel variable has an underlying logical structure which can be expressed by a set of logical equations of the following form :

$$Y_i = y_{i-1} \cdot f_i + y_{i+1}$$

$$i = 1, \ldots, p$$

$$\text{with } y_0 = 1 \text{ and } y_p = 0$$

Each equation includes two terms.

The first term $(y_{i-1} \cdot f_i)$means that Y_i is on (transition from $i-1$ to i or maintenance at the level i) if a function f_i (depending on the logical structure of the system) is fulfilled AND the system is at least at the level y_{i-1} .This condition guarantees that it is not possible to reach the level y_i without being at least at the level y_{i-1} (continuity condition).

The boolean function f_i is the part of the logical model describing the transitions between (i-1) and (i) levels. This function may be different for different values of i and depends in general on other variables in the model. It makes it possible to act by a specific function on each transition of the multilevel variable and gives a greater versatility to the model structure.

For example if $f_i = f_2 = ... = f_p = f$, the multilevel variable is always performing a transition and cannot maintain itself on a level, for instance k , for which $f_j = 1$ $(j < k)$, $f_k = 1$, and $f_{k+1} = 0$. The second term (y_{i+1}), which is "ORed" with the first, ensures that y_i cannot turn off before $y_{i+1} = 0$. The continuity condition is thereby fullfilled for down transitions.

In summary, the functions f_i related to each level of every multilevel variable must be equal to 1 or 0 for the corresponding transition (up or down) to occur ; but the transition will in fact occur only if the continuity conditions are fullfilled.

When the model maker uses the normalized equation written above he must then only care for the functions f_i because the system automatically includes the continuity conditions.

6. What does a logical model with multilevel variables look like ?

Suppose we have to work with a model in which several variables, say a, b, c, have more than two levels.

The formal equation sets for these variables are shown below :

$$\begin{cases} A_1 = f_1^a + a_2 \\ A_2 = a_1 f_2^a + a_3 \\ \vdots \\ A_p = a_{p-1} f_p^a \end{cases} \qquad p+1 \text{ levels for } (A,a)$$

$$\begin{cases} B_1 = f_1^b + b_2 \\ B_2 = b_1 f_2^B + b_3 \\ \vdots \\ B_q = b_{q-1} f_q^b \end{cases} \qquad q+1 \text{ levels for } (B,b)$$

$$\begin{cases} C_1 = f_1^c + c_2 \\ C_2 = c_1 f_2^c + c_3 \\ \vdots \\ C_r = c_{r-1} f_r^c \end{cases} \qquad r+1 \text{ levels for } (C,c)$$

,where (A_i, a_i) ; (B_j, b_j) ; (C_k, c_k) are pairs of secondary functions and variables like the (Y_i, y_i) pair.

For each level one has a special function f_i^a, f_j^b or f_k^c .

One has to write up to (p+q+r) functions to describe the dynamic behaviour of a three multilevel variables system. It is interesting to note that besides the fact that the functions f_i represent the model of interacting levels, the normalized multilevel variable logical equations may sometimes suggest underlying structure in the actual system. In such case, the model- maker may focus his attention on this special problem : is there a one-to-one correspondance between the signification of the levels (for example, those values of the variable are successive important thresholds in the system behaviour) and a new view of the levels in terms of internal mechanisms which produce the thresholds of some observable quantity only as a consequence.

7. A FIRST EXAMPLE WITH THREE LEVELS

Let us suppose that we are modelling a single self-inhibitory element R.

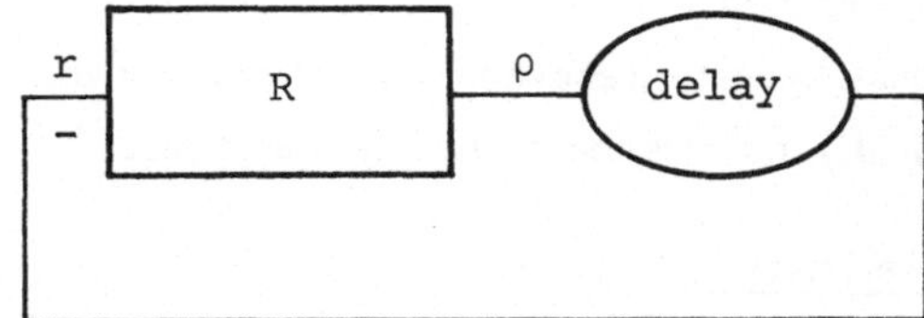

r = activity of gene R

ρ = concentration of the product of gene R

Figure 5 : A single self-inhibitory element.

If we write the associated Boolean equations, we find :

$r = \bar{\rho}$

The temporal behaviour of ρ is shown in Figure 6.

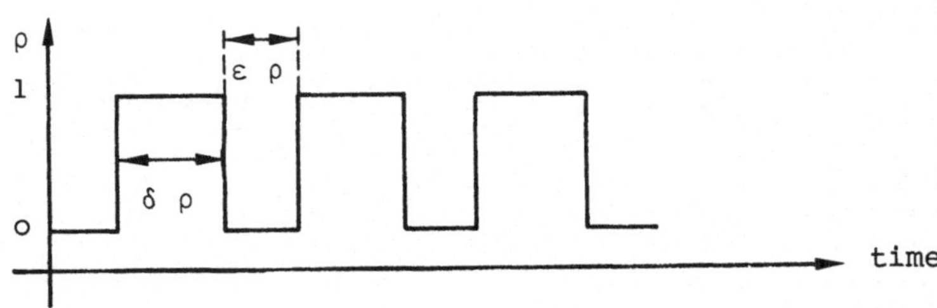

Figure 6 : Temporal behaviour of ρ.

The simple boolean treatment represents the oscillation of ρ in all-or-none terms, whatever the actual amptitude of the oscillation of the concentration (C_ρ) (see Fig.7.).

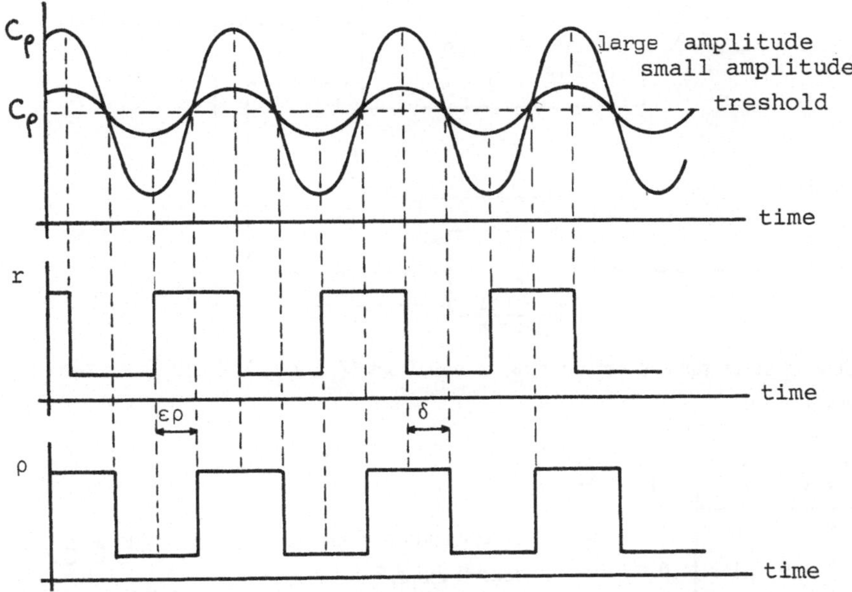

Figure 7 : The behaviour of (r, ρ) does not give any information on the amplitude of an oscillation around the threshold.

C_ρ refers to the concentration of substance ρ.

Let us now use the following model :

$$r_1 = f_1 + \rho_2 \qquad (7.1)$$

$$r_2 = f_2 \cdot \rho_1 \qquad (7.2)$$

with

$$f_1 = f_2 = \overline{\lambda}$$

$$\ell = \overline{\lambda} \cdot \rho_1 \qquad (7.3)$$

The variable (r, ρ) is now treated as a three level variable following equations (5.1) and (5.2). We have here $f_1 = f_2 = \overline{\lambda}$. This supplementary variable (ℓ, λ) has the same equation as (r_2, ρ_2). In fact, ρ_2 and λ differ in their delays and their physical meaning : $\rho_2 = 1$ means that the concentration of ρ, is high, $\lambda = 1$ means, for example, that a molecule of repressor is actually fixed on the operator.

We have chosen this example in view of the great variety of resulting behaviours, more than for genetic reasons.

The states $\rho_1\rho_2$ = 00, 10 and 11 represents the three levels of product ρ concentration. The supplementary variable λ is the feedback variable which insures the negative control.

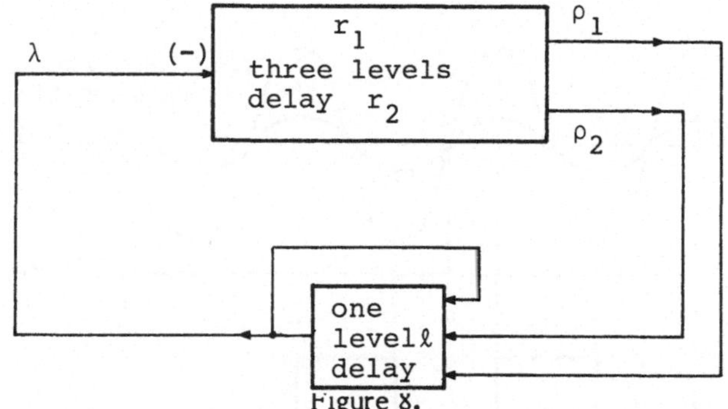

Figure 8.

The following state table corresponding to equations (7.1, 7.2, 7.3) gives the next state for each present state

ρ_1	ρ_2	λ	r_1	r_2	ℓ
0	0	0	1	0	0
0	0	1	0	0	0
1	0	0	1	1	1
1	0	1	0	0	0
1	1	1	1	0	0
1	1	0	1	1	1
Present state			next state		

(in our coding, remember that the state $\rho_1\rho_2$ = 01 doesn't exist.⁺

One can see below the graph of the possible behaviours.

On each arrow we have indicated the conditions on the delays which permits that transition.

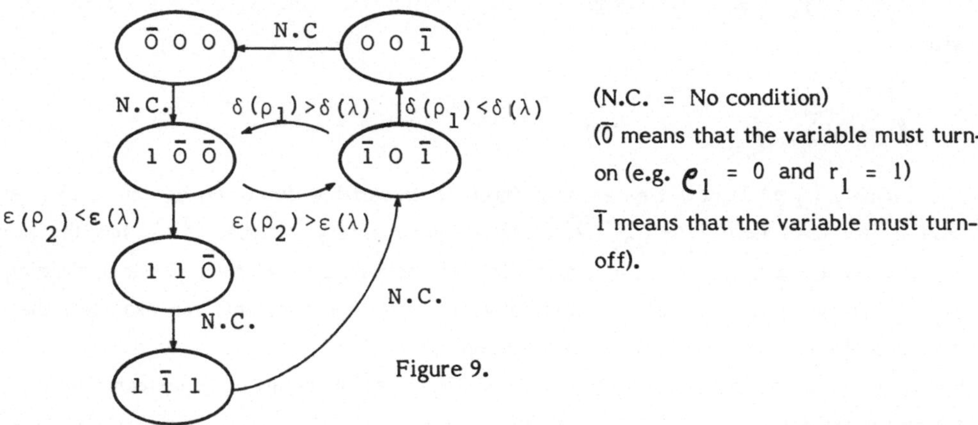

(N.C. = No condition)
($\bar{0}$ means that the variable must turn-on (e.g. ρ_1 = 0 and r_1 = 1)
$\bar{1}$ means that the variable must turn-off).

Figure 9.

⁺ In fact, the continuity conditions, formally inserted in the normalized form of the equations (by y_{i-1} and y_{i+1}), ensure that the system will not go through such a state).

If we have :
$$\begin{cases} \varepsilon(\rho_2) < \varepsilon(\lambda) \\ \delta(\rho_1) > \delta(\lambda) \end{cases}$$

The behaviour is (000, 100, 110, 111, 101, 100 a.s.o.) :

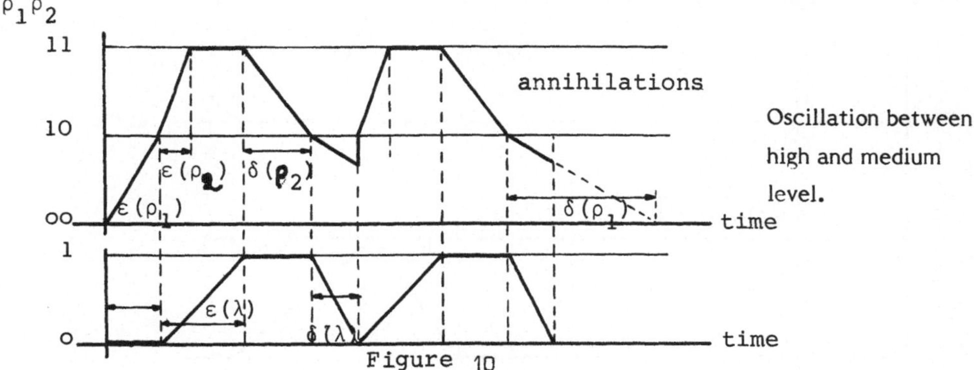

Figure 10

If we have $\begin{cases} \varepsilon(\rho_2) < \varepsilon(\lambda) \\ \delta(\rho_1) < \delta(\lambda) \end{cases}$ (λ is a "slow" variable)

The behaviour is (000, 100, 110, 111, 101, 001, 000, a.s.o.)

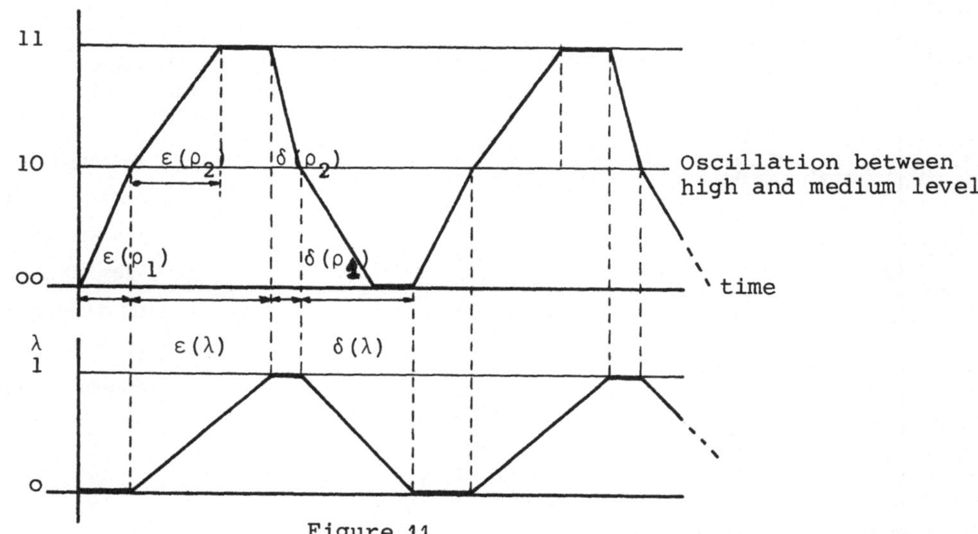

Figure 11

If we have $\begin{cases} \varepsilon(\rho_2) > \varepsilon(\lambda) \\ \delta(\rho_1) < \delta(\lambda) \end{cases}$

The behaviour is : (000, 100, 101, 001, 000, a.s.o.)

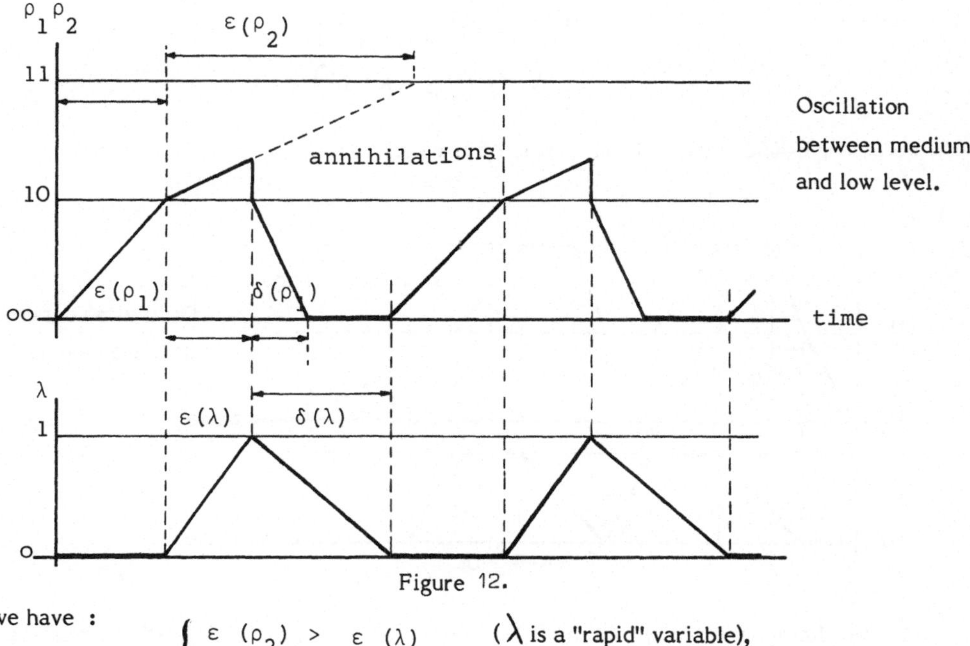

Figure 12.

If we have :

$$\begin{cases} \varepsilon\ (\rho_2) > \varepsilon\ (\lambda) \\ \delta\ (\rho_1) > \delta\ (\lambda) \end{cases}$$

(λ is a "rapid" variable),

the behaviour is : (000, 100, 101, 100, a.s.o.)

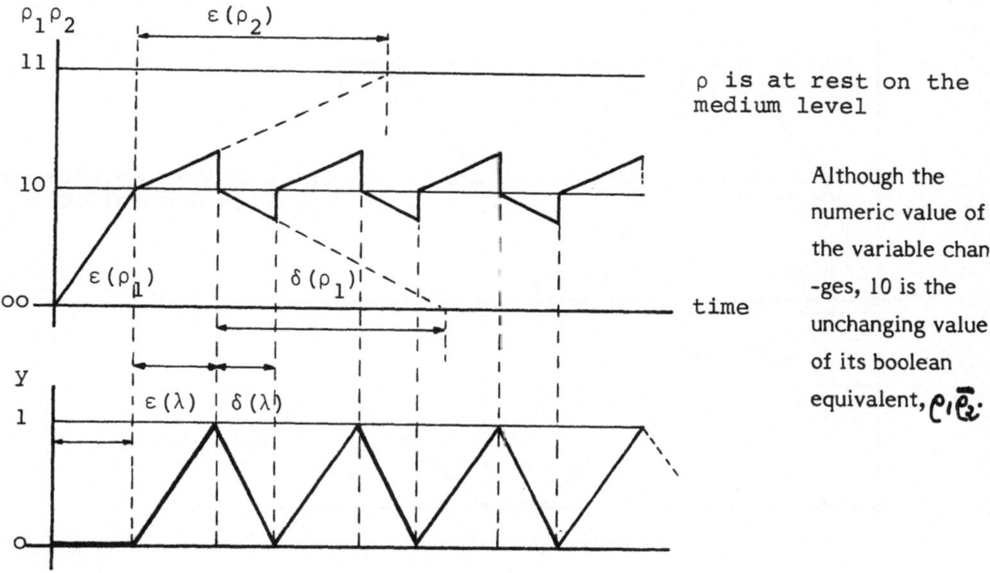

Figure 13.

The four cases just described show that provided certain conditions on the delays are fullfilled the product of the auto-repressive gene R may keep at on intermediate level. The conditions are, of course : the mechanism which performs repression must be faster than the delay to go from medium to high level. Furthermore, the destruction of repression action must also be faster than the delay to go from medium to low level for ρl.

8. A second example taken from the lactose operon.

In the lactose operon (see Chapters I, II, XI, XIII and XVII, 5), the synthesis of the enzymes β-galactosidase and permease is repressed unless the repressor is inactivated by a derivative of lactose (inducer) . The internal concentration of lactose required for induction can be ensured by a moderate concentration of external lactose ($\bar{L}$) if permease is already present, or by a much higher external concentration (L) in the absence of permease.

As permease is involved in the penetration of lactose which is itself necessary for the synthesis of permease, one deals with a positive loop . The situation is complicated by two negative loops.

On the one hand, internal lactose (necessary for the synthesis of β-galactosidase) is destroyed by this enzyme ; on the other hand, this process produces glucose, which might result in catabolic repression of the synthesis of the enzymes (see Fig. 10 of Chapter XVII).

A detailed formal analysis of this regulation can be found in Nicolis and Sanglier (1976). (see also this book, Chapter XI by Nicolis) A preliminary boolean treatment can be found in Thomas (1978) and in this book (Chapter XVII , 5).

The logical equations proposed by Thomas are :

$$y = Y \cdot \lambda \cdot \bar{\gamma} \quad (1)$$

$$l = (L + \upsilon) \cdot \bar{\zeta} \quad (2)$$

$$g = \lambda \cdot \zeta \quad (3)$$

$$z = Z \cdot \lambda \cdot \bar{\gamma} \quad (4)$$

In this simplified model, one reasons as if, whenever β-galactosidase is present, the rate of

destruction of internal lactose was higher than the rate of entry of lactose, so that the concentration of internal lactose is committed to vanish. An alternative possibility would be that whenever $L+\upsilon=1$ the rate of entry of lactose is sufficient to compensate the hydrolysis by β-galactosidase. If this were the case, one would have to write
$\ell = L+\upsilon$ instead of $\ell = (L+\upsilon)\cdot\bar{\zeta}$

One can also refine the analysis as follows. The rate of destruction of internal lactose depends on the concentration of β-galactosidase. One can thus introduce two levels of the enzyme (ζ_1 and ζ_2), such that ζ_1 is sufficient to produce significant level of glucose but only for the level ζ_2 is the destruction of the internal lactose faster than its entry.

thus : $\ell = (L+\upsilon)\cdot\bar{\zeta}_2$
and : $g = \lambda\cdot\zeta_1$

From the initial equation $z = \lambda\cdot\bar{\zeta}$, one derives, using the normalized form of multilevel variable (continuity conditions)

$$z_1 = \lambda\bar{\gamma} + \zeta_2$$
$$z_2 = \lambda\bar{\gamma}\cdot\zeta_1$$

Let us first consider the situation $L = 0$, that is, the external concentration of lactose is low and lactose penetrates only if permease is present :

$$y = \lambda\cdot\bar{\gamma} \qquad (10)$$
$$l = \upsilon\cdot\bar{\zeta}_2 \qquad (11)$$
$$g = \lambda\cdot\zeta_1 \qquad (12)$$
$$z_1 = \lambda\cdot\bar{\gamma} + \zeta_2 \qquad (13)$$
$$z_2 = \lambda\cdot\bar{\gamma}\cdot\zeta_1 \qquad (14)$$

This system was analyzed with the program PRAN2 (described in chapter IX). The simulations given below use as the initial state 01100 and delays of a same average value 50.00, randomly chosen within the interval $\pm$ 80 % around the mean.
In this case the system usually (953 out of 1000 simulations) reaches the "dead" stable state 00000 ; whenever permease has been lost lactose cannot penetrate any more and it is finally exhausted. However, in the other 47 simulations, the system follows cycles as 11000 / 11010 / 11110 / 11100 more conveniently described by the decimal equivalents 24/26/30/28 . In this cycle, β-galactosidase oscillates between the low level 00 and the intermediate level 10 , which produces enough glucose to inhibit further synthesis of β-galactosidase, etc.....

```
STRUCTURF =  1      DONNEES =  1

ETAT INITIAL =
C1100*************************
50.00 50.00 50.00 50.00 50.00 -0.00 -0.00 -0.00 -
40.00 40.00 40.00 40.00 40.00  0.00  0.00  0.00
50.00 50.00 50.00 50.00 50.00 -0.00 -0.00 -0.00 -
40.00 40.00 40.00 40.00 40.00  0.00  0.00  0.00
-0.00 -0.00 -0.00 -0.00 -0.00 -0.00 -0.00 -0.00 -
-0.00 -0.00 -0.00 -0.00 -0.00 -0.00 -0.00 -0.00 -
0-0-0-0-0-0-0-0-015-0-0-0-0-0-0-0-0-0-0-0-0-0-(
NOMBRE D ECHANTILLONS = 1000ESSAI:= 111
     0      000000000000000000000        953
  24  26  30  28
CIRCUIT  1=   7
   8  24  26  27  31  30  14  12
CIRCUIT  2=   6
   8  10  26  30  14  12
CIRCUIT  3=   3
   8  24  26  30  28  12
CIRCUIT  4=  14
   8  24  26  30  14  12
CIRCUIT  5=   9
   8  24  26  27  31  30  28  12
CIRCUIT  6=   2
  24  26  27  31  30  28
CIRCUIT  7=   2
   8  24  26  27  31  23  22  30  14  12
CIRCUIT  8=   1
  18  26  27  31  23  22
CIRCUIT  9=   1
   8  10  26  27  31  23  22  30  28  12
CIRCUIT 10=   1
   8  10  26  30  28  12
CIRCUIT 11=   1
CYCLE =    0
```

If we consider the situation L = 1 (the external concentration of lactose is sufficient to ensure penetration even in the absence of permease), the logical equations become :

$y = \lambda.\overline{\gamma}$ (15)

$l = \overline{\zeta}_2$ (16)

$g = \lambda.\zeta_1$ (17)

$z_1 = \lambda.\overline{\gamma} + \zeta_2$ (18)

$z_2 = \lambda.\overline{\gamma} \bullet \zeta_1$ (19)

The program PRAN2 used in the same conditions as above, shows no stable state but 43 different cycles :

```
STRUCTURE =  1      DONNEES =  1

ETAT INITIAL =
01100*****************************
50.00 50.00 50.00 50.00 50.00 -0.00 -0.00 -0.00 -0.00 -0.00 -0.00
40.00 40.00 40.00 40.00 40.00  0.00  0.00  0.00  0.00  0.00  0.00
50.00 50.00 50.00 50.00 50.00 -0.00 -0.00 -0.00 -0.00 -0.00 -0.00
40.00 40.00 40.00 40.00 40.00  0.00  0.00  0.00  0.00  0.00  0.00
-0.00 -0.00 -0.00 -0.00 -0.00 -0.00 -0.00 -0.00 -0.00 -0.00 -0.00
-0.00 -0.00 -0.00 -0.00 -0.00 -0.00 -0.00 -0.00 -0.00 -0.00 -0.00
0-0-0-0-0-0-0-0-015-0-0-0-0-0-0-0-0-0-0-0-0-0-0-0-0-0-0
NOMBRE D ECHANTILLONS =   100ESSAI := 111
   8  10  14  12
CIRCUIT   1=     6
   8  24  26  30  28  12
CIRCUIT   2=    14
   8  24  26  27  31  15  14  12
CIRCUIT   3=     2
   8  24  26  27  31  30  14  12
CIRCUIT   4=     4
   6  14  12   8  10  26  27  31  23  22
CIRCUIT   5=     1
   2  10  11  27  31  15   7   6
CIRCUIT   6=     1
   2  10  11  15   7   3
CIRCUIT   7=     3
   2  10  26  27  31  15   7   3
CIRCUIT   8=     1
   8  24  26  30  14  12
CIRCUIT   9=    12
   2  10  11  27  31  23  22   6
CIRCUIT  10=     1
   8  10  26  27  31  30  14  12
CIRCUIT  11=     1
  18  26  27  19
CIRCUIT  12=     2
   2  10  26  27  31  15   7   6
CIRCUIT  13=     1
   0   8  10  11  15   7   3   2
```

```
CIRCUIT 14=    1
  24  26  30  28
CIRCUIT 15=    3
  0   8  24  26  27  19   3   2
CIRCUIT 16=    3
  8  10  26  30  14  12
CIRCUIT 17=   10
  2  10  11  27  31  23  19   3
CIRCUIT 18=    1
  0   8  24  26  27  31  23  19  18  16
CIRCUIT 19=    2
  6  14  12   8  24  26  27  31  23  22
CIRCUIT 20=    2
  2  10  11  27  31  23  22  18
CIRCUIT 21=    1
  0   8  10  11  27  31  23  22  18   2
CIRCUIT 22=    1
  2  10  26  27  19  18
CIRCUIT 23=    3
  2  10  26  27  19   3
CIRCUIT 24=    1
  0   8  10  26  27  19   3   2
CIRCUIT 25=    1
  4  12   8  10  11  27  31  15   7   6
CIRCUIT 26=    1
  8  10  26  30  28  12
CIRCUIT 27=    2
  0   8  10  11  27  19   3   2
CIRCUIT 28=    1
  0   8  10  26  27  31  23  22  18   2
CIRCUIT 29=    1
  4  12   8  24  26  27  31  23  22  20
CIRCUIT 30=    1
  6  14  12   8  24  26  27  31  15   7
CIRCUIT 31=    2
  0   8  10  26  27  31  23  19   3   2
CIRCUIT 32=    1
  2  10  11   3
CIRCUIT 33=    3
  8  24  26  27  31  23  22  30  14  12
CIRCUIT 34=    1
  0   8  24  26  27  31  23   7   3   2
CIRCUIT 35=    1
  2  10  11  15   7   6
CIRCUIT 36=    1
  0   8  10  26  27  31  23  19  18   2
CIRCUIT 37=    1
  2  10  11  27  31  23   7   6
CIRCUIT 38=    1
  2  10  11  27  19   3
CIRCUIT 39=    1
  6  14  12   8  24  26  27  31  23   7
CIRCUIT 40=    1
  0   8  10  26  27  31  23   7   3   2
CIRCUIT 41=    1
  0   8  10  26  27  31  23  22   6   2
CIRCUIT 42=    1
  8  10  26  27  31  30  28  12
CIRCUIT 43=    1
CYCLE =     0
```

If we focus attention on circuit n° 14 :
(0/8/10/11/15/17/3/2) or,in binary form,
(00000/01000/01010/01011/01111/00111/00011/00010), the delays are such that all the variables oscillate, giving an interesting behaviour. We have shown below a graphical representation of this cycle.

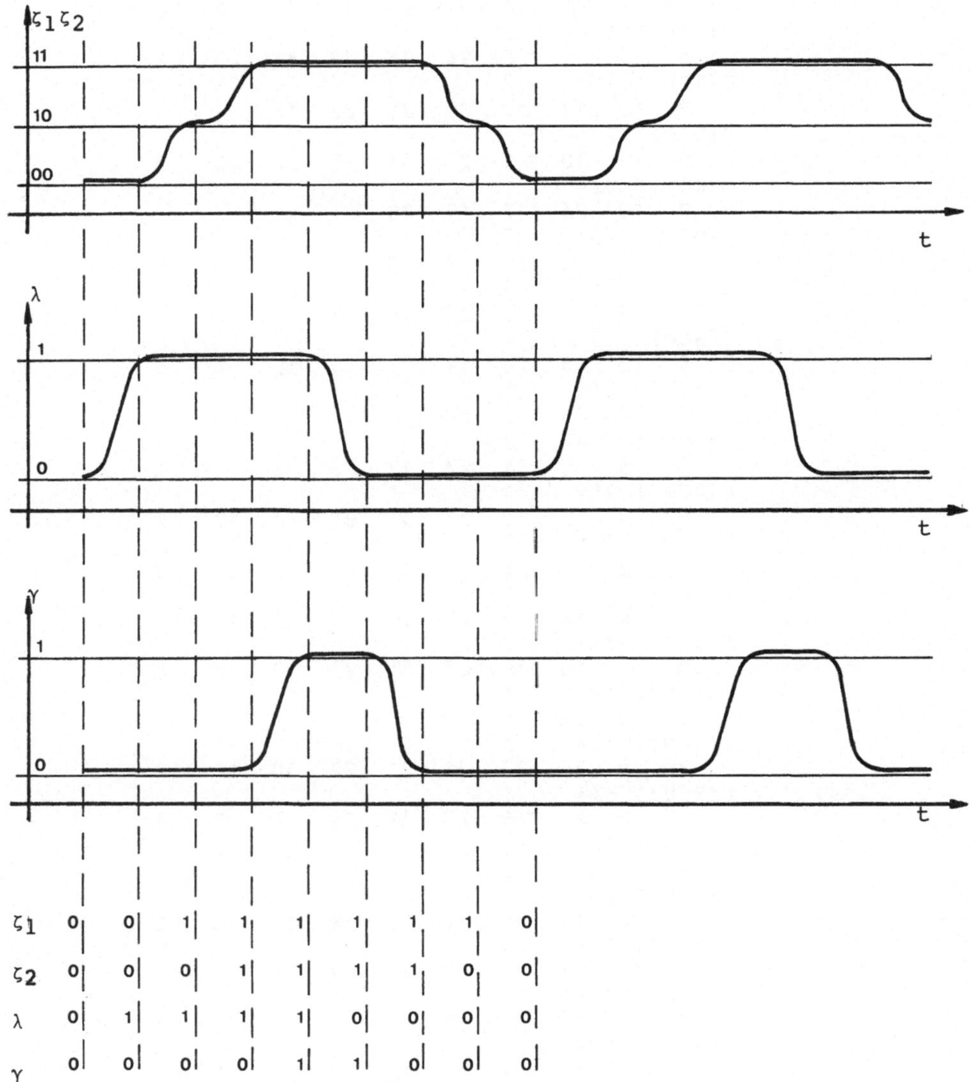

It is of course possible to use in addition intermediate levels in glucose and internal lactose concentrations. The interactions between those different levels make it possible to describe more complex behaviours in the lactose operon. For example, with three levels (low, medium and high) for λ and γ, we have 9 possible configurations of the two products acting on β -galactosidase expression. It is of course possible to further refine the model without leaving a logical approach, by means of multilevel logical variables.

CONCLUSION

The use of multilevel logical variables with an imposed continuity hypothesis is a natural extension of the binary case when more than two levels of the same variable have physical qualitative significance. The logical equations formalizing the model are easy to elaborate and keep their simplifying character.

All the previous methods of analysis (Van Ham, 1974, 1975, 1977) developped for logical models can be used without modification.

Moreover, the variables of a multilevel delay (such as ρ_1, ρ_2) may indicate that there exists an underlying structure previously described by only one variable which, in fact, split into a few other interconnected phenomena.

ACKNOWLEDGEMENT

I wish to thank Professor R. Thomas for many fruitfull discussions.

REFERENCES

GLASS, L. & KAUFFMAN, S.A. (1972). J. Theor. Biol. 34, 219.

NICOLIS, F. & SANGLIER, M. (1976). Biophys. Chem. 4, 113.

SUGITA, M. (1963). J. Theor. Biol. 4, 179.

THOMAS, R. (1973). J. Theor. Biol. 42, 563.

THOMAS, R. & VAN HAM, Ph. (1974). Biochimie, 56, 1529.

THOMAS, R. (1978). J. Theor. Biol. 73, 631.

VAN HAM, P. (1974) Journal A, XV., 2.79.

VAN HAM, P. (1975). Ph. D. Thesis, Université libre de Bruxelles.

VAN HAM, P. (1975) Symposium : Applied aspects of Automata theory, Varna, 2, 728, Edit. Bulgarian Academy of Sciences.

VAN HAM, P. (1977). IFAC Symposium on Discrete Systems, 5, 27.

CHAPTER XVI

R. THOMAS, G. NICOLIS, J. RICHELLE & P. VAN HAM

General discussion on the simplifying assumptions in methods using logical, stochastic or differential equations ; the ranges of applicability and the complementarity of the approches.

The systems we are dealing with can be described in terms of differential, stochastic or logical equations. Although the descriptions usually agree on the essential points, in somes cases there are serious discrepancies ; this is not too surprinsing, in view of the very different types of simplifying assumptions used in the different methods. Some of them are extremely obvious, other are more subtle but no less important. The main purpose of this discussion was to find out, and outline as clearly as possible, which are in each case the crucial simplifying assumptions.

I. Two types of systems.

We felt convenient to focus, within the field of chemistry and biology, on two rather extreme types of systems.

a) systems involving chemical reactions in solution. Biological catalysts may be involved, but they are given to the system rather than produced in it, and usually they do not appear explicitly as elements of the system.

b) populations of living cells with little intercellular interactions (as in dilute bacterial cultures).

For the sake of simplicity, we shall refer to these two types of systems as "chemical" and "bacterial" systems, respectively.

What are the main differences relevant to our discussion between the two types of systems ?

1) Stoichiometric vs catalytic role of the elements of the system in the production of each other.

In the "chemical" systems, the interactions considered are usually the transformation of substances into one another :

$Y_1 \rightarrow Y_2$ means that substance Y_2 is formed at the expense of Y_1. Biological catalysts (enzymes) are frequently involved, but they are usually not used as variables of the system . Their concentration and specific properties are taken into account indirectly, in the value of kinetic constants and the mathematical form of the equations.

In the "bacterial" systems there are of course also substances which transform into one another. However, an important part of the systems consists of macromolecules which control (positively or negatively) the synthesis of one another without being used in the process, in other words, in a catalytic way. In these cases, the catalysts are treated as elements of the system, whose synthesis and disappearance is considered explicitly.

2) Time delays.

Suppose we start a classical chemical reaction at time 0. The concentration of the product α , initially 0, may increase gradually until at time t_α it reaches a chosen value. Even though in the boolean treatment we reason as if substance α were absent before time t_α , we know perfectly that the concentration of α is not nil during this period ; there is no need at the time scale used to consider an "absolute" time delay during which substance α is still completely absent. The situation is quite different in systems which involve the synthesis of macromolecules. When a gene is switched on it takes minutes before the very first protein molecule resulting from this decision has been assembled. This is because hundreds or thousands of nucleotides have to be sequentially condensed into a specific mRNA, and hundreds of amino acids have to be condensed into the proper polypeptide chain. In these cases, there is thus a lag, an absolute delay which has to be taken into consideration, not only in the boolean description (in which it is included in the formal delay) but also in continuous descriptions (in which it should be present explicitly as a delay in the differential equations).

3) One large system or many small systems ?

In the "chemical" situation, one really deals with one system, in the sense that the constituents interact freely with each other. Each of the constituents of the system is present in large numbers, and there is no question as to whether the laws of large numbers are applicable. In diluted bacterial cultures (a typical case of "bacterial" system), one often deals in fact with a large number of practically independant entities. Each one is a system by itself, but this sytem is quite small, and some of the elements involved in regulation may be present in very small numbers (down to 1 or 0). In this case, the laws of large numbers cannot be applied within each small system. Their application to the whole of the constituents of all cells would imply an important simplifying assumtion ; in fact, one would reason as if the cell walls were abolished and the content of the cells were summed to constitute a unique system. It may be helpful to reason as if both "chemical" and " bacterial" systems were composed of many subsystems and consider the local and global behaviours. A difference between the two systems is that in the latter the subsystems are closed and thus independant of each other whereas in the former the subsystems communicate with one another through free diffusion.

Now that some differences between our two types of systems have been briefly discussed, we can see how the various methods can be used in the two cases.

II. Short description of the approaches.

1. Stochastic approach (NICOLIS)

As pointed out in Chapters XI, XII and XIV, in problems involving regulation - whether of the "chemical" or of the "bacterial" type - one deals with statistical phenomena. If the state variables are chosen to be the numbers of molecules of the different constituents, then the latter can change only by discrete (integer) jumps. The basic picture that emerges is therefore that of a random process. If the Markovian assumption* can be accepted, then this process will be governed by a master equation of the form (see e.g. Chapter XI, eq. (28)):

$$\frac{dP(X,t)}{dt} = \sum_{X'} W_{X'X}\, P(X', t) \tag{1}$$

where X denotes, collectively, the numbers of molecules of the various constituents and P(X,t) defines the probability that the system be in state X at time t. According to eq. (1) the evolution can be described as follows. At t_0 the system is at state X_0. During a certain period of time distributed according to a well-defined law expressed in fact in terms of $W_{X'X}$, the system will stay in this state X_0. After a certain interval Δt_0 however, it will transit to a new state X_1, connected to X_0 through a transition matrix which can be deduced uniquely from $W_{X'X}$. Thereafter it will stay on X_1 during a new time interval Δt_1 before performing the next transition to state X_2 and so forth.

We have therefore two phenomena :

(i) "residence", or "holding" times of the system in certain states,

(ii) transitions between states,

both occuring under well-defined rules.

Obviously, this picture can be applied equally well to chemical as to "bacterial" systems as it does not require the validity of any asymptotic law like the law of large numbers or the central limit theorem.

Naturally, in each case one has to construct a transition matrix $W_{X'X}$ reflecting the structure of the particular system under consideration.

---

* A process is Markovian (in the strict sense) if the probability of performing a transition to X at time t depends only on the state X, at the time t' immediately preceeding the transition.

2. Phenomenological approach in terms of differential equations without delays. (NICOLIS)

Having a stochastic master equation of the form (1), one can always construct an equation for the statistical average X by multiplying through by X and summing over all values run by this variable.

As $\langle X \rangle_t = \sum_X X\, P(X,t)$, one obtains

$$\frac{d\langle X \rangle}{dt} = \sum_{X'X} W_{X'X}\; XP(X',t) \tag{2}$$

In general however, the right hand side will not be a function of X alone but will contain, instead, the effect of the variances :

$$\frac{d\langle X \rangle}{dt} = f(\langle X \rangle)\left[1 + g\left(\frac{\langle \delta x^2 \rangle}{\langle x \rangle^2}, \frac{\langle \delta x^3 \rangle}{\langle x \rangle^3}, \ldots\right)\right] \tag{3}$$

If the system is macroscopic and exhibits a single stable state, then the effet of the variances can be neglected and one obtains a closed differential equation for X :

$$\frac{d\langle X \rangle}{dt} = f(\langle X \rangle) \tag{4}$$

which is the basis of the continuous formalism of chemical kinetics. Note the enormous contraction in the description afforded by eq (4) : instead of describing the state by the probability P(x,t) or by the infinite hierarchy of its moments $\langle X^n \rangle$ (n = 1, 2, ...), one now deals with macrostates characterized by $\langle X \rangle$ alone. Nevertheless, this passage is legitimate in all usual chemical systems. An exception is the case of bistable kinetics (Chapter XI, section 5). For these systems, eq. (4) must be reinterpreted in terms of the most probable value rather than in terms of the statistical average. As regards "bacterial" systems, the passage is compromised because of the small numbers of molecules that may be involved in each cell which give rise to variances (δX^n) of the same order as $\langle X \rangle^n$.

One possibility would be to consider a population of cells rather than the dynamics of an individual cell. Let X_i stand for the number of molecules of species X in cell i (i = 1; ..., n). The variable

$$X = X_1 + \ldots + X_n \tag{5}$$

is a sum of n identically distributed independent random variables. Hence, it obeys to the central limit theorem and has a mean value and variance given by :

$$\langle X \rangle = n \langle X_i \rangle \equiv nm$$

$$\langle \delta x^2 \rangle = n \langle \delta x_i{}^2 \rangle \equiv n\sigma^2$$

Hence $\langle \delta X^2 \rangle / \langle X \rangle^2 \simeq \frac{1}{n}$ is negligible for large n.

What is the significance of this collective variable X ?
In a way, considering X one argues as if the cell walls were abolished and the content of each cell dispersed at random in the total cellular volume. On the other hand, for all its crudeness, this may already give an idea of the state of the population as far as, say, the turnover in the induction of a protein (reduced per unit of dry bacterial mass) is concerned.

As mentioned above, systems in which marcomolecular syntheses take place do involve absolute time delays, which can, and should, be taken into account in the various formalisms. On the other hand, delays can be introduced (even in the absence of absolute lags) as an easy way to reduce the number of explicit variables. In either case, the consideration of a delay of a sufficient lenghth can modify the qualitative picture of the process (see Chapter XIV by Richelle).

3. Logical equations. (THOMAS & VAN HAM)

The immediately obvious particularity of the boolean approach is its all-or-none character, with its inherent advantages and drawbacks : simplicity, which permits the treatment of very complex systems indeed, but crudeness. We shall not insist here on the ways to reduce this crudeness (see for instance, Chapter XV, by Van Ham). Rather, we would like to point out additional aspects of the boolean approach, which in fact also result from the discontinuous character of the method.
a) Treating causes as totally inefficient below a certain treshold and fully efficient above this threshold amounts to treat each interaction in the system as if it were infinitely non linear ; in many cases, it amounts to represent a sigmoid curve by a staircase step. This admittedly caricatural attitude in fact provides an extremely convenient way to take into account the non-linear character of a system. Whether this way is not only convenient but also justified, depends on the correctness of the following statement : beyond a certain level of non - linearity, what is essential for a qualitatively correct prevision is that the non-linearity be taken into account, not that it be accurately described.
b) Another manifestation of the discontinuous character of the method is that the n - dimensional space of the variables is cut into 2^n boxes, corresponding each to a boolean state ; each state thus represents a domain within many "microscopic" state are lumped, rather than a point. A trajectory is viewed as the sequence of adjacent boxes successively occupied by the system.
c) Frequently, a state can be followed by either of two or more different next states. This situation is found each time the values of two or more boolean functions have changed but the corresponding internal variables still have their former value - to be more concrete, each time there is more than one decision already taken but not yet executed. One or more of the decisions may have been taken 2, 3, ... n states ahead, and in order to forecast which one will be executed first, one has to take into account not only the preceeding boolean state, but also

states 2, 3, ... n steps removed (see, for instance, Chapter VI, appendix 2). In other words, our boolean description seems "non-Markovian" (see footnote p.3) if the description of each state of the system is limited to the boolean values of the functions and variables ; this non-Markovian character disappears when the description includes the carry-overs of the time delays.

How will the image of a system be altered by the simplifying assumtions involved in the boolean formalism ? The answer is very different for the two types of systems chosen in this discussion. As regards the typically "chemical" systems, we have seen in section 2 that the classical approach using differential equations is an almost perfect tool, because in these systems one does not have to take delays or septation into account. In contrast (see Chapter VII, by Thomas, Section 9), the plain logical formalism is poorly adapted, if only because it is not obvious how to take stoichiometry into account. It may in fact be convenient to use the boolean formalism at the exploratory stage of the study of a chemical model, because it provides with a clear view of the logical structure of the model (including the "hidden" loops) and of the types of behaviours allowed by this logical structure. But given a dynamic behaviour compatible with the logical structure, one must check whether it will actually take place in the conditions of the system - for instance, does it take place even in the absence of absolute time delays ?
Let us now turn to the other type of system considered, a population of non-interacting living cells, here denoted "bacterial" system. In this case, the boolean formalism is certainly helpful, because it takes into account aspects which are usually not considered in terms of classical differential equations : the occurence of absolute time delays and of septation (one deals with a great number of small systems rather than with a large system). In fact, these two factors often operate in opposite directions : for instance, the occurence of absolute time delays vastly widen the domain in which oscillations actually take place ; on the other hand, septation into small non-interacting system prevents any collective behaviour, thus preventing any sustained periodicity at the level of the population. (see Richelle, Chapter XIV).

The <u>multilevel</u> logical approach, as described by Van Ham in Chapter XI, is NOT an attempt to shift from a qualitative to a quantitative analysis. Rather, the purpose is to introduce additional levels where this appears necessary for a correct qualitative analysis. For instance, if the repressor of λ exerts a positive control on its own synthesis at low concentrations and a negative control at high concentration, one has to consider <u>three</u> ranges of concentration rather than two. Another important sophistication of the boolean methods consists of treating the time delays as functions of the state of the system (Van Ham, in preparation). If these improvements are used with caution the logical approach may become much more versatile and supple without loosing its essential simplicity.

III. Comparison of the results of the various methods. (RICHELLE)

One can find in Chapter XIV (by Richelle), a concrete comparison of the various methods in the case of negative loops comprizing less than three elements.

Intuitively, one expects that a system involving a negative loop will oscillate. In fact, it is advisable to consider the situation both at the local and at the global scale. The regulation brought about by the negative feedback loop might be such that locally the concentration of a macromolecule oscillates imperceptibily around an average value, like the temperature in a very efficient thermostat ; in practice, one would then deal with a steady state both locally and globally, and the "chemical" and "bacterial" systems would not be expected to behave differently. Alternatively, the oscillation might be well marked locally. If one deals with a population of non-interacting cells, this will not be apparent at the global scale (unless the cells are synchronized) while in "chemical" systems the interaction of the local behaviours may result in the emergence of a collective behaviour.

The stochastic approach can be applied to the local and global scales, and give a correct description in both the "chemical" and "bacterial" systems.

The analysis using differential equations with delays where required predicts, depending upon the case, a stable steady state, damped oscillations or sustained oscillations. Unless each cell contains a large number of the relevant regulatory molecules, regulation sites and templates this analysis deals with global behaviours. Clearly, an analysis which would predict sustained oscillations in the case of a "bacterial" system would be inappropriate (see above).

The boolean analysis predicts oscillations at the local scale ; this is usually correct, although practically irrelevant whenever the oscillations are imperceptible. A summation of many boolean simulations gives a simulation of a population of independant subsystems, provided the simulation takes into account the dispersion of the individual values of the delays ; this simulation correctly predicts a stable steady state or damped oscillations at the global level .

IV. Conclusion.

In all cases, the stochastic approach seems to provide the most exact description. Unfortunately, it is extremely heavy, and in practice it cannot be adopted for the routine investigation of complex systems.

The continuous approach based on differential equations is especially useful for systems of the "chemical" type. However, it can be used for systems of the "bacterial" type, provided one takes into account that in reality sustained oscillations at the level of the population require proper interactions between cells.

The boolean approach is especially appropriate for systems of the "bacterial" type. However, it may be really useful to apply it, even for systems of the "chemical" type, in the exploratory stage of the study of a model. This is because it provides a clear view of the various types of behaviours allowed by the logical structure of the system (including the "hidden loops" : see Thomas, last section of Chapter VII). Especially useful is the possibility to know from the logical structure of a system whether it can account at all for oscillations, for multiple steady states, etc... in proper conditions. The general statement is that the presence of negative loop(s) is a necessary condition for oscillations and that the presence of positive loop(s) is a necessary condition for multiple steady states.

PART IV

Applications

(boolean and non-boolean)

CHAPTER XVII

Some biological examples

R. THOMAS

17.1. The immunity of phage λ

As described in chapter II, temperate bacteriophages code for a repressor protein whose presence prevents the synthesis of virtually all the other viral proteins. Whether or not this immunity will be actually established, depends on complex gene interactions, which include several feedback loops. Here I shall present two different formal analyses of this system. The first one is described in more detail in Thomas & Van Ham (1974) ; its predictions have been checked experimentally (Thomas, Gathoye and Lambert, 1975).

17.1.1. We will try here to formalize the system described in chapter II in terms of the interactions between genes N, cI, cII and cro. The symbols used can be found in Table 1.

Usual name for the gene	symbols used here for the gene	its expression	its product
N	N	n	ν
cI	R	r	ρ
cII	C	c	γ
cro	D*	d	δ
(= tof			
= fed			
= etc...)			

Table 1.

Thus, we associate with gene N a boolean variable N, whose value is 1 if the system contains a genetically correct copy of the gene, a boolean function n which takes the value 1 if gene N is on, and a boolean variable ν which takes the value 1 when ν , the product of gene N, is present at an efficient concentrations ; and similarily for the other genes.

* This symbol has been chosen because it is one of the rare letters which correspond to neither of the many synonymous used in the litterature for gene cro!

In the experimental system decribed in chapter II, one uses a mutant of phage λ whose repressor is reversibly inactivated at high temperature. For this reason, "immunity" is thus not identical with "presence of the repressor" but more specifically with "presence of the repressor and low temperature".

Thus, $\quad im = \rho . \overline{T}$

and $\quad \overline{im} = \overline{\rho} + T$.

The model we want to formalize first can be described as follows in verbal terms :

- the synthesis of product ν is prevented by immunity and by product δ
- the synthesis of product δ is prevented by immunity (in a more elaborate model one can take into account the fact that it is also prevented by δ itself).
- the synthesis of γ (the product of gene C) is prevented by immunity and requires the presence of ν (in a more sophisticated version one can take into account an additional negative control exerted by δ) :
- the synthesis of the repressor ρ can be achieved by the establishment mechanism (which requires product γ) or by the maintenance mechanism ; in this case, one way to interprete the experimental situation is to say that the process requires preexisting repressor OR no δ product, and that it does not take place at high temperature.

This can be formalized :

$$\begin{cases} n = N.\overline{im}.\overline{\delta} \\ d = D.\overline{im} \\ c = C.\overline{im}.\nu \\ r = \gamma + (\rho + \overline{\delta}).\overline{T} \end{cases} \qquad \text{with } \overline{im} = \overline{\rho} + T \ (1)$$

Note that the various loops mentioned in chapter II are indeed present in these logical equations. For instance, r (the synthesis of ρ) is under positive control of γ; in turn c (the synthesis of γ) is under negative control of immunity and under positive control of ν ; n (the synthesis of ν) is under negative control and ρ and δ ; d (the synthesis of δ) is under negative control of ρ, etc...

Such a set of logical equations is slightly too complex to be easily treated "by hand". It can be, and has been treated with the computer programs described by Van Ham in chapter IX. However in the simplified experimental system chosen by Eisen, N = 0 (one deals with a N^- mutant), thus n and $\nu = 0$; since $\nu = 0$, c and $\gamma = 0$. Thus the system simplifies into :

$$\begin{cases} r = (\rho + \overline{\delta}).\overline{T} \\ d = D.\overline{im} \end{cases}$$

or, more explicitly :

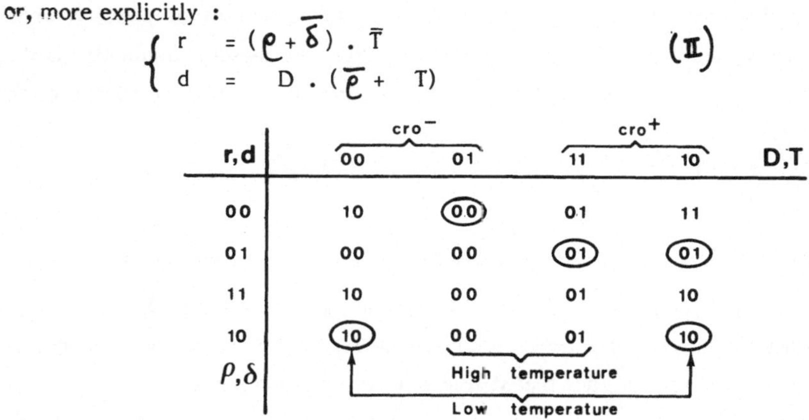

$$\begin{cases} r = (\rho + \bar{\delta}) \cdot \bar{T} \\ d = D \cdot (\bar{\rho} + T) \end{cases} \qquad \text{(II)}$$

r,d \ ρ,δ	00 (cro⁻)	01 (cro⁻)	11 (cro⁺)	10 (cro⁺)
	D,T			
00	10	(00)	01	11
01	00	00	(01)	(01)
11	10	00	01	10
10	(10)	00	01	(10)

Table 2. The state table for the system of equations (II).

It is seen that at high temperature, whether (column 11) or not (column 01) the prophage is cro$^+$, there is a single stable steady state. At low temperature, there is a single stable steady state for the cro$^-$ prophage but <u>two stable steady states for the cro$^+$prophage</u> (column 10). The state table thus clearly visualizes the fundamental result of Eisen et al that in cro$^+$ lysogens the same genome, in the same conditions (low temperature) can display in a permanent way either of two completely different phenotypes.

From the state table, one can easily reconstitute the thermal cycles described by Eisen et al. For a cro$^-$ prophage :

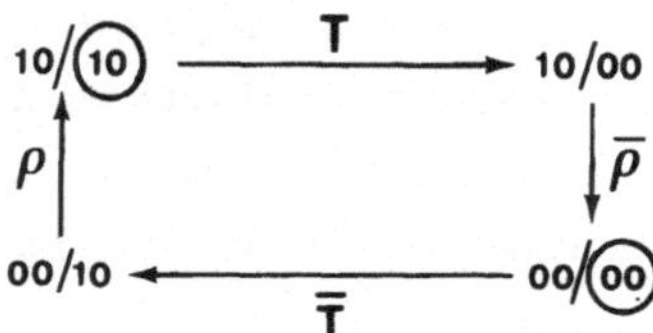

From the stable state 10 /(10) at low temperature (column 00), increasing the temperature (T), a change in an input variable, visualized by an horizontal shift, leads to column 01 , state 10 / 00 . This unstable state leads to 00 /(00), and this situation does not change as long as temperature remains high. If one decides to lower the temperature ($\bar{T}$) , the systems proceeds back from column 01 (state 00 /(00)) to column 00 (state 00 / 10) ; from this unstable state the system returns spontaneously to the initial state 10 / (10) . In this system, temperature thus induces a reversible change from the immune state 10 /(10) to the non-immune state 00 /(00) and vice versa.

For a cro$^+$ prophage the situation is as follows :

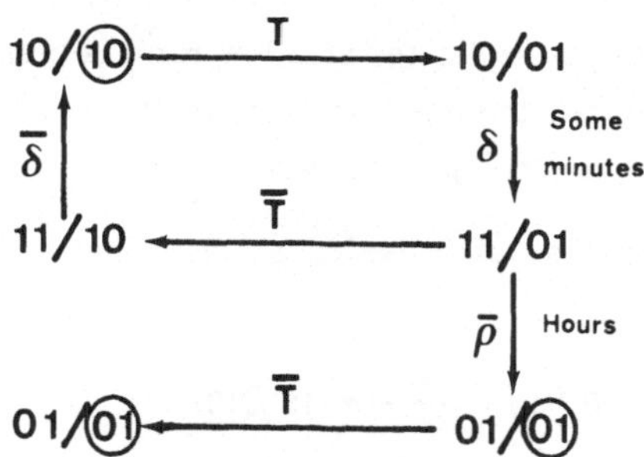

From state 10 / (10), a shift to high temperature leads to state 10 / 01 (or T $\bar{0}$), which might in principle lead to 00 / 01 or to 11 / 01 . But the decay of the repressor takes many hours whereas the synthesis of the cro product takes only some minutes ; thus the sequence is 10 / 01 $\xrightarrow{\delta}$ 11 / 01 . If one now returns to low temperature, the sequence is 11 / 01 $\xrightarrow{\bar{T}}$ 11 / 10 $\xrightarrow{\bar{\delta}}$ 10 / (10) ; but if one waits for some hours before to go back to low temperature, the sequence is 11 / 01 $\xrightarrow{\bar{\rho}}$ 01 / (01) $\xrightarrow{\bar{T}}$ 01 / (01) . Thus, if the temperature shift is short, the loss of immunity is reversible like in the case of a cro$^-$ prophage. But if the system remains long enough at high temperature the loss of immunity becomes irreversible. One can thus have cultures of the same lysogenic strain, in the same conditions (low temperature) which display in a stable way two radically different phenotypes depending on the past history of the system.

What preceeds deals with lysogens. Can one, from the state table (Table 1) interprete known facts about the behaviour of phages λN^- cro$^+$ and λN^- cro$^-$ <u>at infection</u>, and perhaps find predictions which could be subjected to experimental verifications ? In the preeceeding section, we started from the stable steady states to see how they can be perturbed by temperature shifts. In the case of phage infection, the initial state is the one for which the viral products are still absent (row 00 of table 1). For instance, in the case of infection with a λN^- cro$^+$ at low temperature, the initial state is found at the intersect of row 00 and of column 10 ; for this state the values of ρ, δ / r , d are : 00 / 11 . Here, there are in principle two possibilities :

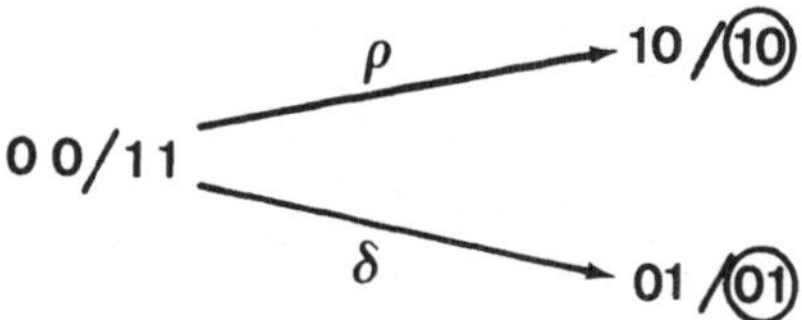

In the first case, immunity would be established. We know that this would prevent the phage from replicating ; and since N^- phage is known not to integrate, it should be lost by dilution. In the second case, immunity is not established ; the phage can thus replicate at the low rate permitted for a N^- phage. This situation is well known (Signer, 1969 ; Lieb , 1970); the phage, which cannot integrate but replicates at a low rate, is perpetuated as a plasmid. In the context of the present model, the high incidence of this response would mean that the delay t_δ is, on the average, shorter than the delay t_ρ.

In the case of infection at low temperature with a λN^- cro^-, the state table gives :

$$00 / 10 \xrightarrow{\rho} 10 / \textcircled{10}$$

Thus, here the matrix predicts only one pathway ; the corresponding pathway was not frequently followed in the case of a λN^- cro^+ phage because it was competed by a faster one. The prediction is that a λN^- cro^- ,in spite of its N^- character, will be able to establish immunity not only occasionally but systematically.

An experimental program was set up to check this prediction (Thomas, Gathoye & Lambert, 1975). It was found indeed that in spite of its N^- character, λ N^- cro^- establishes immunity at a frequency near to 100 % .

If this was all, and the phage was unable to integrate, it would result in a progressive dilution of the phage ; the bacteria would first be all immune, but as they multiply they would progressively lose their phage, and later their immunity. What is found in fact is that not only is immunity efficiently established by λN^- cro^-, but in addition this phage can efficiently integrate, so that, in striking contrast with λ N^- cro^+ , it lysogenises very efficiently.

The ability of λN^- cro^- to integrate can be interpreted in terms of the fact that function int is a catalytic one (while the excision function xis apparently acts in a stoichiometric way).

This gives an opportunity to discuss the problem of how to write the equations for a function which is under more than one control.

Genes Int and Xis are both in the N operon, which is regulated as follows, taking into account that in the presence of the cro product the rate of initiation is reduced by some factor-say 5- and that in the absence of the N product only a limited fraction -say 1 / 10- of the transcriptional waves proceed beyond the terminator.

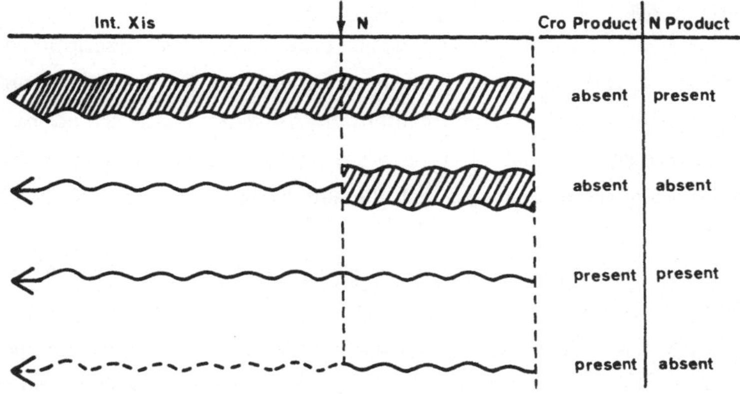

Fig.1. Regulation of the N,... Int, Xis operon by the products of genes N and cro (in the absence of immunity .This schema will have to be revised in view of the recent finding (Katzir et al,1976; Chung & Echols, 1977 ; Court et al, 1977) that Int, but not Xis, can be synthesized by a mechanism involving positive regulation by the products of genes cII and cIII; in practice, whatever the mechanism, Int is expressed if $\gamma + \overline{\delta}$ and Xis is expressed only if $\gamma . \overline{\delta}$.

Assuming these rough estimations, one gets (Table 3) :

γ, δ	Relative rates of transcription
1 0	1 0 0 → $\gamma . \overline{\delta}$ (1 0 0, 2 0, 1 0: $\gamma + \overline{\delta}$)
1 1	2 0
0 0	1 0
0 1	2

Table 3 : Relative rates of transcription of the genes beyond the terminator, based on the crude assumptions given in the text.

Suppose now that a function is "stoichiometric" and requires more than 20 % of the normal transcription in order to be effective. In this case, only the first situation ($\gamma . \overline{\delta}$) is permissive. For a "catalytic" function, 5 % of the normal transcription rate may be sufficient. In this case the three first conditions in table 1 are permissive ; one can thus write that the conditions for the expression of this catalytic gene are $\gamma + \overline{\delta}$.

Genes Int and Xis are under the same controls. That integration, but not excision, is very efficient in λN^- cro^- phage can be interpreted in terms of different quantitative requirements for products Int and Xis. More specifically, one could write :

$$i = I \, . \, \overline{im} \; (\gamma + \overline{\delta})$$

$$x = X \, . \, \overline{im} \; . \, \gamma \, . \, \overline{\delta}$$

The problem of the relationship between integration and the establishment of immunity will be treated formally in section I7.3.

17.2. Another analysis of the control of immunity.

The preceeding section provided a formal description of a moderately complex model. A rather detailed analysis was given only for the much simplified situation chosen by Eisen et al in their experimental work (N^- mutants). In the normal (N^+) situation, the system is already too complex to be conveniently treated "by hand" ; however this, and more complex ones, can by analyzed with the automatized methods described by P. Van Ham in chapter IX.

Now, I would like to give another analysis of the decision to establish or not immunity, based on a simplified model, which can be treated "by hand" even in the normal (N^+) situation. This analysis is given in order to show how one can proceed by using first a model deliberately much too simple, adding progressively the required interactions and looking what is and what is not changed in the operation of the system. Of course, this has been a current practice ; the only difference is the formalized character of the process as it is used here.

The essential feature of the immunity system is clearly the decision between two stable* states, one in which immunity is established and persists, one without immunity. As we already know (see chapter VII) the simplest way to account for this situation is a simple positive feedback loop, and the simplest positive feedback loop corresponds to autocatalysis (in genetic examples, α , the product of gene A, is required for its own synthesis : $\overset{+}{\circlearrowleft}\alpha$).

One easily understands that such a system can be blocked permanently in the "on" or in the "off" position ; but how was the decision taken ? In this extremely simplified case, it looks like a vicious circle. However we know that if one grafts a negative control on a positive loop with a OR connection, the system still displays two stable steady states :

$\rho \xrightarrow{-} \alpha$ (or; $+$ autocatalytic loop on α)

$$\begin{cases} r = 1 \\ a = \bar{\rho} + \alpha \end{cases}$$

ρ, α	r, a
0 0	1 1
0 1	1 1
1 1	(1 1)
1 0	(1 0)

$\bar{0}\bar{0} \xrightarrow{\rho}$ (10); $\bar{0}\bar{0} \xrightarrow{\alpha} \bar{0}1 \xrightarrow{\rho}$ (11)

Now, the decision appears as a race between the expression of genes R and A ; if R wins, the system is blocked in the (10) state, if A wins, it is blocked in the (11) state. Note that in both cases gene R is expressed in the final state ; but in the second case it has been expressed too late to prevent the loop from being switched on.

Can one apply this type of model to the case of immunity ? We know that cI is under positive control of cII for the establishment and of itself for the maintenance, and that cII is under negative control of the cro product. Leaving for the moment any other interaction, this leads to the provisional model of Fig. 2a (in which ρ, $\alpha 1$ and $\alpha 2$ represent the products of genes cro, cII and cI, respectively.

* The non-immune and the immune states can be treated as stable states only as long as one is primarily interested in the decision for or against the establishment of immunity. As soon as one becomes interested in subsequent events, one has to take into account that the non-immune situation will normally lead to phage replication and maturation and to cell lysis, and that the immune situation will be stable only if the phage is integrated.

In fact, the interactions are notably more complex. In addition to those just mentioned, one must take into account :

- that the cI product prevents expression of genes cII and cro (here, $\alpha 2$ prevents the synthesis of ρ and $\alpha 1$: Fig.2b)
- that in addition the cro product prevents the maintenance mechanism of immunity (here, product ρ exerts a negative control on the autocatalytic synthesis of $\alpha 2$: Fig.2c)

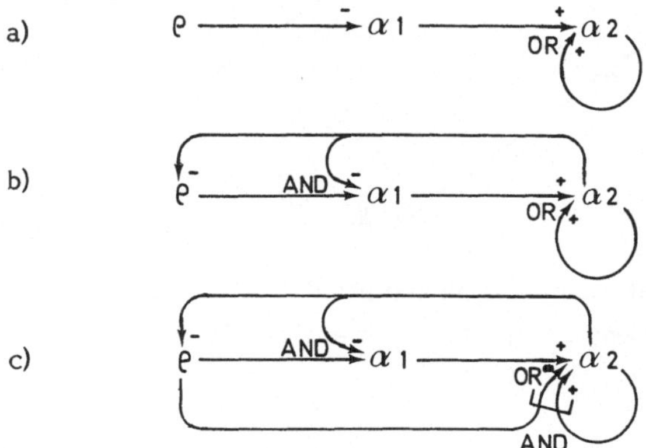

Fig.2. Three models taking into account an increasing number of interactions between ρ, $\alpha 1$ and $\alpha 2$ (which symbolize here the products of genes cro, cII and cI, respectively).

If one introduces successively these two additional interactions, the diagram becomes more and more complex. An explicit diagrammatic description may even become difficult ; I doubt if the situation of $\alpha 2$ in diagram 2 is clear to everybody in the absence of additional comments. It is comforting, however, to notice that in contrast with the diagrams, the equations are only slightly more complicated in cases (b) and (c) than in case (a) and that the state tables for these different cases are different, of course, but of the same complexity.

(a)

$$\begin{cases} r = 1 \\ a1 = \bar{\rho} \\ a2 = \alpha 1 + \alpha 2 \end{cases}$$

$\rho, \alpha 1, \alpha 2$	r, a1, a2
$\bar{0}\,\bar{0}\,0$	1 1 0
$\bar{0}\,\bar{0}\,1$	1 1 1
$\bar{0}\,1\,1$	1 1 1
$\bar{0}\,1\,\bar{0}$	1 1 1
$1\,\bar{1}\,\bar{0}$	1 0 1
$1\,\bar{1}\,1$	1 0 1
1 0 1	1 0 1
1 0 0	1 0 0

(b)

$$\begin{cases} r = \overline{\alpha 2} \\ a1 = \bar{\rho} \cdot \overline{\alpha 2} \\ a2 = \alpha 1 + \alpha 2 \end{cases}$$

$\rho, \alpha 1, \alpha 2$	r, a1, a2
$\bar{0}\,\bar{0}\,0$	1 1 0
0 0 1	0 0 1
$0\,\bar{1}\,1$	0 0 1
$\bar{0}\,1\,\bar{0}$	1 1 1
$1\,\bar{1}\,\bar{0}$	1 0 1
$\bar{1}\,\bar{1}\,1$	0 0 1
$\bar{1}\,0\,1$	0 0 1
1 0 0	1 0 0

(c)

$$\begin{cases} r = \overline{\alpha 2} \\ a1 = \bar{\rho} \cdot \overline{\alpha 2} \\ a2 = \alpha 1 + \bar{\rho} \cdot \alpha 2 \end{cases}$$

$\rho, \alpha 1, \alpha 2$	r, a1, a2
$\bar{0}\,\bar{0}\,0$	1 1 0
0 0 1	0 0 1
$0\,\bar{1}\,1$	0 0 1
$\bar{0}\,1\,\bar{0}$	1 1 1
$1\,\bar{1}\,\bar{0}$	1 0 1
$\bar{1}\,\bar{1}\,1$	0 0 1
$\bar{1}\,0\,\bar{1}$	0 0 0
1 0 0	1 0 0

Table 4.

Table 4. The equations and state tables for the models of Fig.2.

The simple model 2a leads to the graph :

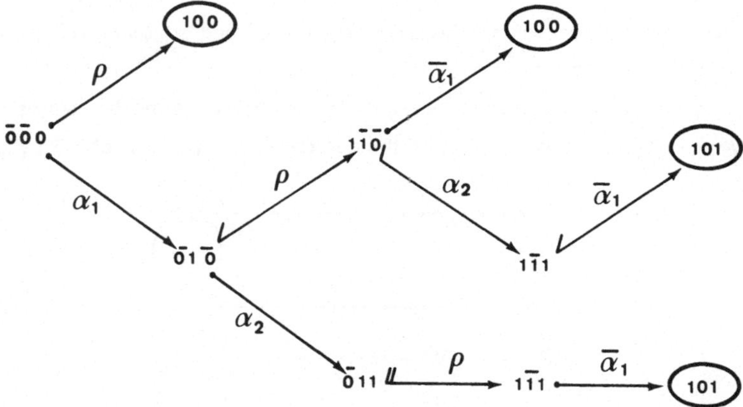

One can now ask : what factors decide whether the system will finally reach state (100) or state (101) ? A treatment similar to the one given at the end of chapter VI gives the following result :

If

$$m \equiv t\rho < t\alpha_1$$

$$n \equiv t\rho < t\alpha_1 + t\alpha_2$$

$$p \equiv t\rho + t\bar{\alpha}_1 < t\alpha_1 + t\alpha_2$$

the system will reach (100) iff $m + p$; it will reach (101) iff $\overline{m}.\overline{p}$

The slightly more complex model 2b leads to the graph :

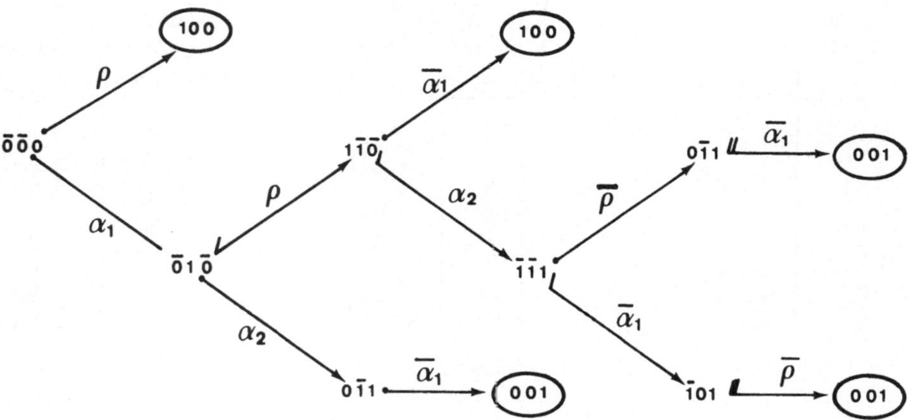

This modified model takes into account the fact that in the immune state gene cro (here symbolizcd R) is off. In spite of the increased complexity of the system, the conditions which lead to the two stable states are the same as in case (a); the immune state (001) will be reached iff $\bar{m}.\bar{p}$.

System (c) leads to the following graph :

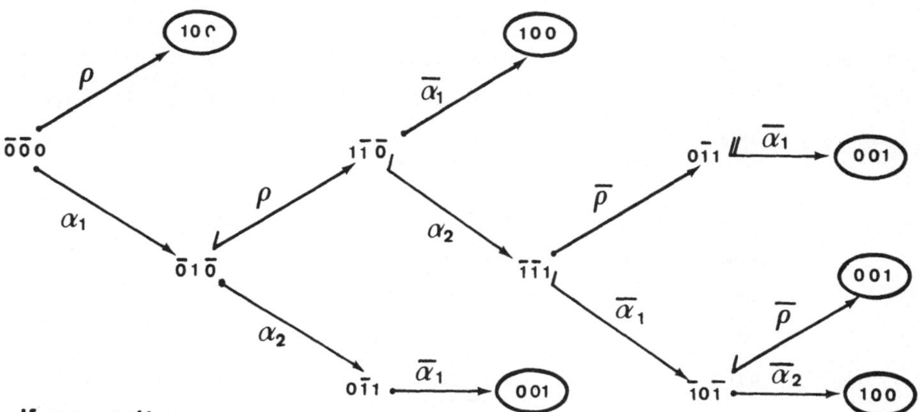

If we write:

$$m \equiv t\rho < t\alpha_1$$

$$n \equiv t\rho < t\alpha_1 + t\alpha_2$$

$$p \equiv t\rho + t\bar{\alpha}_1 < t\alpha_1 + t\alpha_2$$

$$q \equiv t\alpha_1 + t\alpha_2 + t\bar{\rho} < t\rho + t\bar{\alpha}_1$$

$$r \equiv t\alpha_1 + t\alpha_2 + t\bar{\rho} < t\rho + t\bar{\alpha}_1 + t\bar{\alpha}_2$$

The conditions leading to the immune state (001) are :

$$\bar{m}\ (n\bar{p}q + n\bar{p}\bar{q}r + \bar{n})$$

The conditions leading to the non-immune state (100):

$$m + \bar{m}n\ (p + \bar{p}\bar{q}\bar{r})$$

Fortunately, these expressions can be greatly simplified in view of the constraints between variables m , n , p , q and r .

(see the end of chapter VI)

These constraints are :

m implies n , or $\overline{m\bar{n}}$

$\bar{n}$ implies $\bar{p}$, or $\overline{\bar{n}p}$

p implies $\bar{q}$, or $\overline{pq}$

q implies r , or $\overline{q\bar{r}}$

If these unacceptable states are indicated by dashes on the state table, and the situations leading to the immune state (001) and to the non-immune state (100) , respectively by 1 and 4 , it comes (Table 5)

p,q,r \ m,n	00	01	11	10
000	1	4	4	–
001	1	1	4	–
011	1	1	4	–
010	–	–	–	–
110	–	–	–	–
111	–	–	–	–
101	–	4	4	–
100	–	4	4	–

Table 5. The conditions (combinations of values of m , n , p , q , r) which lead to state (001) or to state (100). States (001) and (100) are represented here by their decimal (or octal) equivalents, 1 and 4.

The simplified expressions for the conditions are now :

the system proceeds to (001) if $\bar{n} + \bar{m}\,\bar{p}\,r$ and to (100) if $m + p + n\bar{r}$ *

---

* Note that the two expressions are not strictly complementary (see chapter IV, section 4.7, last §).

If one compares the conditions leading to the non-immune state (001) , in models (a) and (b) on the one hand, (c) on the other hand, one sees that in the last model there is an additional possibility to escape immunity : n $\bar{r}$, this is,

$$t_\rho < t_{\alpha 1} + t_{\alpha 2}$$

and

$$t_{\alpha 1} + t_{\alpha 2} + t_{\bar{\rho}} > t_\rho + t_{\overline{\alpha 1}} + t_{\overline{\alpha 2}}$$

17.3. <u>Integration-excision in lambdoid phages ; its relation with immunity.</u>

"Normal" lysogenization requires both the establishment of immunity and the integration of the phage genome into the bacterial chromosome ; but immunity interferes with the synthesis of the products involved in the integration-excision process. In this section, I present an analysis focused on the interactions between immunity and integration-excision rather than on the details of the controls exerted on these processes.

17.3.1. But I must first give some background about the processes of integration and excision in lambdoid phage. The general principles were discovered by Campbell (1962). In the mature phage, the chromosome is a linear DNA molecule with "sticky ends" (⤚⟶). Soon after infection this structure circularizes (⟲ see Fig.3). A region of the bacterial chromosome and a region of the viral chromosome somehow recognize each other and a reciprocal genetic exchange takes place between these regions ; this results in the integration of the phage genome (then called prophage) in the bacterial genome. Excision is supposed to take place by an inverse mechanism. Abnormal excision (very rare) results in a so-called <u>transducing</u> phage particle, which carries some bacterial genes adjacent to the prophage, usually at the expense of viral genes at the other end of the prophage ; for instance, λgal phage carry the bacterial operon involved in galactose catabolism, but they are usually defective because the "right" part of the prophage is missing.

It was thought initially that the regions of the bacterial and viral chromosome involved in the processes of integration and excision were homologous (that is, had identical or closely similar nucleotide sequences). However, it was soon realized (Guerrini, 1969) that this is not the case. There is probably a very short (some nucleotide pairs) common segment, whithin which the recombination takes place, flanked by regions which are symbolized BB' on the bacterial chromosome, PP' on the phage chromosome ; and B , B' , P and P' are different from each other. Normal integration, normal excision and illegitimate excision are schematized in Fig.3. Other situations are described in Fig.4., 5 and 6. For a well-documented description, see Gottesman § Weissberg (1971).

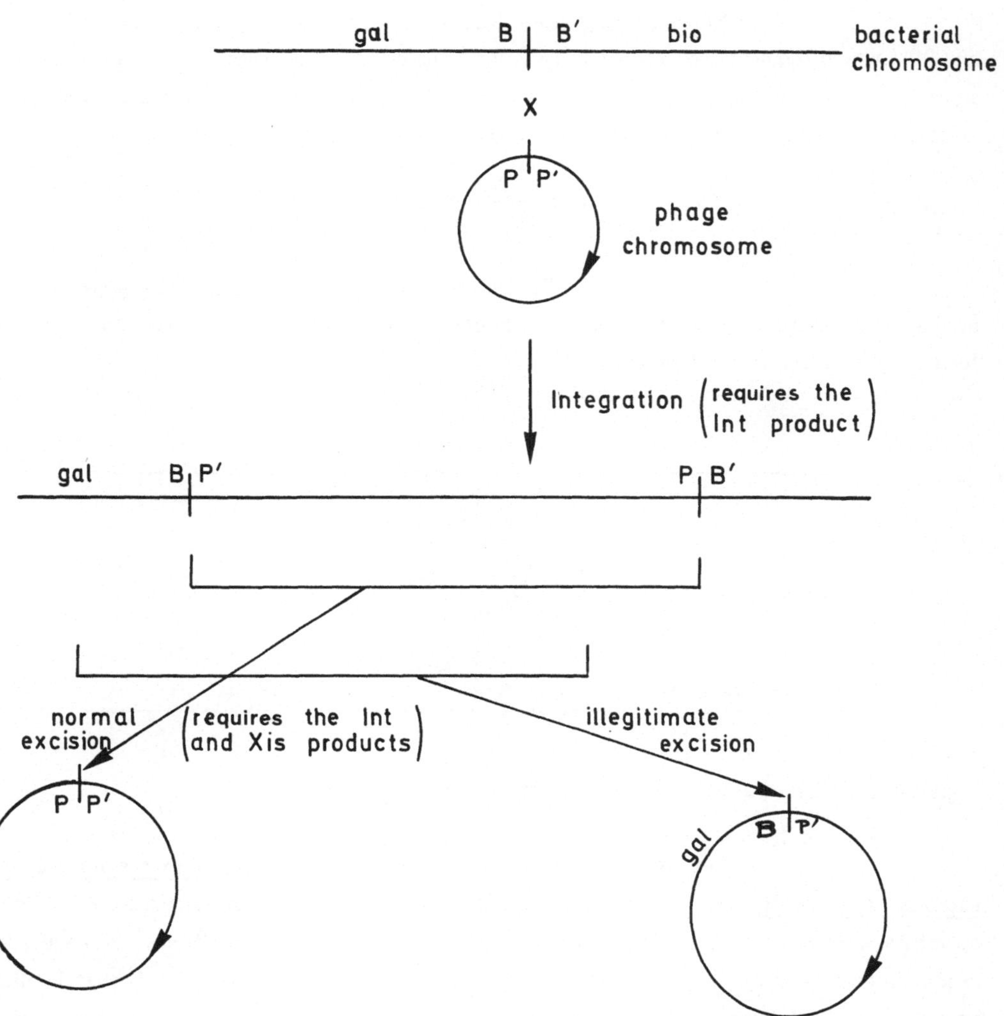

Fig.3. Integration ; normal and illegitimate excision. It is seen that illegitimate excision can produce a phage carrying the gal operon from the bacterial host ; in these particles the zone relevant to the integration-excision process is BP' instead of PP' .

Normal integration (Fig.3.) involves an interaction between BB' and PP', which requires the product of gene Int. The ends of the resulting prophage have the structures : BP' and PB'. Normal excision involves an interaction between these new structures, which requires the products of genes Int and Xis. Thus :

$$\text{B B'} + \text{P P'} \underset{\text{Int, Xis}}{\overset{\text{Int}}{\rightleftharpoons}} \text{B P'} + \text{P B'}$$

In the case of a <u>Xis</u> mutant (or of a N^- cro^- mutant), gene Int is normally expressed but gene Xis is not (or not efficiently). Since the Xis product is required for excision, integration is irreversible :

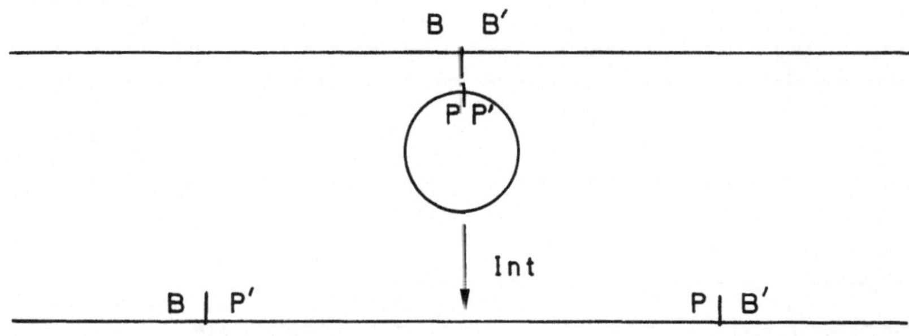

Fig.4. Integration of a Xis^- (or a N^- cro^-) mutant.

Let us now turn to the integration of <u>a λgal to the left of an established prophage</u>. As shown above (Fig.3), a λgal phage has the structure BP'. This structure does not interact efficiently with the normal bacterial site (BB'). However, it readily interacts with the left end of a prophage (BP'). One can see (Fig.5) that in this particular case the interacting structures are the same for integration and for excision ; experiments show that the Int product, but not the Xis product, is required :

$$\text{B P'} + \text{B P'} \underset{\text{Int}}{\overset{\text{Int}}{\rightleftharpoons}} \text{B P'} + \text{B P'}$$

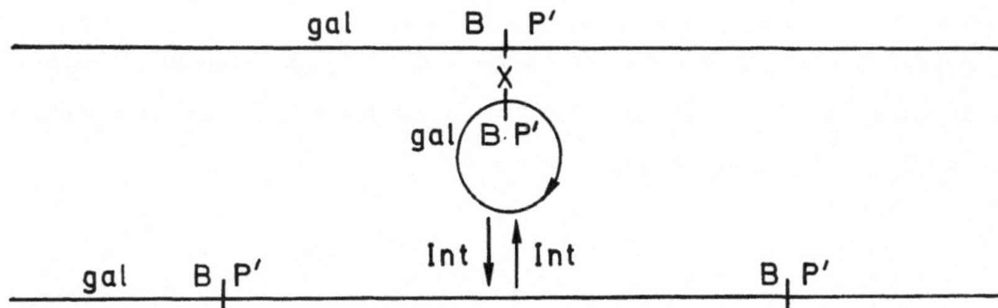

Fig.5. Integration of a λgal to the left of an established prophage.

Finally, we shall consider the interaction of a λgal with the right end of an established prophage. In this case, (Fig.6) the interaction which results in integration is PB' x BP' (the same interaction results in excision in the "normal" case) ; conversedly, the interaction which results in excision is PP' x BB' (the same interaction results in integration in the "normal" case). Accordingly, when a λ gal interacts with the right end of a prophage, integration would require the Int and Xis products (like excision in the normal case), while excision would require only the Int product (like integration in the normal case). Why this combination is inefficient, is briefly discussed below.

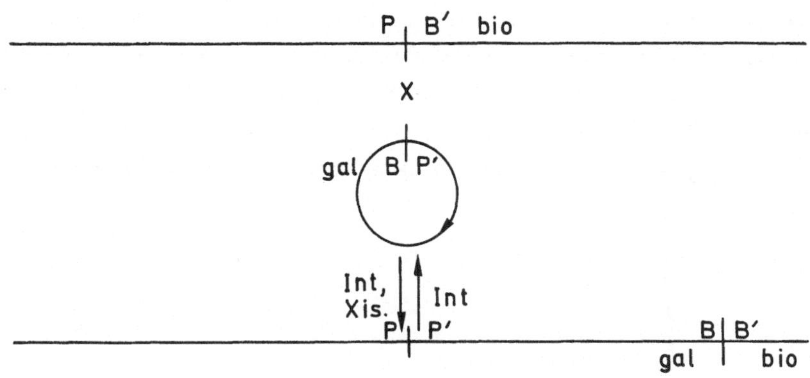

Fig.6. Interaction of a λgal with the right end of an established prophage.

17.3.2. Formal analysis.

We already know that the establishment of immunity and the process of integration-excision are each subject to complex controls. It would be possible to proceed to a complete analysis in which all the known controls are taken into account ; in view of the number of variables,this could not be done "by hand", but such models can be treated with the logical machine "Delphine" (see chapter VIII by Van Ham) or with computer programs (see chapter IX by Van Ham). Here I choose rather to treat the other controls in a simplified way and to focus a) on the interaction between immunity and integration-excision and b) on the structural aspects described in 17.3.1. To be more specific, I consider only the fraction of the bacterial population in which immunity is, at least transiently, established, because the other cells will lyse and they have thus no opportunity to become lysogens. If one takes this attitude, one can provisionally ignore the factors of the decision to establish or not immunity, and simply reason as if gene cI (here called R)were expressed constitutively ; $r = 1$, and the repressor ρ, first absent, will appear after a delay t_{ρ} whose average value depends on the genetic and physiologic circonstances. In the formalism used (see chapter VI) R, I and X represent the genes (cI, Int, Xis, respectively), r, i and x represent the state of expression (on or off) of these genes and ρ, ι and ξ represent the presence of the gene products.

In these cells, after a transitory period during which the phage can integrate and excise conceivably several times, there is finaly a decision : either the phage is and remains integrated, or it is and remains non-integrated (see below). The first situation really leads to a stable state (the lysogenic bacterium) in which immunity is present but since the phage is integrated it is passively replicated and transmitted to the progeny. The alternative situation does not really lead to a stable state, since in the presence of immunity a non-integrated phage is not replicated ; it will thus be diluted out and immunity itself will be eventually lost because the cells contains no more copy of the phage genome. This could very well be formalized. However in this analysis I do not consider the late-occuring process of the loss of the phage and of immunity. Finally, I assume that in the absence of immunity the various positive controls involved in the synthesis of the enzymes Int and Xis are expressed, and that in practice only the establishment of immunity will interfere with the synthesis of these enzymes. Both for the establishment of immunity and for the control of integration-excision, the simplification amounts to replace a detailed analysis of the expression of the genes by a characteristic delay between infection and the expression of each gene.

With these simplificative assumptions, it comes :

$$\begin{cases} r = R \\ i = I \cdot \overline{\varrho} \\ x = X \cdot \overline{\varrho} \end{cases}$$

As regards the process on integration-excision itslef, let λ mean that there is an integrated copy of the phage under study in the bacterium, $\overline{\lambda}$ that there is no integrated copy. Function ℓ, of which λ is the memorization variable, has the value 1 if there is not yet a phage integrated but conditions are such that a phage can integrate, or if there is an integrated copy and the conditions are such that it cannot excise. In the four cases considered above, the equations for ℓ will be, respectively :

a) Normal case : $\ell = \overline{\lambda}\iota + \lambda(\overline{\iota} + \overline{\xi})$

The first term of ℓ considers the situation in which there is not yet an integrated phage but enzyme Int is present so that the condition for the integration are fulfilled ; the second term considers the situation in which there is a phage integrated and it will not excise because enzyme Int or enzyme Xis or both are absent.

b) Xis$^-$ (or N$^-$) phage :

$$\ell = \overline{\lambda}\iota + \lambda \text{ (or, more simply, } \ell = \iota + \lambda)$$

c) λ gal to the left of a prophage :

$$\ell = \overline{\lambda}\iota + \lambda\overline{\iota}$$

d) λgal to the right of a prophage :

$$\ell = \overline{\lambda}\iota\xi + \lambda\overline{\iota}$$

Simplified as it is, the model leads to an extremely rich behaviour.

I shall first consider situation (b), (Xis^-, or N^- phage) which is the simplest one :

$$r = R \qquad i = I \cdot \overline{\rho} \qquad \ell = \iota + \lambda$$

or, if R and I = 1

$$r = 1 \qquad i = \overline{\rho} \qquad \ell = \iota + \lambda$$

a logical structure which can be represented : $\rho \xrightarrow{-} \iota \xrightarrow[\text{OR}]{+} \lambda$ (with a positive loop "+" on λ)

This logical structure is already familiar to us (see this chapter section 17.2.,model a). We know that such a structure provides a choice between two stable steady states, in this case, one with a phage stably integrated, one with no phage integrated. The matrix, the graph and the analysis of the conditions leading to the two stable states are the same as in section 17.2, model a. Let us simply mention that the conditions to get a lysogen (stable state (1 0 1) are : $\overline{m}\ \overline{p}$, this is :

$$t_{\rho} > t_{\iota}$$

$$t_{\rho} + t_{\overline{\iota}} > t_{\iota} + t_{\lambda}$$

Thus, in order to obtain a stable lysogen following infection with a Xis^- (or a N^- cro^-) phage, two conditions must be fulfilled :

- the establishment of immunity must take more time than the production of the Int enzyme.
- once immunity is established gene Int is switched off and its product begins to decay , the sum (establishment of immunity plus decay of the Int product) must be longer than the sum of the delays leading to the appearance of the Int product and subsequently to effective integration.

One may ask why this model involves a positive loop, formally equivalent to an autocatalytic process. This is because the maintenance of the integrated state does not require the continuous presence of the Int product ; once realized, integration persists, unless the Int and Xis product are both present (and this never happens in a Xis^- phage).

Let us now examine the integration of a λgal to the left of a prophage (case c).

$$r = R$$

$$i = I \cdot \overline{\rho}$$

$$\ell = \overline{\lambda}\,\iota + \lambda\,\overline{\iota}$$

If genes R and I are normal, this becomes :

$$r = 1$$

$$i = \overline{\rho}$$

$$\ell = \overline{\lambda}\,\iota + \lambda\,\overline{\iota}$$

The state table is found in table 6(a). Note that in this system (as well as in the others in the group) the behaviour of r/ρ is especially simple ; r = 1 , and ρ , initially 0 , will reach the value 1 after a delay t_ρ and keep this value. In such cases, it is often convenient to treat ρ as if it were an input variable (table 6 ,b).

a)

r, i, ℓ	... R,I = 1 ...
0 0 0	1 1 0
0 0 1	1 1 1
0 1 1	1 1 0
0 1 0	1 1 1
1 1 0	1 0 1
1 1 1	1 0 0
1 0 1	(1 0 1)
1 0 0	(1 0 0)
ρ, ι, λ	

b)

i, ℓ	R, I = 1: ρ = 0	R, I = 1: ρ = 1
0 0	1 0	(0 0)
0 1	1 1	(0 1)
1 1	1 0	0 0
1 0	1 1	0 1
ι, λ		

Table 6 : state tables for the integration of a λgal to the left of a prophage.

From table 6 ,b, it is apparent that as long as immunity is absent (ρ = 0) the graph is :

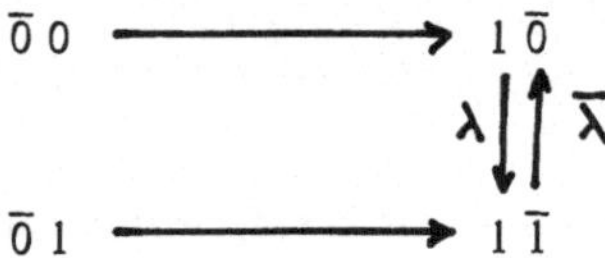

which may involve repeated integrations and excisions. But as soon as immunity is present (ρ = 1) the synthesis of the Int product is blocked and the Int product present will slowly decay ; the system moves towards one of two stable steady states.

An interesting representation of the situation is given by the cubic graph (Miller, 1965, Kunzmann, 1965 ; for systems of biological interest, Glass, 1975 ; Thomas & Van Ham, 1974) (Fig.7). A difference with the graphs of Glass is that there may be two arrows on the same edge of the cube ; this is because we admit the possibility of self-input. Here, λ appears in the equation of ℓ . As a result, the system can oscillate ($\lambda \rightleftarrows \bar{\lambda}$), either before ($\bar{0}\,1\,\bar{0} \rightleftarrows \bar{0}\,1\,\bar{1}$) or after ($1\,\bar{1}\,\bar{0} \rightleftarrows 1\,\bar{1}\,\bar{1}$) immunity has been established.

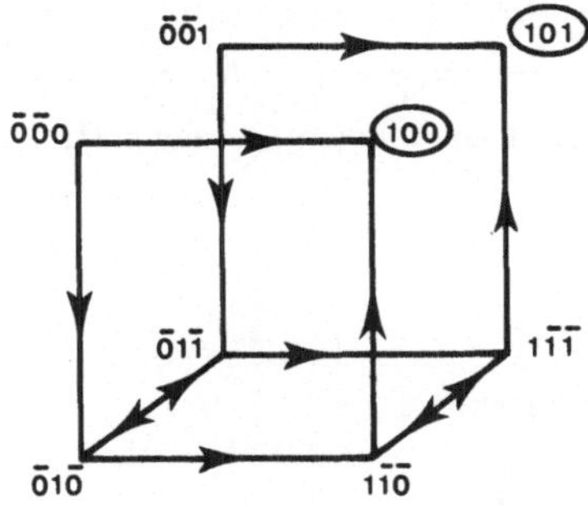

Fig.7. The cubic graph of the system:interaction of a λ gal to the left of an integrated prophage.

The two remaining cases (normal situation, and integration of a λgal to the right of a prophage) are more complex, as not only Int, but also Xis are involved.
Table 7 and Fig.8 give, respectively, the states table and the cubic graph in the case R, I, X = 1 , this is, genes cI , Int and Xis are normal. Variable ρ has been formally treated as an input variable for sake of clarity.

		R,I,X = 1
i,x,ℓ	ρ=0	ρ=1
000	110	(000)
001	111	(001)
011	111	001
010	110	000
110	111	001
111	110	000
101	111	001
100	111	001
ι,ξ,λ		

Table 7. The states table for the integration and excision of a normal prophage.

In this perspective, Fig.8 shows the situation as two cubes corresponding, one to the situation $\rho = 0$, the other to the situation $\rho = 1$. As long as $\rho = 0$ the trajectory of the system is on the cube to the left, but at any moment ρ may shift to one and then the system jumps to the cube to the right.

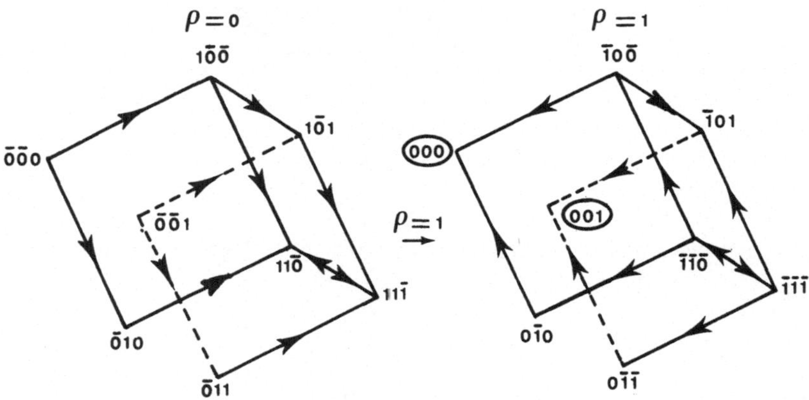

Fig.8 The cubic graph for the integration and excision of a normal prophage. This graph derives from Table 7.

One can proceed to the complete logical analysis of such a system, and find the exact conditions (relative values of the time delays or linear combinations of them) which lead to each of the possible pathways. Another line of reasoning is the following. In each individual case (in each bacterial cell) each delay has a well-defined value, and the set of values of the delays impose a well-defined pathway. However, if one considers the whole population, the value of a delay will vary from cell to cell ; each delay can be characterized, not by a precise value, but by an <u>average value</u> and a <u>distribution function</u>. One way to analyze the system consists thus of giving a computer :

- the set of logical equations defining the logical structure of the system.
- the average value ascribed to each time delay.
- the distribution of the values of the time delays.

One can then make a number of simulations, each representing an individual cell, and count how often each of the possible pathways has been followed.

It was felt that in many cases it would be reasonable to use a normal distribution with a dispersion roughly proportional to the duration of the delay itself. These ideas have been materialized in the program PRAN2 of Van Ham (see Chapter IX). In this program, for sake of simplicity,a square distribution is used, in which the actual value chosen for a delay of average value d is drawn at random in the interval $d \pm 0.2\,d$ or $d \pm 0.1\,d$ *

Here, the problem was treated as follows. I was interested in how the frequency of the integrated state would be affected by varying the time required for the establishment of immunity, in the four different cases discussed above. For that, a reasonable (I hope) mean value was ascribed to each delay, except for the delay t_{ϱ} , for which a whole set of values was used. In addition, two cases were considered, one in which the decay of the Xis product is rapid (4 arbitrary time units) as it is presumably the case (Weisberg and Gottesman, 1971), one in which it is much slower (17 time units). A number of simulations were performed in each of the four situations described above, with two different values of t_{ξ} and a number of different values of t_{ϱ} . For each set of values of the delays, one measured the fraction of the simulations which lead to stable integration.

Fig.9. a, b, c and d give the frequencies of integration as a function of t_{ϱ} , the time required for the establishment of immunity ; for each figure there are two curves, one for $t_{\xi} = 4$ time units, the other for $t_{\xi} = 17$ time units . In Fig.9 b and c, the curves corresponding to $t_{\xi} = 4$ time units and $t_{\xi} = 17$ time units of course coincide (for the simple reason that in these two cases x does not appear in the equations). The difference between b and c is that in the first case the frequency of integration tends to 1 as the appearance of immunity is delayed whereas in case c this is not so.

An interesting case is shown by Fig.9 d,. When t_{ξ} is short (as it should) there is 0 % stable integration. This is because in this combination, once the Xis product has disappeared the phage cannot integrate any more but it can still excise as long as there is Int product left. (Weisberg & Gottesman, 1971).

---

* It must also be mentioned that in this program the value of each delay is chosen for each individual simulation, but if a given delay appears more than once in a simulation, the same value is used. This attitude amounts to consider that the actual value of each delay behaves essentially as a property or each individual cell. Another extreme attitude consist of saying that the dispersion of the values of a delay is as wide for the successive occurence of a delay within a same cell as its dispersion within the cell population. This point is discussed by Richelle in chapter XIV.

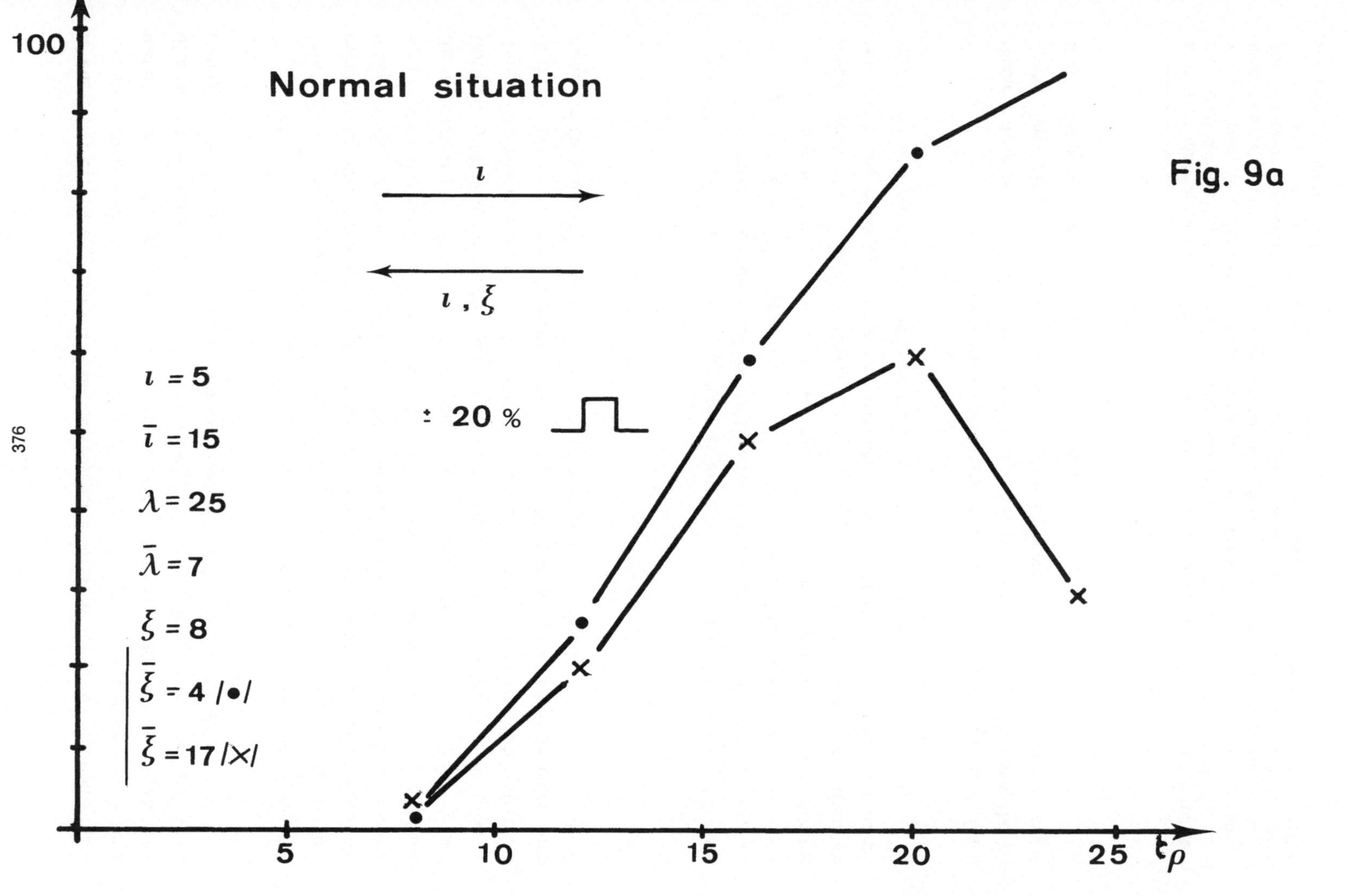

100
Normal situation
ι
ι, ξ
ι = 5
ῑ = 15
λ = 25
λ̄ = 7
ξ = 8
ξ̄ = 4 /•/
ξ̄ = 17 /×/
± 20 %
5
10
15
20
25
t_ρ

Fig. 9a

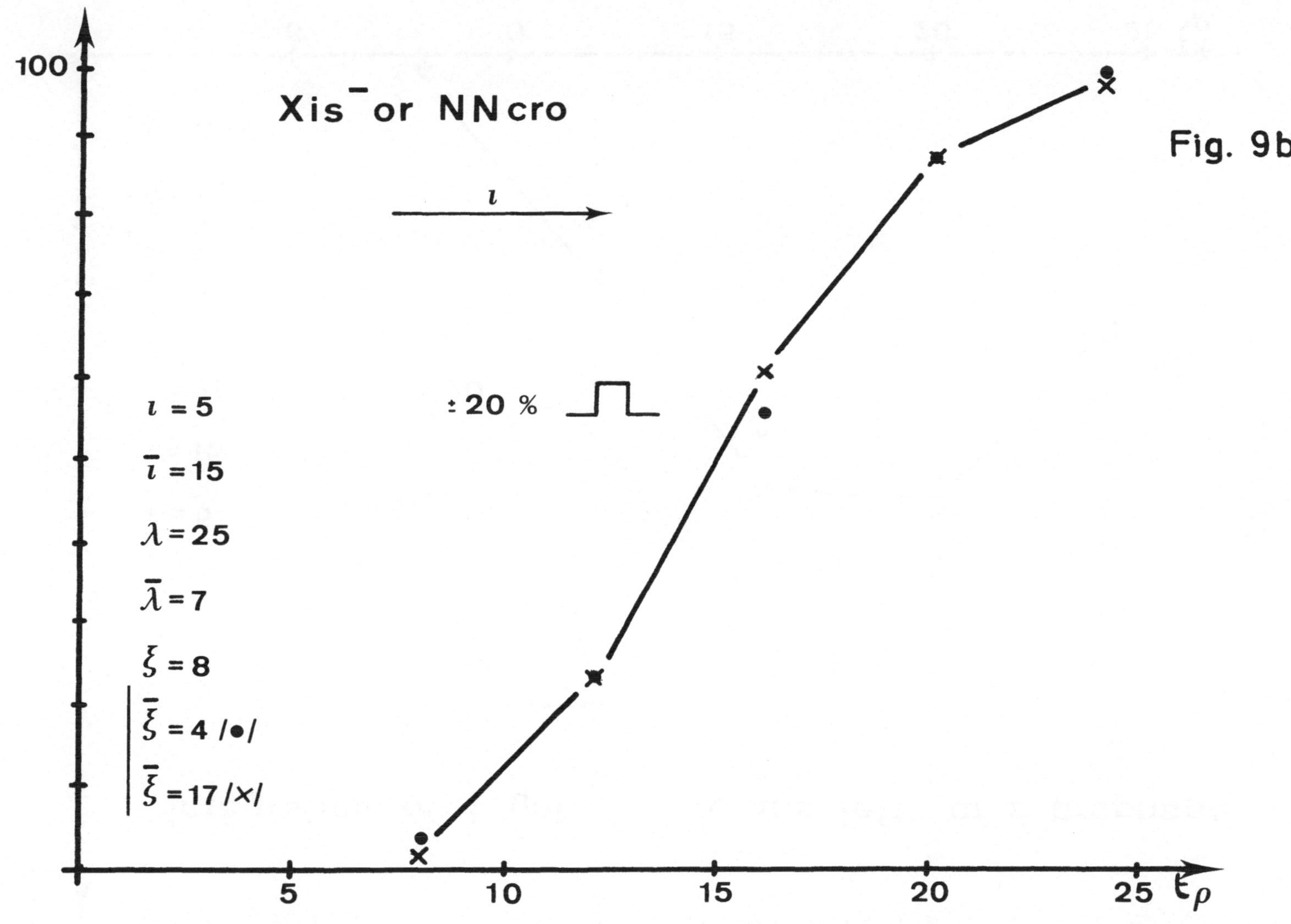
100
Xis$^-$ or NN cro
ι
$\pm$ 20 %
$\iota = 5$
$\bar{\iota} = 15$
$\lambda = 25$
$\bar{\lambda} = 7$
$\xi = 8$
$\bar{\xi} = 4$ /•/
$\bar{\xi} = 17$ /×/
5
10
15
20
25
ε_ρ

Fig. 9b

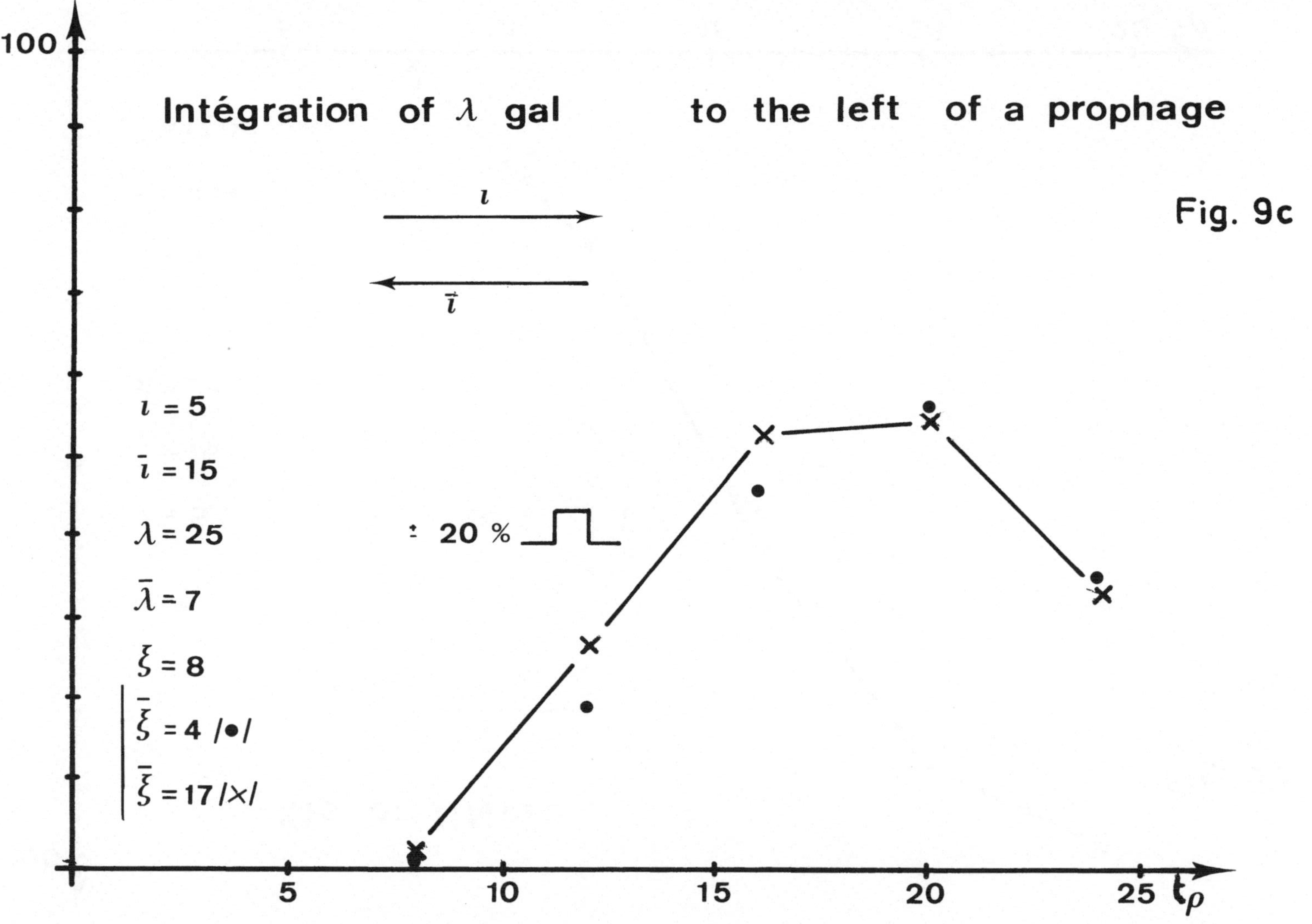

Intégration of λ gal to the left of a prophage
Fig. 9c
ι
ῑ
ι = 5
ῑ = 15
λ = 25
λ̄ = 7
ξ = 8
ξ̄ = 4 /•/
ξ̄ = 17 /×/
± 20 %
100
5
10
15
20
25
t_ρ

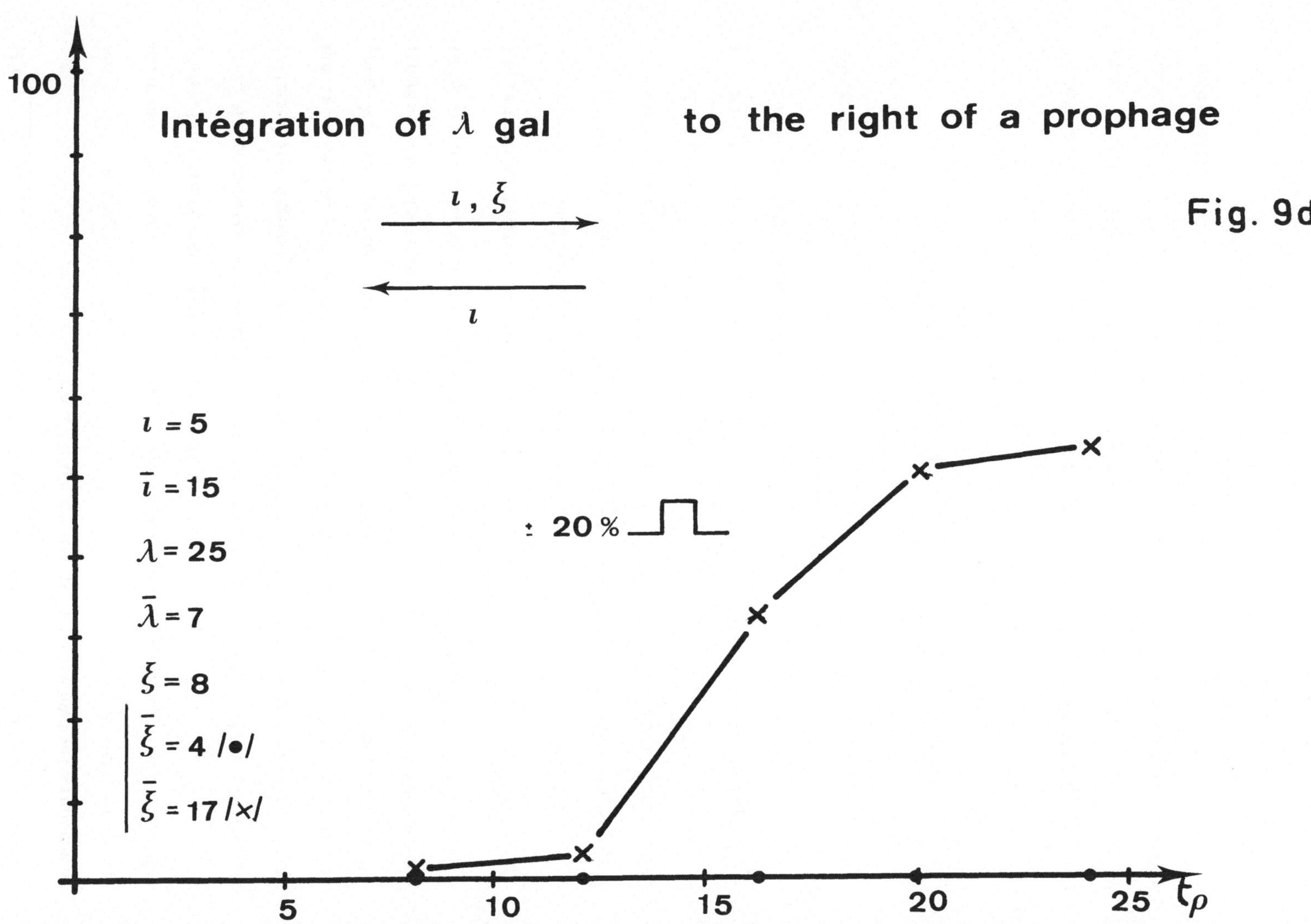

100
Intégration of λ gal to the right of a prophage
ι, ξ
ι
Fig. 9d
ι = 5
ῑ = 15
λ = 25
λ̄ = 7
ξ = 8
ξ̄ = 4 /●/
ξ̄ = 17 /×/
± 20 %
5
10
15
20
25
τ_ρ

17.4. The control of immunity in bacteriophage P22. (see Levine 1972 ; Botstein et al 1975).

In bacteriophage P22, the control of immunity involves a refinement not present in phage lambda. A gene called Ant produces an "antirepressor", which can inactivate or destroy the repressor. However, gene Ant is under negative control of a closely linked gene, called Mnt (for maintenance).

Thus, the persistance of immunity requires the operation of gene c2 (which produces the repressor) and of gene Mnt (whose product prevents the synthesis of the antirepressor). There are two types of virulent mutants. The vir_B mutants, comparable to the vir mutants of phage λ, are operator mutations which render the lytic functions insensitive to the repressor. Vir_A mutations inactivate the operator of gene Ant, thus rendering the synthesis of the Ant product constitutive (insensitive to the product of gene Mnt).

Let us try to formalize the following ("classical") model :

- genes c2 (symbolized C) and Mnt (symbolized M) are expressed constitutively.
- the lytic functions (symbolized L) are expressed in the absence of repressor ($\overline{\gamma}$) or if there is a vir_B mutation ($\overline{0}_L$, that is, the operator of the Lytic functions is inactivated) or if the product of gene Ant is present (α).
- gene Ant is expressed if there is no Mnt product ($\overline{\mu}$) or if there is a vir_A mutation ($\overline{0}_A$, that is, the operator of gene Ant is inactivated).

This gives

$$c = C$$

$$\ell = L(\overline{\gamma} + \overline{0}_L + \alpha)$$

$$m = M$$

$$a = A(\overline{0}_A + \overline{\mu})$$

Table 7 shows that this logical structure is consistent with a single stable state (1 0 1 0), in which immunity is present and the lytic functions are off. One can check that there is no stable cycle (in fact, no cycle at all) and that a system endowed with this logical structure would necessarily lead to the immune state.

c, l, m, a	"classical" model	model with an additional interaction ($m = \bar{\lambda}$)
0 0 0 0	1 1 1 1	1 1 1 1
0 0 0 1	1 1 1 1	1 1 1 1
0 0 1 1	1 1 1 0	1 1 1 0
0 0 1 0	1 1 1 0	1 1 1 0
0 1 1 0	1 1 1 0	1 1 0 0
0 1 1 1	1 1 1 0	1 1 0 0
0 1 0 1	1 1 1 1	1 1 0 1
0 1 0 0	1 1 1 1	1 1 0 1
1 1 0 0	1 0 1 1	1 0 0 1
1 1 0 1	1 1 1 1	(1 1 0 1)
1 1 1 1	1 1 1 0	1 1 0 0
1 1 1 0	1 0 1 0	1 0 0 0
1 0 1 0	(1 0 1 0)	(1 0 1 0)
1 0 1 1	1 1 1 0	1 1 1 0
1 0 0 1	1 1 1 1	1 1 1 1
1 0 0 0	1 0 1 1	1 0 1 1
$\gamma, \lambda, \mu, \alpha$		

Table 7. State tables in the case of a normal phage (C, L, M, A ; 0_L and 0_A = 1). In this case the equations for the "classical" model reduce to : $c = 1$; $\ell = \bar{\gamma} + \alpha$; $m = 1$; $a = \bar{\mu}$.

Thus, the model as such does not account for the fact that phage P22 can choose between the lytic and immune responses. In order to permit such a choice, the system should comprize an additional interaction in order to display a positive feedback loop. Simple possibilities would be that gene c2 or Mnt is directly or indirectly repressed by one of the lytic functions ($c = \bar{\lambda}$ or $m = \bar{\lambda}$) ; this would create a positive loop* ($\overset{-}{\rightleftarrows}_{-}$) since the lytic functions are themselves under negative control of gene c2 and, indirectly, under negative control of gene Mnt. Another possibility would be that the lytic response has an autocatalytic component ($\ell = \cdots + \lambda$).

---

* A positive loop is a loop with an even number of negative interactions. This structure leads to two stable (and an unstable) steady states (see chapter VI).

Any of these additional interactions would provide the system with the experimentally observed choice between two stable states* (see Table 7).

Various experimental situations can be formalized as follows :

- let the input variable S_A symbolize superinfection with a vir_A mutant. If one considers this possibility,

$$a = A\bar{\mu} + S_A$$

If $S_A = 1$ the system has only one stable state. (1 1 0 1) which represents the lytic pathway

- let the input variable S_B symbolize superinfection with a vir_B mutant. If one considers this possibility,

$$\ell = L(\bar{\gamma} + \alpha) + S_B,$$

and again if $S_B = 1$ one finds only the stable state (1 1 0 1).

- If gene <u>Mnt</u> carries a thermosensitive mutation, one can write :

$a = \bar{\mu} + T$, and if $T = 1$ (high temperature) one gets again a single stable state (1 1 0 1).

- In the case of a prophage deleted for the region <u>Mnt-Ant</u>, only superinfection (S) with a normal phage can provide these genes ; thus :

$$m = \bar{\lambda} \cdot S$$

$$a = \bar{\mu} \cdot S$$

In the absence of superinfection, one finds a single stable state immune (1 0 0 0) while after superinfection a lytic state (1 1 0 1) appears, etc...

* One might argue that the situation in which the lytic functions are on is not really a stable one, as the destruction of the cell results in a block of the lytic functions. However, since the model does not consider these late developments, the lytic situation appears formally as stable.

17.5. Again the lactose operon !

There is an enormous amount of litterature about the lactose operon. (see "The lactose operon", 1970). Much of it deals with a more and more refined genetic and biochemical analysis of transcription from the lac promoter, and control of this transcription by the lac repressor and by the cap protein. Very little deals with the integration into a coherent network of : lactose uptake, induction of β-galactosidase and permease , breakdown of lactose and role of the glucose thus formed in catabolic repression. A beautiful work by Novick and Wiener (1957) and by Cohn and Horibata (1959) is described in a simplified way in this book (Chapter II). This work established clearly the fundamental result that in proper conditions an inducible strain can be blocked permanently in one of two well distinct phenotypic states (one induced, one non-induced), in the absence of any genetic difference and in a common environment ; which stable state is chosen depends on a detail of the past history of the cultures. The basis for the mechanism of this epigenetic difference, as well as the general importance of the phenomenon have been fully understood by the authors without the help of any formalism. In the meantime, very little experimental work has been devoted to these aspects. It is the great merit of Nicolis and Sanglier (1976) to have undertaken a theoretical analysis in which the complex network is treated as a whole ; a number of interactions are taken into account, each of course in a simplified way. The authors could account for the multiple steady states observed in certain conditions and predict in other conditions multiple cycles.

A boolean analysis of the same system is briefly described in Thomas (1978). Let us simply recall three steps of increasing complexity. For the elementary description of the system,see this book, Chapter II, and Fig.10.

a) One uses a gratuitous inducer, this is, a substance which can induce the expression of the lac operon but is not split by β-galactosidase. In this case, the system can be described by the following set of equations (the symbols can be found in the legend to Fig.10).

$$i = I$$

$$z = Z\ (\bar{\iota} + \bar{0}_L + \lambda)$$

$$y = Y\ (\bar{\iota} + \bar{0}_L + \lambda)$$

$$\ell = \upsilon L_1 + L_2$$

In a strain which produces efficient repressor ($I = 1$) and carries normal Z and Y genes and a normal lactose operator (Z, Y and $0_L = 1$), this reduces to :

$$z = \lambda$$

$$y = \lambda$$

$$\ell = \upsilon L_1 + L_2$$

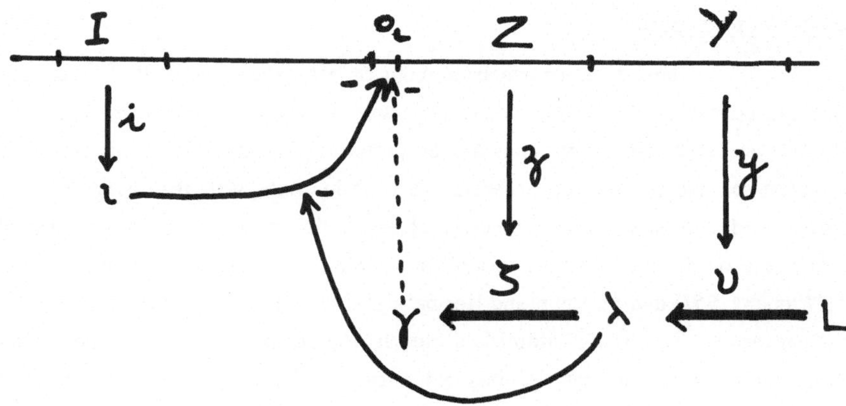

Fig.10. A schema of the lactose operon

0_L is the lac operator.

I, i and ɩ (the gene, its expression, the presence of the product) refer to the repressor of the lac operon.

Z, z and ζ (the gene, its expression, the presence of the product) refer to β-galactosidase.

Y, y and υ (the gene, its expression, the presence of its product) refer to β-galactoside permease.

$L \Longrightarrow \lambda$ represent the conversion of external into internal lactose (or gratuitous inducer, depending on the case) and $\lambda \Longrightarrow \gamma$ represents the formation of glucose at the expense of internal lactose.

An arrow like $Z \rightarrow \zeta$ means that Z takes part in the synthesis of ζ without being consumed in the operation. An arrow like $\lambda \Rightarrow \gamma$ means that γ is produced at the expense of λ . This is seen in the differential equations as a negative term $- k[\lambda]$ in the expression of the time derivative of $[\lambda]$, and in the logical treatment as a negative feedback exerted by λ on itself. (see also the last section of Chapter VII).

For L , the external concentration of lactose , one uses two boolean variables L_1 and L_2 corresponding to two thresholds. If $\bar{L}_1 \bar{L}_2$ (low level) there is no penetration ; if $L_1 \bar{L}_2$ the concentration is such that lactose penetrates in the presence of permease but not in its absence, if $L_1 L_2$ the concentration is such that lactose penetrates even in the absence of permease ; $\bar{L}_1 L_2$ has no physical meaning, and $L_1 L_2$ can thus be written simply L_2.

In this system there is a positive feedback loop between λ and ν ; the permease ν, if present, permits the entry of inducer, and internal inducer (λ) is necessary for the synthesis of permease. <u>This positive loop is the origin of the multiple stable steady states experimentaly observed</u>.

b) <u>One uses lactose itself</u> . Once it has penetrated into the cell even the basal level of β-galactosidase is sufficient to convert a small part of it into a close derivative, allolactose which is the true inducer (Jobes & Bourgeois, 1972) . To the extent that persistance of the inducer is dependant on the presence of internal lactose, we will reason here as if internal lactose were the inducer*. In contrast with the "gratuitous" inducer mentioned in the preceeding §, lactose is destroyed by β-galactosidase. In the equation of ℓ (which tells in which conditions the level of internal lactose increases towards its efficient treshold or remains above it) we have now to take into account not only the penetration of lactose but also the degradation of internal lactose by β-galactosidase. In differential equations, the situation is described by the presence of two terms in the expression of the time derivative of internal lactose concentration, a positive term for the entry of lactose and a negative term for its breakdown by β-galactosidase. In crude boolean terms, we simply ask whether in the presence of β-galactosidase the breakdown of lactose is significantly faster or slower than its entry. If it is slower, one can keep the equation $\ell = L_1\nu + L_2$. If it is faster one must take into account that the absence of β-galactosidase is an additional condition for $\ell = 1$:

$$\ell = (L_1\nu + L_2) . \overline{\mathfrak{z}} \;^{**}$$

Elementary as it is, this formal description takes into account the essential qualitative alteration introduced by the transition from a gratuitous to a metabolizable inducer. There is now an additional interaction, which creates a negative feedback loop ;
β -galactosidase destroys internal lactose, a derivative of which is necessary for the synthesis of β -galactosidase.

---

* To have an idea of the real complexity of the situation, see Jobes& Bourgeois (1973), who show that lactose generates the inducer, allolactose but is itself slightly antagonist of the inducer !

** This treatment will seem too primitive to many. For instance, the rate of entry of lactose is presumably much higher in the case νL_1 (a moderate external concentration of lactose with permease present) than in the case $\bar{\nu} L_2$ (a high concentration of external lactose but no permease present). It would thus be advisable to consider two levels of β-galactosidase $\mathfrak{z}_1$ and $\mathfrak{z}_2$, such that $\mathfrak{z}_1$ is sufficient to overcome the entry by the slow mechanism, but $\mathfrak{z}_2$ is necessary to overcome the entry by the fast mechanism :
$\ell = L_1 \nu \overline{\mathfrak{z}}_2 + L_2 \overline{\mathfrak{z}}_1$.

Such systems have been analyzed in chapter VII. One should consider two situations. If the external concentration of lactose is such that it enters even in the absence of permease ($L_2 = 1$) the equations become :

$$z = Z \,.\, \lambda$$
$$y = Y \,.\, \lambda$$
$$\ell = \overline{\xi} \;,$$

in which permease is not used as a retroaction variable. In this simple case, the logical structure is a simple negative feedback loop : $\xi \underset{+}{\overset{-}{\rightleftarrows}} \lambda$

and the boolean analysis predicts a stable cycle with the following sequence of transitions :

$$\lambda \rightarrow \xi \rightarrow \overline{\lambda} \rightarrow \overline{\xi}$$

If the lactose concentration is lower ($L_1 \overline{L}_2$), the equations are :

$$z = Z \,.\, \lambda$$
$$y = Y \,.\, \lambda$$
$$\ell = \upsilon \,.\, \overline{\xi}$$

and the logical structure is a pair of "tangent" feedback loops, one positive, one negative . As mentioned in chapter VII, this leads to a choice between one stable state and cycle(s). In the present case there is a choice between the stable state (0 0 0) in which the system is "dead" and a cycle (the same as above) :

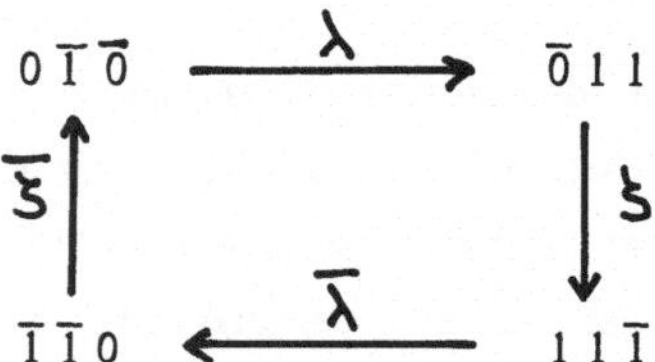

in which the concentrations of internal lactose and of β-galactosidase oscillate. Since galactosidase is very stable, $\overline{\xi}$ must take place by dilution due to the expansion of the system, during cell division, rather than to the decay of the enzyme.

The conditions to reach the cycle and remain in it are :

$$t_{\upsilon} < t_{\overline{\lambda}}$$

and

$$t_{\overline{\upsilon}} < t_{\overline{\xi}} + t_{\lambda}$$

c) Nicolis and Sanglier (1976) also take into account the additional possibility that the glucose liberated by the breakdown of internal lactose might result in catabolic repression of the transcription of the lac operon . In fact this does not seem to be the case (Jobe & Bourgeois, 1973). I shall nevertheless consider the possibility here, just to extend the parallel between the analysis of Nicolis and Sanglier (1976) and ours. It is well known that, in spite of its name, catabolic repression is a positive control ; however, as this positive control is inhibited by glucose, we simply write that the expression of genes Z and Y does not take place when the glucose concentration exceeds a threshold ($\gamma = 1$).

The equations become :

$$z = Z \,.\, \lambda \,.\, \overline{\gamma}$$
$$y = Y \,.\, \lambda \,.\, \overline{\gamma}$$
$$\ell = (L_1 \upsilon + L_2) \,.\, \overline{\xi}$$
$$g = \lambda \,.\, \xi$$

Here again, the situation is very different depending on whether the external lactose concentration is high ($L_2 = 1$) or low ($L_1 \overline{L}_2$). In the first case, the presence of permease does not operate as a retroaction variable, and the equations are :

$$z = Z \,.\, \lambda \,.\, \overline{\xi}$$
$$\ell = \overline{\xi}$$
$$g = \lambda \xi$$

In this case, the boolean analysis predicts no stable state but five elementary cycles :

$$A = \lambda \xi \overline{\lambda} \overline{\xi}$$
$$B = \xi \gamma \overline{\xi} \overline{\gamma}$$
$$C = \xi \gamma \overline{\lambda} \overline{\xi} \lambda \overline{\gamma}$$
$$D = \lambda \xi \gamma \overline{\lambda} \overline{\xi} \overline{\gamma}$$
$$E = \lambda \xi \gamma \overline{\lambda} \overline{\gamma} \overline{\xi}$$

The conditions of stability of the various cycles have been determined in terms of the relative values of time delays or linear combinations of them. If these delays are kept constant, the system remains on a defined "orbit", but it can jump from one orbit to another if a proper change occurs in the value of a delay.

If the external concentration of lactose is low, the boolean analysis predicts a choice between a "dead" stable state and a number of cycles. The possible occurence of more than one cycle had already been mentioned by Nicolis and Sanglier 1976, using differential equations.

17.6. An exercise with neurones.

One begins to understand some locomotory rhythms, like the swimming of the urodela or leech, in terms of networks comprising a small number of neurones with well-defined interconnections. It was extremely tempting to try whether kinetic logic can be applied to this type of work; what follows is an exercise based on the papers of Friesen & Stent (1977, 1978). Boolean analysis has been applied to neuronal circuits already long ago (Rashevsky, 1948), and theapplicability to the work of Friesen and Stent is mentioned by Glass & Pasternack (1978). In Friesen and Stent's work, periodicity is accounted for basically by a recurrent cyclic inhibition implying an odd number of elements*, of which the simplest case is : neurone A is inhibited by neurone B, which is inhibited by neurone C, which is inhibited by neurone A (Székely, 1965 ; Friesen & Stent op. cit.). As described by these authors : "This network consists of an inhibitory ring formed by three tonically excited neurones (A-C), of which each makes inhibitory synaptic contact with and receives inhibitory synaptic input from one other cell. If, as indicated in Figure 2a, cell C happens to be in a depolarized, impulse-generating state, its postsynaptic cell, B, must be in a hyperpolarized, inactive state, while its presynaptic cell, A, is recovering from past inhibition. As soon as cell A has recovered from inhibition and reached its impulse generation threshold cell C becomes inhibited,thus desinhibiting cell B and allowing the latter to enter its recovery phase. Once cell B has recovered, it inhibits cell A, thus allowing cell C to begin recovery; and once cell C has recovered, so that cell B enters its inactive phase and cell A its recovery phase, one cycle of the oscillation has been completed". (see this Chapter, Fig.2a).

The most schematic representation of such situations is a phase diagram (Fig.1b), in which each bar indicates the duration of the impulse burst of the cell. This simple level of description can be formalized simply by associating with neurones A, B, C boolean variables α , β , γ which take the value 1 when the neurone is producing an impulse burst, 0 otherwise. One can thus associate with each step of the sequence ("state") a binary number formed by the values of α, β and γ (Fig.1,c).

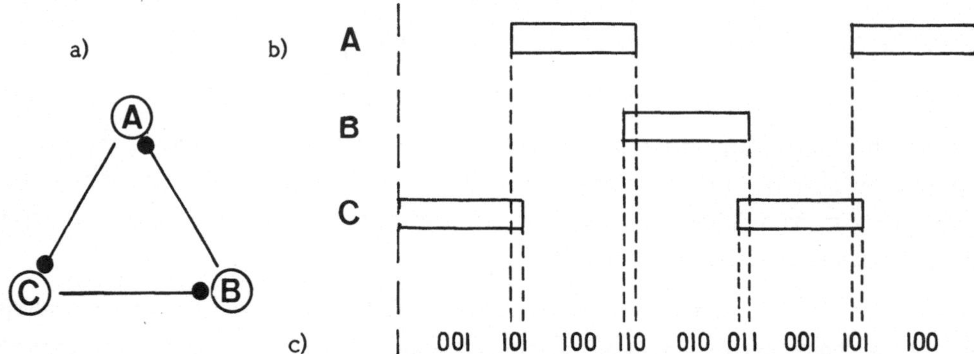

Fig.1. A simple network with recurrent cyclic inhibition (Friesen & Stent, 1978)
a) Three-cell network ; filled small circles designate inhibitory synaptic contacts
b) The corresponding phase diagram
c) Boolean description

---

* A particular case of negative loop, see Chapter VII.

Note that in this network the change from $\alpha = 0$ to $\alpha = 1$ is the cause of the change from $\gamma = 1$ to $\gamma = 0$, which it must thus preceede ; and for this reason 1 0 1 is the logical transition, between 0 0 1 and 1 0 0 . This is why the bars have been represented slightly overlapping. In practice, the duration of states 1 0 1 , 1 1 0 and 0 0 1 is so short that it is not taken into account by Friesen & Stent.

A slightly more sophisticated description (Friesen & Stent, 1978) is shown in Fig.2a.

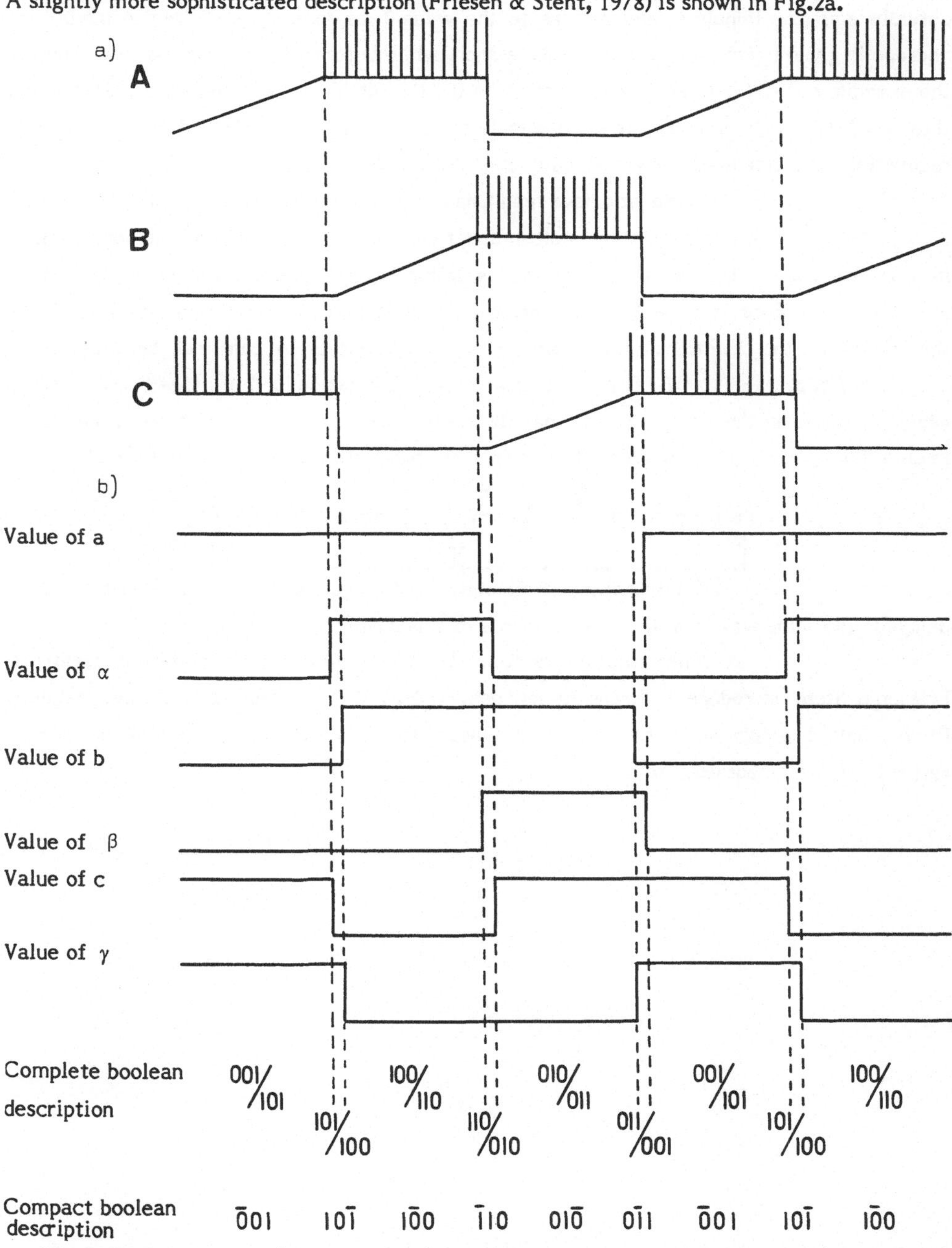

Fig.2. Same network a) Traces representing the membrane potential and impulse burst activity of individual neurones.

b) Representation of the values of the boolean function and variable associated with each neurone.

One way to formalize this description is as follows. In addition to the already mentioned boolean variables α , β , γ ,... which tell whether each neurone is actually emitting an impulse burst, one defines boolean functions a, b, c , ... which take the value 1 as soon as the corresponding neurone is desinhibited* (even though it is not yet actually emitting impulses) and returns to the value 0 as soon as the order to inhibit the neurone is given (even though it may take a short while before the order is obeyed). Thus, in the example of Fig.1-2, when β drops from 1 to 0 , the inhibition on neurone A disappears ; thus a = 1 (the cell is asked to emit impulses) even though α is still 0 (the cell which is recovering inhibition does not yet actually emit impulse (see Fig.2b).

Those who have read Chapter VI have realized that a, b, c, ... and α , β , γ , ... are, respectively, boolean functions and their associated memorization or internal variables. In the first state of Fig.2. the values of the variables, α , β and γ are 0 0 1 , but since neurone A is recovering from inhibition, function a has the value 1 and the values of the functions a, b, c, are 1 0 1 . The situation can thus be represented α , β, γ / a, b, c = 0 0 1 / 1 0 1 . The compact representation is $\bar{0}\,0\,1$ (see Chapter VI), in which the dash on the first 0 means that the corresponding neurone is not yet producing an impulse burst, but is already ordered to do so. The complete sequence can be formalized :

$$\bar{0}\,0\,1 \xrightarrow{\alpha} 1\,0\,\bar{1} \xrightarrow{\bar{\gamma}} 1\,\bar{0}\,0 \xrightarrow{\beta} \bar{1}\,1\,0 \xrightarrow{\bar{\alpha}} 0\,1\,\bar{0} \xrightarrow{\gamma} 0\,\bar{1}\,1 \quad (\xrightarrow{\bar{\beta}} \text{back to } \bar{0}\,0\,1)$$

The transitions from state to state can be treated in terms of time delays in the same way as in the other sections of this chapter.

When the interactions involve neurones from different segments, Friesen & Stent introduce a "transmission delay" H, 2 H , ... depending on the distance. Provisionall, I simply add these delays to those already considered. This may has been refined in more elaborate attempts.

---

* or activated, if one deals with a positive control.

Logical analysis as applied to this type of systems.

a. Given a network, find the resultant phase diagram by theoretical analysis.

One example from Friesen & Stent (1978) is shown in Fig.3a. As already shown by Székely (1965) this network produces periodic activities of the four neurones. Our way to reach this result is shown in Fig.3b and c. Note that simple recurrent cyclic inhibition between the four elements(a positive loop : see Chapter VII)would result in two stable steady states; the cross connections provide the negative loops responsible for the periodic activity.

a)

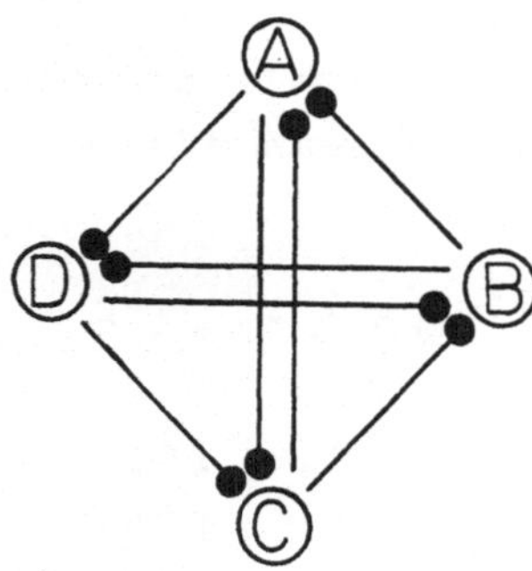

b)

$$\begin{cases} a = \bar{\beta} \cdot \bar{\gamma} \\ b = \bar{\gamma} \cdot \bar{\delta} \\ c = \bar{\alpha} \cdot \bar{\delta} \\ d = \bar{\alpha} \cdot \bar{\beta} \end{cases}$$

c)

$\alpha\ \beta\ \gamma\ \delta$	a b c d
$\bar{0}\ \bar{0}\ \bar{0}\ \bar{0}$	1 1 1 1
$\bar{0}$ 0 0 1	1 0 0 1
0 0 $\bar{1}$ 1	0 0 0 1
0 0 1 $\bar{0}$	0 0 1 1
0 $\bar{1}$ 1 0	0 0 1 0
0 $\bar{1}\ \bar{1}\ \bar{1}$	0 0 0 0
0 $\bar{1}$ 0 $\bar{1}$	0 0 0 0
0 1 $\bar{0}$ 0	0 1 1 0
$\bar{1}$ 1 0 0	0 1 0 0
$\bar{1}\ \bar{1}$ 0 $\bar{1}$	0 0 0 0
$\bar{1}\ \bar{1}\ \bar{1}\ \bar{1}$	0 0 0 0
$\bar{1}\ \bar{1}\ \bar{1}$ 0	0 0 0 0
$\bar{1}$ 0 $\bar{1}$ 0	0 0 0 0
$\bar{1}$ 0 $\bar{1}\ \bar{1}$	0 0 0 0
1 0 0 $\bar{1}$	1 0 0 0
1 $\bar{0}$ 0 0	1 1 0 0

d)

$$\begin{array}{ccccccc} 1\bar{0}00 & \xrightarrow{\beta} & \bar{1}100 & \xrightarrow{\bar{\alpha}} & 01\bar{0}0 & \xrightarrow{\gamma} & 0\bar{1}10 \\ \bar{\delta}\uparrow & & & & & & \downarrow\bar{\beta} \\ 100\bar{1} & \xleftarrow[\alpha]{} & \bar{0}001 & \xleftarrow[\bar{\gamma}]{} & 00\bar{1}1 & \xleftarrow[\delta]{} & 001\bar{0} \end{array}$$

Fig.3. Derivation of the dynamic behaviour of a network. a) A network proposed by Székely (1965) and by Friesen and Stent (1978). b) Its description by boolean equations. c) Matrix (or state table). d) The final state : a stable cycle.

First, the network is described in terms of logical equations (Fig.3b) ; $a = \bar{\beta} . \bar{\gamma}$ means that neurone A is desinhibited (a = 1) iff neither neurone B nor neurone C is emitting impulses, etc... . Second, we write the matrix giving the values a, b, c and d for each combination of values of α, β, γ, δ. The complete analysis of the system will not be given here. Suffice to remark that for eight of the sixteen states (combinations of values of α, β, γ and δ) only one of the variables is asked to change its value at a time (only one of the digits is topped by a dash). This means that each of these states has only one possible next state ; and since these states can be arranged in a closed sequence, one deals with a stable cycle. Note that the lengths of the various transitions are not the same. In particular, the transitions denoted $\bar{\alpha}$, $\bar{\beta}$, $\bar{\gamma}$ and $\bar{\delta}$ are almost immediate ; their duration is considered negligeable by Friesen & Stent (1978, Fig.2b).

The kind of models given initially to account for oscillations did not comprize explicit negative loops. An example cited in the same paper of Friesen & Stent is treated in Fig.4.

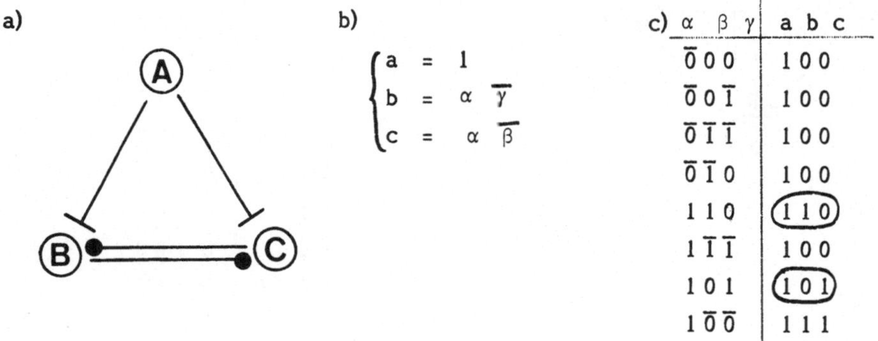

α β γ	a b c
$\bar{0}00$	100
$\bar{0}0\bar{1}$	100
$\bar{0}\bar{1}\bar{1}$	100
$\bar{0}\bar{1}0$	100
110	(110)
$1\bar{1}\bar{1}$	100
101	(101)
$1\bar{0}\bar{0}$	111

Fig.4. A model quoted by Friesen and Stent (1978) but not favoured by them. The model intends to account for permanent oscillations but our analysis indicates rather a choice between two stable states.

The matrix immediately shows that this logical structure gives the choice between two stable states ; in addition a short inspection shows that no oscillations are generated by this system as such. If one wants to get oscillations, one has to introduce a negative loop such as •A↩ (or, in terms of logical equations, $a = \bar{\alpha}$). See the discussion in Friesen & Stent (1978), and also the second part of Chapter VII.

Significantly more complex networks can be treated "by hand" as above. However, as the number of internal variables increases, it becomes more and more necessary to use ad hoc computer programs (see Van Ham, Chapter IX) or a logical simulator (see Van Ham, Chapter VIII, and the next section).

b. Given a proposed logical structure, find the resultant trace(s) by simulation.

The "full interneuronal network" of Friesen and Stent (1977) and even more complex ones, should be amenable without major problems to simulation with the logical machine "Delphine" (see Chapter VIII by Van Ham), which can handle 24 pairs (function - internal variable), and thus 24 neurones. It is now possible to formalize and simulate not only the normal but also the "rectifying electrical junctions" described by Friesen and Stent.

So far, we have simulated the networks of Friesen and Stent's (1977) figures 1b, 5b and 7. In the first and second case the final state is a stable cycle ; the networks oscillate exactly in the predicted way whatever the values of the delays, except of course in the very artificial case in which two or more delays are exactly identical* In the third case, there is no stable cycle but the required sequence can be obtained for proper values of the delays. A detailed analysis has not yet been performed.

c. Given the phase diagram (or the trace) of a group of neurones, find networks, as simple as possible, which account for it.

Assuming that the observed behaviour of a set of neurones can be understood in terms of the interactions between these same neurones, one can try to derive the network from the phase diagram. When one operates in that way (from the dynamic behaviour towards the logical structure) the solution is seldom, if ever, unique.
Let us first treat in reverse the second example of section a. Given the sequence :

$$\begin{array}{ccccccc} 1000 & \xrightarrow{\beta} & 1100 & \xrightarrow{\bar{\alpha}} & 0100 & \xrightarrow{\gamma} & 0110 \\ \bar{\delta}\uparrow & & & & & & \downarrow\bar{\beta} \\ 1001 & \xleftarrow{\alpha} & 0001 & \xleftarrow{\bar{\gamma}} & 0011 & \xleftarrow{\delta} & 0010 \end{array}$$

find simple logical structures consistent with this sequence. The problem can be understood in two ways.

α) One may want to find a network whose operation involves **the sequence as a stable cycle ; in other words, once a state of the sequence has been reached the cycle is followed whatever the values of the time delays. This amounts to say that each state in the sequence has only one possible next state, a situation in which the states forming the sequence are symbolized by numbers with only one digit topped with a dash (like in Fig.3d).

* When all the "on" delays are identical and the same for the "off" delays, the system can pulse between two extreme states like $\bar{0}\,\bar{0}\,\bar{0}\,\bar{0} \rightleftarrows \bar{1}\,\bar{1}\,\bar{1}\,\bar{1}$. This artificial situation disappears when the time delays are, if only slightly, different from each other, or if one takes as the initial state one of the states of the stable cycle.

** I refer intentionally to the sequence as "involved" as a stable cycle rather than "implied" by the logical structure. This is because one can construct a system in which the stable cycle is not the only possible final state. For instance, one might want to have the choice between the above mentioned sequence as a stable cycle, and a stable state, say (0 1 0 1). In that case, one would specify one more state in the matrix 4a : 0 1 0 1 / (0 1 0 1) . The simplest system is : $a = \bar{\beta}.\bar{\gamma}$; $b = \bar{\gamma}\,(\beta + \bar{\delta})$; $c = \bar{\alpha}.\bar{\delta}$; $d = \bar{\alpha}.(\bar{\beta} + \delta)$. One can show that this system has indeed the choice between the stable state (0 1 0 1) and the stable cycle of Fig.3d. Which one is reached depends on the initial state and time delays,; but if the cycle is reached it is followed indefinitely whatever the time delays.

From the graph, one can derive a partial state table. Remember that $1\bar{0}00$ is a compact notation for $\alpha\ \beta\ \gamma\ \delta$ /a b c d = 1 0 0 0 / 1 1 0 0 , that is : neurone α is "on", neurones β , γ and δ are "off", but one of them (β) is committed to become active after a short time delay, etc ... Thus, in the matrix we can write that for the state 1 0 0 0 of the variables, the values of a, b, c, d, are 1 1 0 0 respectively ; and similarily for the other states present in the sequence. As for the other combinations of values of the variables the corresponding values of a, b, c, d, are "indifferent", that is, a "0" or a "1" would satisfy the requirements (Chapter IV); for these states we write thus in the matrix : 0 0 0 0 /- - - - etc ... (see Fig.5,a).

For a system comprising the elements A, B, C and D only, the set of the 0 and 1 digits in this matrix describes the conditions necessary and sufficient for the system to involve the sequence defined above as a stable cycle.

Fig.5.

a)

$\alpha\ \beta\ \gamma\ \delta$	a b c d
0 0 0 0	- - - -
0 0 0 1	1 0 0 1
0 0 1 1	0 0 0 1
0 0 1 0	0 0 1 1
0 1 1 0	0 0 1 0
0 1 1 1	- - - -
0 1 0 1	- - - -
0 1 0 0	0 1 1 0
1 1 0 0	0 1 0 0
1 1 0 1	- - - -
1 1 1 1	- - - -
1 1 1 0	- - - -
1 0 1 0	- - - -
1 0 1 1	- - - -
1 0 0 1	1 0 0 0
1 0 0 0	1 1 0 0

b)

a	00	01	11	10 $\alpha\beta$
00	-	0	0	1
01	1	-	-	1
11	0	-	-	-
10	0	0	-	-
$\gamma\delta$				

b	00	01	11	10 $\alpha\beta$
00	-	1	1	1
01	0	-	-	0
11	0	-	-	-
10	0	0	-	-
$\gamma\delta$				

c	00	01	11	10 $\alpha\beta$
00	-	1	0	0
01	0	-	-	0
11	0	-	-	-
10	1	1	-	-
$\gamma\delta$				

d	00	01	11	10 $\alpha\beta$
00	-	0	0	0
01	1	-	-	0
11	1	-	-	-
10	1	0	-	-
$\gamma\delta$				

c)

a	00	01	11	10 $\alpha\beta$
00	1	0	0	1
01	1	0	0	1
11	0	0	0	0
10	0	0	0	0
$\gamma\delta$				

b	00	01	11	10 $\alpha\beta$
00	1	1	1	1
01	0	0	0	0
11	0	0	0	0
10	0	0	0	0
$\gamma\delta$				

c	00	01	11	10 $\alpha\beta$
00	1	1	0	0
01	0	0	0	0
11	0	0	0	0
10	1	1	0	0
$\gamma\delta$				

d	00	01	11	10 $\alpha\beta$
00	1	0	0	0'
01	1	0	0	0
11	1	0	0	0
10	1	0	0	0
$\gamma\delta$				

Fig.5. a) State table built from the sequence of Fig.3,d.

b) The same, but the values of functions a, b, c and d are written separately in order to facilitate the implementation of the simplest logical equations consistent with the requirements.

c) Complete matrices showing the functions chosen as the simplest ones.

With a little training one finds immediately the simplest equations from matrix 5b.

$$a = \bar{\beta} \,.\, \bar{\gamma} \; ; \quad b = \bar{\gamma} \,.\, \bar{\delta} \; ; \quad c = \bar{\alpha} \,.\, \bar{\delta} \; ; \quad d = \bar{\alpha} \,.\, \bar{\beta}$$

which are in fact those given in section 1, Fig.3b.

This choice amounts to replace "indifferences" by "1" on "0" as indicated in Fig.5c.

β) In the preceeding paragraph, the constraints were that, for each state in the cycle, one well-defined variable, and that only, be asked to change its value. More generally, one might search for simple networks which would be consistent with a desired sequence of states without, however, necessarily involving it as a stable cycle. In this option, the constraint would be simply that the variable in question be asked to change its value, without excluding that other variables be asked to change their values as well. Thus, instead of the well-defined assignement $\alpha\beta\gamma\delta$ / a b c d = 1 0 0 0 / 1 1 0 0 , one would have $\alpha\ \beta\ \gamma\ \delta$ / a b c d = 1 0 0 0 / - 1 - - , which means that variable β is asked to commute from 0 to 1 ; nothing is told regarding the other variables, which may or may not be asked to commute. The matrix (not shown) comprises matrix 5a as a particular case. It provides one with a choice of very simple equations (three for each a, b, c and d). The logical structures described by any choice of one of these expressions for each a, b, c and d , all generate the desired sequence as one of their possible pathways. But in general the sequence will be followed only for well-defined relative values of the time delays.

A second example is taken from the same authors. The "trace" of Fig.5b in Friesen & Stent (1977) provides the sequence* :

$$\begin{array}{ccccccccc} 0\bar{1}011 & \xrightarrow{\bar{\beta}} & \bar{0}0011 & \xrightarrow{\alpha} & 100\bar{1}1 & \xrightarrow{\bar{\beta}'} & 10\bar{0}01 & \xrightarrow{\alpha'} & 1010\bar{1} \\ \gamma' \uparrow & & & & & & & & \bar{\gamma}' \downarrow \\ 0101\bar{0} & \xleftarrow{\bar{\alpha}'} & 01\bar{1}10 & \xleftarrow{\beta'} & 011\bar{0}0 & \xleftarrow{\bar{\alpha}} & \bar{1}1100 & \xleftarrow{\beta} & 1\bar{0}100 \end{array}$$

Reasoning in the same way as in the first example, one constructs the matrices of Fig.6. For instance, one finds in the sequence, state $0\bar{1}011$; this means that for the state 0 1 0 1 1 of variables α , β , α' , β' , γ' , the corresponding functions a, b, a', b', c' must have the values 0 0 0 1 1 , respectively . This requirement and those related to the other states of the sequence are found in the matrices.

* In the trace given by Friesen and Stent, such transition as the initiation of the emission by neurone B and the termination of the emission by neurone A are treated as simultaneous. I assume that they occur in that order, with a very short time delay.

a ($\alpha'\beta'\gamma'$ \ $\alpha\beta$)	00	01	11	10
000	-	-	-	-
001	-	-	-	1
011	1	0	-	1
010	-	0	-	-
110	-	0	-	-
111	-	-	-	-
101	-	-	-	1
100	-	0	0	1

$a = \overline{\beta}$

b ($\alpha'\beta'\gamma'$ \ $\alpha\beta$)	00	01	11	10
000	-	-	-	-
001	-	-	-	0
011	0	0	-	0
010	-	1	-	-
110	-	1	-	-
111	-	-	-	-
101	-	-	-	0
100	-	1	1	1

$b = \overline{\gamma}'$

a' ($\alpha'\beta'\gamma'$ \ $\alpha\beta$)	00	01	11	10
000	-	-	-	-
001	-	-	-	1
011	0	0	-	0
010	-	0	-	-
110	-	0	-	-
111	-	-	-	-
101	-	-	-	1
100	-	1	1	1

$a' = \overline{\beta}'$

b' ($\alpha'\beta'\gamma'$ \ $\alpha\beta$)	00	01	11	10
000	-	-	-	-
001	-	-	-	0
011	1	1	-	0
010	-	1	-	-
110	-	1	-	-
111	-	-	-	-
101	-	-	-	0
100	-	1	0	0

$b' = \overline{\alpha}$

c' ($\alpha'\beta'\gamma'$ \ $\alpha\beta$)	00	01	11	10
000	-	-	-	-
001	-	-	-	1
011	1	1	-	1
010	-	1	-	-
110	-	0	-	-
111	-	-	-	-
101	-	-	-	0
100	-	0	0	0

$c' = \overline{\alpha}'$

Fig.6. One wants to find which connections between the neurones A,B,A', B' and C' would generate the above-mentioned sequence as a stable cycle. The matrices give the constraints which must be respected.

The simplest expressions for a, b, a', b', c' are :

$$a = \overline{\beta}$$

$$b = \overline{\gamma}'$$

$$a' = \overline{\beta}'$$

$$b' = \overline{\alpha}$$

$$c' = \overline{\alpha}'$$

They correspond exactly to the logical structure from which Friesen & Stent have derived the trace analyzed here ; their model is thus the simplest one which would produce the sequence in question.

Note that in the treatment outlined in Fig.6, I have asked which simple connections of elements A, B, A', B' and C' would provide the desired sequence as a stable cycle. In this particular case, when one tries to find the simplest connections which would permit the sequence, but not necessarily as a stable cycle, one gets the same set of logical equations.

A third example is that of the generation of the pyloric rhythm in the lobster (described in Friesen & Stent (1978, Fig.3d) after Maynard & Selverston (1975)).

The experimental phase diagram can be formalized :

$$100 \xrightarrow{\bar{\alpha}} 000 \xrightarrow{\beta} 010 \xrightarrow{\gamma} 011 \xrightarrow{\bar{\beta}} 001 \xrightarrow{\alpha} 101 \xrightarrow{\bar{\gamma}} 100$$

The circuit proposed by the original authors and discussed by Friesen & Stent is shown in Fig.7a. According to our analysis, this circuit gives a choice between two stable states,but it does not provide for oscillations (see Fig.7a). As remarked by Friesen & Stent, oscillations are made possible if one of the elements, for instance neurone A, pulses by itself (a = $\bar{\alpha}$) ; but the cycle thus obtained does not reproduce the desired sequence (Fig.7b).

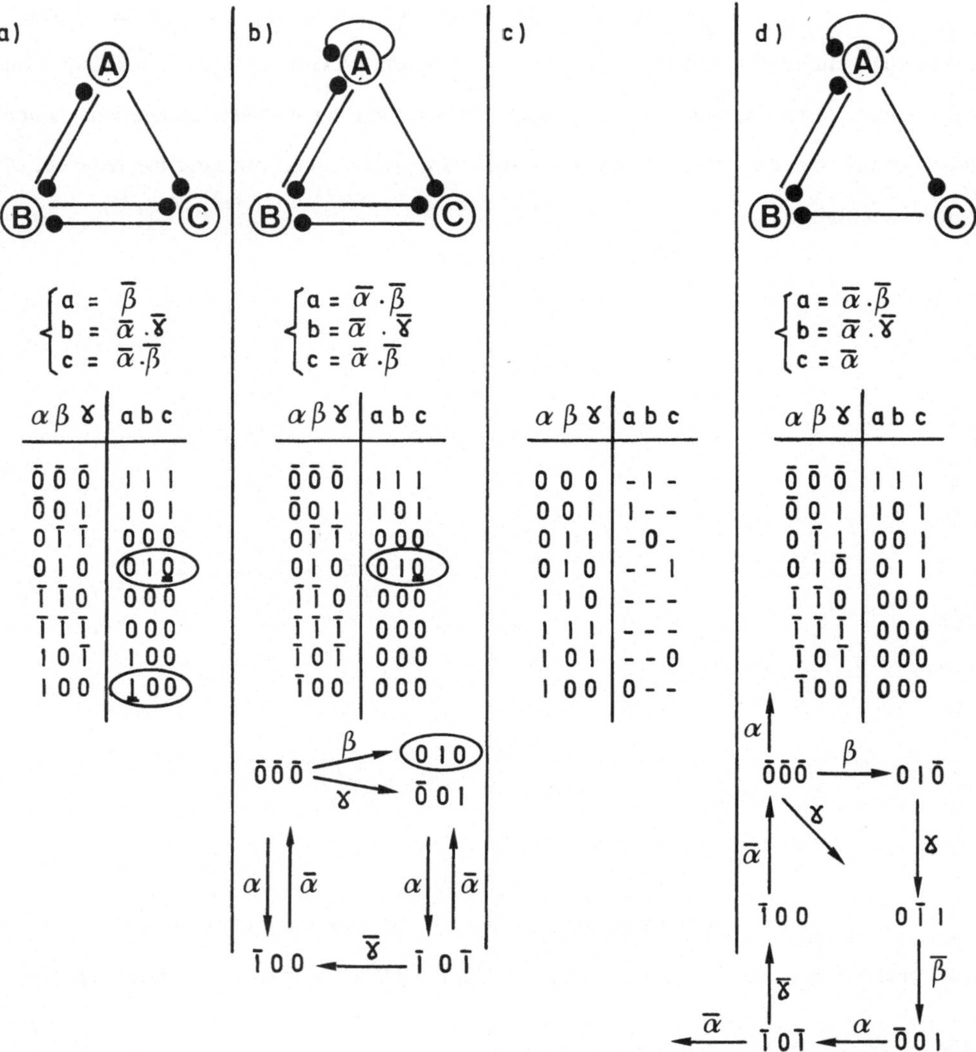

Fig.7. a) A circuit proposed by Maynard & Selverston (1975).

b) A modification proposed by Friesen & Stent (1978).

c) The constraints to respect if one wants to have the above - mentioned sequence.

d) A simple suggestion.

One can thus ask : given three elements A, B, C, which interactions between these elements would be consistent with the occurence of the experimental sequence? Using the same reasoning, we can tell that in state 1 0 0 variable α is asked to commute from 1 to 0 ; whether or not variable β or variable γ , or both, are also asked to commute at this stage, is irrelevant. Thus, for the values 1 0 0 of α , β and γ , the values of a, b and c are 0 - - . Applying this reasoning to each state of the sequence required leads to the matrix 7c. A comparison of this matrix with 7a shows that the conditions described in 7c are violated at two places in 7a : for the state 0 1 0 of the variables α , β and γ , the value of c is 0 instead of 1 , and for the state 1 0 0 of the variables the value of a is 1 instead of 0 . The simplest way to modify 7a in order to render it compatible with the constraints described in 7c seems to insert a negative loop on A (as already suggested) and in addition to suppress the inhibition exerted by B on C . This gives :

$$a = \bar{\alpha} \cdot \bar{\beta}$$

$$b = \bar{\alpha} \cdot \bar{\gamma}$$

$$c = \bar{\alpha}$$

This structure provides the sequence asked, if the time delays fulfil certain conditions ($t_\beta < t_\alpha$; $t_\beta < t_\gamma$ and $t_{\bar{\gamma}} < t_{\bar{\alpha}}$).

This kind of applications should not be taken for more than amusing exercises as long as they are developped by outsiders like me ; but I feel that the methods described here would be really useful in the hands of experts in the field.

References

Beckwith, J.R. & Zipser D. (1970)
The lactose operon.
Cold Spring Harbor, Laboratory New York.

Botstein, D. Lew, K.K., Jarvik, V. & Swanson, C.A.
J. Mol. Biol., 91, 439.

Campbell, A. (1962)
Advan. Genet. 11, 101.

Chung, S. & Echols, H. (1977)
Virology, 79, 312.

Cohn, M. & Horibata, K. (1959)
(Chapter I)

Court, D., Adhya, S., Nash, H. and Enguist, L. (1977)
in DNA Insertion Elements, plasmids and Episomes.
Bukhari,A.I., Shapiro, J & Adhya, S. eds.
Cold Spring Harbor Laboratory, New York, p.389.

Eisen, H., Pereira da Silva , L. & Brachet, P. (1970)
Proc. Natl. Acad. Sci. US, 66, 855

Friesen, W.O. & Stent, G.S. (1977)
Biol. Cybernetics, 28, 27-40.

Friesen, W.O. & Stent, G.S. (1978)
Ann. Rev. Biophys. Bioeng. 7, 37-61.

Glass, L. & Pasternack, J.S. (1978).
Bull. Math. Biol. 40, 27-44.

Glass, L. (1975) J. Theor. Biol., 54, 85.

Gottesman, R.A. & Weisberg, M.E. (1971)
in The bacteriophage lambda, A.D. Hershey, ed., Cold Spring Harbor Laboratory, New York.

Guerrini, F. (1969)
J. Mol. Biol. 46, 523.

Jobe, A. & Bourgeois, S. (1973)
J. Mol. Biol. 75, 303.

Jobe, A. & Bourgeois, S. (1972)
J. Mol. Biol. 69, 397.

Katzir, N., Oppenheim, A., Belfort, M and Oppenheim, A. (1976)
Virology, 74, 324.

Levine, M. (1972)
Current topics in microbiol. and Immunol. 58, 135.

Lieb, M. (1970)
J. Virol. 6, 218.

Maynard, D.M. & Selverston, A.I. (1975)
J. Comp. Physiol. 100, 161-182.

Miller, R.F. (1965)
Switching theory, J. Wiley & Sons.

Nicolis, G. & Sanglier, M. (1976)
Biophys.Chem. 4, 113.

Novick, A. & Wiener, M. (1957)
Proc. Natl. Acad. Sci.US, 43, 553.

Rashevsky, N. (1948)
Mathematical Biophysics, The University of Chicago Press.

Signer,E. (1969)
Nature, 223, 158.

Székely, G. (1965)
Acta Physiol. Acad. Sci. Hung. 27, 285-289.

Thomas, R. (1978)
J. Theor. Biol. 73, 631.

Thomas, R. & Van Ham, P. (1974)
Biochimie, 56, 1529.

Thomas, R., Gathoye, A.M. & Lambert, L. (1975)
Eur. J. Biochem. 71, 211.

CHAPTER XVIII

BOOLEAN FORMALISM AND URBAN DEVELOPMENT

Françoise Boon and André de Palma

Faculté des Sciences - Service de Chimie-Physique II
Université Libre de Bruxelles
1050 Brussels, Belgium

Summary

Our aim in this paper is to evaluate if the boolean formalism could be successfully applied to the study of the city and if so, if it could be of some help in the process of making planning decisions.

In the introduction, we try to point out what makes the evolution of the city so difficult to forecast : the complexity of the urban relationships and the "snowball" effects that arise from it. The theoretician who wants to take these effects into account, has the choice between two approaches which are complementary, as we try to show.

As a first trial of the boolean formalism in this field, we have chosen - as a specific urban problem - the residential location of city dwellers. A simple model allows us to show the existence of a variety of stable combinations of groupings of four social classes whose differenciated behaviours could be explained, for instance by differences in income and nationality. The nature of these combinations depends on :
(i) the perception that each social class has of the "physical" characteristics of the city (population density, neighbourhood quality and home-to-work travel time) considered as input variables;
(ii) the presence or the absence of the different types of residents, as well as the housing price, these being considered as the internal variables of the system. The results, projected onto the space of a theoretical city, qualitatively conform to the reality of a big city. The comparison is made with the Agglomeration of Brussels, Belgium.

The formalism is then analyzed from the point of view of its potential contribution to the making of planning decisions. It is clear that it could be of some help in the search for coherence between goals and means, in the search and evaluation of possible temporal strategies and, with the help of simulations, in the estimation of probabilities of occurrence of the different stable combinations of population grouping.

These encouraging results make us think that the application of the boolean formalism to the study of urban systems should be pursued further.

\+ \+

\+

Cities have grown so rapidly over recent decades that it becomes more and more necessary to guide their development. But, at the same time, their complexity has become such that the planner's task is far more difficult today than ever before.
With these considerations in mind, we have asked ourselves the following questions : could the boolean formalism be applied to the study of the city and if so, could it be of some help in the process of making planning decisions ?

Our first attempt in modelling [1)] trys to show the possible application to urban problems of the boolean formalism, as it has been used by R. Thomas in Genetics [2)] and to discuss its relation to the ideas that the Brussels School of Physical Chemistry of I. Prigogine has developped with a formalism based on continuous variables [3)].

THE CITY, A COMPLEX SYSTEM

In the complex system that is the city, let us consider two scales of phenomena : the scale of individual relationships between the constituting elements of the city that we will call the economic and social agents, and the scale of collective structures wich arise from all the individual interactions.

1 - Complexity of individual relationships [4)]

Let us imagine a small commercial town, located in a rural area, and let us introduce a large factory into it. Such an event sets in motion several chains of reactions.

The initial creation of a certain number of industrial jobs allows an almost equivalent number of families to make a living. Furthermore some proportion of these families settle down in the town. Thus from the beginning one registers a demographic jump.
The arrival of these families induces a substantial increase of the daily consumption of goods and services and thus a growth of the tertiary economic sector. In the short run, convenience shops and services, such as grocery stores and banks, start to multiply. In the medium run, new tertiary functions appear in town because the demand for scarcer goods and services has risen and crossed the threshold of their economic profitability. An essential step is made : the induction of a ter-

tiary development from the initial demographic growth.

In the meantime another chain of reactions has been set in motion. The functioning of the factory gives rise to a local demand for related industrial activities, suppliers and buyers of the factory. A series of new industrial units may come to the town, continuing the industrial development.

These two chains of reactions do not stay independent from each other; they interact. For instance, the need for new houses and shops promotes the growth of the building industry and, in the opposite way, the creation of industrial jobs induces a growth in the tertiary sector.

The induction of all these jobs contributes to the further attraction of new families to what is slowly becoming a city, especially as its power of attraction is increasing owing to the diversification of its urban functions. This development cycle is a self-supporting spiral, with the population bigger at the end of each cycle than at the beginning.

On the other hand, the multiplication of interactions between more and more numerous and diversified individuals increases the possibilities of technological progress, adoption of more efficient financial institutions, increases as well the speed of diffusion of local and imported new ideas. When they have reached maturity, these innovations are implemented by dynamic entrepreneurs. Employment grows as well as population. Another self-supporting development spiral goes on, of a psycho-sociological nature this time.

So the multiplication of the rumber and sorts of socio-economic agents in the city makes the intra-urban relationships more and more complex with time.

2 - Relative simplicity of individual behaviours

Such a tangle of relationships seems to be balanced by a simplicity in the agent's behaviour.

Let us take the example of a retail-shop. In order to make this activity profitable, the shopper must gather enough clients around him. These clients are very different : people belonging to various social classes, other retail-shops, repair-shops,... The relationships that the shop

maintains with the city are complex. Despite this however, its behaviour is quite simple : below a critical number of clients, the shop will disappear; above it, it will last and even develop. Its behaviour can then be summarized as "to be or not to be" and the main influencing factor of this behaviour will be the number of clients, expressed in the form of "above or below the threshold".
Many economic agents respond to population thresholds (all the retail-shops and population services for instance); industrial firms more often need the presence or the absence of specific factors varying with the type of activity (a good transport system and a large and flat piece of land; or the proximity of the central business district,...); as to people, they seem to choose their residential location as a function of a certain number of precise factors, varying with the type of household (a house in a quite neighbourhood with a backyard for the children; or an appartment in the middle of the city,...) [5].

As we see it, the factors influencing individual behaviours cannot always be quantified. Many non-quantifiable factors, difficult to detect and to evaluate but yet very present, do indeed influence individual behaviour in the city. They are often related to the quality of the environment (the attractiveness of the architecture, the presence of trees in the streets, the quietness of the neighbourhoods,...) or to the nature of the groups present (cooperativity between groups or rejection, ...) or to other characteristics of the city.

3 - Collective structures and non-linearities

The sum of firms' and residents' individual behaviours leads to the rise of structures at a collective scale. These structures give to the city a specific spatial organization, abundantly described in the litterature [6]. Schematically, they are :

- a network of unequally important commercial and service centers, relatively well distributed in space;
- a more random spatial distribution of industrial aggregates;
- and finally, the residence, filling up the rest of the urban space, structured itself as a function of the socio-economic status of the dwellers, their position in their life-cycle and their identification to minority groups.

In liberal societies, the connection between these two scales of phenomena (individual and collective) is made through the price attached to

a given piece of land. Coming directly from individual demands, the price of the land is a powerful regulator in the process of spatial patterns formation.
Let us take the example of the central business district [7]. When the town is small, the center is not very dense and not very diversified economically. Any new function may easily take root in it. With time, the center grows in importance and in attractiveness for economic activities as well as for residents, so that it becomes rapidly impossible to locate everybody in it. The growing demand for space in the center makes the price of the land go up and a selection begins to operate on the basis of income. The first to leave the center are the residents and the poorest among the economic activities - small businesses and local services. In the course of its growth, the town is then shaped first by centripetal forces and next by growing centrifugal forces which will accentuate their importance with time. Urban space gets structured little by little. A hierarchy of business centers appears and the center becomes the "Central Business District" with its luxury shops, offices and administration. In the residential areas, there is a high demand for pleasant neighbourhoods which attract wealthy residents who can afford high housing prices; old and depreciated areas are left to poor families.
With time, the urban space shows an increasing tendency to specialization (rich, poor, middle class neighbourhoods, industrial, office, entertainment areas,...) which tend to accentuate different individual behaviours. For example, although we may continue living with a small number of undesirable agents, the growth of these beyond some threshold may make the situation unbearable. Specialization accelerates; contrasts become striking. The big modern city confronts socio-economic extremes within a reduced space. It is the richest in invention but also in violence.

One can ask ourselves why, in the course of its growth, the city does not keep that almost homogeneous internal structure it shows at the small village stage. It is because the increasing complexity of the relationships makes the behaviours more and more interdependent, i.e. more and more non-linear, so that "snow-ball" effects start to appear : a mere nothing is enough to see the entire population of a neighbourhood being replaced by a new type.
The best example of this phenomenon is the spatial expansion of the negro ghetto in an american city [8]. As long as the negro ghetto stays at a few blocks distance from a white neighbourhood, the white dwellers,

who meet many Negroes in their daily trips, do not fear this proximity. But one needs no more than 5% of negro residents in a block (at least in the city of Seattle) to see the white residents leave or sell their houses. From this critical point, the replacement of the population may occur very quickly; the neighbourhood is deeply disturbed for some time and then finds a new equilibrium again.

4 - The surprises of urban planning

It often happens that the summation of individual behaviours leads to overall structures which finally go against individual benefits. It is one of the paradoxes of urban development we witness every day as urbanites.
A good example of this is the urban expressway policy. The best way to decrease the downtown traffic, decision makers said, is to build large automobile expressways. But the unpredicted result of this new network was to accelerate the migration of families to the suburbs and the use of automobiles as daily means of transportation. More and more drivers are using expressways today so that we can say that the building of urban expressways has accelerated the blocking of the centers.

The degree of unpredictability of city growth is a function of the complexity of the city. Thus it increases with city size. This is why the evolution of big modern agglomerations has become so difficult to control.
Consequently it is of a great importance to conceive mathematical models capable of anticipating these massive instabilities in order to stem them or to divert them.

We will not call to mind here the different types of models of urban development that have been conceived in Social Science. We will only refer to two parallel and complementary approaches, born in different scientific fields, external to Social Science, which are begining to be applied to the study of cities.
One of them is rooted in the theory of bifurcations, developped by I. Prigogine's School of Physical-Chemistry in Brussels [3]. Applied to the internal structure of the city, with the help of a formalism based on continuous variables [9], it leads to a quantitative description of neighbourhood growth and development in time. Its long term vision of the city's evolution attempts to point out the conditions in which massive instabilities develop : individual migrations suddenly cease to

compensate each other; random nucleations appear and grow; contrasting group behaviours appear leading to a new spatial order.
On the contrary, the second approach takes for granted the existence of non-linearities at the scale of social groups, with the hypothesis that group behaviours are contrasted by definition. This simplification in the collective behaviours, introduced by the use of binary variables, allows to consider a large variety of groups, without the model and the mathematics becomming too complex. This last formalism is suitable for describing very complex systems; it has already been used to good effect in Genetics by R. Thomas [2].
These two approaches are dynamic. The way time is considered in both of them is basically the same. Indeed a logical equation - as it is conceived in this paper - expresses with discrete variables, the variation in time of the demand at a particular location, in a similar way to that of a differential equation with continuous variables.

What we know about cities makes us think that collective instabilities in them are numerous and that individual behaviours often respond to thresholds. Let us accept this as a premise and we will justify the use of the boolean formalism. We will end up with a qualitative description of a complex system in time, a description nevertheless limited to the short and medium terms.
This is what we will try to show with the example of a simple model of residential location.

A SIMPLE MODEL OF RESIDENTIAL LOCATION

In its present version, the model introduced here is aimed at showing the possibilities of the method rather than accurately simulating some aspects of reality : it is not based on field work; it is guided more by our intuition of individual behaviours than by actual knowledge of people's attitudes towards urban space. It is thus speculative. Nevertheless it draws its inspiration from the situations with which we have been confronted in the city we personally know best : the Agglomeration of Brussels, Belgium.
We will start by describing the various individual relationships that maintain the different socio-economic agents (these are the internal variables : 4 classes of population and the housing price). The behavioural equations that follow express each group's conditions of immigration into and emigration out of any neighbourhood. Some of the be-

havioural factors found in these equations describe the characteristics of the urban environment that are relevant to the problem (these are the input variables : population density, quality of the neighbourhood and home-to-work travel time). After that we will build up the spontaneous sequences of immigration and emigration and show the collective structures that arise from them.

1 - The behavioural equations

Let us imagine a hypothetical city in which live four classes of dwellers whose different behaviours can be explained, for instance, by cultural and socio-economic differences :

- high income residents ($\mathcal{A}$)
- middle income residents ($\mathcal{B}$)
- low income autochthonal residents ($\mathcal{C}$)
- low income foreign minority group ($\mathcal{D}$)

Each class chooses its residential location as a function of a series of constraints and requirements which are characteristic and define each class with respect to the others. Among these are the relevant "physical" characteristics of the city : population density(D), quality of the neighbourhood(H) and home-to-work travel time(T). Four behavioural equations follow (the functions a, b, c, d), expressing each group's potential demand for residential location. If the system, that is the neighbourhood, meets the requirements of one or several classes, the residential immigration can take place and this leads to the satisfaction of the potential demand (the memory variables $\alpha, \beta, \gamma, \delta$). At this stage, we suppose that the housing supply is instantaneously adjusted to the demand, i.e. that one immediately builds all types of housing necessary to satisfy the demand. In reality, there can be a long time delay between the expression of a demand and the building of the corresponding housing type (it is often the case with social housing). The opposite can however also happen : the supply comes before the demand and the buildings stay empty. In any case, the method could be modified in order to describe situations where demand and supply are not necessarily in equilibrium.

The time delay that passes between the moment where the potential demand is expressed and the moment where it is satisfied is a characteristic delay we will call "migration delay". One of the advantages of the boolean formalism is that one may differentiate the migration kinetics, by giving different values to a definite group's migration delay

for immigration and emigration, but we will not do so here.
Let us point out that it can happen with this model a "catch-back" phenomenon, characteristic of all sequential systems [10]. It may happen that the period of time during which a potential demand of immigration is expressed is too short to allow effective immigration (fig.1).
This would correspond to the following case. Let us imagine that one builds a new subway line in a dense neighbourhood, located close to the central business district and occupied by a low income minority group. The demand for residences along the line increasing, the price of land starts going up. Low income families leave the area little by little but are still numerous enough to prevent the immigration of middle income families. At a certain time, the pourcentage of people belonging to the minority group falls low enough to allow the middle class potential demand to rise. Nevertheless some time is required for middle income families to realize this and react positively to it. During this time, the price of land continues to rise so that it oversteps the acceptable range for middle income families. Finally then, the period of time during which the neighbourhood has been accessible to middle income families has been too short to allow massive immigration to take place.

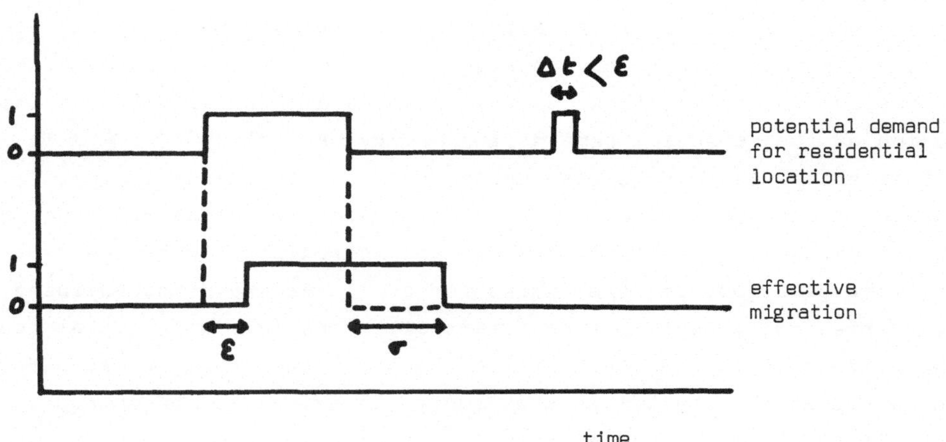

Fig 1 - A "catch-back" phenomenon

To the four internal variables already cited (the four groups of people), we will add a fifth one of another nature : the housing price ($\mathcal{P}$). The potential variation of the housing price depends on several conditions; their formulation leads to a fifth behavioural equation (function p). When these conditions are met in a neighbourhood, the housing price can rise (memory variable π).

Now what are the location constraints and requirements of each class of population and what are the conditions of variation of the housing price ?
Since we are in the process of writing urban planning fiction, let us imagine that the relationships between people are tainted by a peculiar tendency towards discrimination : the first three classes reject the foreigners belonging to the minority group (. $\bar{\delta}$.). The foreigners stick together (. δ .) for cultural reasons but also because they are rejected by the other groups (δ is partially due to the $\bar{\delta}$ of the other groups). As to the three first classes, they are willing to mix, except when the population density is high : in that case, each of them rejects the one which is just below in the income scale ($\overline{D.\beta}$ and $\overline{D.\gamma}$).
Besides these general characteristics, each class of population has requirements reflecting a value system of its own.

Let us imagine for instance that high income residents ($\mathcal{A}$) give a great importance to the quality of the environment (.H.). On the other hand, their attitude towards autochthonal low income people is less drastic than theirs towards middle income people : they accept to mix with the former, even at high population density, providing residents of their own class have already migrated in the neighbourhood before them ($\alpha + \bar{\gamma}$). This bridge connecting the two ends of the economic scale can be interpreted as a tendency of some high class people to be interested in low income people.
Their hypothetical behavioural equation is then the following :

$$a = \bar{\delta}.H.(\overline{D.\beta}).(\alpha + \bar{\gamma})$$

Middle income residents ($\mathcal{B}$) are relatively sensitive to the cost of urban life. The urban space is not entirely accessible to them for financial reasons : they cannot afford at the same time expensive housing and high home-to-work travel time or cost ($\overline{\pi.T}$); but they can afford the three other possibilities ($\bar{\pi}.T$, $\bar{\pi}.\bar{T}$, $\pi.\bar{T}$). Yet there is a rider to add here, revealing a wish to climb the rungs of the social ladder :

they like to live in the vicinity of wealthy people even if it is costly ($+\Pi.T.\alpha$).
After simplification, their equation is the following :

$$b = \bar{\delta}.(\overline{D.\gamma}).(\overline{\Pi.T} + \alpha)$$

Autochthonal low income residents ($\mathcal{C}$) are highly sensitive to the cost of urban life. Their possibilities of choosing a residential location are even more limited : they can afford to live in the city providing they have cheap housing and a low home-to-work travel time or cost ($\overline{\Pi.T}$).

$$c = \bar{\delta}.\bar{\Pi}.\bar{T}$$

Foreign low income residents ($\mathcal{D}$) respond to the same economic constraints as the autochthonal low income residents. They differ from them however, by the necessity or choice of living together.

$$d = \delta.\bar{\Pi}.\bar{T}$$

To these four behavioural equations, let us add now the housing price ($\mathcal{P}$) equation. In a first trial, we considered the housing price as an input variable, but further on we have transformed it into an internal variable [11], because decision makers generally cannot control it in a free entreprise society. We have made it essentially sensitive to the nature of the residents : it goes down when low income people of any kind immigrate into the neighbourhood; when these are missing, the housing price goes up when wealthy people come in or when the neighbourhood is characterized by a good quality environment and a low density.

$$p = \bar{\gamma}.\bar{\delta}.(\alpha + H.\bar{D})$$

At this stage of modelling, each economic and social variable is very simply represented by one binary variable, the value of whose threshold in the system has not been determined. Field work would reveal a greater complexity in the behaviours. For instance, the critical percentage of the foreign residents in a neighbourhood, leading to some reaction on the part of the other social classes - i.e. the "social efficiency threshold" of the $\mathcal{D}$ variable - could well be different from one social class to another. One would need to use then several binary variables to express these nuances between classes or a delayed multi-

level logical representation [12].
What is true for an internal variable is true too for an input variable : each social class may perceive differently the physical characteristics of the urban space.

The equations we have just set up are purely deterministic : they describe the behaviour of the average person of each social class. Chance - i.e. the existence of some irrationality in the behaviours or the local presence of other factors than the ones that have been introduced in the equations - this chance does not affect the individual behaviours. Nevertheless it does influence the final states by its intervention in the choice of the initial states.
At the end of this paper, we will moderate the deterministic nature of the individual behaviours by introducing a random variation around the mean of migration time delays. From given input and initial states, we will finally end up with an estimation of the probabilities linked to each stable state (see the results of the simulations on fig.13 and in reference 17).

2 - The decision variables (input variables)

The input variables express the influence of the outside world on the system. They immediately emerge from the behavioural equations of which they are parameters. They characterize the "physical" nature of the neighbourhood. For the urban planner, for instance,they are decision variables, the channel by which he can make the urban structure change.

Population density (D) is indeed often determined by local land use plans.

Neighbourhood quality (H) is also largely influenced by land use plans. One could consider it as a function of other input variables, such as the type of housing, the type of environment,... that we will not define here.

Home-to-work travel time (T) is largely determined by the location of jobs and transport networks ; it can be interpreted also as the cost of transportation. It is typically the kind of variable that the urban planner can control.
In the model, the three binary variables associated with the three de-

Internal variables

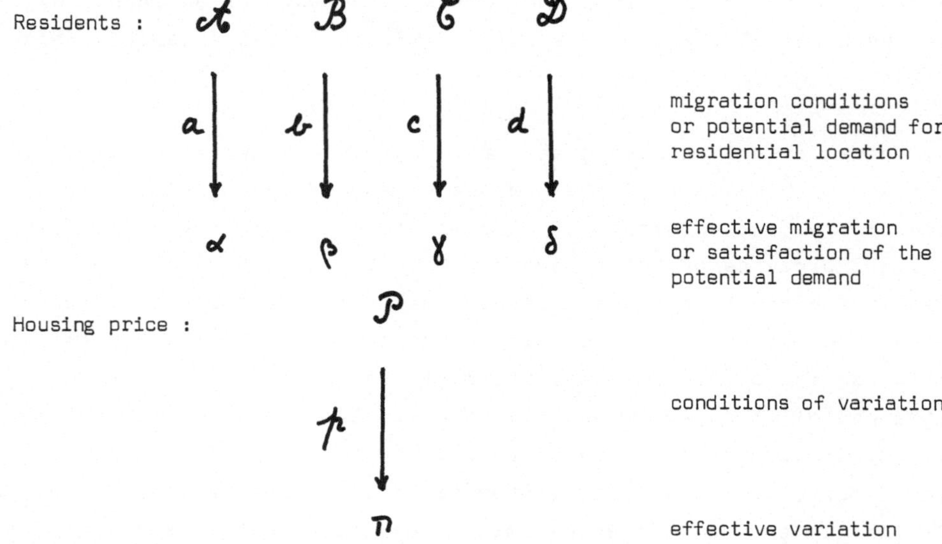

Input variables

D : population density

H : neighbourhood quality

T : home-to-work travel time or cost

A_{ij}	π	α	β	γ	δ
π	0	0	-	-	-
α	+	+	+	0	0
β	0	-	0	0	0
γ	-	-	-	0	0
δ	-	-	-	-	+
D	-	-	-	0	0
H	+	+	0	0	0
T	0	0	-	-	-

A_{ij} : Effect of the variable in the row i on the variable in the column j.

Fig. 2 - Functionning of the model

Fig.3 - State table

pabcd (π,α,β,γ,δ)	D=0 H,T 0 0	0 1	1 0	1 1	D=1 H,T 0 0	0 1	1 0	1 1
00000	00110	00100	11110	11100	00110	00100	01110	01100
00001	00001	00000	00001	00000	00001	00000	00001	00000
00010	00110	00100	00110	00100	00010	00000	00010	00000
00011	00001	00000	00001	00000	00001	00000	00001	00000
00100	00110	00100	11110	11100	00110	00100	00110	00100
00101	00001	00000	00001	00000	00001	00000	00001	00000
00110	00110	00100	00110	00100	00010	00000	00010	00000
00111	00001	00000	00001	00000	00001	00000	00001	00000
01000	10110	10100	11110	11100	10110	10100	11110	11100
01001	00001	00000	00001	00000	00001	00000	00001	00000
01010	00110	00100	01110	01100	00010	00000	01010	01000
01011	00001	00000	00001	00000	00001	00000	00001	00000
01100	10110	10100	11110	11100	10110	10100	10110	10100
01101	00001	00000	00001	00000	00001	00000	00001	00000
01110	00110	00100	01110	01100	00010	00000	00010	00000
01111	00001	00000	00001	00000	00001	00000	00001	00000
10000	00100	00000	11100	11000	00100	00000	01100	01000
10001	00000	00000	00000	00000	00000	00000	00000	00000
10010	00100	00000	00100	00000	00000	00000	00000	00000
10011	00000	00000	00000	00000	00000	00000	00000	00000
10100	00100	00000	11100	11000	00100	00000	00100	00000
10101	00000	00000	00000	00000	00000	00000	00000	00000
10110	00100	00000	00100	00000	00000	00000	00000	00000
10111	00000	00000	00000	00000	00000	00000	00000	00000
11000	10100	10100	11100	11100	10100	10100	11100	11100
11001	00000	00000	00000	00000	00000	00000	00000	00000
11010	00100	00100	01100	01100	00000	00000	01000	01000
11011	00000	00000	00000	00000	00000	00000	00000	00000
11100	10100	10100	11100	11100	10100	10100	10100	10100
11101	00000	00000	00000	00000	00000	00000	00000	00000
11110	00100	00100	01100	01100	00000	00000	00000	00000
11111	00000	00000	00000	00000	00000	00000	00000	00000

Input variable :

D : population density
H : neighbourhood quality
T : home-to-work travel time or cost

Internal variables :

π : housing price
α : high income residents
β : middle income residents
γ : autochthonal low income residents
δ : foreign low income residents

Potential demands :

p : variation of the housing price
migration of :
a : high income residents
b : middle income residents
c : autochthonal low income residents
d : foreign low income residents

Nature of the states

00000 : unstable
<u>00000</u> : stable

cision variables take the value 1 when they overstep a threshold arbitrarily chosen.

Figure 2 synthesizes the functionning of this hypothetical model.

3 - The collective structures, projected into a theoretical city

The state table (or flow table - fig.3) describes all the possible situations in which the system can be.
The system - as we have defined it - is the neighbourhood. Consequently the state table gives us a complete theoretical description of all types of neighbourhoods : each column represents a "physical" situation and each line a social content. Nevertheless the whole table does not represent a city because the neighbourhoods do not interact with each other [13]. It would be an important improvement to the method to introduce interactions between first neighbours; the model would then give a global description of the city instead of a local one.
In this state table, almost all states are unstable; only a few are stable.

It is then important to define the concept of stability when it is applied to the city which is characterized by change. This will lead us to specify the nature of the urban problems which will be the most suitable for boolean treatment, as well as to define the time scale in which we operate.
The stability of the final state is such that neither input variables, nor internal variables will be allowed to change value. This implies a certain permanence of the urban structures which will not be allowed to vary enough to make any variable pass from one logical state to another. This constrains the neighbourhoods to remain qualitatively unchanged but does not exclude quantitative modifications, such as people's and firms' migrations, on the condition that the qualitative balance be untouched. Then we see that the boolean formalism suits urban development problems (qualitative changes) better, by definition, than urban growth problems (quantitative changes), because of its intrinsic sensitivity to the former.
As to the time scale, it has to include both the characteristic time required to go from the initial state to the final state, as well as a time sufficient to show the stability of the final state. This implies that the input variables be unchanged, and consequently the "physical" structures of the city be relatively permanent. Since these physical

structures (means of transportation, firms' technical requirements,...) are largely dependent on the technico-economic system, the time scale is then reduced to the short and middle term which, by definition, exclude the possibility of any modification of the technico-economic system. Nevertheless it is possible to lengthen the prediction term of this type of model by defining a temporal sequence of input states. This sequence has to be arbitrarily chosen since it cannot be predicted by the model.

One can give a spatial image to this state table by imagining a theoretical city which would show the main tendencies observed in reality. A real city may be cut up on the basis of criteria such as the ones we have defined as input variables (population density, neighbourhood quality and home-to-work travel time), since we observe that :

- the population density decreases exponentially to the periphery and increases again in the surrounding satellites; in the center, two cases are possible : the density reaches its maximum or decreases to form what is called a density crater;
- the main employment areas are the central business district (C.B.D.) and one or several industrial areas, generally aranged along an axis;
- the neighbourhood quality is generally mediocre in the C.B.D. and around industrial areas; good in the periphery, in the historical center and, in some cases, in that part of the C.B.D. that has been recently renewed.

In order to define the specific theoretical city we are presenting, we have had to make two types of choice.
On the one hand, we have chosen its physical characteristics so as to make it look very much like the Agglomeration of Brussels, Belgium, so that we would finally be able to compare theoretical results with the real situation. Nevertheless the theoretical city is more complete than Brussels in order to make it match better the state table.
On the other hand, we have been forced to locate the social efficiency thresholds in an arbitrary manner, since we have not determined their values.

The resulting cutting is inevitably arbitrary with the following characteristics (fig.4) : the industrial areas are arranged along a S.W-N.E rapid transit axis, going through the C.B.D.; the factories located at the S.W. end of the axis have in fact closed down. The population density forms a central crater, diminishes to the periphery and increases again in three residential satellites located in the urban fringe.

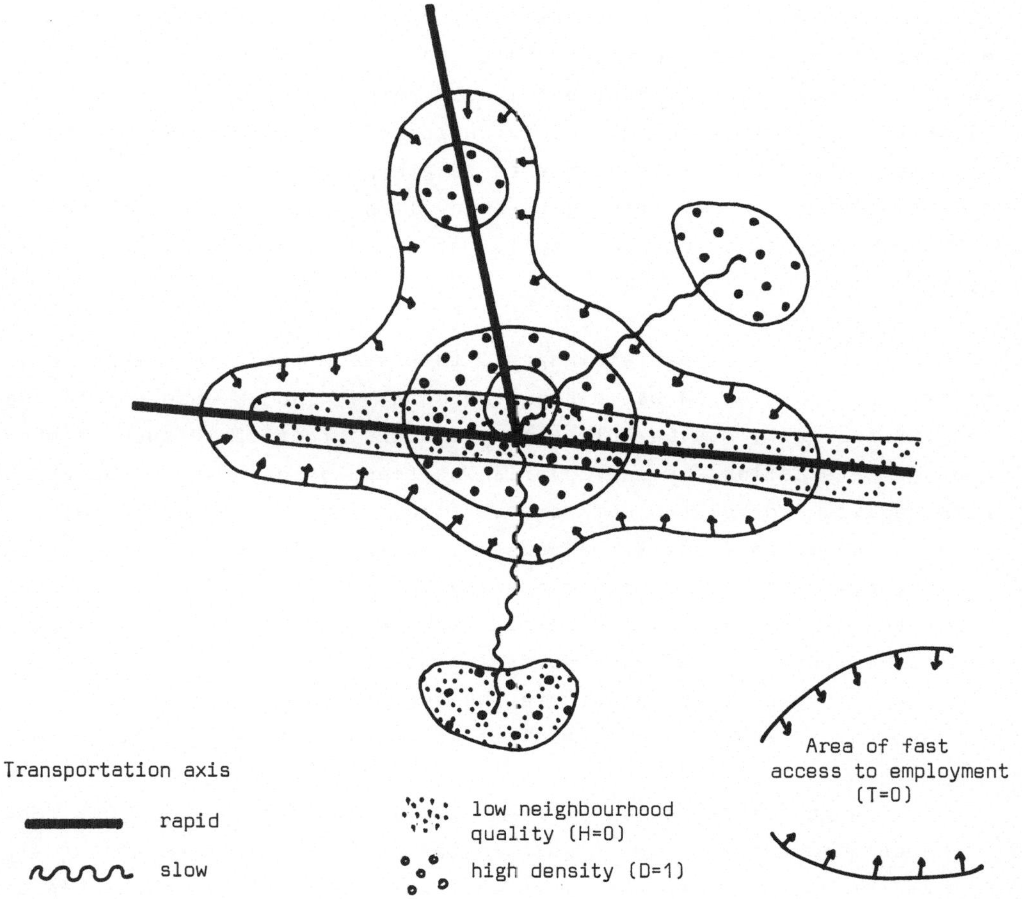

Area of fast
access to employment
(T=0)
Transportation axis
rapid
slow
low neighbourhood
quality (H=0)
high density (D=1)

Fig. 4 - Theoretical city - Social efficiency thresholds of the input variables

Fig. 5 - Theoretical city - Input states map

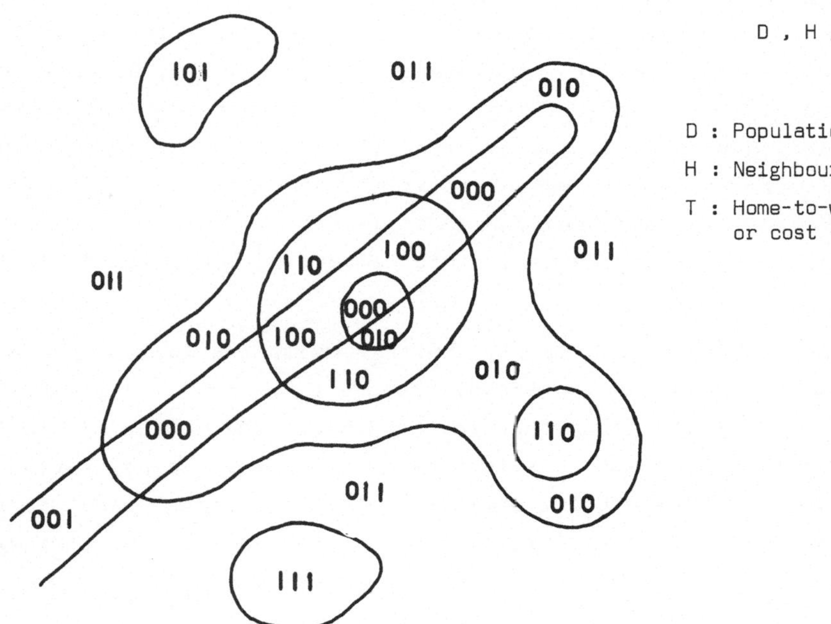

101
011
010
000
100
110
011
011
000
010
010
100
110
010
000
110
011
010
001
111
D , H , T
D : Population density
H : Neighbourhood quality
T : Home-to-work travel time
or cost

One of them only is well linked to the C.B.D. by a rapid mean of transport. The neighbourhood quality is mediocre along the industrial axis and in the Northern satellite.

Let us build up now the input state map (fig.5). If one takes n criteria to each of which is associated a binary variable (as it is the case in this study), each portion of the city will be characterized by one of the 2^n possible combinations of values of the n binary variables. In our case, each area of the theoretical city will be defined by three numbers, corresponding to the columns of the state table.
If D represents the population density, H the neighbourhood quality and T the home-to-work travel time, each input state will have the following meaning(s) :

D	H	T	
0	0	0	- depopulated industrial center - secondary pole of industrial employment
0	0	1	- low quality suburb (residential area mixed with abandoned factories for instance)
0	1	0	- depopulated historical center, renewed center or C.B.D. - pleasant peripheral residential area, well linked to the employment areas
0	1	1	- pleasant residential suburb, badly linked to the employment areas
1	0	0	- part of the first dense fringe, surrounding the C.B.D., where residences and factories are mixed
1	0	1	- industrial satellite, badly linked to the employment areas, showing a dominant dormitory function
1	1	0	- part of the first dense ring, surrounding the C.B.D., showing a high neighbourhood quality (renewed or historical populated area) - pleasant residential satellite, with a dominant dormitory function, well linked to the employment areas by a rapid mean of transport
1	1	1	- pleasant residential satellite, with a dominant dormitory function, badly linked to the employment areas

Looking at the state table (fig.3), we notice that a neighbourhood can reach different stable states even though it is in the same physical state (column). Indeed, the behavioural equations, while assuming an average mecanism of interaction at the individual scale, do not determine in an unequivocal manner the final state of the system at a collective scale, i.e. the distribution of people and housing price in the urban space. The factors influencing behaviour do not intervene solely; the choice of the initial state - that is the history of the system - may considerably modify the final state. A small random fluctuation,

changing the value of one internal variable, is sometimes enough to make a very different structure arise at the collective scale.
The following example illustrates the influence on the system of a change of value of the internal variable δ . Two very different stable states appear :

Input state (D,H,T) : 000

Initial state (π, α, β, γ, δ) : $00\bar{0}\bar{0}0 \xrightarrow{\beta} 001\bar{0}0 \xrightarrow{\gamma} \underline{00110}$

$\qquad\qquad\qquad\qquad\qquad\qquad\quad \searrow^{\gamma}\ 00\bar{0}10 \xrightarrow{\beta} \underline{00110}$

Initial state : $\underline{00001}$

The system - as it has been defined - can reach seven different stable states (π, α, β, γ, δ) :

- at low housing price :
  three situations of one class dominance :
  00001 foreign low income residents
  00010 autochthonal low income residents
  00100 middle income residents
  two possibilities of mixing two groups :
  00110 middle income and autochthonal low income residents
  01010 high income and autochthonal low income residents
  one possibility of partial integration :
  01110 mixing of the three autochthonal classes
- at high housing price, there is only one possible stable state :
  11100 the mixing of the two highest income classes

The projection in space of the stable states (fig.6), on the basis of the input states map (fig.5), gives a image which is a bit confused of the spatial organization of the theoretical city. However, the maps showing the possible locations of each social class at stable state (fig. 7a,b,c,d) are somewhat clearer. We will comment on these later on.

Doing the same operation for the housing price, we get a map showing the spatial distribution of housing price at stable state (fig.8), which is qualitatively consistent with the real situation of a city like Columbus, Ohio (fig.9) [14)].

4 - The comparison of the theoretical city with a real city

Before comparing the theoretical results with the last census of

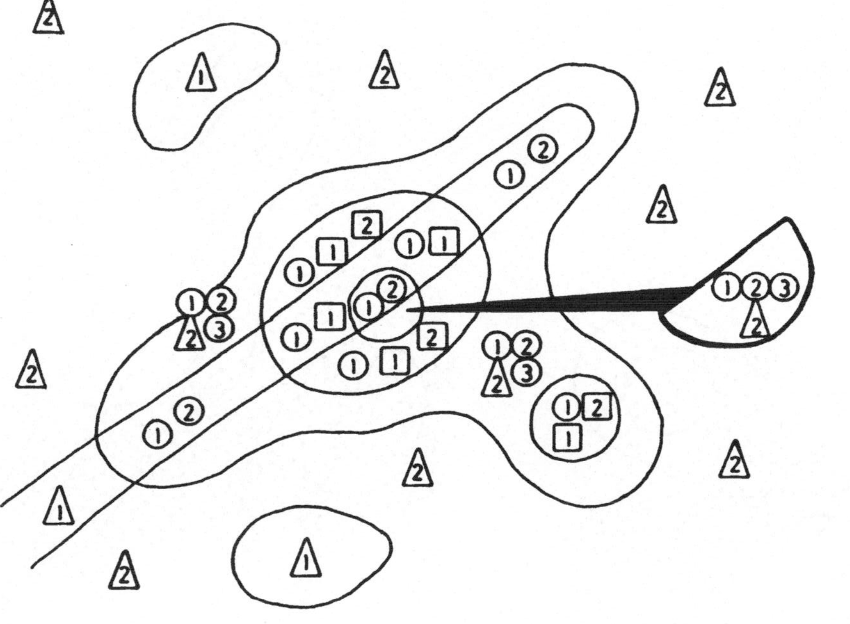

Fig. 6 - Theoretical city -
Stable states map

Predominance of one social group		Cohabitation of two social groups		Cohabitation of three social groups	
1	00001	2	00110	3	01110
1	00010	2	01010		
1	00100	2	11100		

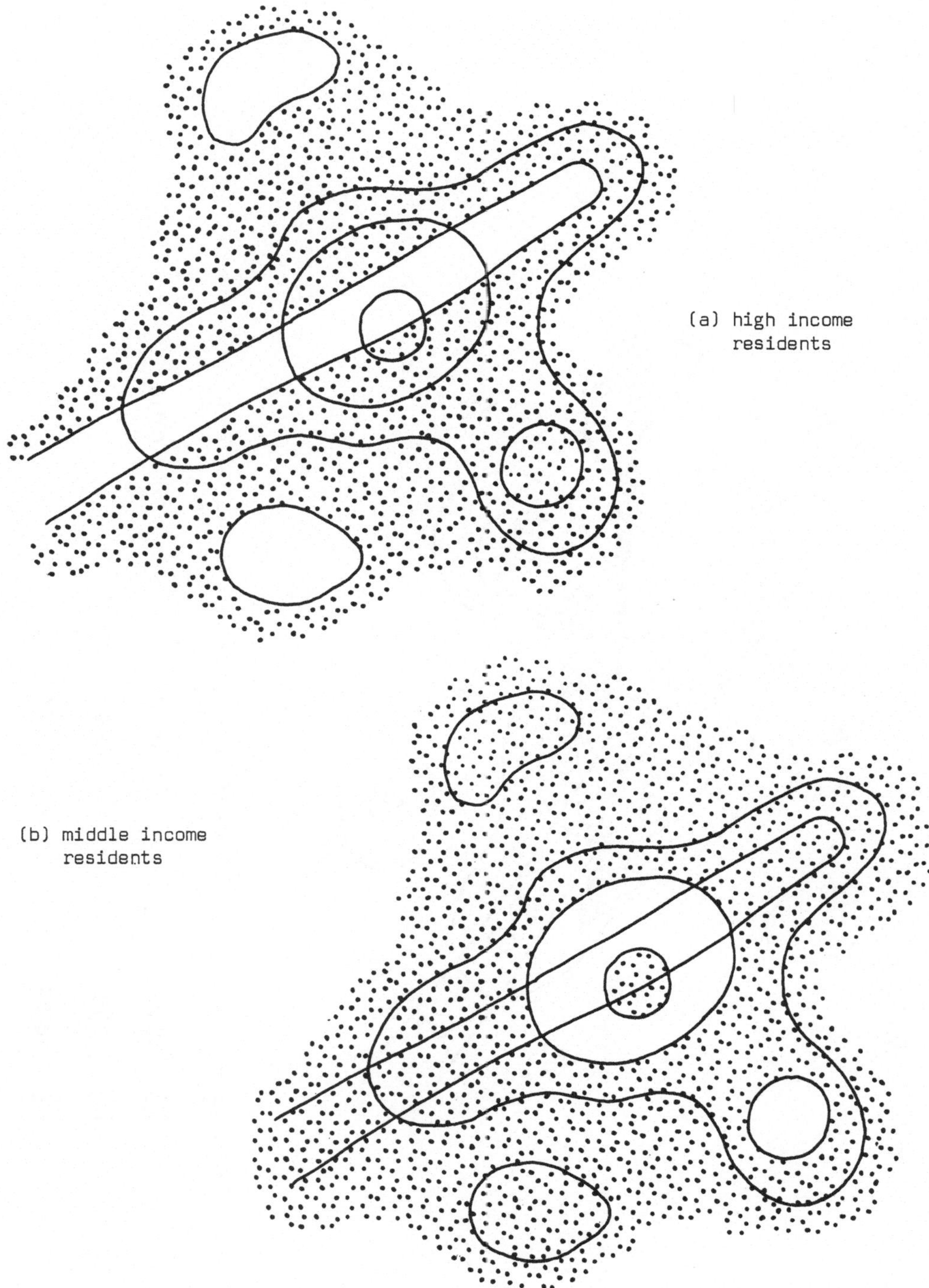

Fig. 7a,b - Theoretical city - Possible locations of the different social groups at stable state : (a) high income residents, (b) middle income residents

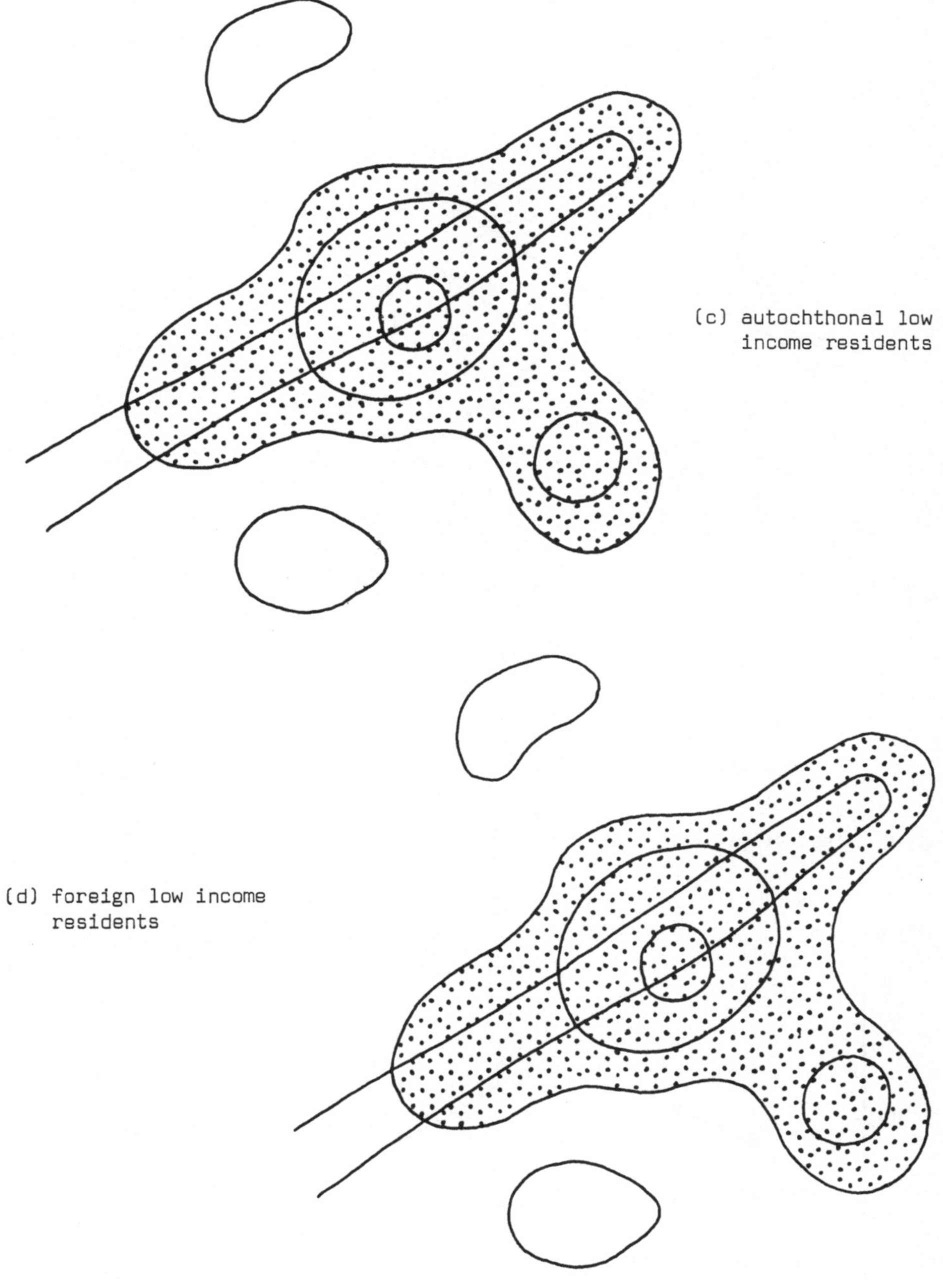

Fig. 7c,d - Theoretical city - Possible locations of the different social groups at stable state : (c) autochthonal low income residents, (d) foreign low income residents

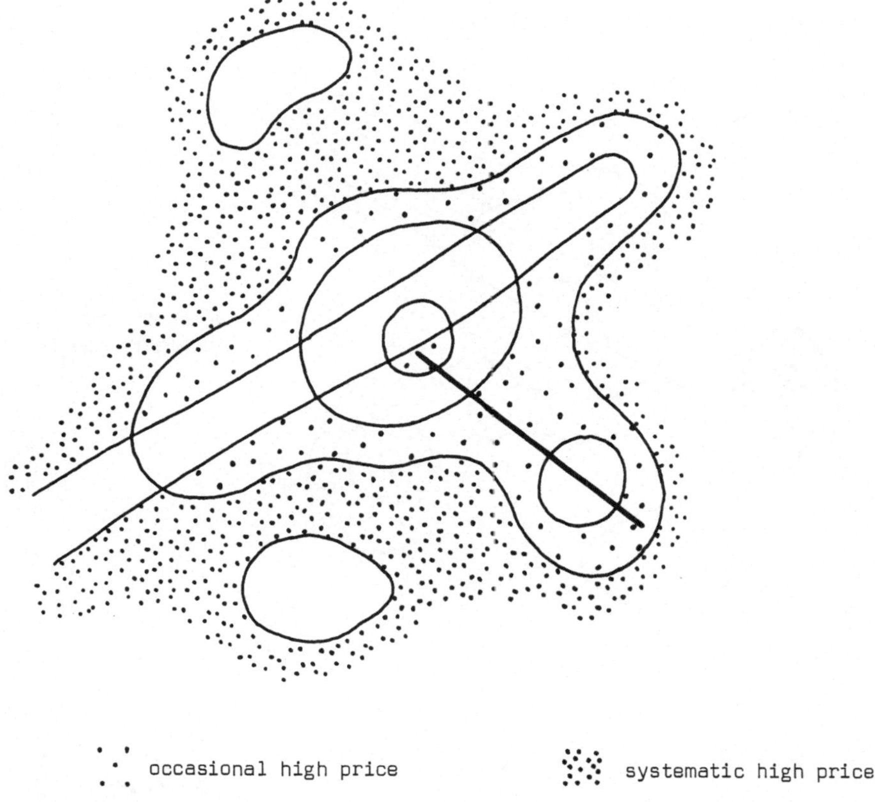

Fig. 8 - Theoretical city - Housing price at stable state

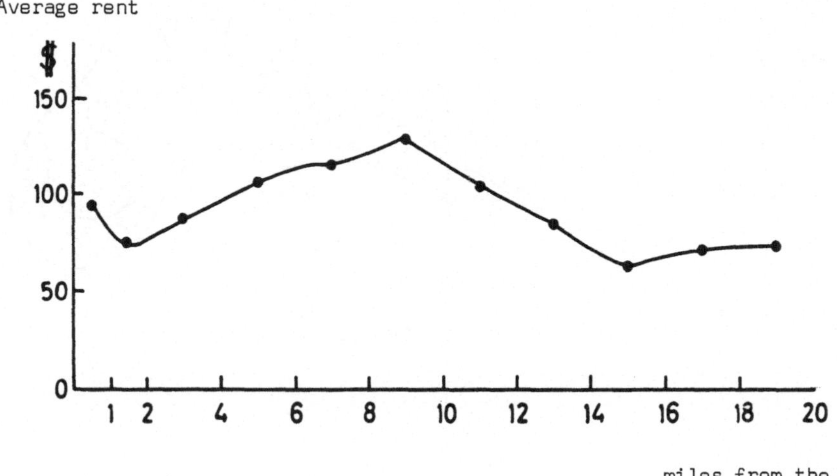

Fig. 9 - Columbus, Ohio - Average rent as a function of the distance to the city center - this profile corresponds to the axis drawn on fig.8

the Brussels' population, it is necessary to describe the "physical" characteristics of the Agglomeration (fig.10) [15).]
Divided into two parts by a S.W-N.E canalized river, hugged by a railroad line, the Agglomeration of Brussels is crossed today by an industrial axis passing through its center (this center is symbolized on all maps by the pentagon that used to form the walls of the ancient city). Small factories and often damaged old houses are intimately mixed along this axis. The ancient city, depopulated today, has developped a classical central business district on the eastern bank of the valley. Around this area, there is a ring of densely populated old neighbourhoods which is the zone corresponding to the first expansion of the city outside its walls. The population density decreases to the periphery sothat one proceeds little by little from a dominance of compact appartment buildings to a dominance of single family houses with much open space. To the South, a beautiful old beech-grove is one of the most attractive spots of the Agglomeration.

For Brussels, there is no information available about the critical values of the <u>input</u> variables corresponding to the social efficiency thresholds of the model. Thus we are forced to leave it to the imagination of the reader to interpret the features of the map described above by trying to locate estimates of these thresholds in space.

We have no information either about the corresponding values for the <u>internal</u> variables. As a matter of fact, the only information we have is the spatial distribution of the socio-economic characteristics of the Brussels' population. This can be interpreted as a spatial image of the values reached by the behavioural equations of the model. Consequently, here again, we are forced to choose arbitrarily the real values corresponding to the theoretical thresholds.

Because of the limitations of the 1970 population census, we have approximated income classes with occupational classes :

- high income residents correspond to professionals, employers and high level employees;
- middle income residents to the other types of employees;
- low income residents to workers.

Figures 11a,b,c show the census tracts having a proportion of each of these occupational classes higher than its average proportion in the Agglomeration (percentage of the total active population). As to the (active and non active) foreigners, the census gives only their propor-

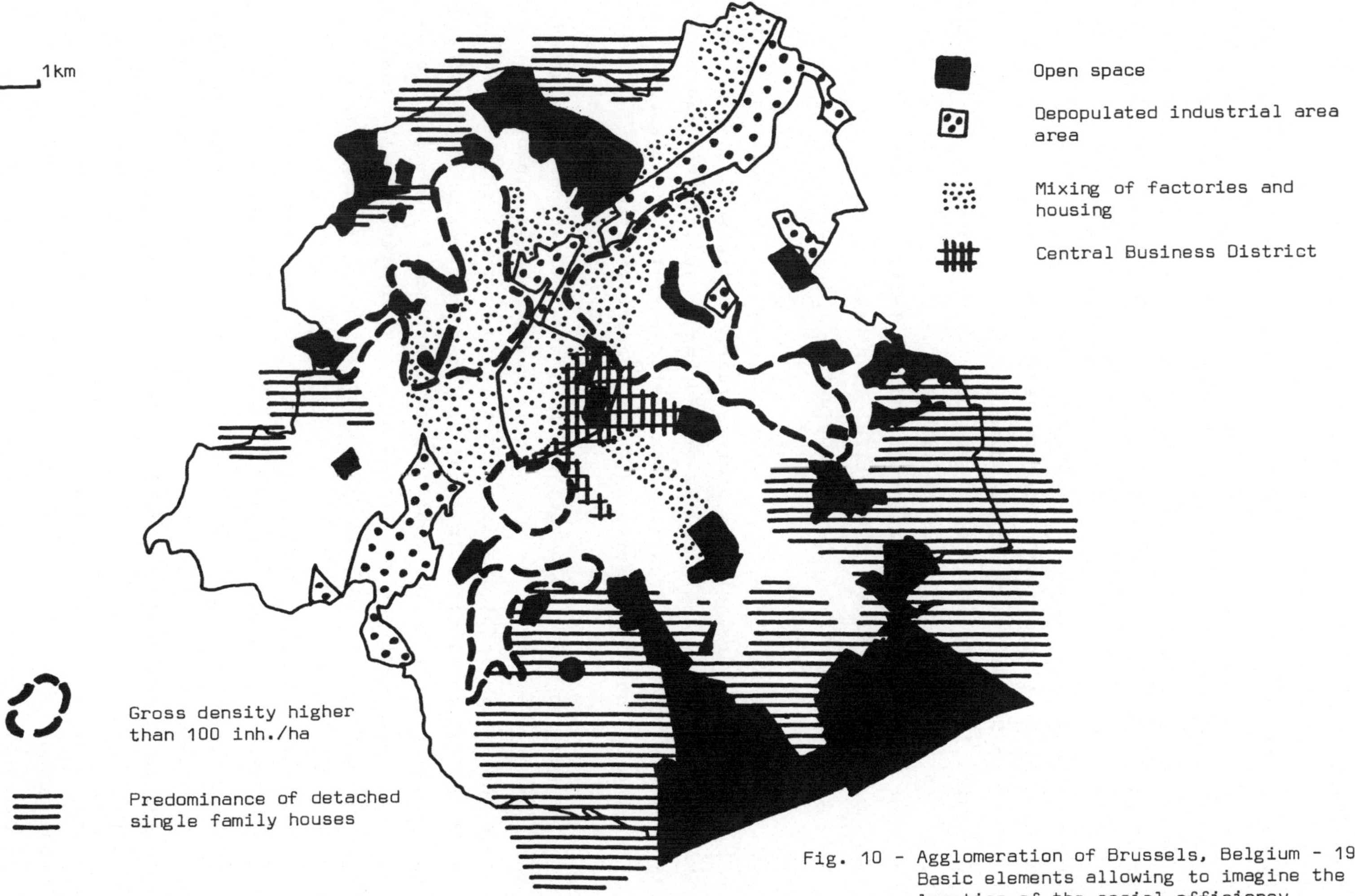

Fig. 10 - Agglomeration of Brussels, Belgium - 1975
Basic elements allowing to imagine the location of the social efficiency thresholds of the input variables

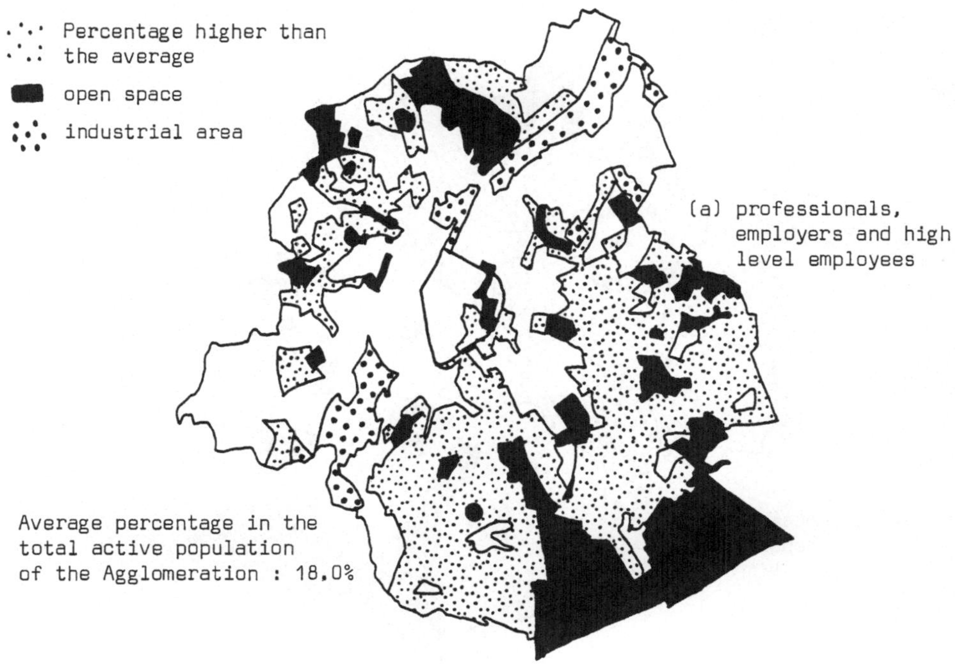

Fig.11a,b - Agglomeration of Brussels - Location of residents by occupational categories - 1970 : (a) professionals, employers and high level employees, (b) other types of employees

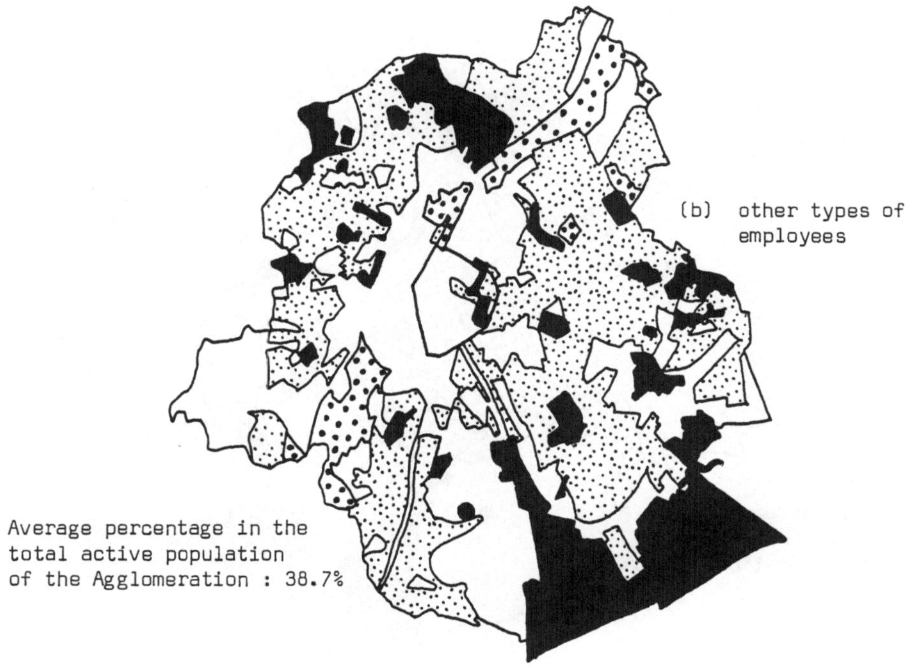

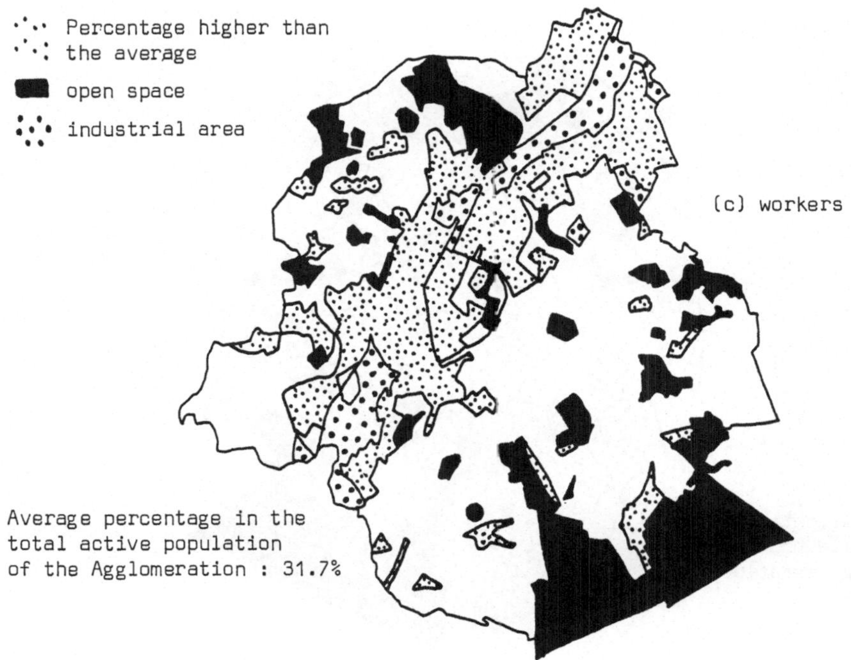

Fig.11c - Agglomeration of Brussels - Location of residents by occupational categories - 1970 : (c) workers

Fig.11d - Agglomeration of Brussels - Location of residents by nationality - 1970 : (d) foreigners

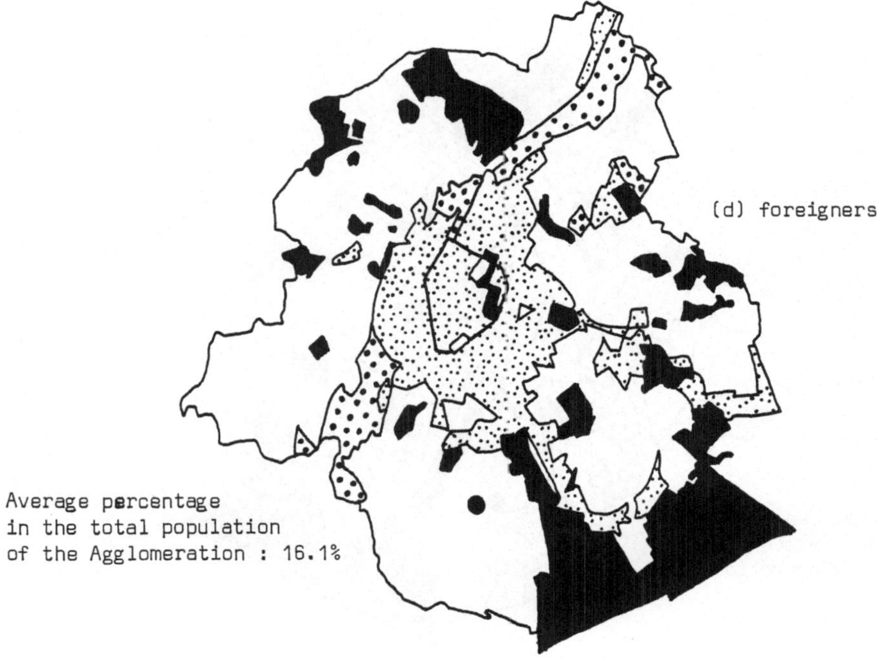

tion in the total population, whatever their activity may be (fig.11d).

The internal structure of the theoretical city is qualitatively rather consistent with the reality of the Agglomeration of Brussels which does not have any peripheral satellite inside its administrative limits.
Indeed high income families (fig. 7a and 11a) carefully avoid the industrial axis on both maps; in Brussels, they seem to concentrate around public parks and in the peripheral neighbourhoods where single family houses are predominant. One finds them also in the restored and renewed parts of the center. It seems that they are more sensitive to density than indicated in the model.
The middle income residents (fig. 7b and 11b) mix with industrial activities only where the population density is relatively low; they effectively seem to be rather sensitive to population density.
The workers in Brussels (fig.11c) are clearly limited to the areas close to the industrial axis. A comparison with the foreigners' map (fig. 11d) shows that the foreign workers concentrate around the C.B.D., while the autochthonal workers have a tendency to spread along the industrial axis. This difference in the spatial distribution of workers, according to their nationality, does not appear on the theoretical maps (fig. 7c and 7d) which, in other respects, reproduce quite well the overall tendency of workers to locate close to the industrial areas.

With the same data, let us now draw a map of the 1970 residential cohabitations (fig.12).
Arbitrarily again, two classes of population cohabit in a neighbourhood when the proportion of each of them in the neighbourhood oversteps its average in the Agglomeration, diminished by 20%.
These cohabitations in 1970 are not always stable states. Some areas - we know it - have started to undergo a deep mutation process. It is the case, for instance, of the vast South-Western area where high income families mix with workers : it is the last rural area included in the Agglomeration limits which has undergone the classical process of urbanization by expansion of the urban fringe. Other areas have been stable for some time : the Southern wealthy neighbourhoods, for instance, and the central foreign neighbourhoods which will stay foreign and poor for a long time, even though they are submitted to a strong demographic pressure and because of that changing quickly. But informations are missing about this problem of neighbourhood stability.

Fig.12 - Agglomeration of Brussels - Occupational cohabitations - 1970

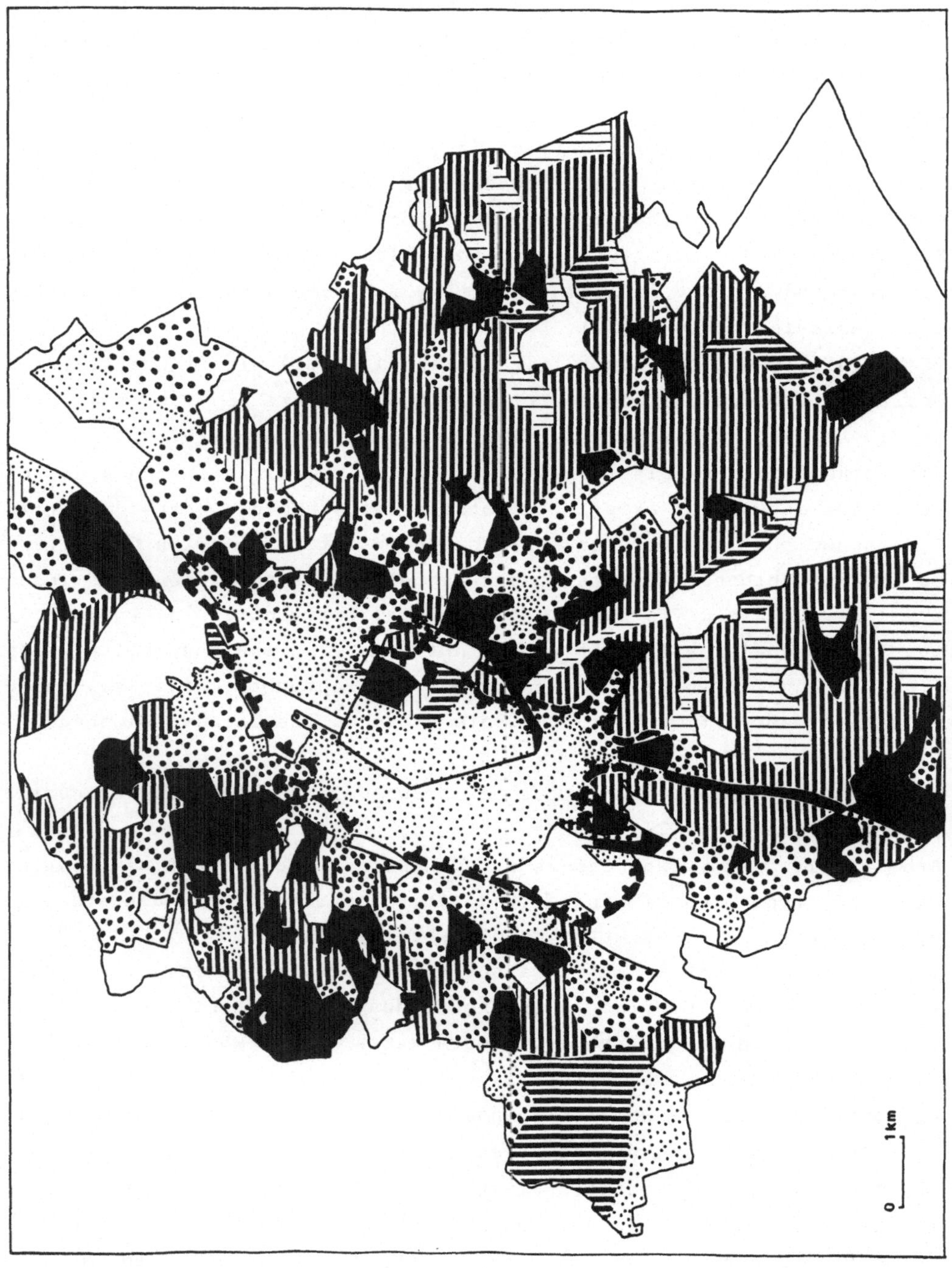

Caption of figure 12

Uninhabited area

High percentage of foreigners in the total population

Occupational cohabitations

P Professionals, Employers and high level Employees

E Other types of Employees

W Workers

W

W - E

W - E - P

E - P

E

When we compare the map of all the possible stable states in the theoretical city (fig.6) with the map of the existing states in the Agglomeration of Brussels (fig.12), we notice that the real situation looks much simpler than the theoretical one. Whatever the area, a selection seems to operate among the possible stable states in order to promote one or two of them. In other words, in any area, each possible stable state does not have the same frequency of occurrence.
In order to make this type of model more realistic, we should then associate a probability to each possible stable state. This is what we have done with the help of computer simulations, as we will see later on.
For the moment, let us analyze in detail the dynamics of the model in order to appreciate how the boolean formalism could help in making planning decisions.

BOOLEAN FORMALISM, A TOOL FOR DECISION MAKING ?

The boolean formalism, when it is applied to the study of the city, allows a complex situation to be clarified, and according to which one can adopt two types of attitudes :

- on the one hand, one can wish to maximize one's individual interest: as for example a household looking for the best residential location or an entrepreneur trying to maximize his profit;
- on the other hand, one can wish to promote the collective interest of the entire urban community, knowing well that there is often some conflict between the individual and the collective points of view. This will be the goal of the urban planner and sometimes of the politician.

Even though the boolean formalism allows us to adopt either of these points of view, we will consider here only the second case : the case of the urban planner who is confronted with the necessity of adapting the urban space to the development constraints.

1 - A general development policy or the search for coherence between goals and means

Let us imagine the case of an urban planner, taking as a goal, the integration of the different social classes in the theoretical city. Implicitly he tries to reach his goal with a strategy which would eliminate a maximum of undesirable effects. Furthermore he wishes his goal to be maintained in time : so this has to be a stable state. Consequently he wants to reach a stable state which is as close as possible to 01111 or 11111. The tendency of the three autochthonal classes being to exclude foreigners, it is not surprising to find that in the state table (fig.3),there is no stable state integrating all social classes. The stable state being the closest to this goal is the one that integrates the three autochthonal classes : 01110. It can be reached only in a low density and nice neighbourhood which is close to the employment areas (column 010) and where housing prices are low.

In these conditions, the urban planner's reasoning will be the following :
For the integration of the foreigners, there is only one solution : change the behaviour of the three autochthonal classes with regard to them and this will lead to a change in their behavioural equations.
For the integration of the three other classes, there are a lot of possible solutions, dominated by these two requirements : put the neighbourhood in the input state (column) 010 and let the housing price be low (π = 0).

- fix a maximum to the population density which is lower than the threshold for which the inter-class rejection takes place. Employment continuing to increase spontaneously in the center, the home-to-work travel time will tend to increase and make the variable T turn to 1;
- there are two possibilities of maintaining T at the value 0 :
  1/ if one wants to keep centralized the internal structure of the city (which is its spontaneous tendency), one should stop the growth of the city and carry forward the overall urban growth on other cities of the region;
  2/ on the contrary, if one does not want to stop the city growth, it becomes necessary to change its internal structure. Here again there are two possibilities :
     a/ decentralize the employment to the periphery
     b/ or improve the transportation system
- this being ensured, maintain the housing price at a low level. A new choice appears :
  1/ stop the land speculation
  2/ or increase the income of the low income classes

Obviously the means suggested by the state table are not all equivalent; some of them are feasible for the date assigned to the planner, others are not. Everything depends on the room he has for manoeuver in

a specific socio-economic system. It falls to him to appreciate his constraints.
This information about the coherence existing between goals and means is of a crucial importance for the planner.

2 - The planning of a specific neighbourhood or the search for the best strategy of investments

Maintaining the same goal of social integration - 01110 - let us imagine a nice neighbourhood (H=1), close to the employment areas (T=0) and let us consider this case at low and high densities, starting from the initial state 00000.
How could a planner program the right sequence of investments ?

a/ at low density

The detailed description of all the possible paths that the system can follow gives the entire spectrum of possible investment strategies. All paths do not lead to the goal 01110 and, among the ones that reach it, some are better than others.

Input state (D,H,T) : 010
Initial state (π, α, β, γ, δ) : 00000

```
00000 -π- 10000 -α- 11000 -β- 11100
                -β- 10100 -α- 11100
      -α- 01000 -π- 11000 -β- 11100
                -β- 01100 -π- 11100
                              -γ- 01110
                -γ- 01010 -β- 01110      goal
      -β- 00100 -α- 01100 -γ- 01110
                              -π- 11100
                -π- 10100 -α- 11100
                -γ- 00110
      -γ- 00010 -β- 00110
```

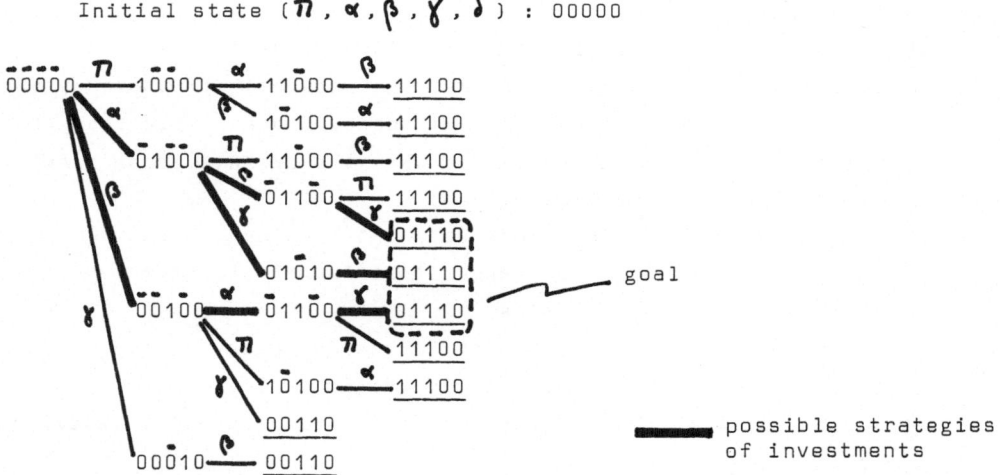

possible strategies of investments

Indeed it is only the building of wealthy or middle income family residences first that will lead to the goal. The building of low income family houses first would definitely divert the system from the goal. But in these conditions the decision maker is confronted with a possible increase in housing price during the execution of his project,and because

of this, the system can be diverted from the goal at any time.
How could the planner evaluate the chances of success of the chosen strategy ?

Computer simulations are able to give some elements of the solution (for details on these simulations, see reference 17).
Let us suppose that the immigration (ε) and emigration (σ) delays of the social groups, on the average and in any neighbourhood, obey the following relationships :

$$\varepsilon(\delta) < \varepsilon(\beta) < \varepsilon(\alpha) < \varepsilon(\gamma)$$

$$\varepsilon(\alpha) = \sigma(\alpha)$$
$$\varepsilon(\beta) = \sigma(\beta)$$
$$\varepsilon(\gamma) = \sigma(\gamma)$$
$$\varepsilon(\delta) = \sigma(\delta)$$

Thus we assume that the most mobile residents in space are the foreign low income families and the most stable ones the autochthonal low income families. Let us maintain constant these migration delays (arbitrary chosen as 20, 30, 40 and 50 with a random variation of 20% around these averages) while allowing the rise delay of the housing price to vary (turn-on delay).
The probability for the occurrence of a stable state strongly varies as a function of the housing price rise delay (fig.13). When the land speculation is intense and the price rise rapid, the threshold value is quickly exceeded and the only possible state is 11100. The wealthy families are the only ones who can afford land speculation of such intensity and middle income families who are intent on climbing the social ladder. When land speculation becomes weaker, the rise in the price of housing takes longer and the most probable stable state is still 11100. Two other stable states appear however : 01110 (the goal) and 00110 (the mixing of middle income families with autochthonal low income families). A further lengthening of the delay strongly increases the probability of occurrence of the goal (01110), gives the 00110 state with a low probability and gives zero probability for the state 11100.

In this theoretical example, we see that the probability of occurrence of the goal is a function of the rapidity with which the price of housing rises. In reality (fig.12), the frequency of occurrence of a state depends on the location of the neighbourhood in the city. In other

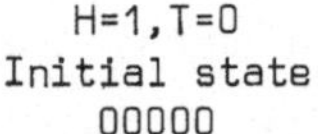

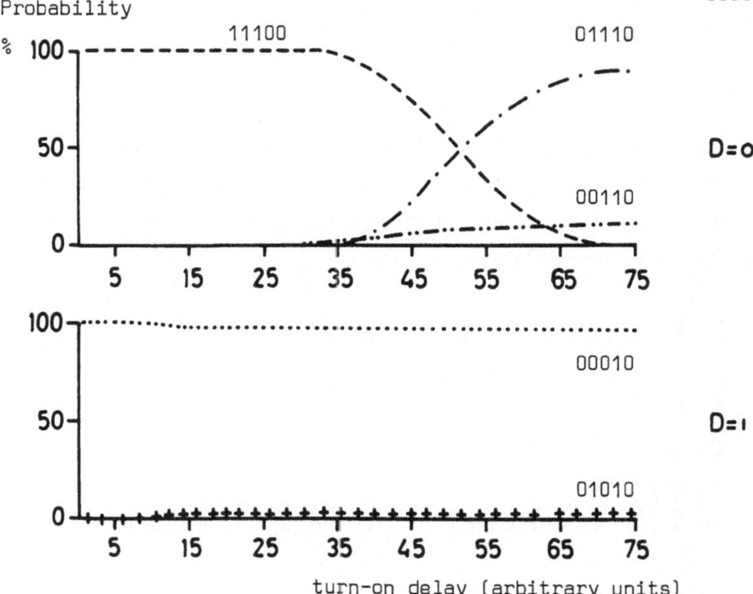

Fig 13 - Probability of occurrence of the stable states as a function of the housing price rise delay (turn-on delay)

Fig 14 - Probability of occurrence of the stable states as a function of the housing price fall delay (turn-off delay)

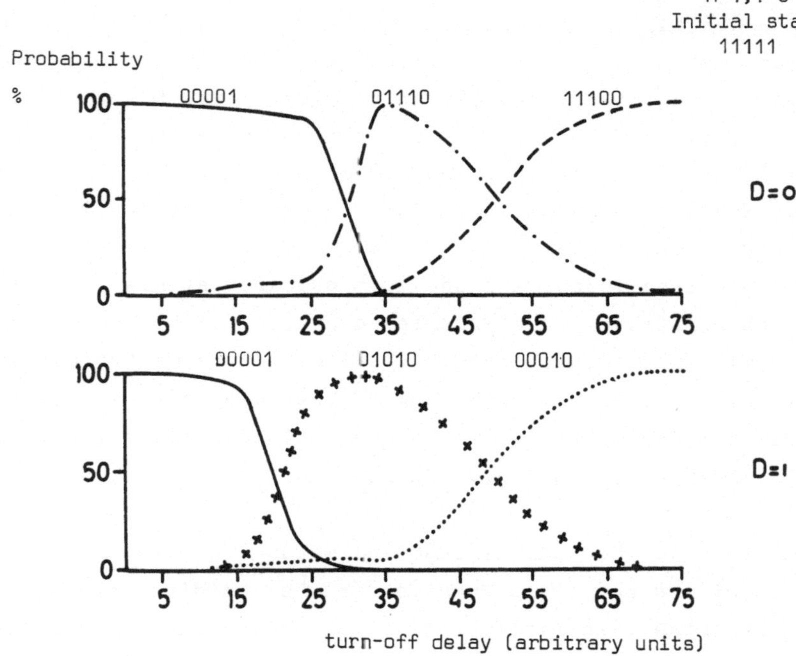

words, this would mean that the intensity of land speculation varies in the urban space and that computer simulations of that kind would probably be able to give some information about the spatial distribution of the speed of variation of the housing price.

Investment strategy should include a variable fraction of its costs used to cut down on land speculation. Here again, the urban planner is free to chose the best way to do so, according to the economic system involved. He could even try to make a cost-benefit analysis of different development strategies.

Other initial conditions - 11111 - still at low density (fig.14), give a very different probability distribution of the goal 01110 which can be interpreted as the influence of history on the final state of the system. In this case, the influencing factor is no longer the rate at which the price of housing rises, but that of its fall (turn-off delay). The temporal strategy will be modified. On the other hand, it is only from this initial state that one sees the formation of ethnic neighbourhoods. Foreign immigrants must be imposed on the system explicitly; they do not appear spontaneously in the neighbourhood by the internal dynamics of the system.

b/ at high density

At high density, whatever the initial state may be, the goal 01110 is unattainable (fig.13 and 14).

Input state (D,H,T) : 110

Initial state (π, α, β, γ, δ) : 00000

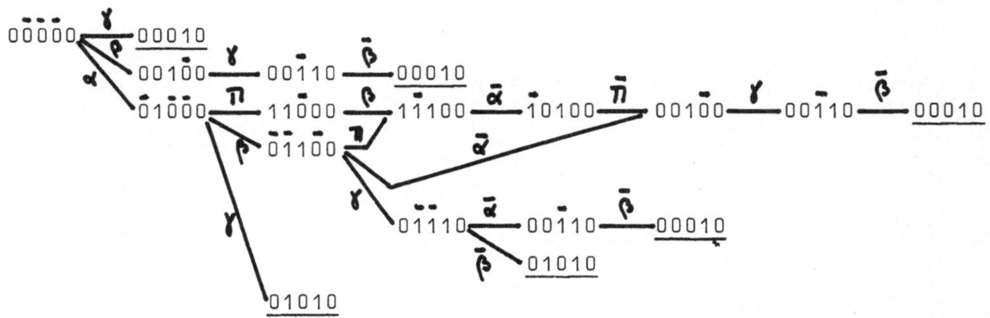

Residential segregation greatly increases, as one would expect from the equations. Here again, the initial conditions strongly influence the probability of occurrence of the stable states (fig.14).

CONCLUSION

The boolean formalism has the great advantage of being easily adaptable to the dynamic analysis of complex systems. It remains manageable when a continuous formalism, although attractive because of its greater analytical power, becomes too complex to handle. Its flexibility, its simplicity and the rapidity of analysis that it allows could make it a precious help to decision making in socio-economic matters.

Acknowledgment

We are grateful to Professors Jean-Pierre Boon and René Thomas for their patience in convincing us to apply the boolean formalism to the study of the city. These ideas have had a chance to develop, thanks to our participation in the studies of population dynamics realized by the Brussels School of Professor Prigogine. We are indepted to him for his interest and encouragements. This work has been pursued beyond our initial goal thanks to fruitful discussions with and suggestions by Philippe Van Ham, Jean-Louis Deneubourg, Peter Allen and Isabelle Stengers. We take pleasure in warmly acknowledging them.

This work was supported by the Department of Transportation, Transportation System Center, Cambridge, USA (D.O.T. - T.S.C. - 1185), la Commission des Communautés Européennes (DG XII-B-4), les Instituts Internationaux de Physique et Chimie fondés par E. Solvay and Minna-James-Heineman Stiftung.

Notes and references

1) This work has been accomplished during the semester following the EMBO course "Formal analysis of Genetic Regulation". Université Libre de Bruxelles - September 6-17, 1977.
2) Thomas (R), Boolean formalization of genetic control circuits, J. Theor. Biol (1973) 42, 563-585
   Florinne (J), La synthèse des machines logiques et son automatisation - Presses Académiques Européennes, Dunod, 1964
3) Nicolis (G), Prigogine (I), Self organization in non equilibrium systems, Wiley, 1977
4) In this section, we follow the approach by Allen Pred presented in his book : The spatial dynamics of U.S. Urban-Industrial growth, 1800-1914 - Interpretive and Theoretical Essays, M.I.T. Press, 1966
5) see e.g. Hoover(E.M.), The location of Economic Activity, Mc Graw-Hill, 1948
   Castells(E), Ahtik(V), Touraine(A), Zygel(S), La mobilité des entreprises industrielles dans la région parisienne, Cahiers de l'I.A.U.R.P., vol 11, Paris 1968

Berry(B.J.L.), Geography of market centers and retail distribution, Prentice Hall, 1967

6) see e.g. Bourne(L.S.), ed., Internal Structure of the city - Readings on space and environment, Oxford University Press, 1971
Berry(B.J.L.), Horton(F.E.), Geographic perspective on urban systems - with integrated readings, Prentice Hall, 1970

7) Vance(J.E.), Focus on Downtown, in Internal Structure of the city, op. cit.

8) Morrill(R.M.), The Negro Ghetto - Problems and Alternatives, in Internal Structure of the city, op. cit.

9) Allen(P.M.), Deneubourg(J.L.), Sanglier(M), Boon(F), de Palma(A), Dynamic Urban growth models - Final report prepared for the Department of Transportation - Transportation System Center, Cambridge - USA - 1977

10) Philippe Van Ham (private communication)
Van Ham(P), Ph.D. Dissertation, Modèles discrets à actions différées, Université Libre de Bruxelles - 1975

11) This modification was suggested by René Thomas

12) Philippe Van Ham (private communication)
Van Ham(P), Delayed multilevel logical representation of functions, to be published in J. Theor. Biol.

13) see different studies on spatial diffusion of socio-economic phenomena. Morrill(R.L.), The spatial organization of society, Duxbury Press, 1970

14) Bronitsky(L), Costello(M), et al., Urban data book, U.S. Department of Transportation, 1975

15) Agglomération de Bruxelles, Carte de Situation Existante au 1/1/1975

16) Institut National de Statistique - Recensement de la Population au 31/12/1970, Bruxelles, Belgique

17) The computer program used for the simulations is the PRAN 2 program, realized by Philippe Van Ham.
Van Ham(P), A random simulation of deferred actions logical systems dynamics, IFAC, Congres on discrete systems, Dresden (DDR), march 1977, vol.5, p.27-35.

Details of the simulations

Initial state 00000

Each point of the curves is the average calculated from eight series of one hundred simulations. These series are characterized by variable turn-off delays of the housing price (arbitrary units from 5 to 85). For each simulation, the values of the delays are randomly chosen in a range of 20% around the values.

Initial state 11111

The simulation method is the same but here the average is computed from variable turn-on delays of the housing price.

CHAPTER XIX

NUMERICAL MODELS OF OSCILLATING GENETIC SIGNALS

A.Rörsch, M.A.E.Groothuis and A.M.H.Schepman

Department of Biochemistry, Leiden State University, The Netherlands.

INTRODUCTION.

The study of genetic regulation is one of the most intriguing subjects in molecular biology. It is in itself of great importance to understand the regulatory mechanism of cell metabolism, of cyclic processes such as cell division and of cell differentiation. A new dimension was added recently when it was postulated that the evolution of multicellular organisms would be the evolution of genetic regulatory circuits, rather than the evolution of their individual cell constituents (King & Wilson,1975). Thus in order to learn about the pathways of the evolution of higher organisms, we must learn more about the qualities of particular regulatory circuits and how Darwinian selection can act upon them.

The operon model of Jacob and Monod (1961) for the regulation of gene function in bacteria, is still a basic and most popular model. Over the last 15 years it has undergone further refinement by the experimental identification of repressor proteins, operator, promotor and termination sites, without having its elegance affected (for review see Beckwith and Sipser, 1970).

Also for the regulation of genes in eukaryotic cells a model was proposed by Britten and Davidson (1971). These authors postulated a regulatory function for RNA, rather than for a protein, synthesized under the control of an RNA molecule. Unfortunately little is known about the structural elements involved in the genetic regulations in eukaryotes. An essential role has been attributed to low molecular weight affecter molecules, for example steroids, which probably induce conformational changes in macromolecular constituents, leading to changes in the stability of such molecules and changes in protein-nucleic acid interactions. It is hoped that by isolation of individual, operon-like structures from eukaryotes by molecular cloning soon some light will be shown on the mechanisms of their regulatory circuits.

Since we expect the basic principles of genetic regulation in eukaryotes not to differ essentially from those established in their ancestors, the prokaryotes, and since we know quite a bit of the properties of the primary and secondary genetic signals,RNA and protein, it is time now to develop, parallel to the experimental approach, a theoretical one, in order to establish hypotheses which can be tested and challenged experimentally. To understand regulation fully it is not sufficient to identify the structural elements involved in regulation. It is a phenomenon in which quantitative aspects are of utmost importance to understand also its qualitative properties.

The structural model of Jacob and Monod induced Goodwin (1965) to present a quantitative description of it in mathematical terms. This was probably the first attempt to attribute to a genetic regulatory circuit a quality which is based on quantitative properties. Taking into consideration rates of synthesis and stability of mRNA and protein, as well as the interaction of a repressor molecule with a promotor-operator (OP) site, Goodwin presented equations for alterations of mRNA and protein levels with time. A simulation on an analogue computer would have indicated that a negative feedback loop, would lead to an oscillating behaviour. The Goodwin model was challenged by others (Fraser and Tiwari, 1974; Tiwari et al.,1974) and refined with respect to the algebraic description of the processes which occur in between transcription and feedback to the gene. However, the basic concept of Goodwin was not challenged so far. He assumed that the interaction of a repressor protein with an OP site, will determine a rate of transcription which can be described by a similar function as derived for the rate of substrate conversion by an enzyme, in the presence of an inhibitor: $V = a/(b+K.(I))$ (Goodwin, 1976).

Several years ago Thomas (1973) presented a very different approach to describe models for genetic regulatory circuits, using the principles of Boolean algebra. His basic concept is that the transcription of a gene is an all-or-none phenomenon; a gene must considered to be either 'on', subject to transcription, or 'off'. Although the primary and secondary signal generated by a gene in terms of numbers of mRNA and protein molecules might fluctuate strongly, he stated that it would be sufficient to consider two conditions for the signals, 'on' and 'off' only, in order to describe the mutual influences of different genes in a regulatory circuit. As Thomas wrote in one of his first papers (1973): *There is a rather narrow range of concentration above which a substance is fully efficient ('present' or 'on') and below which it is essentially inefficient ('absent' or 'off').*

This approach allowed Thomas to symbolize complicated genetic regulatory circuits in a rather simple way and to predict whether such circuits would have one or more stable states or would be oscillating. There is an obvious parallel between the approach of Thomas and the one used by chemical and electronic engineers to design regulatory circuits and no doubt the engineers' experience with regulating industrial processes can be used to 'model' also biological regulation. The merit of the Boolean approach is no doubt that it can be used in principle to describe at least the patterns of complicated regulatory circuits which become very difficult to survey without such formalization. There is nevertheless still much to say in favour of the continuous approach (Goodwin, 1976) because it takes in consideration the nature of the genetic signals (mRNA and protein) in terms of levels varying with time. The production of mRNA and protein is an energy consuming process (for a protein of 100 amino acid residues, 300 high energy phosphate bonds and for its corresponding mRNA another 600). We must assume that next to specificity, the strive for economy will be of great importance as a selective force in evolution of regulatory circuits. In order to understand how particular circuits came into being we must closely follow the mRNA and protein signals. From the literature we learn, by doing so, we soon end up with complicated algebraic equations, difficult to understand by the non-mathematician. We experienced, however, that if we borrow from the discontinuous, (Boolean) approach, the idea of a certain setpoint level (at which a particular signal has an effect, below which it does not signify) and use it in the continuous description, the algebraic equations remain rather simple. We still need a digital computer to solve those equations numerically. The programming, however, is simple and straightforward and can be broken down to a limited number of different operations, which in various combinations can be used to simulate a variety of oscillating genetic circuits.

One should look upon this approach as similar to the one used by Thomas, with the extension that the primary and eventually secondary genetic signals (mRNA and protein) are quantified, which allows the determination of amplitudes and frequencies of oscillation in relation to protein and mRNA stability, the differential rate of their synthesis and changes in setpoint level.

Here we restrict ourselves mainly to the description of the application of the method to an oscillating circuit consisting of two genes, r and a, of which the gene product of the first has an inhibitory effect on the transcription of the latter and the gene product of the second gene has a stimulating effect on the transcription of the first one.

SYMBOLS

a, b, c etc.	genes with a stimulating effect on others
r, s, t etc.	genes with an inhibitory effect on others
A,B,C,R,S,T,X	number of RNA molecules from genes a,b,c,r,s,t,x
a,b,c,r,s,t,y	number of protein molecules produced by the corresponding RNA molecules
A_0, a_0	number of molecules at time t=0 when the gene is switched on
A_1, a_1	number of molecules at time t_1 when gene gives a signal (= setpoint level)
A_2, a_2	number of molecules at time t_2 when gene is switched off
A_3, a_3	number of molecules at time t_3 when 'a' reaches again a_1 (usually $a_3=a_1$)
a_4	number of molecules at time t_4 when A=1
F_a	rate of synthesis of RNA molecules of gene a
G_a	rate of breakdown of RNA molecules of gene a
N_a	$= F_a/G_a$
H_a	rate of synthesis of protein molecules per mRNA molecule 'A'
I_a	rate of breakdown of protein molecules
P_a	$= H_a/I_a$
k_a	$= I_a/G_a$ relative stability protein/mRNA
g_a	$= H_a/F_a$ relative rate of synthesis protein/mRNA
p	$= G_r/G_a$ relative stability of two mRNA molecules

BASIC CONCEPTS IN THE APPROACH

A genetic signal which is produced, signifies because it is also subject to breakdown. Obviously, because if this was not the case - if the gene product was stable for ever - it could hardly be considered a signal.

Here we assume that primary and secondary gene products (mRNA and protein respectively) have a stability of their own and are subject to decay according to the target theory (Lea 1955) and that the decay can be described as a single hit process. Consequently the number of molecules present at time 't' - in the absence of new synthesis - is given by the first term of the normal distribution :

(1.1) $X_t = X_0 . e^{-G.t}$

in which X_0 the number of molecules at time t_0 and G the constant for the stability of the molecule of type 'X'.

The rate of breakdown is given by :

(1.2) $\delta X/\delta t|_S = -G.X_o.e^{-G.t} = -G.X$

Thus, G is defined as

(1.3) $G = -X^{-1}.\delta X/\delta t|_S$

Next we assume that synthesis of product X, occurs at a constant rate, independently from the number of molecules, previously produced :

(1.4) $+\delta X/\delta t|_B = F$

Obviously the rate of synthesis is directly proportional to the number of templates which are subject to transcription (or translation).

For the sake of simplicity we assume that in the case of transcription one gene is involved. We also assume that the rate of protein synthesis is directly proportional to the number of mRNA molecules :

(1.5) $\delta y/\delta t|_B = H.X$

in which 'X' the number of mRNA molecules.

If a gene is subject to active transcription we must combine (1.2), (1.4) :

(1.6) $dX/dt = \delta X/\delta t|_B - \delta X/\delta t|_S = F - G.X$

We introduce the number N = F/G which is also the maximum number of mRNA molecules that ever can be reached (dX/dt = 0).

(1.7) $dX/dt = G.(F.G^{-1} - X) = G.(N-X)$

(1.8) $\frac{dX}{N-X} = G.dt$

(1.9) $\frac{d(N-X)}{N-X} = -G.dt$

(1.10) $N - X = (N - X_0).e^{-G.t}$

If the gene is switched off, the signal 'X' will decay according to (1.1).

Similarly we can consider the synthesis of protein 'y' on mRNA 'X' :

(1.11) $dy/dt = H.X - I.y$

We introduce the value P = H/I, the maximum number of protein molecules which can be produced on a single mRNA molecule ($dy/dt_{X=1} = 0$).

(1.12) $dy/dt = I.(H.I^{-1} - I.y) = I(P.X - y)$

(1.13) $$\frac{dy}{P.X - y} = I.dt$$

(1.14) $P.X - y = (P.X - y_0).e^{-I.t}$

Next we assume that a primary genetic signal (RNA) raises during oscillation from a minimum X_0 to a maximum X_2 and again is reduced to X_0 in the absence of transcription. During synthesis a critical value X_1 will be passed at which the signal has an effect on another gene. This we name the setpoint level X_1. During synthesis the signal is effectively "off" between X_o and X_1 and "on" between X_1 and X_2, according to the concept of Thomas (1973). See figure 1. Similarly such a critical value will be passed during decay of the signal, in the absence of transcription, from X_2 to X_o and we indicate this value X_3. (Note that we adopt for X_3 a value, not necessarily equal to X_1, of which the significance will be explained later)

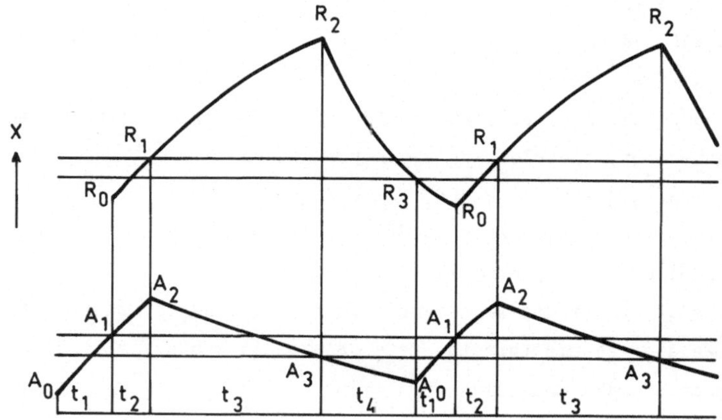

Figure 1.
The basic model for an oscillating regulatory circuit.
Abscis: time. Ordinate: level of mRNA molecules 'A' and 'R'.

To summarize, the time course of the raise and fall of the primary signal can be broken down als follows :

From X_0 to setpoint level X_1 (from 1.10) :

(1.15) $(N-X_1) = (N-X_0).e^{-G.t_1}$

From X_1 to maximum level X_2 (from 1.10) :

(1.16) $(N-X_2) = (N-X_1).e^{-G.t_2}$

From the maximum level X_2 to the setpoint level X_3, during breakdown (from 1.1):

(1.17) $X_3 = X_2.e^{-G.t_3}$

and from the setpoint level X_3 to the minimum level X_0 (from 1.1) :

(1.18) $X_0 = X_3.e^{-G.t_4}$

From (1.15) to (1.18) we can also derive equations for the time the <u>signal</u> is considered as to be "on" (above setpoint level) and "off" (below setpoint level) and also for the frequency of the oscillation :

The signal is "on" :

(1.19) $$T_{on} = t_2 + t_3 = G^{-1}.\ln((N-X_1).(N-X_2)^{-1}.X_2.(X_3)^{-1})$$

and the signal is "off"

(1.20) $$T_{off} = t_1 + t_4 = G^{-1}.\ln((N-X_0).(N-X_1)^{-1}.X_3.(X_0)^{-1})$$

and the period of the oscillation is represented by

(1.21) $$T = T_{on} + T_{off}$$

T_{on} and T_{off} and the period T are directly proportional to G^{-1}. For the sake of simplicity we shall express time usually as G.t units, or in other words, adopt a value G=1.

THE BEHAVIOUR OF SECONDARY GENETIC SIGNALS

The raise and fall of the RNA signal is fully described by equation (1.10) and (1.1). The raise and fall of the protein signal by equation (1.14). Note that when the last intact mRNA molecule has decayed (X=0) (1.14) is reduced to :

(2.1) $$y = y_0.e^{-I.t}$$

in which y_0 is the number of protein molecules at the time when the last mRNA molecule disappeared (indicated further as y_4)

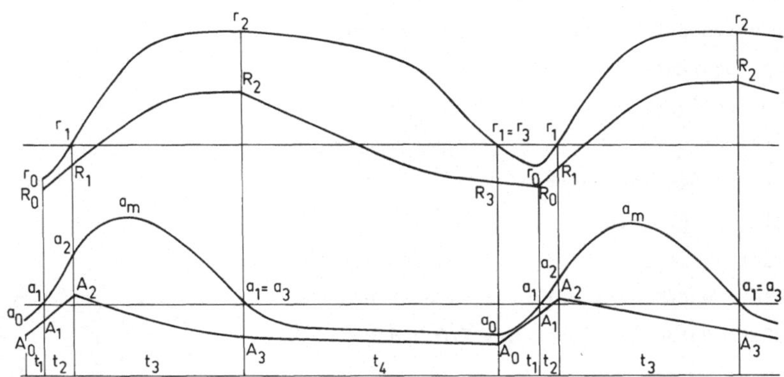

Figure 2.
The protein mediated model for an RA system.

In order to estimate the number of protein molecules at any time during the raise and fall of the primary RNA signal (see fig.2), we assume firstly that only complete and intact mRNA molecules contribute to net protein synthesis. Thus (1.14) is in fact a step function, which can be described in general as :

(2.2) $$y_t = P.X - (P.X - y_{t=0}).e^{-I.t}$$

in which y_t the number of protein molecules at time 't' after time t=0, when for the first time 'X' complete RNA molecules were available for translation. And next, if t becomes the space of time required to synthesize one more mRNA molecule than X, we write :

(2.3) $$y_{(X+1)} = P.X - (P.X - y_X).e^{-I.t}$$

Here we introduce the constant k = I/G, the relative stability of protein and nucleic acid, in order to relate 'y' directly to 'X'.

From (1.15) and (2.1) we can write during transcription

$$e^{-I.t} = e^{-k.G.t} = ((N-X)/N-X_0))^k$$

and in (2.3) :

(2.4) $$y_{(X+1)} = P.(X) - (P.(X) - y_X).((N-X-1)/(N-X))^k$$

$$y(X+2) = P.(X+1)-(P.(X+1)-y_{(X+1)}).((N-X-2)/(N-X-1))k$$

(2.4a) $$y_{(X+i+1)} = P(X+i) - (P(X+i)-y_{(X+i)}).((N-X-i-1)/(N-X-i))^k$$

In the absence of mRNA synthesis we can write from (1.1) and (2.1) :

$$e^{-I.t} = e^{-k.G.t} = (X/X_o)^k$$

Assume that the gene is switched off at X_2 = (X+i+1) than :

(2.5) $$y_{(X_2-1)} = P.(X_2-1) - (P.(X_2-1) - y_{X_2}).((X_2-1)/X_2)^k$$

$$y_{(X_2-j)} = P.(X_2-j) - (P.(X_2-j) - y_{(X_2-j+1)}).((X_2-j)/(X_2-j+1))^k$$

In FORTRAN IV a programme was developed to calculate X = f(G.t) and y = f(X), for the various input values P,N,k,X_o,y_o and X_2.

In figures 3 - 11 the results of a number of runs with this programme are graphically presented. It is of importance to keep in mind how the primary constants, which define the system are related to the working constants P, N and k.

F = promotor activity

G = stability of mRNA

H = rate of protein synthesis

I = stability of protein

In addition to k it is useful to employ an additional ratio g = H/F, the relative rate of protein and mRNA synthesis. Since N = F/G and P = H/I, we can write :

P = g.F/k.G = g.N/k or

(2.6) k.P = g.N.

We have limited our preliminary studies to a limited number of values for P.N. and k. Their relationship to G,F,I,H and g can be read from Table I.

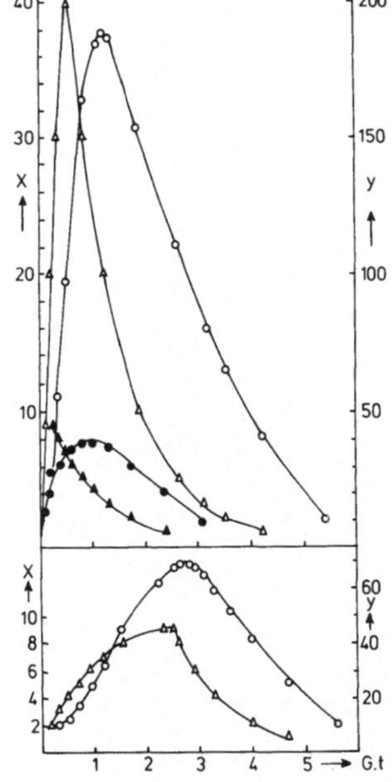

Figure 3.
The signals X(mRNA) (Δ) and y(protein) (o) as a function of time.
Upper fig. N=100,P=10,k=1.0
X_o=0,y_o=10(o Δ)X_2=40
(● ▲)X_2= 9
Lower fig. N=10,P=10,k=1.0
X_o=0,y_o=10,X_2=9.

In reality, k and g may vary over a wider range of values than we considered so far. It is well known that various proteins may have very different stabilities, which are determined by secondary, tertiary and quaternary structure and consequently difficult to predict. We can also assume that this stability is subject to conformational changes under the influence of other, including low molecular weight, molecules.

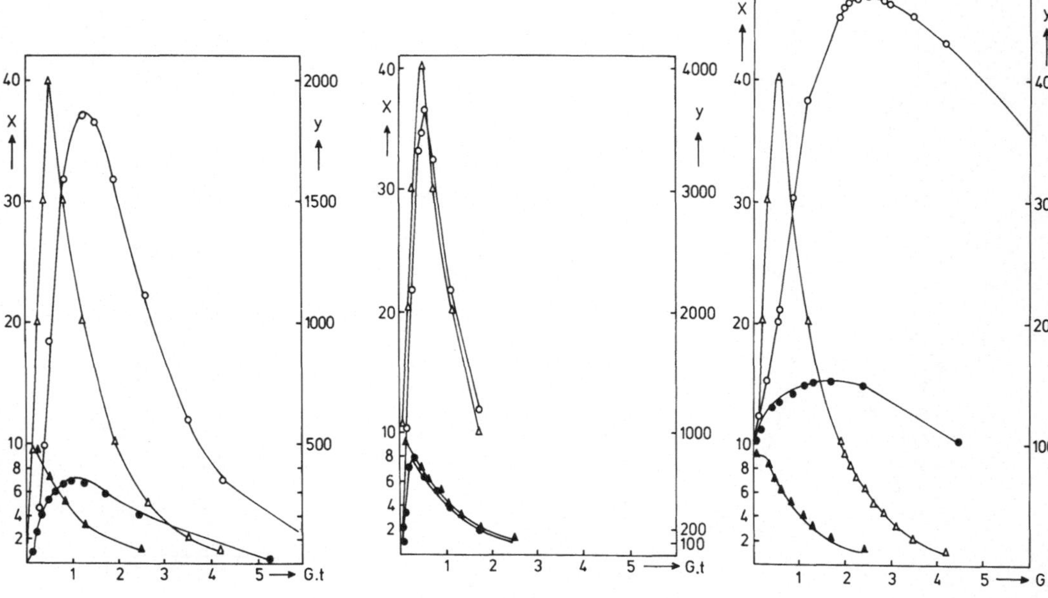

Figure 4. P=100,N=100,k=1
$X_o = 0, y_o=10$
(o Δ) $X_2=40$
(● ▲) $X_2= 9$

Figure 5. P=100,N=100,k=10
$X_o= 0,y_o=100$
(o Δ) $X_2=40$
(● ▲) $X_2=9.1$

Figure 6. P=100,N=100,k=0.1
$X_o= 0,y_o=100$
(o Δ)$X_2=40$
(● ▲)$X_2=9.1$

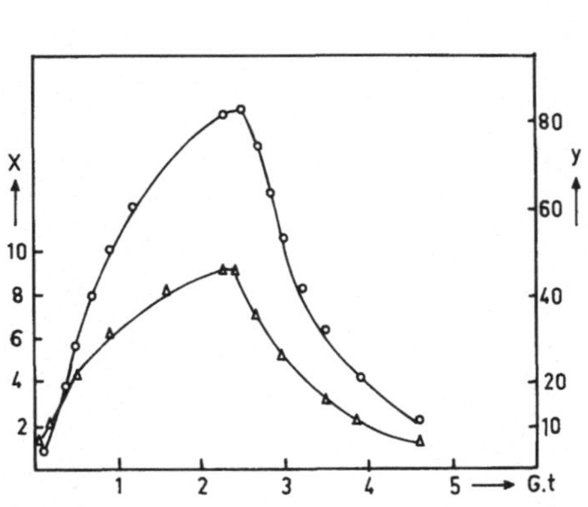

Figure 7. N=10,P=10,k=10
$X_o = 0,\ y_o = 10;\ X_2 = 9$

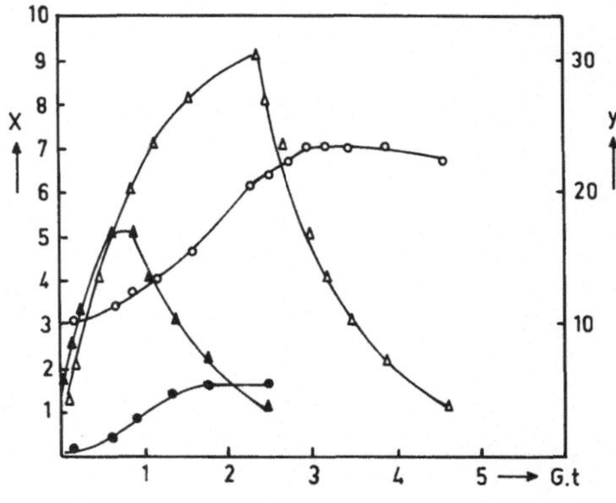

Figure 8. N=10,P=10,k=0.1. $X_o=0$
(● ▲) $y_o = 0;\ X_2 = 5.1$
(o Δ) $y_o = 10; X_2 = 9$

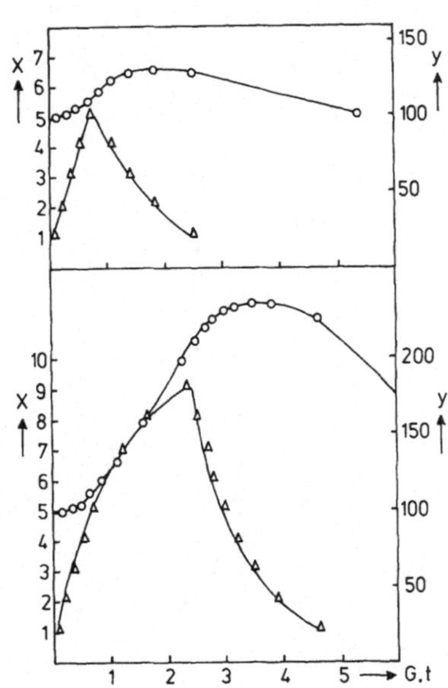

Figure 9. N=10,P=100,k=0.1
X_o=0, y_o=110
Upper figure: X_2 = 5
Lower figure: X_2 = 9

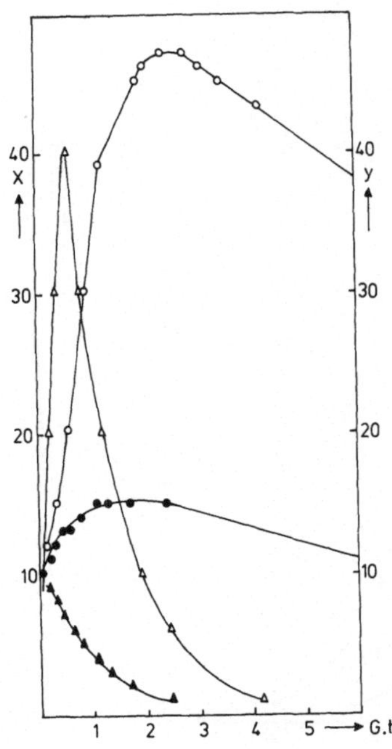

Figure 10. N=100,P=10,k=0.1
X_o = 0, y_o = 10
(o Δ) X_2 = 40
(● ▲) X_2 = 9

TABLE I

G	F	I	H	N	P	k	g
1	10	0.1	1	10	10	0.1	0.1
1	10	0.1	10	10	100	0.1	1.0
1	100	0.1	1	100	10	0.1	0.01
1	100	0.1	10	100	100	0.1	0.1
1	10	1	10	10	10	1	1
1	10	1	100	10	100	1	10
1	100	1	10	100	10	1	0.1
1	100	1	100	100	100	1	1
1	10	10	100	10	10	10	10
1	10	10	1000	10	100	10	100
1	100	10	100	100	10	10	1
1	100	10	1000	100	100	10	10

THE BASIC MODEL FOR A REGULATORY CIRCUIT

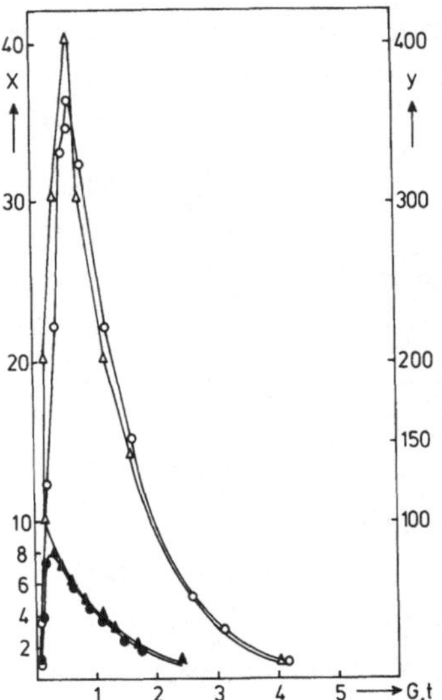

Figure 11. N=100,P=10,k=10
$X_o = 0$, $y_o = 10$
(o Δ) $X_2 = 40$; (● ▲) $X_2 = 9$

The relatively simple algebraic formulation we adopted for the raise and fall of a genetic signal, easily allows a numerical description of regulatory circuits. Here we consider firstly a system, presented in Fig.1, in which we assume that the primary signal from a gene has itself a regulatory effect on another. The gene product 'A' of gene *a* has above the setpoint level A_1 a stimulating effect on the transcription of gene *r*. The resulting product 'R' has above the level R_1 an inhibiting effect on the transcription of gene *a*. Consequently when 'R' reaches the level R_1, gene *a* is switched off, no more 'A' is produced and 'A' decays. Synthesis of R continues until breakdown of product 'A' has proceeded to setpoint level A_3. Below the level A_3, gene *r* is switched off, no more 'R' is produced and 'R' decays. When the setpoint level R_3 is reached, 'R' has no longer an inhibitory effect on the transcription of gene *a* and synthesis of 'A' is resumed. R decays further from R_3 to R_0 until 'A' reaches again setpoint level A_1 at which again the synthesis of 'R' from gene *r* is stimulated. This is actually a simple model for a regulatory circuit, based on the Britton and Davidson model for eukaryotes. Later we shall argue that this model has also significance if not the primary gene product but the secondary gene product (protein) exerts a regulatory effect. Note in anticipation that we have assigned a different value to setpoint levels during synthesis (A_1 and R_1) than to the setpoints during decay of the primary signals (A_3 and R_3).

To describe the behaviour of the system quantitatively we need only equations (1.15) and (1.18).

When A has reached its setpoint level A_1 after time t_1 it will continue to grow until A_2 is reached and we write (from(1.16)):

(3.1) $$(N_a-A_2)/(N_a-A_1) = e^{-G_a \cdot t_2}$$

In the same period t_2 'R' will increase from R_0 to R_1 according to (1.15) :

(3.2) $$(N_r-R_1)/(N_r-R_o) = e^{-G_r \cdot t_2}$$

Combining (3.1) and (3.2) gives :

(3.3) $$- t_2 = G_a^{-1} . \ln((N_a-A_2)/(N_a-A_1)) = G_r^{-1} . \ln((N_r-R_1)/(N_r-R_0))$$

Now we introduce the constant p

(3.4) $$p = G_r/G_a$$

Then we can write instead of (3.3) :

(3.5) $$((N_a-A_2)/(N_a-A_1))^p = (N_r-R_1)/(N_r-R_0)$$

Similarly we can write, by chosing the appropriate combinations of equations (1.15) to (1.18) for t_3, t_4 and t_1 :

(3.6) $$-t_3 = G_a^{-1}.\ln(A_3/A_2) = G_r^{-1}.\ln((N_r-R_2)/(N_r-R_1))$$

(3.7) $$-t_4 = G_{a_1}^{-1}.\ln(A_0/A_3) = G_r^{-1}.\ln(R_3/R_2)$$

(3.8) $$-t_1 = G_a^{-1}.\ln(N_a-A_1)/(N_a-A_o)) = G_r^{-1}.\ln(R_0/R_3)$$

If we consider A_1, A_3, N_a, R_1, R_3, N_r and p as constants which determine the system, we have in fact four equations with four unknowns, from which the unknowns can be solved in principle :

from (3.5) :

(3.9) $$A_2 = N_a - K_1(N_r-R_0)^{-1/p} \qquad K_1=(N_a-A_1).(N_r-R_1)^{1/p}$$

from (3.6) :

(3.10) $$R_2 = N_r - K_2(A_2)^{-p} \qquad K_2 =(A_3)^p.(N_r-R_1)$$

from (3.7) :

(3.11) $$A_0 = K_3(R_2)^{-1/p} \qquad K_3 =(A_3).(R_3)^{1/p}$$

from (3.8) :

(3.12) $$R_0 = K_4(N_a-A_o)^{-p} \qquad K_4 =(N_a-A_1)^p.(R_3)$$

Substitution of the unknowns in a single equation, leads, however, to a very complicated one, for example for A_2 :

(3.13) $$A_2 = N_a-K_1.(N_r-K_4.(N_a-K_3.(N_r-K_2.(A_2)^{-p})^{-1/p})^{-p})^{-1/p}$$

It is far more easy to search for a numerical solution, with a simple computer programme, that can be outlined as follows :

1. Input values are chosen for p, N_r, N_a, R_1, R_3, A_1 and A_3 and also initial values for R_0 and A_0, for example $R_{0(1)} = A_{o(1)} = 0$.
2. From (3.9) a value $A_{2(1)}$ is calculated
3. From (3.10) a value $R_{2(1)}$ is calculated, substituting $A_{2(1)}$ from step (2)
4. From (3.11) a new value $A_{0(2)}$ is calculated, substituting $R_{2(1)}$ from step (3)
5. From (3.12) a new value $R_{o(2)}$ is calculated, substituting $A_{0(2)}$ from step (4)
6. From (3.9) a new value $A_{2(2)}$ is calculated, substituting $R_{0(2)}$ from step (5)
7. From (3.10) a new value $R_{2(2)}$ is calculated, substituting $A_{2(2)}$ from step (6)

and so on, until

$A_{0(n+1)} = A_{o(n)}$, $R_{o(n+1)} = R_{o(n)}$, $A_{2(n+1)} = A_{2(n)}$ and $R_{2(n+1)} = R_{2(n)}$.

THE PROTEIN MEDIATED MODEL

The model presented above can be further sophisticated if not 'A' and 'R' but their secondary genetic signals 'a' and 'r' exert a regulatory effect (see figure 2). Gene *a* produces a mRNA 'A' which produces a protein 'a', which at level a_1 induces transcription of gene *r*. The mRNA 'R' produces a repressor protein 'r', which at level r_1 inhibits further production of mRNA from gene *a*. The mRNA 'A' decays, although synthesis of protein 'a' will continue, according to equation (2.5) until the last 'A' molecule has disappeared. If 'a' reaches the value a_3 (and now we assume

$a_3=a_1$, the setpoint level at which 'a' has an effect on gene *r*) production of 'R' stops and therewith, when 'R' has decayed to R=1, the production of repressor molecules 'r'. When 'r' reaches the value $r_3(=r_1)$ gene *a* is derepressed and synthesis of 'A', and therewith of 'a' is resumed.

Again we can search for a numerical description of the system, using essentially equations (1.15) to (1.18) and (2.4) and (2.5).

The computer programme runs as follows :

1. Input constants are chosen for P_a, N_a, k_a, a_1, P_r, N_r, k_2 and r_1 and input values are chosen for A_o, a_o, R_o and r_o (for example 0)
2. A value A_1 is calculated from step function (2.4)
   It is essentially a search for a value 'X' in a subroutine when y_X reaches the value a_1.
3. Similarly, a value R_1 is calculated from step function (2.4).
4. A value A_2 is calculated from (3.9), substituting A_1 and R_1 obtained from (2) and (3) and the initial value R_o.
5. A value a_2 is calculated from step function (2.4), substituting A_2 from (4).
6. Next we need a value A_3 to be calculated. We need to distinguish here between two possibilities, illustrated in figure 12.
   6a. The value $a_4(=y_4)$ is lower than the setpoint level $a_3(=a_1=y_3)$
   6b. The value a_4 is higher than a_1.
   We need firstly to determine the value a_4, that is the level of 'a' when the last mRNA molecule would disappear, using stepfunction (2.5) substituting for $X_2-j = 1$.
7. a. If a_4 is lower than the setpoint level a_1 we must use step function (2.4) to search for a value $A_3(= (X_2-j)$ when $y_{(X_2-j)}$ equals a_1 (see figure 12)
   b. If a_4 is higher than the sepoint level a_1, 'a' decays in the absence of mRNA according to equation (2.1). We must calculate then a hypothetical value for A_3 (lower than 1) from :

(4.1) $$y_1=y_4.e^{-I.t}= y_4.e^{-k.G.t}$$
$$=y_4.(X_3/1)^k$$

(4.2) $$X_3 = (y_1/y_4)^{1/k}$$

8. Next a value for R_2 follows from A_3, R_1 and A_2, previously calculated in step (7), (3) and (4) and substituted in (3.10).
9. A corresponding value for r_2 is calculated from the step function (2.4) similarly to step (5).
10. Next we have to consider a value for R_3 and again we have to distinguish the two cases, presented already for A_3 in step(6). Thus first of all a value r_4 is determined, using step function (2.5), when 'R' reaches the value 1.
11. a. If r_4 is lower than the setpoint level r_1, we must use again step function (2.5) to search for a value $R_3(= X_2-j)$ when $y_{(X_2-j)}$ equals r_1.
    b. If r_4 is higher than r, we can find R_3 from equation (4.2) (see step 7b) substituting for X_3, y_1, y_4 and k, resp. R_3, r_1, r_4 and k_r.

12. Then we obtain a (possible hypothetical) new value for A_o from equation (3.11) substituting R_2, A_3 and R_3, previously calculated in step (8), (7) and (11) respectively. (A_0 has a hypothetical value if lower than 1). To obtain the corresponding value a_0 we must again clearly distinguish between two possibilities, A_o is higher than 1 or A_0 is lower than 1.
13. a. If A_0 is higher than 1, a value a_o must be derived from step function (2.5) substituting for X_2-j = A_o.
    b. If A_o is lower than 1, its value can be found from equation (4.1) substituting for y_1, y_4, X_3 and k, resp. a_o, a_4, A_3 and k_a.
14. A new value for A_1 can now be obtained, as described in step (2).
15. Similarly to step (12) we obtain a (possibly hypothetical) new value for R_o from equation (3.12) substituting A_o, A_1 and R_3, previously obtained in step (12), (14) and (11) respectively.

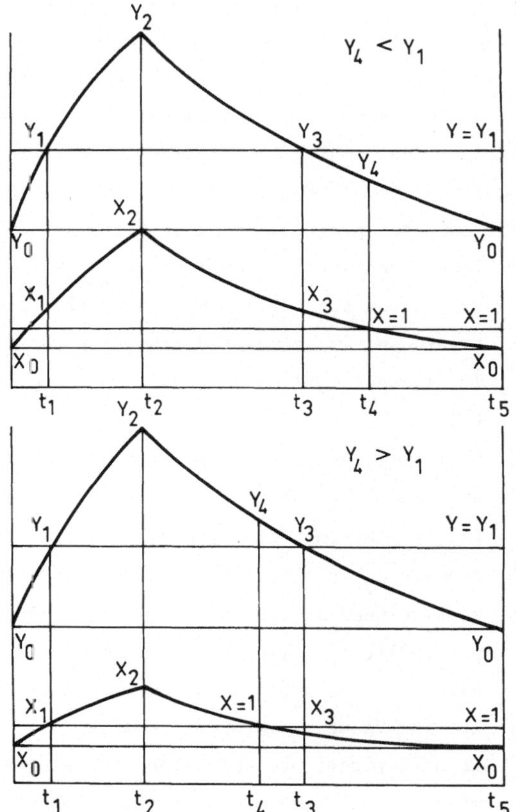

Figure 12. The decay of y and X with time.
Upper figure, X_3 is greater than 1
Lower figure, X_3 is smaller than 1

16. Similarly to step (13) we obtain a new corresponding value a_0, either from step function (2.5) or from (4.1).
17. A new value for R_1 follows, as described in step (3)
18. and the calculation sequence is continued from step (4) to step (18) until $A_{0(n+1)} = A_{o(n)}$, $A_{2(n+1)} = A_{2(n)}$ etc.etc. and also constant values for a_o, a_2, r_o, r_2 are obtained.

    Note that the two step functions (2.4) and (2.5) fulfill multifold purposes; (2.4) is used to determine values for A_1, R_1 and a_2 and r_2. (2.5) is used to determine values for a_4 and r_4 and sometimes to determine A_3, R_3 and a_0, r_0. The calculations to be performed with equations (2.4) and (2.5) are combined into two subroutines, indicated as SYN(THESIS) and BR(EAK)DOWN and those subroutines can be used in any other main programme to describe any other regulatory circuit, if appropriate equations are obtained to relate various RNA levels from the genes in the circuit.

THE SIGNIFICANCE OF VARIOUS VARIABLES

The numercial approach allows a description of oscillating behaviour in terms of frequency and amplitude as a function of setpoint levels and the stability of the molecules involved. Here we shall present firstly some calculations on the most basic model (fig.1) presented with the assumption that RNA exerts a regulatory effect. As we shall demonstrate later, also a protein mediated model (fig.2) for a RA system leads to stable osciallations, both in the RNA and protein signal. Consequently, the basic (RNA) model is also representative to a certain extend for the more sophisticated but more complicated model and because of its greater surveyability, the basic model is presented first.

In anticipation we note that the protein mediated model, always induces a setpoint level (X_3) which is considerably lower than (X_1). Therefore we also studied the effect of various ratios A_3/A_1 and R_3/R_1 with the basic model.

A value $X_3 < X_1$ originates from a variety of causes :

1. In the protein mediated model this is directly correlated to $k = I/G$.
   A high k (=10) indicates an unstable protein, a low k (=0.1) a stable protein. We shall see that the more stable the protein is, the lower the ratio X_3/X_1 will be.
2. If there is a delay before the molecular signal reaches its destination, (an OP site) the signal will be reduced below its actual setpoint level before it exerts an effect.
3. The signalling molecule may have a greater stability when coupled to the OP site than when it is free. Consequently the bound molecule will decay later than when the setpoint level is reached.

   Further, the study of the effect of various setpoint levels during generation of the signal (X_1) is of significance particularly with respect to post synthesis modifications of the signalling molecules - by co-repressors and co-activators - which alter the affinity of the signalling molecules for its receptor site.

   Both A_1 and A_3 are of interest to study the behaviour of a particular oscillating system itself. It needs no argument that the study of other characteristics - rate of synthesis and breakdown of signals - is of interest when comparing different systems, although we do not want to exclude beforehand that in theory alterations in rate of transcription and translation, may have also a regulatory effect in a single system itself (Ames and Hartman, 1973).

THE BASIC MODEL

First of all it is noteworthy that if $A_1 = A_3$ and $R_1 = R_3$, for any initial value of A_o and R_o and any value of p, the amplitude of the oscillation decreases in each subsequent cycle and the system reaches finally a steady state in which $A_o = A_1 = A_2$ and $R_0 = R_1 = R_2$.
Note that this is obviously one solution of equations (3.9) to (3.12).
Apparently there is no other solution under the above condition, as we proved by taking all possible input values for $A_1 = A_3$, $R_1 = R_3$, N_a, N_r and p since never a stable oscillation was observed. In fact, under the condition $A_1 = A_3$ and $R_1 = R_3$, the system is an ideal one in order to maintain a fixed level of 'A' and 'R'.
We observed, however, that if we reduce A_3 or R_3 just slightly (1% is sufficient) with respect to A_1 or R_1, a stable oscillation is obtained after a limited number of cycles. Some results are collected in table II.

TABLE II
The stability of oscillations in the RA model system as a function of A_3 and R_3

A_1	A_3	R_1	R_3	A_2	A_o	R_2	R_o	number of cycles after which stable oscillation is observed
0.1	0.100	0.1	0.100	0.100	0.100	0.100	0.100	-
0.1	0.100	0.1	0.099	0.103	0.075	0.131	0.096	15
0.1	0.099	0.1	0.100	0.103	0.075	0.133	0.097	15
0.1	0.099	0.1	0.099	0.105	0.066	0.148	0.095	14
0.1	0.090	0.1	0.090	0.116	0.027	0.305	0.083	5

In this table and the following graphs (fig.13-19) we express A_1, A_3, R_1, R_3, A_o, A_2, R_o and R_2 as a fraction of resp. N_a and N_r, to simplify the figures. If both A_3/A_1 and R_3/R_1 are 0.9, stable oscillation is observed after the 5th cycle. With lower values, stable oscillation is already observed in the 3rd cycle.
In figure 13 the effect of various A_3 and R_3 values are presented. In this and subsequent figures, full lines represent 'R' values and dotted lines 'A' values. In the upper figures the actual values of A_o, A_2, R_o and R_2, in the lower figures the period 'T' of the oscillation (in G.t units) is presented and also the fraction of the time that the 'A' and 'R' signals are below setpoint level ("off"). Note that alterations in A_3 have an effect on A_o only whereas alterations of R_3 have an effect on both A_o and R_o. Both have a strong effect on the period T.
Next we studied the effect of the setpoint levels A_1 and R_1 under conditions for various A_3/A_1 and R_3/R_1 ratios.
In figure 14 these ratios are both 0.9 (which corresponds in a protein mediated model, to be discussed below, with a high value of k). It should be noted that the changes of setpoint level A_1, especially influence A_o and R_o but most important, the time that 'R' is "off". The changes of setpoint level R_1 also influence strongly A_o and R_o but also both for 'A' and 'R', the fraction of time the signals are "off".

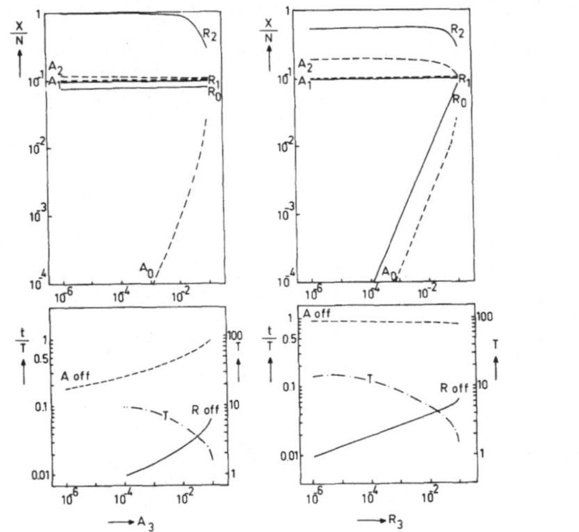

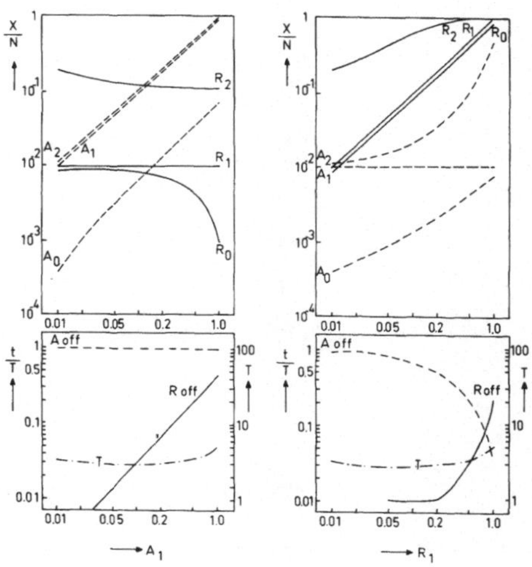

Figure 13. A_o and A_2, R_o and R_0, T and T_{off} as a function of A_3 and R_3.
Left : A_1=0.1, R_1=0.1, R_3=0.09 p=1.0
Right: A_1=0.1, R_1=0.1, A_3=0.09 p=1.0

Figure 14. A_0, A_2, R_o, R_2 as a function of A_1 and R_1.
Left : A_3/A_1=0.0,R_1=0.01,R_3/R_1=0.9 p=1
Right: A_3/A_1=0.9,A_1=0.01,R_3/R_1=0.9 p=1

Figure 15 illustrates the effect of changes in setpoint level R_1 for various values of A_3/A_1 and R_3/R_1 and similarly figure 16 the effect of changes in setpoint level A_1.

In figure 17 the effect of alterations in p (relative sensitivity of RNAs to breakdown) are illustrated for several A_3/A_1 and R_3/R_1 ratios. These alterations of course influence strongly the period (T) of the oscillation.

In figure 18 and 19 again the effect of alterations in setpoint levels are described, but now for two values of p, 2 and 0.5 respectively.

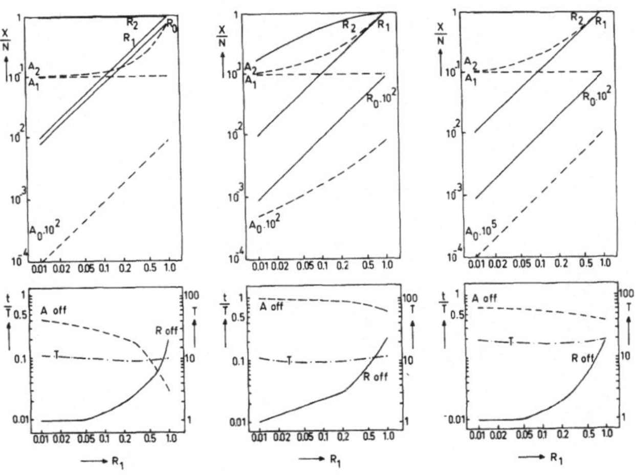

Figure 15. A_o,A_2,R_o,R_2 as a function of R_1
A_1 = 0.1 p=1.
Left: A_3/A_1 = 0.001 R_3/R_1 = 0.001
Middle: A_3/A_1 = 0.9 R_3/R_1 = 0.001
Right: A_3/A_1 = 0.001 R_3/R_1 = 0.9

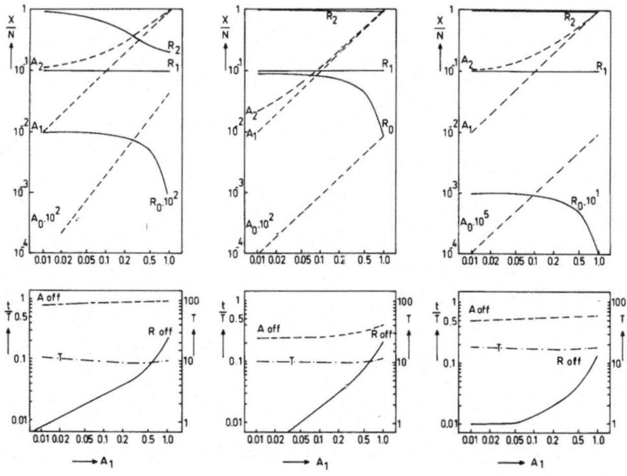

Figure 16. A_o, A_2, R_o, R_2 as a function of A_1
$R_1 = 0.1$ $p = 1.0$
Left : $A_3/A_1 = 0.001$ $R_3/R_1 = 0.001$
Middle: $A_3/A_1 = 0.001$ $R_3/R_1 = 0.9$
Right: $A_3/A_1 = 0.9$ $R_3/R_1 = 0.001$

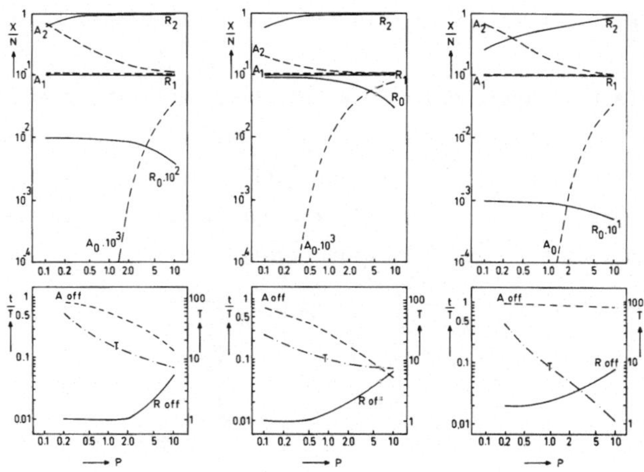

Figure 17. A_o, A_2, R_o, R_2 as a function of p.
$A_1 = 0.1$ $R_1 = 0.1$
Left : $A_3/A_1 = 0.001$ $R_3/R_1 = 0.001$
Middle: $A_3/A_1 = 0.001$ $R_3/R_1 = 0.9$
Right: $A_3/A_1 = 0.9$ $R_3/R_1 = 0.001$

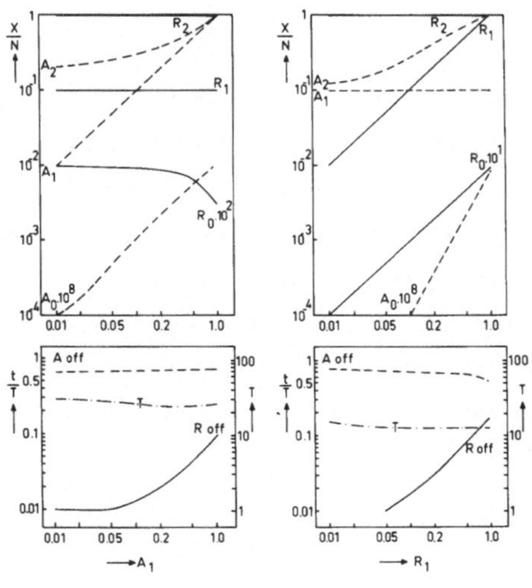

Figure 18.
A_o, A_2, R_o, R_2 as a function of setpoint levels at p = 0.5
$A_3/A_1 = R_3/R_1 = 0.001$

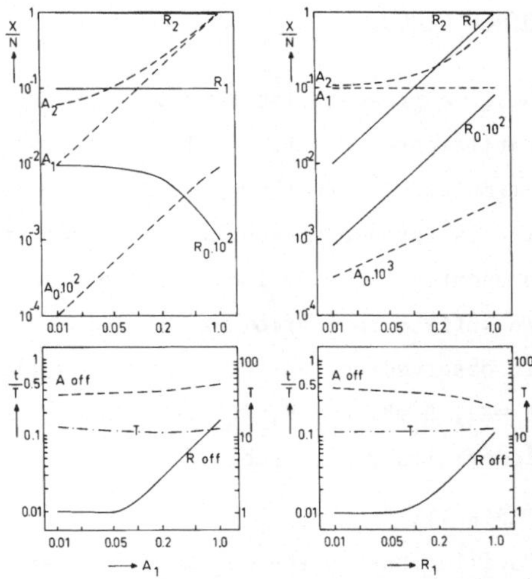

Figure 19.
A_o, A_2, R_o, R_2 as a function of setpoint levels at p = 2
$A_3/A_1 = R_3/R_1 = 0.001$

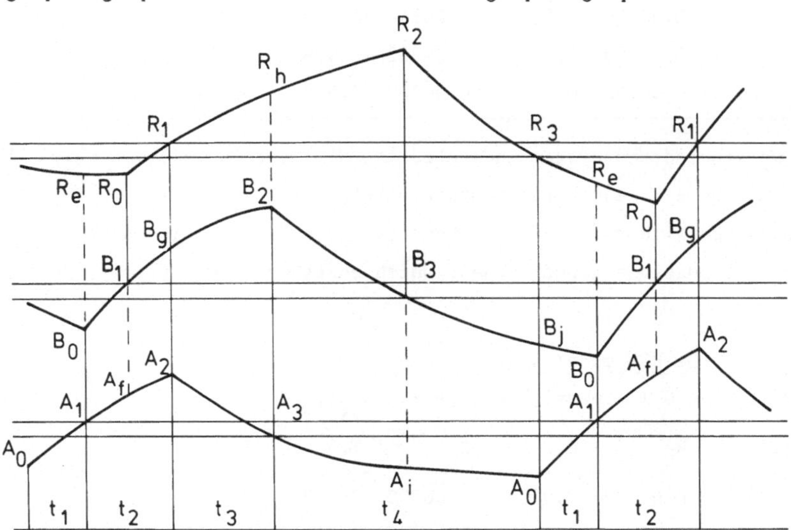

Figure 20. Oscillation scheme for the ABR system.

(5.1) $A_2 = 1-K_1(1-R_o)^{-1/pq}.(1-B)^{-1/p}$ $\quad K_1 = (1-R_1)^{1/pq}.(1-B_1)^{1/p}.(1-A_1)$ $\quad p=G_a/G_b \quad q=G_b/G_r$

(5.2) $B_2 = 1-K_2(1-R_o)^{-1/q}.(A_2)^{-p}$ $\quad K_2 = (1-R_1)^{1/q}.A_3{}^p.(1-B_1)$

(5.3) $R_2 = 1-K_3(B_2)^{-q}.(A_2)^{-pq}$ $\quad K_3 = (B_3)^q.(A_3)^{pq}.(1-R_1)$

(5.4) $A_0 = K_4(B_2)^{-1/p}.(R_2)^{-1/pq}$ $\quad K_4 = (B_3)^{1/p}.(R_3)^{1/pq}.A_3$

(5.5) $B_0 = K_5(1-A_0)^{-p}.(R_2)^{-1/q}$ $\quad K_5 = (1-A_1)^p.(R_3)^{1/q}.B_3$

(5.6) $R_0 = K_6(1-B_0)^{-q}.(1-A_o)^{-pq}$ $\quad K_6 = (1-B_1)^q.(1-A_1)^{pq}.R_3$

(All levels of 'A', 'B' and 'R' expressed as ratios of N_a, N_b and N_r respectively)

OTHER CIRCUITS

Here we present also some preliminary results with other circuits in which three genes participate. Figure 20 illustrates the oscillations in a system in which gene *a* has a stimulating effect on gene *b*, which itself has a stimulating effect on gene *r*, and the latter has an inhibitory effect on the first gene *a*. In the legenda also are presented the equations for the relationships among the various RNA levels.
We suffice to mentionhere that in this particular system, a stable oscillation quickly is observed for any of the input variables and that even when the setpoint levels $A_1=A_3$, $B_1=B_3$ and $R_1=R_3$ are adopted, unlike in the RA system, oscillation continues indefinitely. See table III.

TABLE III

Oscillations in the ABR model system

$A_1=A_3=B_1=B_3=R_1=R_3=0.1$ p=q= 1.0 $B_{0(1)}=R_{0(1)}=0$

Round	A_2	A_o	B_2	B_o	R_2	R_o
1	0.271	0.00149	0.701	0.00000	0.953	0.0000
2	0.198	0.00196	0.555	0.00946	0.918	0.0819
3	0.198	0.00196	0.555	0.00982	0.918	0.0819
4	0.198	0.00196	0.555	0.00982	0.918	0.0819

Similarly we studied the system indicated as RST, (see fig.21), in which the genes *r*, *s* and *t* all have a sequential inhibiting effect.
Again we observed with this model system that quickly a stable oscillation is established for all possible input values, which continues indefinitely. See Table IV.

TABLE IV

Oscillations in the RST model system

$R_1=R_3=S_1=S_3=T_1=T_3=0.1$ p=q=1 $R_{2(1)}=1$ $S_{0(1)}=0$

Round	R_2	R_0	S_2	S_0	T_2	T_o
1	1.000	0.00979	0.9109	0.00000	0.9190	0.00997
2	0.910	0.00999	0.9101	0.00999	0.9101	0.00999
3	0.910	0.00999	0.9101	0.00999	0.9101	0.00999

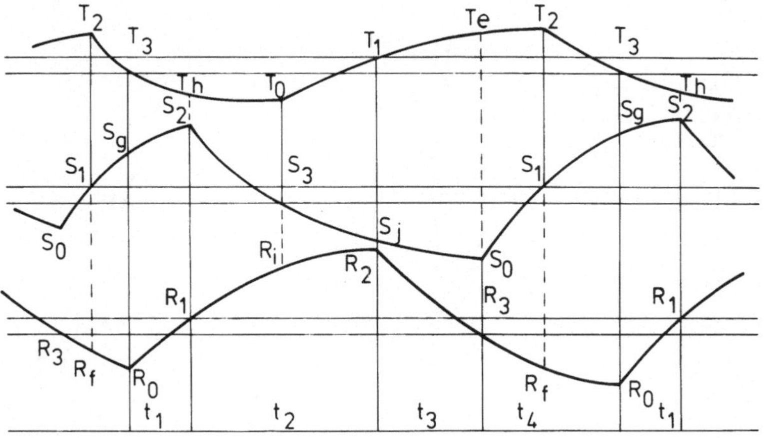

Figure 21.
Oscillation scheme for the RST system. $p=G_r/G_s$ $q=G_s/G_t$

(6.1) $T_2=1. - K_1(1-S_0)^{-q}.R_2^{-p.q}$ $K_1= (1-S_1)^q.R_3.(1-T_1)$

(6.2) $R_o= K_2(1-S_0)^{-1/p}.(T_2)^{-1/pq}$ $K_2= (1-S_1)^{1/p}.(T_3)^{1/pq}.R_3$

(6.3) $S_2= 1 - K_3(1-R_0)^{-p}.(T_2)^{-1/q}$ $K_3= (1-R_1)^p.(T_3)^{1/q}.(1-S_1)$

(6.4) $T_0= K_4(1-R_0)^{-pq}.(S_2)^{-q}$ $K_4= (1-R_1)^{pq}.(S_3)^q.T_3$

(6.5) $R_2= K_5(1-T_0)^{-1/pq}.(S_2)^{-1/p}$ $K_5= (1-T_1)^{1/pq}.(S_3)^{1/p}.(1-R_1)$

(6.6) $S_0= K_6(1-T_0)^{-1/q}.(R_2)^{-p}$ $K_6= (1-T_1)^{1/q}.(R_3)^p.S_3$

(All levels of 'R', 'S' and 'T' expressed as ratios of resp. N_r, N_s and N_t)

THE PROTEIN MEDIATED MODEL OF THE RA SYSTEM

We are currently undertaking a systematic study of this model for setpoint alterations (a_1 and r_1) and variance of P_r, P_a, N_r, N_a, k_r, k_a and p. We observe under all conditions, independent of input values for A_0, a_0, R_0 and r_0 a quickly established stable oscillation which continues indefinitely.
The results of two representative computer simulations are presented in table V and VI. Note that in the first case the system is stable at the second round and in the second case in the third round. The actual raise and fall of the signals 'R', 'r', 'A' and 'a' are presented in figure 22 and 23.

Fig.22

The behaviour of the primary and secundary signals of the genes in the RA model (Table V) at stable oscillation.

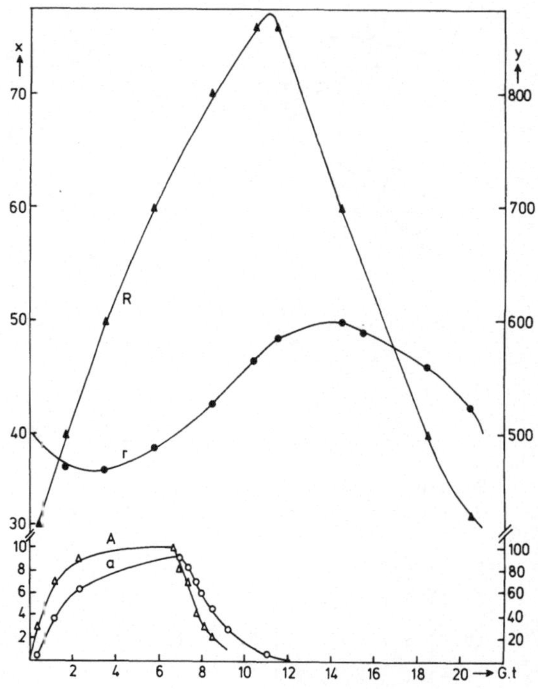

TABLE V Oscillation of a R-A system

$P_R = 10$, $N_R = 100$, $K_R = 1.0$, $r_1 = 500$

$P_A = 10$, $N_A = 10$, $K_A = 1.0$, $a_1 = 5$, $p = 0.1$

Input values : $A_o = 0$ $a_o = 0$
$R_o = 0$ $r_o = 0$

T is oscillation time in G.t. units. R_{off} and A_{off} is t_{off}/T

ROUND	1	2	3	4
PROTEIN				
r_0	0	493	489	489
r_2	606	572	573	573
a_0	0	0	0	0
a_2	100	89	89	89
NUCLEIC ACID				
R_0	0.00E+00	0.28E+02	0.30E+02	0.30E+02
R_1	81.00	66.00	66.00	66.00
R_2	87.30	77.18	77.18	77.18
R_3	0.29E+02	0.13E+02	0.31E+02	0.31E+02
A_0	0.00E+00	0.29E-05	0.20E-04	0.20E-04
A_1	3.00	3.00	3.00	3.00
A_2	10.00	10.00	9.99	9.99
A_3	0.18E+00	0.19E+00	0.19E+00	0.19E+00
R_3/R_1	0.36E+00	0.47E+00	0.47E+00	0.47E+00
A_3/A_1	0.60E-01	0.62E-01	0.62E-01	0.62E-01
FREQUENCY				
T_R	2.29	2.07	2.07	2.07
R_{off}	0.343	0.367	0.367	0.367
T_A		20.97	20.70	20.70
A_{off}		0.452	0.458	0.458

TABLE VI Oscillation of a R-A system

$P_R = 10$, $N_R = 10$, $K_R = 0.1$, $r_1 = 5$

$P_A = 10$, $N_A = 10$, $K_A = 0.1$, $a_1 = 5$

$p = 1.0$

Input values : $A_o = 0$ $a_o = 0$
$R_o = 0$ $r_o = 0$

T is oscillation time in G.t. units. R_{off} and A_{off} is t_{off}/T

ROUND	1	2	3	4
PROTEIN				
r_0	0	4	4	4
r_2	58	38	38	38
a_0	0	0	0	0
a_2	12	7	7	7
NUCLEIC ACID				
R_0	0.00E+00	0.30E-10	0.14E-08	0.14E-08
R_1	7.00	4.00	4.00	4.00
R_2	10.00	9.97	9.97	9.97
R_3	0.10E-09	0.47E-08	0.47E-08	0.47E-08
A_0	0.00E+00	0.38E-14	0.16E-10	0.16E-10
A_1	7.00	7.00	7.00	7.00
A_2	9.10	8.20	8.20	8.20
A_3	0.38E-03	0.35E-01	0.35E-01	0.35E-01
R_3/R_1	0.14E-10	0.12E-08	0.12E-08	0.12E-08
A_3/A_1	0.54E-04	0.49E-02	0.49E-02	0.49E-02
FREQUENCY				
T_R	37.14	28.65	28.65	28.65
R_{off}	0.046	0.060	0.060	0.060
T_A	37.83	28.65	28.65	28.65
A_{off}	0.701	0.791	0.791	0.791

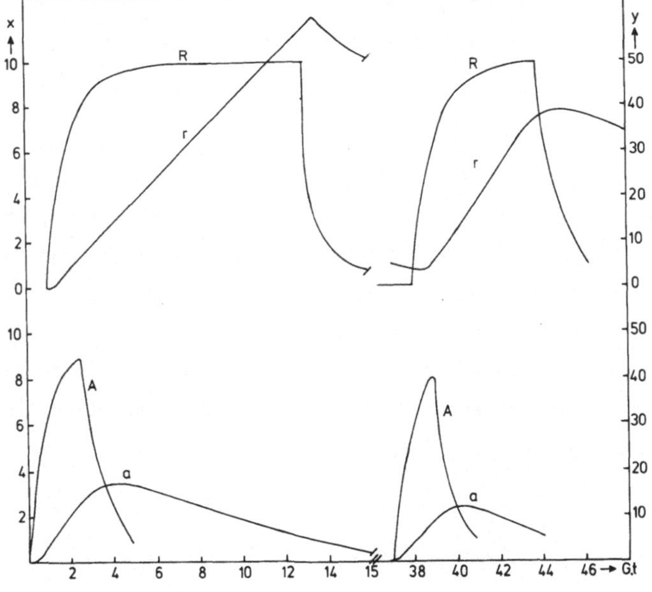

Figure 23.
The behaviour of the primary and secondary signals of the genes in the RA model (see Table VI) at the first and second round of oscillation.

DISCUSSION

The numerical approach presented here gives a surveyable picture of the behaviour of primary and secondary genetic signals in oscillating circuits.

The essential difference with the approach of others, is the introduction of the concept of a setpoint level, which simplifies the necessary equations and make them more apprehensible. The method should be considered as an extension of the Boolean approach (Thomas 1973) which allows the quantification of the signals involved. This quantification we want to study in the future in more detail, in order to attribute certain qualities to oscillating circuits, for example to maintain particular enzyme levels or to guide cyclic processes. Sofar we have studied only three of the most simple genetic circuits, indicated as RA, RAB and RST. We observed a strong influence of the ratios X_3/X_1 on oscillatory behaviour. We observed with the protein mediated model that this ratio is mainly determined by the stability of the protein involved. If we read for a low X_3/X_1 value a high protein stability, even the simplified RNA models give some valuable information about the behaviour of more sophisticated models.

A very general observation is, that the level of the signal from a repressor gene, tends in most cases to raise to its maximum level, where the rate of synthesis balances breakdown (N=F/G). This is of course, from the point of view of energy economy, wasteful. We know, however, that nature has also developed systems for the autoregulation of genes in order to limit transcription above the setpoint level. The Boolean approach predicts that for example in an RA system, if R is selfregulation, still oscillation will occur.

By combining the various elements of our computer programme in a variety of ways, we can now quantify the signals in a variety of regulatory circuits. We have the intention to study in the future particularly the energy cost of specific regulatory (model) systems, in relation to their functions and the evolutionary aspects thereof.

SUMMARY

The Boolean description of models for genetic oscillating systems is extended by a numerical description of the molecular signals from the genes in such systems. When the Boolean approach predicts a cycle of states for those genes, the raise and fall of the corresponding signalling molecules is described by simple equations, which allow, by the use of a digital computer, the determination of amplitude and frequency of oscillation, as a function of the stability of the molecules involved, the rate of transcription and translation and the setpoint levels at which those signalling molecules interact with regulatory elements of the genes.

REFERENCES

Ames, B.N. and P.E.Hartman (1973) Cold Spring Harbor Symp.Quant.Biol.XXVIII, 349.

Beckwith, J.R. and P.Sipser (1970). The lactose operon. Cold Spring Harbor Laboratory.

Britten, R.J. and E.H.Davidson (1971) Quart.Rev.Biol.46, 111.

Fraser, A. and J.Tiwari (1974), J.Theor.Biol.47, 397.

Goodwin, B.C. (1965), in Adv.Enzyme Regulation 3.

Goodwin, B.C. (1976). Analytical physiology of cells and developing organisms. Academic Press.

Jacob, F. and J.Monod (1961) Cold Spring Harbor Symp.Quant.Biol.XVI, 193.

King, M.C. and A.C.Wilson (1975) Science 188, 107.

Lea, D.E. (1955). Action of radiation on living cells. Cambridge University Press.

Thomas, R. (1973), J.Theor.Biol. 42, 563.

Tiwari, J., A.Fraser and R.Beckman (1974), J.Theor.Biol.45, 311.

CHAPTER XX.

A MODEL FOR THE ACHIEVEMENT OF ACCURACY IN BIOLOGY AND ECONOMY

Alain Ghysen and André Farber.

Laboratoire de Génétique and Centre d'Economie Mathématique et d'Econométrie, Université Libre de Bruxelles, Brussels, Belgium.

I. INTRODUCTION.

High levels of accuracy are difficult to achieve, yet they are critically important for the correct function of living systems. In molecular biology, where the problem is mostly one of accurate transfer of information, various systems have been developed to reach the required level of accuracy. An interesting case is provided by the mechanism of DNA replication. It appears that in bacteria, the enzyme which polymerizes the nucleotides into a new chain of DNA also presents an exonuclease activity which selectively excises mismatched nucleotides (Brutlag and Kornberg, 1972). In eukaryotes, where the purified DNA polymerases have no nuclease activity, the "proofreading" seems to be done by a separate exonuclease which is responsible for the removal of incorrect nucleotides (Spadari, Villani and Radman, in prep.). This suggests that the replication of DNA is a two-step process : first the DNA polymerase catalyzes the addition of a new nucleotide on the growing DNA chain ; then a nuclease checks the fidelity of the process, thereby increasing the initial level of accuracy. A "proofreading" mechanism also seems to operate during the synthesis of proteins (Ninio, 1974).

Another way to reach high levels of accuracy is the combinatorial use of moderately specific steps to yield a completely specific result. Thus, the aminoacylation of tRNA's is done in two steps. Each step displays a limited specificity, while the complete reaction is highly specific (Ebel et al, 1973, Ghysen and Celis, 1974). In a different field, the accuracy with which insect sensory neurons find their targets is apparently obtained by an "adress" system which combines two moderately large repertoires to generate a much larger set of specificities (Ghysen, 1978). In more complex nervous systems, a "proofreading" mechanism could be used to keep a high level of accuracy in the connectivity (Changeux et al, 1973).

These various examples point towards a general tendency : the achievement of high levels of accuracy is obtained by the accumulation of less accurate steps or subsystems. This tendency may be expected to be operative at the supracellular level as well. A model has been developed, which accounts for the development of very accurate patterns during the development of an insect, Drosophila. In this model, many moderately accurate responses are automatically averaged out, and a highly accurate result is generated. The process of averaging out depends on the existence of a system of intercellular communication.

We suggest that a similar mechanism may be operative in economy, e.g. in the determination of market prices, and we give some preliminary evidence which supports this suggestion.

II. THE PATTERN OF BRISTLES IN DROSOPHILA

A. Development of the pattern.

One of the most striking cases of precise multicellular pattern is the positioning of bristles on the epidermis of some insects. In adult Drosophila, large bristles (macrochaetes) are found at highly reproducible locations on the body. The dorsal thorax bears eleven pairs of these macrochaetes. The accuracy in the location of these bristles is remarkable, as shown by the almost perfect symetry of the thoracic pattern even though the left and right halves of the thorax develop as separate entities and fuse along the midline long after the position of the macrochaetes has been determined.

The macrochaetes are mechanoreceptor sensory organs. Histological examination reveals that these sensory structures are contributed by four cells : one synthetizes the bristle shaft, a second produces the socket, a third differentiates into the underlying sensory neuron and the fourth becomes a glial cell. Two questions have to be answered before the formation of the pattern can be modeled. First, how are the four cells that contribute a bristle structure assembled ? Second, do some or all of these cells differentiate at the appropriate location, or do they differentiate elsewhere and subsequently migrate to the final location (Ursprung, 1967) ?

As for the first question, histological studies have shown that the various cells involved in the development of one bristle derive from an ordinary epithelial cell (Wigglesworth, 1953). This mother cell undergoes two rounds of differentiative divisions ; the resulting four sister cells become the four elements of the bristle structure. There is thus no need for a special mechanism to assemble these cells since they are generated side by side.

The second question has been solved by the use of genetic mosaics (Stern, 1968). If the mother cell migrated after being specified to develop into a bristle, one should observe that bristles are not clonally related to the surrounding tissue. This may indeed happen in the case of the evenly spaced bristles which cover the abdomen of the fly. However in the case of the precisely located macrochaetes on the thorax, the bristles which develop within marked clones are always marked. Thus, migration is clearly not involved in the precise positioning of bristles.

These data on the origin of precisely located macrochaetes can be summarized as follows : at a given time during development, the cells which occupy defined positions in the epithelial sheet embark in a differentiative program which results in the development of a bristle.

Many observations in developmental biology indicate that some sort of "positional information" (Wolpert, 1969) allows cells in different regions of a tissue to behave differently. The epidermis of insects has been a favorite system for the experimental analysis of the concept of positional information (for a review, see Lawrence,1973).

It appears that whatever property is used by the cells to read their position, behaves formally as a monotonic gradient. This gradient might correspond to a gradient of concentration of a diffusible substance (Crick, 1970) or to any property which could be continuously modified from one edge of the tissue to the other.

We assume that a similar system exists in the developing imaginal wing disc, the sheet of epithelial cells which will differentiate into a wing and a half-thorax of the adult fly. Any point in the tissue could in theory be specified by its coordinates along two orthogonal axes, e.g. by two values of concentration along two orthogonal concentration gradients. However it is unlikely that the reading of positional information is so accurate as to allow the recognition of a specific position by the cell which occupies that position, at a time when the disc comprises one or two thousand cells. Furthermore, various observations have shown that if a bristle does not develop at the appropriate site, e.g. because the site is occupied by mutant cells, then a bristle may be formed at a slightly displaced location by a normal cell (Stern 1954). Thus the ability to differentiate into a bristle is not restricted to the cell which occupies the presumptive site, but to a diffuse region which includes the presumptive site and extends over a few cell diameters. How is it then that in normal conditions, only one bristle is found in each region, and that this bristle is accurately positioned ?

B Description of the model.

A model has been developed (Richelle and Ghysen) in which the determination of a bristle occurs in two contrasting steps. The first step is the response of the cells to their position in the tissue. This response is the synthesis of a specific inducer ; it is triggered by a given set of positional values (e.g. a given set of concentrations if the positional information is conveyed by orthogonal gradients of concentration of two diffusible substances). This response is probabilistic rather than accurate. By this it is meant that cells which are close to a particular set of positional values are more likely to engage in the synthesis of the inducer than cells further away. The level of accuracy required at this step is rather low : the probability of synthetizing the chaetogen is taken to change from 0 to 1 over a distance of 10 cell diameters, while the field is approximately 50 x 50 cells at that time. This means that cells which are up to 9 cell diameters (about a fifth of the length of the time) away from the presumptive site (the position defined by a set of positional values) can still respond to that set of positional values by turning on the synthesis of the inducer. However the probability that they will do so decreases as the distance from the presumptive site increases.

The inducer is a small molecule, e.g. an oligopeptide of about 1.000 MW, and can therefore diffuse freely across specialized cell junctions (Loewenstein, 1966) which interconnect the epithelial cells in imaginal discs (Poodry and Schneiderman, 1970). Early after the synthesis has begun, the distribution of concentration of the inducer over the field will present a maximum at the position of each synthetizing cell in the cluster that surrounds a given presumptive site. Later, diffusion leads to a progressive merging of these small peaks. Eventually a single maximum will be found for the whole cluster, which may comprise about 100 inducer-synthetizing cells. Simulations have shown that the position of this maximum will be located at, or adjacent to, the presumptive site, and will be essentially unaffected by large statistical fluctuations in the exact arrangement of synthetizing cells.

The second step of the process is threshold-dependent : when the concentration of the inducer in any epithelial cell reaches a given value (threshold), that cell is induced to differentiate into a bristle apparatus. It is assumed that when an epithelial cell is induced, it prevents neighboring cells from being induced too. This assumption is based on observations made on the bristle pattern of another insect, Rhodnius, which indicate the presence of a zone of inhibition around each newly induced bristle (Wigglesworth, 1940). The result of the second step is that one , and only one , cell will be induced to differentiate into a bristle apparatus at the positions where the concentration of inducer is maximum, i.e. at the presumptive sites.

The whole process ensures that bristles will develop at the appropriate locations, even though the accuracy of this positioning largely exceeds the ability of any cell in the tissue.

C Analysis of the model.

The details of the model outlined above will be published elsewhere (Richelle and Ghysen). Computer simulations have shown that this model is highly successful in reproducing various patterns found on the adult fly, as well as abnormal patterns found on mutants (Ghysen and Richelle) and bizarre patterns observed on flies mosaic for normal and mutant tissues (Stern, 1954, Gottlieb, 1964). Here we would like to analyze this model in terms of discontinuous vs. continuous description of biological phenomena.

The model combines the two approaches, each in its own domain. First, the decision of the cells to engage in the synthesis of the inducer is typically discontinuous : it is an all-or-none, stable decision. This type of behavior is generally thought to operate throughout development : indeed developmental genetics points towards a sequence of binary, heritable decisions as one of the main mechanisms of genetic control of development. (Garcia-Bellido, 1973). More generally a boolean approach seems appropriate for the analysis of the main features of genetic control mechanisms (Thomas, this volume). Indeed, the number of elements involved is small : repressors are present in small numbers in the regulatory circuits of phages and bacteria (Gilbert and Müller-Hill, 1966), the number of cells involved in each compartmentalization in Drosophila is of the order of 50-100 (Garcia-Bellido, 1973) and the number of cells involved in the location of one bristle, according to the model summarized above, is of the same order. This order of magnitude further precludes the use of a continuous approach based on differential equations. On the other hand it is suitable for a "simulation" approach which is likely to provide the best approximation of the actual phenomena (Richelle, this volume).

The next step is very different in nature from the first one. It involves the diffusion of large numbers of inducer molecules in a sheet of cells which can be assimilated to a 2 dimensional homogeneous space. A continuous approach is clearly required to describe this part of the process. Biological values taken from the literature were given to the various parameters of the general equation of diffusion. The equation was integrated relative to time and space in order to decribe the evolution of the system, assuming that each inducer-synthetizing cell behaves as a point source with a constant rate of synthesis. It is interesting that the shape of the distribution of concentration around the source fits the requirements of the model in the sense that it allows the development of a single maximum of concentration for a scattered cluster of about 100 inducer-synthetizing cells, yet it allows the development of eleven distinct maxima (corresponding to as many scattered and partly overlapping clusters) over the whole field. It should be emphasized that this adequacy of the diffusion process was in no way "imposed". As stated above, the shape of the concentration around a source is derived from the classical diffusion equation.

II. THE PATTERN OF PRICES IN ECONOMY

A. The rationality of the stock market.

An economy is a system in which scarce resources (manpower, energy, raw material) have to be allocated among alternatives uses. Prices are generally considered to indicate the relative scarcity of the various resources, and the price system has been widely studied by economists as a regulative system whereby resources are oriented to their best allocation. A central tenet of these studies is that prices are accurate, and the question arises naturally of how this accuracy is achieved.

The stock market is interesting to look at in order to analyze this question. Its structure is indeed simple compard to other markets (such as the labor market or the real estate market) and data on its functioning are numerous and can be accessed easily. In the early 60's several papers have shown (Osborne 1959, Fama 1965) that stock price behavior can be described by a random walk model

$$P_t = P_{t-1} + e_t$$
$$E(e_t) = 0$$
$$\mathrm{cov}(e_t, e_{t-s}) = 0 \qquad S \neq 0$$

where p_t is the price of a stock on day t, e_t is the change in the price from $t - 1$ to t, E and cov are the expected value and covariance operators respectively. The model essentially indicates that the history of past prices does not add any information to the current price for predicting future prices. The above model has indeed the property that the current price is the best predictor of any future price.

The result has been generalized to show that any public information is reflected in the current price and that the knowledge of that information does not improve the forecasting ability compared to a naïve forecasting technique that would take the current price as the best forecast. The stock market is therefore considered "rational", in the sense that the current price summarizes all the available information.

It is not clear how this rationality is achieved. The price quoted on any day results from the comparison of the orders to buy and to sell given by the different investors. Models developped in finance usually assume that each individual in the market has free access to all the information and that, based on this information, all have the same expectations. The rationality of the market as a whole would simply reflect the rationality of the individuals acting on this market.

This hypothesis does not resist close scrutiny. Information is far from being a free good. This is clearly shown by the existence of a whole profession, the financial analysts, the function of which is to gather and interpret pieces of information (balance sheets, economic reports, political gossips, ...) . Morever, the interpretation is far from being unambiguous and may lead different individuals to different expectations.

A confirmation of this is to be found, for instance, in the different values estimated by various "experts" for the purchase of a fraction of a mine by the Irish Government. Estimated values ranged from Ł 1 million to Ł 104 millions (cf. Irish Times, December 1975). This type of situation, although extreme, doe not seem to be abnormal and practitioners report ranges from 1 to 10 as being fairly frequent. Furthermore the very existence of transactions between buyers and sellers, on which the whole mechanism is based, shows that not all people have the same expectations : indeed the side of the transaction on which they are reflects their expectations, and it will appear a posteriori that one of the two parties was wrong. One is therefore faced to the problem of how to reconcile the rationality of the market as a whole with the limited rationality of the individuals acting on the market.

B. Relevance of the model.

The "paradoxical" rationality of the stock market has similarities with the case of the accurate pattern formation dealt with in section II, where the accuracy of the response appears greater than the accuracy of the components. The model proposed to explain that case relied on a "built - in" communication system which automatically averages a number of moderately accurate responses. The same type of model could clearly explain how rational prices are generated on the basis of a large number of moderately rational individuals.

The most obvious prediction of the model is that the accuracy of the result depends on the number of moderately accurate elements which are used. More specifically, the model suggests that low volumes of transaction should be associated with higher variability of the forecasting error. We shall first briefly explain the two main systems of stock quotation, as it will appear that our prediction will generate different results in each system.

In the first type of organization, orders to buy and sell are pooled together at given times (usually at fixed hours, every trading day) and the price is established in order to balance buy and sell orders. The second type of organization is characterized by continuous trading. A "dealer" stands ready to buy and sell, and acts as a buffer between asynchronous orders by quoting at any moment a purchase and a sale price.

In the first system, the price quoted reflects directly the forecasts of the traders. In this system the prediction of the model is straightforward : the variation of the daily price should be larger when the volume of transcation is low. Cohen et al (1977) have analyzed the relationship between the variance of daily forecasting errors and the values of share outstanding for a random sample of stocks traded at given times on the Tokyo and the Rio de Janeiro Stock Exchanges. They found a significant (P .05) negative relationship between these two variables. A preliminary study made for this paper on the Brussels Stock Exchange also shows that for the stocks of seven companies in the same economic field, the standard deviation of the daily forecasting error is higher for each of the three stocks with small average monthly volume of transaction than for any of the companies with a large volume (P .03). This tendency is illustrated by Fig.1., which shows the histogram of the daily forecasting errors for the three companies with large volumes of transactions (A) and for the three companies with a small volume of transaction (B).

The second system is slightly more complex to analyze. Indeed the "dealer" acts as a buffer so that the variation in the individual offers is not expressed anymore in the stock price. Therefore the fluctuations of the price should not be related to the transaction volume, and in fact they are not (Cohen et al, 1977). On the other hand the "dealer" himself wants to maximize his own profit, and therefore he will set the purchase and sale prices as different as he can while still meeting a large enough proportion of the offer. Therefore the purchase and sale prices quoted at any time by the "dealer" will be a function of the spread of the distribution of individual offers : the wider the distribution, the larger the difference between purchase and sale prices quoted by the "dealer". This question has been analyzed by Tinic and West (1974) for the New York Stock Exchange. They found that the difference between purchase and sale prices offered by the dealer is inversely related to the volume of transaction.

These results certainly fit the expectations derived from the model under examination. However we wish to emphasize that these observations should be taken with care as they could be explained in other ways. A comparison between other aspects of the stock marker and predictions of the model are in order before more definite conclusions can be drawn.

A.G. is Chargé de Recherche au Fonds National de la Recherche Scientifique (Belgium).

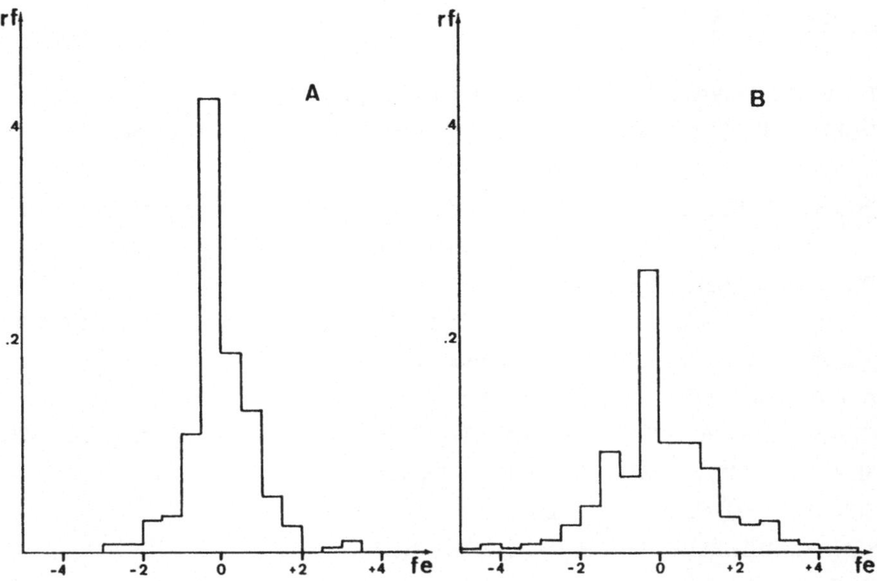

Figure 1. Relative frequency histograms of the daily forecasting errors.

The histograms were obtained by pooling together the daily forecasting errors of the three stocks with large volumes of transcations (A) and of the three stocks with small volumes of transactions (B).

rf : relative frequency ; fe : forecasting error (in percent).

REFERENCES

BRUTLAG. D., KORNBERG, A. (1972) J. Biol. Chem, 247, 241-248.

CHANGEUX, JP., COURREGE, P., DANCHIN, A. (1973). Proc. Nat. Acad. Sci. U.S. 10, 2974-2978.

COHEN, K., MAIER, S., NESS, W., OKUDA, H., SCHWARTZ, R. WHITCANTS, D. (1977) Journal of Banking and Finance, 1, 219-247.

CRICK, F. (1970) Nature 225, 420-422.

EBEL, J.P., GIEGE, R., BONNET, J., KERN, D., BEFORT, N., BOLLACK,C., FASIOLO, F., GLANGLOFF, J., DIRHEIMER, G. (1973) Biochimie, 55, 547-557.

FAMA, E.F. (1965) Journal of Business, 38, 34-105.

GARCIA-BELLIDO, A., RIPOLL, P., MORATA, G. (1973) Nature New. Biol. 245, 251-253.

GHYSEN, A., CELIS, J., (1974) J. Mol. Biol., 83, 333-351.

GHYSEN, A. (1978) Nature, 274, 869-872.

GHYSEN, A., RICHELLE, J. - in press. (Dev. Biol)

GILBERT, W., MULLER-HILL, B. (1966) Proc. Nat. Sci., 56, 1891-1898.

GOTTLIEB, F-J. (1964) Genetics, 49, 139-160.

LAWRENCE, P. (1973) in "Developmental systems : Insects "Vol.II., Counce and Waddington eds., Academic Press, pp. 157-209.

LOEWENSTEIN, W.R. (1976). Cold Spring Harbor Symp. on quantitative Biology, Vol. XL., pp. 49-63.

NINIO, J. (1974) J. Mol. Biol. 297-313.

OSBORNE, M.F.M. (1959) Operations Research 7, 145-173.

POODRY, C.A., SCHNEIDERMAN, H.A. (1970) , Roux'Arch., 166, 1-44.

RICHELLE, J., GHYSEN, A. - in press . (Dev. Biol)

STERN, C. (1954) Caryologia, Vol. Suppl., pp. 355-369.

STERN, C. (1968) Genetic mosaics and other essays, Harvard Univ. Press.

TINIC, S., WEST, R. (1974). The Journal of Finance, 29, 729-746.

URSPRUNG, H. (1969) in "Major Problems in Developmental Biology", Locke ed., Academic Press, pp. 189-197.

WIGGLESWORTH, V.B. (1940) J. Exp. Biol. , 17, 180-200.

WIGGLESWORTH, V.B. (1953) Quart J. Micr. Sci., 94, 93-112.

WOLPERT, L. (1969) J. Theoret. Biol. , 25, 1-47.

CHAPTER XX1

Serial Thresholds and Homeotic Variation

Jonathan M.W. Slack

Imperial Cancer Research Fund,
Mill Hill Laboratories,
Burtonhole Lane,
London, NW7 1AD.

SERIAL THRESHOLDS AND HOMEOTIC VARIATION

Introduction

There is a certain difference between a formal description of a system in which the chemical steps are understood, such as the lac operon of *E.coli*, and a system in which they are not. All the problems of animal morphogenesis and pattern formation, which is my own field of interest, lie in this second class. This does not mean that formalization of data is not possible, and indeed it may be very interesting, but it is important to remember that the data concerned are large scale phenomena arising from mutation, regeneration or grafting experiments. This means that the reasoning is highly abstract, and to a biochemist or geneticist has a rather curious character. The justification for such exercises on paper or on the computer must be *a posteriori*, so if it helps us to find the answers then it was useful, if it does not then it was not useful.

The problem I want to consider in this article is a contradiction which appears to exist at the phenomenological level in animal development. On the one hand there is a large body of data which suggests that the control of pattern formation during embryogenesis is carried out by means of continuous signals or gradients, which will be defined in this article as some function of the cellular metabolism which varies smoothly and monotonically across the early embryo. The simplest type of gradient would be a concentration gradient of a single substance. On the other hand there are in the arthropods various types of metaplastic transformation, in particular homeotic mutations and transdetermination, which seem difficult to reconcile with an anatomy which was originally laid down in response to a gradient.

I want to concentrate attention on the early embryonic development of insects because it is here that the contradiction seems most acute. There are three strong pieces of evidence that the pattern in the long axis of the insect egg is formed in response to a gradient:-

(i) When embryos of various species are divided by transverse constriction, a gap is created in the series of structures which later develops. This gap is wider the earlier the developmental stage at which the constriction was made. This "gap phenomenon" is interpreted as a result of the accumulation of a morphogen on one side of the constriction and its dissipation on the other side. (Sander 1976).

(ii) In the leafhopper, *Euscelis*, it is possible to induce an ordered secondary set of posterior structures by the influence of posterior cytoplasm on the anterior part of the egg. (Sander 1976).

(iii) In Drosophila, there exists a maternal effect mutant, bicaudal, which gives rise to double-posterior embryos. Different embryos are not all identical but exhibit a range of forms in which different numbers of structures lie on each side of the plane of mirror symmetry (Nüsslein 1977 and Fig. 1). It is believed that the mutation alters the constitution of the oocytes in such a way that a U-shaped gradient is formed

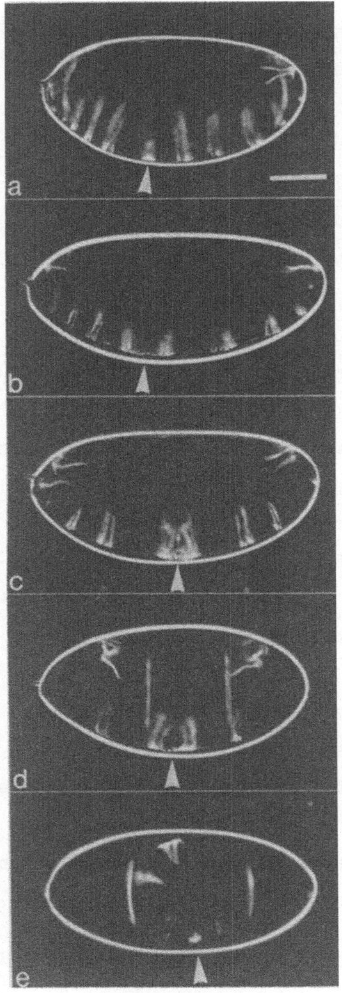

Figure 1. *Drosophila* embryos affected by the maternal mutation *bicaudal*. The embryos a-e form a graded series of mirror symmetrical duplications which contain fewer and fewer structures between the two posterior segments. Taken from Nüsslein (1977).

Figure 2. The homeotic mutant *antennapaedia*. The antenna is converted into a leg. Specimen provided by Sheena Pinchin.

instead of a monotonic one, and the number of segments formed would vary depending on the depth of the minimum of the U.

In the case of holometabolous insects such as Drosophila, the rudiments which are formed in response to the longitudinal gradient are thought to include anlagen for the larval segments and for the imaginal discs which will give rise to the adult structures during metamorphosis.

But it is the properties of the imaginal discs which provide the strongest evidence against a gradient in the egg and in favour of the states of determination being coded by a *combinatorial* arrangement of genes. Certain mutations, called homeotic mutations, convert one disc into another so that it will develop into a different organ from that expected. For example *bithorax* converts anterior metathorax into anterior mesothorax *ophthalmoptera* converts eye into anterior wing, and *antennapaedia* converts antenna into 2nd leg (Morata and Lawrence 1977 and Fig. 2).

In addition, when disc fragments are cultured in the abdomens of adult flies it is found that they sometimes change their states of determination and if they are later implanted into a larva about to undergo metamorphosis, they differentiate into structures characteristic of the adult organ formed from a *different* disc. This is called transdetermination (Gehring 1972). Kauffman (1973, 1975) has analysed this and other data in detail and proposed a coding for the discs which are combinations of *on* and *off* states of four genes. These genes would be the wild type forms of the homeotic mutants and have been called *selector genes* by Garcia-Bellido (1975).

It is very hard to see how a monotonic gradient could yield a sequence of such combinations, eg., 1011, 1001, 1101, 1111, so Kauffman and co-workers (Kauffman, Shymko and Trabert 1978) have argued that pattern formation in the early embryo of *Drosophila* occurs by means of morphogen distribution which arise from a system of coupled reactions with nonlinear kinetics forming spatial dissipative structures of the type analysed by Nicholis and Prigogine (1977). Several different spatial distributions of the morphogen arise in a sequence which is controlled by the variation of a parameter, such as the viscosity of the egg cytoplasm. For each successive pattern, a different gene is turned on in regions of the embryo with high morphogen levels and remains off in regions with low morphogen levels.

This is a most ingenious theory, but it is not in accordance with most of the experimental results on early embryos such as those mentioned above, which suggest that the rudiments for the basic body plan are laid down in response to a continuous and monotonic gradient. What I want to do in the present paper is to suggest that the contradiction may disappear once we consider the nature and the properties of the responses to a gradient in terms of a theory of serial and independent thresholds.

Serial Thresholds

Typically an embryo or an embryonic organ rudiment goes through a regulative phase and then a mosaic phase (Huxley and de Beer 1934). The operational distinction between the two periods is that during the regulative phase the removal of part of the rudiment does not affect the eventual structures which are formed, while during the mosaic phase removal of a part produces a corresponding deletion in the eventual structures. Embryologists regard the regulative phase as being the time before or during the passage of the signals which bring about the regionalization of the field, and the mosaic phase as being the time after the parts have become irreversibly determined.

Let us suppose that the signal is a gradient, and let us also suppose that the cells of the field are competent to respond to this gradient by the possession of n biochemical switches, each of which is turned on at a different level of the signal (see Lewis *et al.*, 1977 for discussion of thresholds). If the lowest threshold is exceeded everywhere, as is required if the most distal territory is to differ from the state it was in prior to the signal, then n discrete territories will be formed,

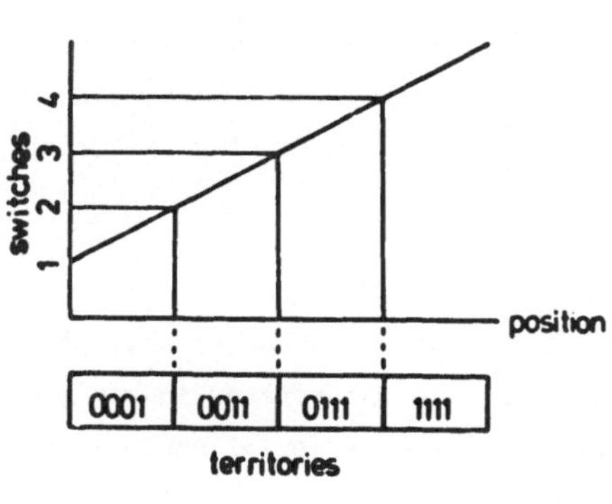

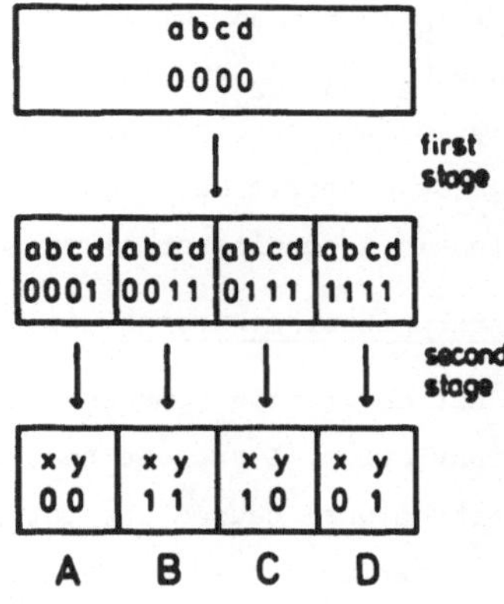

Fig 3

Fig 4

Figure 3. Binary codings arising from independent threshold responses to a monotonic gradient.

Figure 4. An example of pattern formation controlled by a gradient showing the primary responses, the final states of gene activity and the names for the different territories.

each differing from its neighbour by the turning on of one more switch, (Fig. 3). If the states of the switches are written in the same order as their relative thresholds for activation then the coding of a territory will be of the form 0000011111. For the ith territory all the switches from 1 to i are *on* (value 1) and from (i + 1) to n are *off* (value 0). The coding of the i th territory will be symbolised as S_i. Mathematically it is easy to see that the space of all the points S_i has a simple metric:-

$$d(S_i,S_j) = |i - j|$$

and analagous results hold for spaces of higher dimensionality. For example a two dimensional field will consist of territories defined by two independent sets of switches.

$$\begin{matrix} 0000011111 \\ 0001111111 \end{matrix} \equiv S_{5,7}$$

and the metric is:-

$$d(S_{ik},S_{j\ell}) = |i - j| + |k - \ell|$$

The assumption of territorial codings which form a metric space seems inescapable for organs which are capable of regeneration. The imaginal discs are among such organs (Bryant 1975) and although I do not wish to discuss the matter at any length here, I believe that there are two types of process at work. The first is a "filling in" process, called convergence, and the second is a reconstitution of extrema, called divergence (Goodwin 1977). For example S_5 and S_9 when placed in contact will generate

S_6, S_7, S_8 in the gap by convergence, while S_5 when situated at a free edge will generate S_4, S_3, S_2 and S_1 to complete the distalwards sequence by divergence. It is easy to imagine biochemical processes capable of comparing and incrementing the codings of a metric set, but very difficult for an organ composed of territories labelled in a combinatorial manner.

The Interpretation Process

Just as protein synthesis has two chemically distinct phase of transcription and translation, I propose that the embryonic signals are interpreted in two phases. First the signal causes the formation of the territories with serial codings, then in each territory this coding leads to the activation of a certain combination of structural genes. When we examine the histological appearance of the differentiated cells, or measure their biochemical properties, we are looking at the second level of expression. But if we damage the organ so as to obtain a regenerative response then we are looking at the properties of the first level.

Consider a simple example. Suppose that an embryonic signal acts on a field whose competence resides in the capacity of four genes: a, b, c, & d, to be switched on at different thresholds. The resulting serial codings become translated into different combinations of activity of two structural genes, x & y, and the four resulting visible phenotypes are called A, B, C & D (Fig. 4).

Since this scheme is wholly speculative, it may seem rash to say anything much about the second "translation" stage. But some features become quite obvious as soon as the problem is posed. First, there is the question of the relative information content of the primary codings and the final differentiated states. If the competence of the field resides in n independently regulated elements and the character of the final differentiated states in r independently regulated elements, then clearly $r \leqslant n$ since the information content of a primary territory coding will be n bits and the information content of subsequent states can be no greater (Shannon 1948). Secondly, we have assumed that in normal development only the serial codings are used and not the other possible combinations. So it becomes interesting to ask what would happen if an unnatural coding is generated by mutation. This problem can be simply illustrated by means of the Veitch matrix introduced for the study of bacterial gene networks by Thomas (1973). In such a matrix the values of the output variables (in this case x and y) are tabulated for all possible combinations of the input variables (a, b, c, d). Where the formulation of the problem does not include a particular input combination then this position in the table is left blank. Using the example of Fig. 4, we take all 16 possible combinations of a b c d, we fill in the normal outcomes and leave the unnatural ones indeterminate:-

		c,d			
	x,y	0 0	0 1	1 1	1 0
	0 0	---	0 0	1 1	---
a,b	0 1	---	---	1 0	---
	1 1	---	---	0 1	---
	1 0	---	---	---	---

Then we ask, what are the simplest translation rules for generating the normal result? The advantage of the Veitch matrix is that this can be seen at a glance, and is in fact:-

$$x = \bar{a}c$$

$$y = a + \bar{b}c$$

($\bar{a}$ is <u>not</u> a, ac is a <u>and</u> c, a + b is a <u>or</u> b). These rules now define the outcomes of the unnatural combinations as well:

	x,y	c,d: 0 0	0 1	1 1	1 0
	0 0	0 0	0 0	1 1	1 1
a,b	0 1	0 0	0 0	1 0	1 0
	1 1	0 1	0 1	0 1	0 1
	1 0	0 1	0 1	0 1	0 1

and replacing the combinations of x, y with their phenotype names we get a matrix for predicting the metaplastic changes which will arise from mutation or epigenetic failure of one of the switches a, b, c, d.

	state	c,d: 0 0	0 1	1 1	1 0
	0 0	A	A	B	B
a,b	0 1	A	A	C	C
	1 1	D	D	D	D
	1 0	D	D	D	D

For example if gene c is inactive because of mutation, the codings will be 0001, 0001, 0101 and 1101. The anatomy of the creature will become AAAD instead of ABCD. The consequences of each of the four genes being defective are shown as "anatomy 1" in the table below. The translation rules could be more complex, for example:-

$$x = \bar{a}cd + \bar{c}\bar{d}$$

$$y = abc + \bar{b}cd$$

are equally compatible with the normal development, but would give rise to a different set of mutant anatomies (anatomy 2):

Mutant	Anatomy (1)	Anatomy (2)
a	ABCC	ABCC
b	ABBD	ABBD
c	AAAD	AAAA
d	ABCD	CAAD

There are three main points to be made from this kind of analysis. First that a defect in one element can cause several regions of tissue to change their character. Secondly, even though the original signal was a gradient and the territorial codings are serial, there are some "jumps" between non-neighbouring territories. Thirdly, there is one regularity superimposed on the other changes which is the systematic conversion of the i th territory to the i-1 th when the i th element is defective. This of course follows trivially from the existence of a serial coding.

All these three features are found in nature. Mutations frequently affect several parts of the body, and in fact among the vertebrates it is hard to find developmental mutants which have simple and clearcut phenotypes.

The fact that jumps between non-neighbours are found in arthropods and not in vertebrates may perhaps be because there are more genes active in the various terminal cells types in the vertebrates. In the example given above there are no unnatural combinations of the end products, but the more genes there are, the less likely would errors in pattern formation be to convert one part of the body into another.

Systematic conversion of a structure into its neighbour, and the associated loss of one element from a linear series was called "backwards homeosis" by Bateson (1894) who cites various examples in the teeth, vertebrate and ribs of animals and Man. The gap is often accompanied by other defects in other positions and it will be seen that these could arise from the unnatural combinations formed proximal to elements which had been turned off and distal to one which had been jammed on. The opposite phenomenon of the presence of an extra element in the sequence is also sometimes encountered and may perhaps be explained as resulting from the division of a territory into two.

References

1. Bateson, W. (1894) Materials for the Study of Variation, Macmillan, London.
2. Bryant, P.J. (1975) Regeneration and duplication in imaginal discs. p. 71-93 in Cell Patterning", CIBA Symposium 29.
3. Garcia-Bellido, A. (1975) Genetic control of wing disc development in Drosophila, p. 161-182, in "Cell Patterning", CIBA Symposium 29.
4. Gehring, W. (1972) The stability of the determined state in cultures of imaginal discs in *Drosophila*. p. 35-58, in "The Biology of Imaginal Disks", H. Ursprung & R. Nöthiger. Springer-Verlag, Berlin.
5. Goodwin, B. (1977) Analytical physiology of cells and developing organisms. Academic Press.
6. Huxley, J.S. & de Beer, G.A. (1934) The Elements of Experimental Embryology. Cambridge University Press.
7. Kauffman, S.A. (1973) Control circuits for determination and transdetermination. Science, 181, 310-318.
8. Kauffman, S.A. (1975) Control circuits for determination and transdetermination: interpreting positional information in a binary epigenetic code. p. 201-221, in CIBA Symposium 29.
9. Kauffman, S.A., Shymko, R.M. & Trabert, K. (1978) Control of sequential compartment formation in *Drosophila*. Science, 199, 259-270.
10. Lewis, J., Slack, J.M.W. & Wolpert, L. (1977) Thresholds in development. J. Theor. Biol., 65, 579-590.
11. Morata, G. & Lawrence, P.A. (1977) Homeotic genes, compartments and cell deter--mination in *Drosophila*, Nature, 265, 211-216.
12. Nicholis, G. & Prigogine, I. (1977) Self Organization in Nonequilibrium Systems. John Wiley & Sons, New York.
13. Nüsslein-Volhard, C. (1977) Genetic analysis of pattern formation in the embryo of *Drosophila melanogaster*. Characterization of the maternal effect mutant *bicaudal* Wilhelm Roux's Archives, 183, 249-268.
14. Thomas, R. (1973) Boolean formalization of genetic control circuits. J. Theor. Biol., 42, 563-585.
15. Sander, K. (1976) Specification of the basic body pattern in insect embryogenesis. Advances in Insect Physiology 12, 125-238.
16. Shannon, C.E. (1948) A mathematical theory of communication. Bell Systoms Technical Journal 27, 379-423.

CHAPTER XXII

AN APPLICATION OF CATASTROPHE THEORY TO THE STUDY OF A SWITCH IN DICTYOSTELIUM DISCOIDEUM

M. J. BAZIN and P.T. SAUNDERS

Departments of Microbiology and Mathematics.

Queen Elizabeth College
Campden Hill Road
London W8 7AH
England

At the end of the workshop a discussion was held on the relative advantages of representing biological systems by continuous or discrete models. It was recognized that the most suitable formulation could depend not only on the essential nature of the system being studied but also on the level at which it was being considered. For example, the growth of a tissue is well represented as a continuous process if viewed on a sufficiently large scale, but on a microscopic scale it is clearly a sequence of separate cell division events. Conversely, a system which has a dynamic obeying van der Pol's equation will exhibit discontinuous behaviour on the macroscopic scale, even though a microscopic analysis of the system would reveal nothing that was not inherently continuous. We describe here an application in which we have used catastrophe theory to determine what we believe to be the correct critical variables governing a particular biological system. The technique we have employed illustrates the interplay of discrete and continuous modelling in a striking fashion the key observations are of an apparently discrete switch, but the derivation of the conclusions depends (through catastrophe theory) on the assumption that there exists an underlying smooth dynamic.

The mathematical study of the interaction between a predator and its prey began over 50 years ago when d'Ancona (1926) observed out-of-phase variations in the populations of two species of fish in the Adriatic Sea. Volterra (1926) proposed the simple model

$$\begin{aligned} \dot{H} &= \alpha H - \lambda PH \\ \dot{P} &= \mu PH - \beta P \end{aligned} \qquad (1)$$

Here H and P are the prey and predator population densities respectively, α, β, λ, μ are constants, and a dot denotes differentiation with respect to time. These "Lotka-Volterra" equations cannot be solved in closed form, but it can be shown that the solutions are of the right form, i.e. that they are periodic with the phase of H about a quarter period ahead of that of P.

The apparent success of the Lotka-Volterra equations in accounting for the population oscillations observed by d'Ancona (and, later, by other workers in different predator-prey systems) has led to their playing a key role in theoretical ecology. Certainly most theoretical discussions of population variations (and the related problem of the stability of ecosystems) tend to assume that oscillations arise in essentially the way predicted by the Lotka-Volterra equations, i.e., through the predator-prey interaction. (It is, to be sure, generally acknowledged that limit cycles may exist, but much less has been done with these because they are so much harder to work with, as they cannot be adequately analyzed by Liapounov's first method.)

There are, however, some problems with the Lotka-Volterra equations. In the first place, they are clearly only a rough approximation to the actual interactions which take place between and within species. Unfortunately they are also structurally unstable, in that almost any functional perturbation changes the constant amplitude oscillations about a neutrally stable equilibrium point to oscillations which either damp fairly rapidly or else increase without limit. The persistent oscillations characteristic of the L-V equations are therefore not characteristic of most plausible modifications of them. More importantly (since one can always postulate just sufficient perturbations to counteract the damping) whenever population oscillations have been studied in detail they have turned out to be better explained as being driven by variations in the environment or, less commonly, as limit cycles caused by some interaction involving more than just two species in isolation.

Since at least part of the reason that there does not appear to be a well-established example of population oscillations produced by a simple predator-prey interaction may simply be that it is difficult to obtain data which is sufficiently accurate and which covers a sufficiently long period of time, we decided to study an artificial ecosystem consisting of two species of microorganisms, the cellular slime mould amoeba, *Dictyostelium discoideum*, feeding on the common gut bacterium, *Escherichia coli*. Microorganisms have the great advantages that they can be grown in very large numbers, which effectively eliminates statistical fluctuations, and their generation time is a few hours, rather than months or even years. Moreover they can be grown in a relatively well-controlled environment; we used a culture vessel, called a "chemostat" in which the volume of the culture is kept constant by supplying nutrient at the same rate that effluent is being removed. This rate is called the dilution rate, and is customarily denoted by D. Because the system is well-mixed, the concentration of nutrient and organisms in the effluent is the same as that in the vessel, which permits measurements to be made without perturbing the system. In addition, the continuous loss of organisms makes it possible to study an actively growing population without allowing an increase in numbers sufficient to constitute a change in the environment.

Of course we did not expect that the Lotka-Volterra equations as they stand would provide a good description of the system, so we introduced two additional features. First, we included a third equation to represent the variation in the concentration of the nutrient (glucose) on which the bacteria were feeding. Apart from the obvious gain in realism in that the specific growth rate of the bacteria was no longer constant, the extra equation also allows the possibility of limit-cycle behaviour. We also wrote the equations in the form

$$\begin{aligned} \dot{S} &= DI - DS - Hf(S,H) \\ \dot{H} &= Hf(S,H) - DH - Pg(H,P) \\ \dot{P} &= Pg(H,P) - DP \end{aligned} \tag{2}$$

where S is the nutrient concentration and I is the nutrient concentration in the input. The functions f and g were not specified in advance. We hoped to be able to determine them empirically and in the meantime we could try to discover what we could predict about the general behaviour of the system without making any particular choice of functions. One problem which interested us was the so-called "paradox of enrichment" (Rosenzweig, 1971) which states that too great a supply of nutrient can destabilize an equilibrium. We were able to show that this effect was likely to occur if the organisms were limited by saturation (i.e., by having as much food as they needed and being unable to consume any more) but not if they were limited by some form of law of diminishing returns (Saunders & Bazin, 1975). A consequence of this is that one ought to be able to deduce which form of limitation is acting in a given system by increasing the nutrient and observing whether the system destabilizes or becomes even more stable. We have not yet obtained conclusive results for our system, though the evidence suggests that the limitation is of the diminishing returns kind. It was this result, however, that first drew our attention to the fact that it is possible to come to certain conclusions about a system without having a definite model and definite equations, and that when the mechanism of a system is not known in detail it is even preferable to have results which are known not to be critically model-dependent.

When we carried out the experiments (Dent, Bazin & Saunders, 1976) we found, as had Tsuchiya, Drake, Jost & Frederickson (1972) that the system exhibited rapidly damped oscillations, and not the persistent oscillations we had hoped for. It appeared that the predator-prey interaction could make the passage from the initial state to the stable equilibrium oscillatory, but it could not actually produce oscillations. This was not entirely unexpected, but what was surprising was that the results did not seem consistent with any model of the quite general type given by equations (2).

A typical set of results is shown in Figs. 1 and 2. Perhaps the most obvious of the unexpected features is the way in which the specific growth rate of the amoebae remained constant over quite along periods of time (compared to the generation time of these organisms, which is about 3h) even though the prey density was changing by a factor of as much as 100. There was then a rapid change to a new specific growth rate, which was in turn maintained for a considerable time. At about 80h after the start of the experiment all the quantities we were measuring and their derivatives as well appeared to be the same as they had been at 30h, yet at 80h the populations began to recover whereas at 30h they had continued to fall. It is also difficult to explain how during the interval from 20h to 30h the bacterial biovolume density (the number density multiplied by the mean cell volume to obtain an estimate of the bacterial biomass density) declined at a rate greater than the dilution rate D, when at this time the amoebal biovolume was declining at almost precisely the dilution rate, which seems to imply that the amoebae were not feeding. Finally (although this is not shown in the figures), after about 225h the population densities of both organisms, which had remained approximately constant for some 100h, moved smoothly to new values, which were maintened until the ènd of the experiment.

At this stage it seemed clear to us that we were unlikely ever to be able to fit a straightforward differential equations model to the data; the system we had chosen to be a simple example of Lotka-Volterra dynamics in action was turning out to be anything but that. So instead of trying to construct a complete model of the system we decided to see what if anything could be deduced about the interaction between the organisms. As we didn't know what we were looking for we tried a number of approaches, and the one that has so far proved most profitable was based on catastrophe theory.

We were particularly struck by the way in which the amoebae went through alternating phases of exponential growth and exponential decline, so that the best curve through the data points appeared to be a succession of straight lines. This implied discontinuities, or at any rate near-discontinuities, in the specific growth rate, λ, so we decided to try to interpret the behaviour of the amoebae in terms of catastrophe theory with λ as the state variable. From Thom's list (Thom, 1972) we chose the cusp, because it is the simplest which is consistent with the observation that, as the experiment continued, the jumps in λ became smaller and eventually disappeared altogether.

For an introduction to catastrophe theory we refer the reader to Zeeman's article in Scientific American (Zeeman, 1976) or to Poston & Stewart (1978). What we need to know is that it involves making the assumption that the specific growth rate λ is governed by one or more ordinary differential equations and that it is almost always at or near an equilibrium. This is in fact what is usually supposed in the study of systems involving biochemical reactions. The theory then tells us that the simplest mechanism that can produce the observed behaviour will involve two critical "control variables" and that the state of the system is then determined by finding the minima of a Liapounov function which is obtainable from the canonical form

$$V(x) = x^4 - ux^2 + vx$$

by a diffeomorphism (i.e., a smooth transformation) which maps the actual control variables to the canonical control variables u and v and a second diffeomorphism which maps the actual state variable λ to the canonical state variable x. (The second diffeomorphism can involve the control variables if certain conditions are satisfied, but it turned out that we did not need the extra freedom this allows.)

Stationary points of $V(x)$ are of course located by solving the equation $dV/dx = 0$; this is a cubic in x and so will have either one real root or three depending on the values of u and v. The system will correspondingly have either one stable equilibrium or else two stable equilibria separated by an unstable one. In the u-v plane the boundary between the region with on real root and the region with three is a curve along which there are three real roots, two of which are equal, so that V(x) has a minimum and a point of inflexion. The equation of this curve, which is found by solving the equations $dV/dx = d^2V/dx^2 = 0$, is $27v^2 = 8u^3$.

Small changes in the control variables produce small changes in the stationary values of V(x), and in general this implies small changes in the state variable. Sudden changes occur when the system moves from one steady state to another. In principle this can happen anywhere within the region in which there is more than one stable steady state corresponding to a single pair (u,v). It is, however, most likely to occur on the boundary of this region, for if the system happens to be in the stable equilibrium which disappears by coalescing with the unstable one, it will have no alternative but to move to the other stable steady state. Systems which always remain in the equilibrium they are in until it disappears are said to obey the "perfect delay convention".

In Fig.3 we have sketched the function V(x) for different values of u and v. In Fig.4 we show what happens as the control variable V is increased from V_A at A to V_B at B and then decreased from V_A again. Because the jumps occur only if the equilibrium that the system is in disappears, they occur only as the system passes out of the region with two equilibria, i.e., at (vi) when passing from A to B, but at (ii) when passing from B to A. The system therefore exhibits hystersis. Finally, in Fig.5 we have drawn the surface $dV/dx = 0$ and the projection into the u-v plane of the curve $dV/dx = d^2V/dx^2 = 0$. The state of the system is always represented by a point on the surface, and we see that the perfect delay convention implies that the phase point will remain on the upper or lower sheet as long as it can, jumping up or down only when necessary. The middle sheet represents the unstable equilibria, and we do not expect ever to find the phase point on it.

The first step in constructing our model was to choose the variables. We had already fixed on λ as the state variable because it was in this quantity that the sudden jumps occurred. As one control variable we chose the prey biomass density, H, because most previous workers had assumed that this did chiefly determine the specific growth rate. We knew from catastrophe theory that we could not reproduce the observed behaviour without a second control variable, and we chose t, the time measured from the start of the experiment. This appeared a natural selection, particularly as there appeared to be time-dependent changes within the system, but it had the effect of making our model different from most others, which are based on autonomous differential equations.

With this choice of variables, Fig.1 became the projection into the control space of the trajectory of the system. We marked the points at which sudden jumps up and down occurred and sketched a cusp-like curve connecting them; in fact it turned out that a pair of straight lines would serve (Fig.6). In terms of the canonical variables u,v the equation of the curve is $27v^2 = 8u^3$ so we had to construct a diffeomorphism which would carry the equation of the pair of lines into this form. We did this by translating the origin to the point of intersection of the lines, rotating the axes so that the new coordinate axes were the bisectors of the angles between the lines, and then making a further transformation of the form

$$u \longmapsto au \qquad v \longmapsto bv^{\frac{2}{3}}$$

where a and b are constants. This turned out to be sufficient for our purpose but is, of course, not unique.

We could determine x as a function of u and v by using the equation $dV/dx = 0$. Since we had data giving λ in terms of H and t we could obtain an empirical diffeomorphism relating x and λ simply by curve-fitting. This gave us a theoretical relation expressing λ as a function of H and t, and we integrated this numerically to obtain P as a function of t. The result is compared with the data in Fig.7. The agreement is obviously good, although we have to bear in mind that what is significant is the almost straight lines and the decreasing periods and slopes. Once the model reproduces these qualitative features correctly, the close numerical agreement between the observed and predicted values of the slopes can usually be arranged by suitable choice of the diffeomorphisms described above.

There is, however, a problem with this model. At first glance Fig.6 may appear to be the usual sort of picture one draws in catastrophe theory, but in fact there is one important difference: the sudden jumps occur not as the trajectory leaves the cusp, but as it enters it. Now catastrophe theory does not require that the perfect delay convention be obeyed; in some situations, for example, systems follow instead the so-called Maxwell convention, in which the equilibrium that is chosen is the one corresponding to the global minimum of potential. In the present case, however, we appear to have a system which moves to a new steady state as soon as it appears, even though this state is at a higher potential. There seems to be no obvious reason why this should happen unless we postulate some separate mechanism not included in our model. For example, we might suppose that the amoebae are sensitive to the rate of change of the bacterial biomass density. When this derivative is positive the amoebae try to move to the rapid growth mode, but they cannot do this until the bacterial biomass density is great enough that the corresponding equilibrium point of the biochemical reactions governing the behaviour of the amoebae. Consequently as soon as the equilibrium appears the system moves to it. Conversely, the amoebae react to a falling bacterial biomass by attempting to switch to the slow growth mode, but this too cannot be done until the appropriate equilibrium point appears.

This seemed to us to be sufficiently implausible that we decided to look for an alternative explanation. Most models of microbial predation do suppose, as we had here, that the feeding rate of the organisms depends on the density of the nutrient. There is, however, evidence (Curds & Cockburn, 1971, and also some of our own unpublished data) which tends to suggest that, at least in the case of amoebae, the critical variable may be the ratio of food to feeders; this is of course the case for higher organisms, such as man, in which what matters is the number of cattle, say, per head of population. So we repeated the analysis we have just described, except that we replaced H by H/P, and it turned out that this altered the picture considerably. Fig.8 illustrates H/P as a function of time, with the locations of the sudden jumps in λ marked. These points all lie on a straight line, which we may take as the catastrophe set of a cusp catastrophe using the Maxwell convention. We can now obtain a fit just as good as the one in Fig.7 (by choosing diffeomorphisms suitably) but with the important difference that there is no need to postulate a separate switch. We conclude that H/P is far more likely to be the correct critical variable.

The smooth curves drawn in Figs.6 and 8 had been obtained by the method of cubic splines, and we had to consider to what extent our conclusions might be sensitive to small errors in these curves or, indeed, in the original data. It is clear that in Fig.6 no small change in the curves nor difference of an hour or two in the location of the jumps can affect the qualitative picture: the configuration of the jumps will remain the same and neither the perfect delay nor Maxwell conventions will be applicable. In the case of Fig.8 the position is somewhat different, as similar perturbations would almost certainly result in the jumps no longer lying on a single straight line. This will not necessarily affect our conclusions, since the line need not be straight and our argument also applies if the configuration is that associated with the perfect delay convention. All the same, it seemed best to investigate further.

We noticed that the smooth curve in Fig.7, which had been based on the catastrophe theory model, was a better fit to the data than the curve we had obtained by using the cubic spline. This was not surprising; the spline technique is not really suitable for nearly straight lines, and tends inevitably to round the corners too much. So we recalculated the ratio H/P using the new curve. The results are shown in Fig.9. It now appears that it is the perfect delay convention which is applicable, and this is really what we would expect, given the sorts of biochemical equations which are probably involved.

Further confirmation is provided by later results of Owen (1979) which are illustrated in Figs.10 and 11. This work was carried out with a somewhat different aim, and consequently less data relevant to our problem was obtained. Even so, it is clear from this unsmoothed data that the results based on H/P are consistent with the perfect delay convention, whereas those based on H are not. Viewed in this way we also see that the configuration in Fig.10 is also qualitatively different from that in Fig.6, which casts further doubt on the hypothesis that H is the correct critical variable. We also see here an example of how we can use catastrophe theory to draw conclusions from data to which we could have no hope of fitting a model based on differential equations.

Conclusions:

In most mathematical studies in science we postulate a mechanism for the effect we are interested in, translate this into a mathematical model, derive observational consequences, and compare these with the data. What we have done here is quite different, in that we have no particular mechanism in mind. Instead we have argued that if one assumption (that H/P is the correct critical variable) is correct, then a comparatively simple mechanism can account for the behaviour of the amoebae. If, on the other hand, the alternative assumption (that it is H that is the correct critical variable) is true, then a much more complicated mechanism will be required with, in effect, two separate switches to control a single response. We conclude that the former assumption is much more likely to be correct.

Compared with more familiar techniques, our analysis has a number of advantages. It allows conclusions to be drawn on the basis of data which is not as accurate as would be need to fit a model based on differential equations. We do not have to propose a definite mechanism, and, by the same token, the conclusions do not stand or fall with one choice of mechanism or _ad hoc_ equation. And while our analysis has not given us the underlying mechanism, it has provided us with an important clue: we are more likely to be able to determine the nature of the switch if we know the critical variable to which it responds.

We consider that our work provides an example of one way in which catastrophe theory can be used and we expect that similar applications will be found in other fields. Observations of switching behaviour in complex systems can be used to gain information about the underlying processes. Even more importantly, however, we consider that our work illustrates how the ways in which catastrophe theory is used can differ from those in which more familiar techniques of applied mathematics are employed. It may well be that before catastrophe theory makes the impact on science in general and biology in particular that we are confident it will make, much more will have to be done in learning how best to apply it. We hope that our work is a step in this direction.

This research was supported, in part, by the UK NERC.

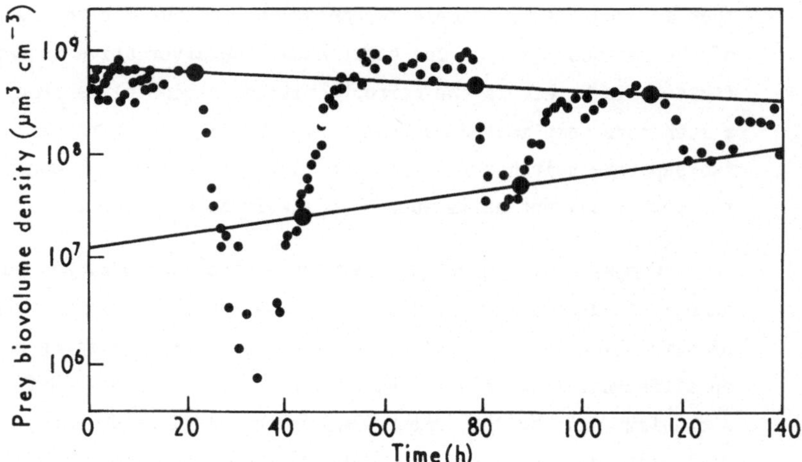

Fig. 1. Change in biovolume density of Escherichia coli grown together with Dictyostelium discoideum amoebae in chemostat culture. The large solid circles indicate the times at which the amoebal specific growth rate changed rapidly. The lines connecting these points were used to define the catastrophe set of a cusp catastrophe as described in the text.

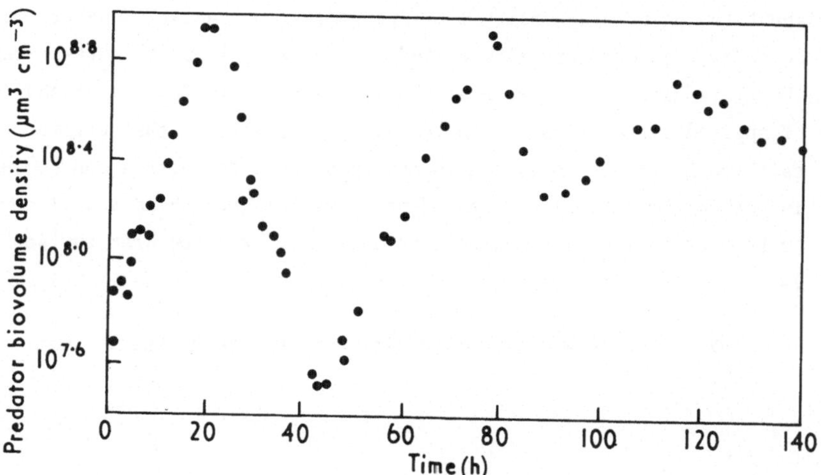

Fig. 2. Amoebal biovolume density during the experiment described in Fig. 1.

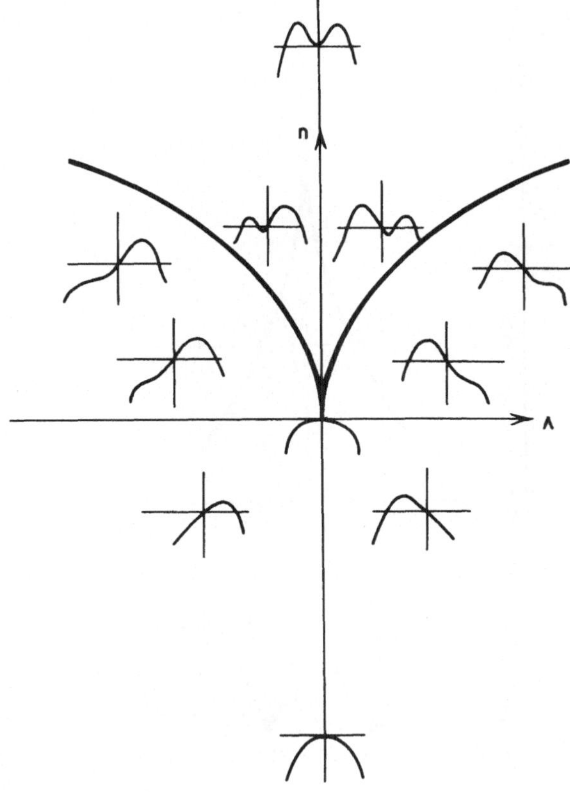

Figure 3.

Fig. 3. The projection of a cusp catastrophe into the u-v plane. The heavy line shows the catastrophe set and the small diagrams represent the number of stable equilibria occurring in the plane. Inside the cusp-shaped region there are two stable equilibria. Sudden changes in the state variable occur when one of these disappears as the system moves out of this region.

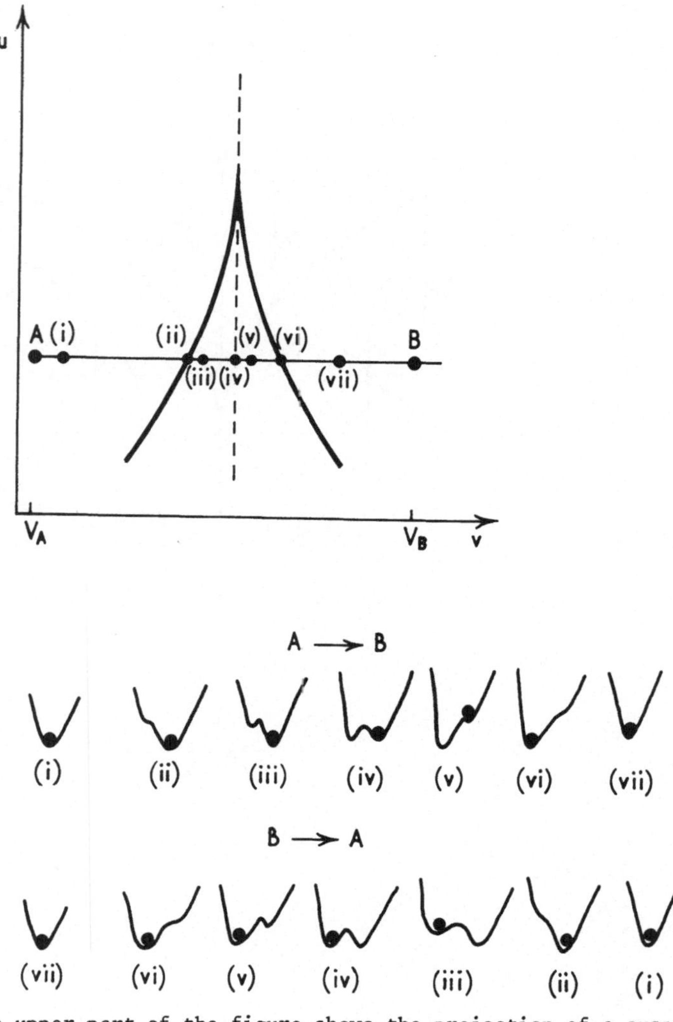

Fig. 4. The upper part of the figure shows the projection of a cusp catastrophe in the u-v plane with the cusp-shaped region bounded by the catastrophe set and bisected by the Maxwell set, indicated by the broken line. In the lower half of the figure the stability of the system is represented by a series of diagrams in which a ball rolls to a local minimum whenever possible. The behaviour of the system as V changes from V_A at A to V_B at B is shown. After (ii) a second equilibrium appears, but only becomes effective after the disappearance of the original equilibrium at (vi). At this point a sudden change in the state variable occurs. In the opposite direction, going from B to A, the jump occurs at (ii) i.e. at a different value of V.

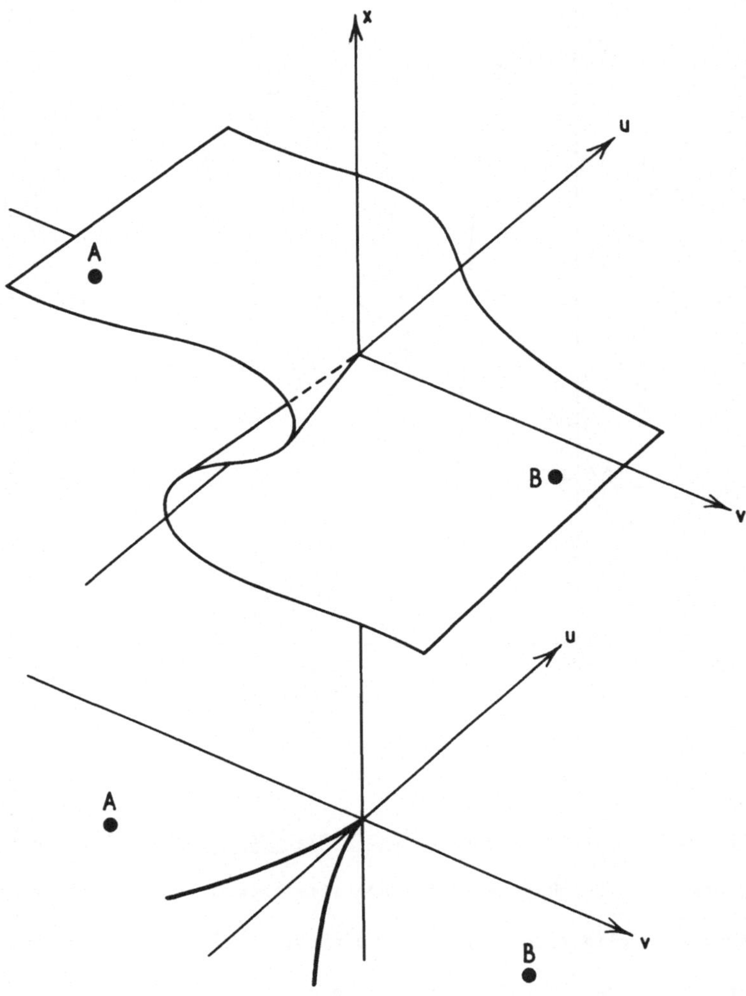

Fig. 5. The manifold of the canonical cusp catastrophe, its projection into control (u-v) space and the points A and B of the previous figure show in this space and on the surface.

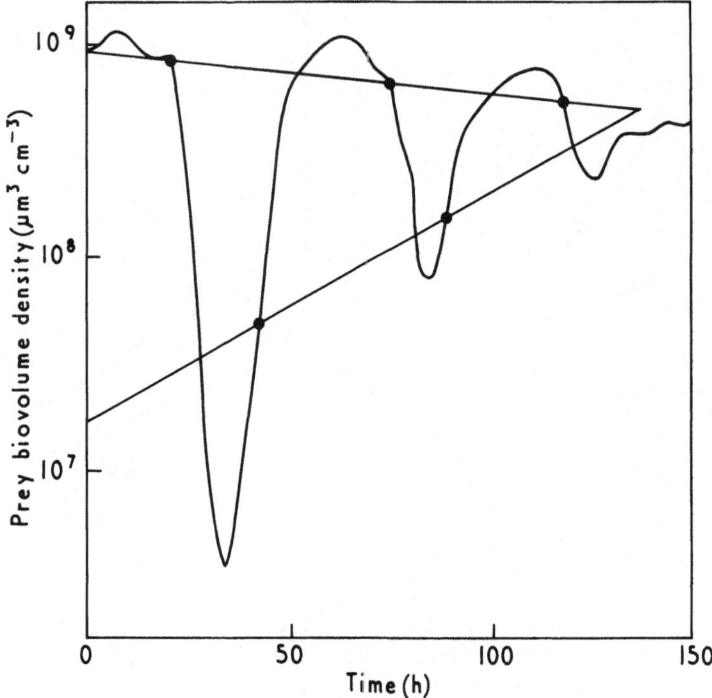

Fig. 6. Biovolume density of the prey, Escherichia coli, showing the points at which changes in the specific growth rate of the predator occurred (solid circles). The two straight lines represent the catastrophe set in the control space of the system.

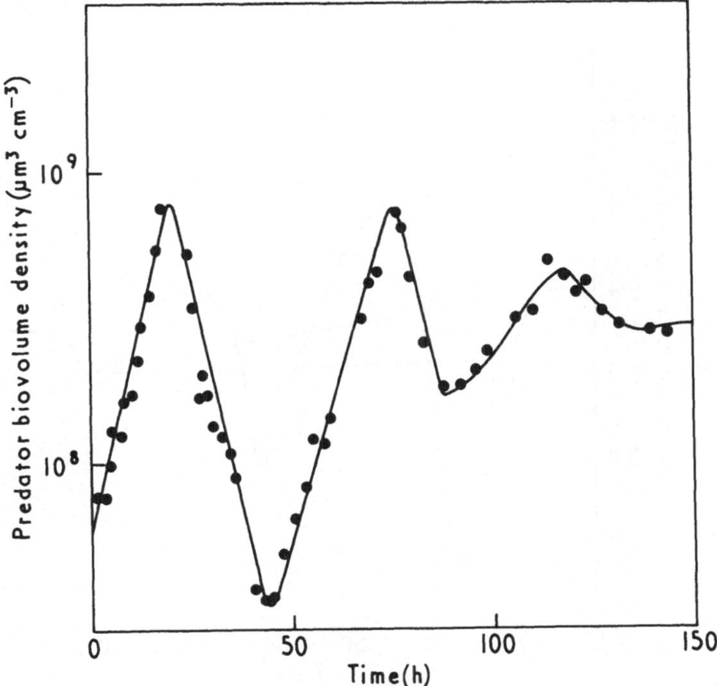

Fig. 7. Biovolume density of D. discoideum. The solid line is a fit obtained by using a cusp catastrophe with H and t as control variables.

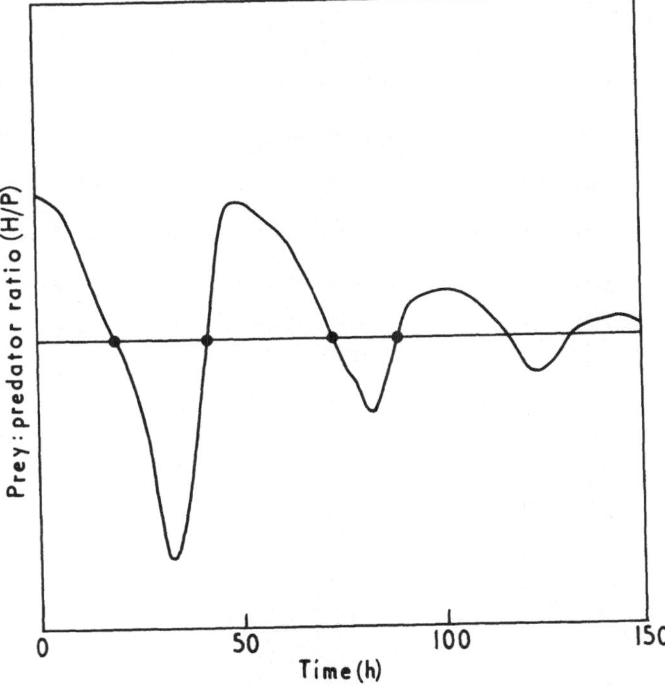

Fig. 8. As in Fig. 6, but with the prey:predator ratio as the ordinate. The solid line can be taken to represent the Maxwell set of a cusp catastrophe. The smooth curve was generated by the method of cubic splices from the data of Fig. 7.

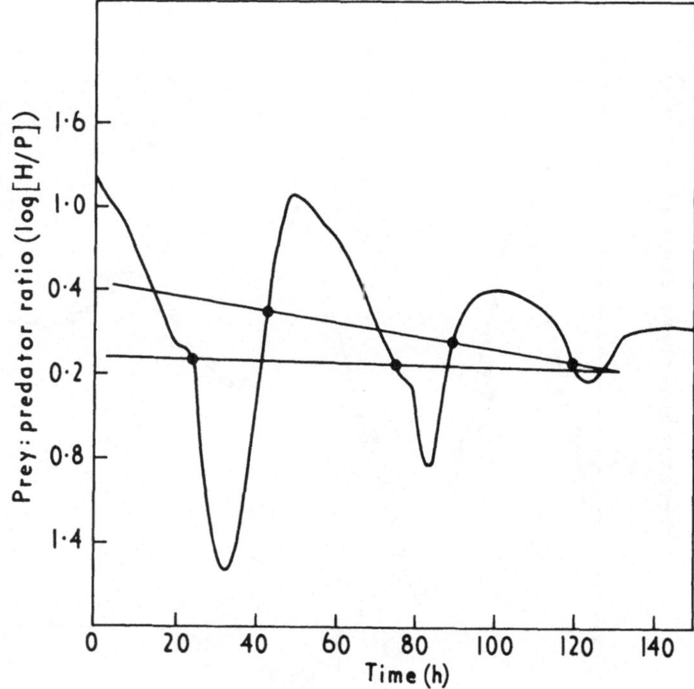

Fig. 9. As in Fig. 8 but using the catastrophe theory model to fit the data, as shown in Fig. 7. In this case sudden jumps in the system outline a cusp-shaped region in control space. In contrast to the situation shown in Fig. 6, the jumps occur as the trajectory moves out of this region and so conforms with the perfect delay convention.

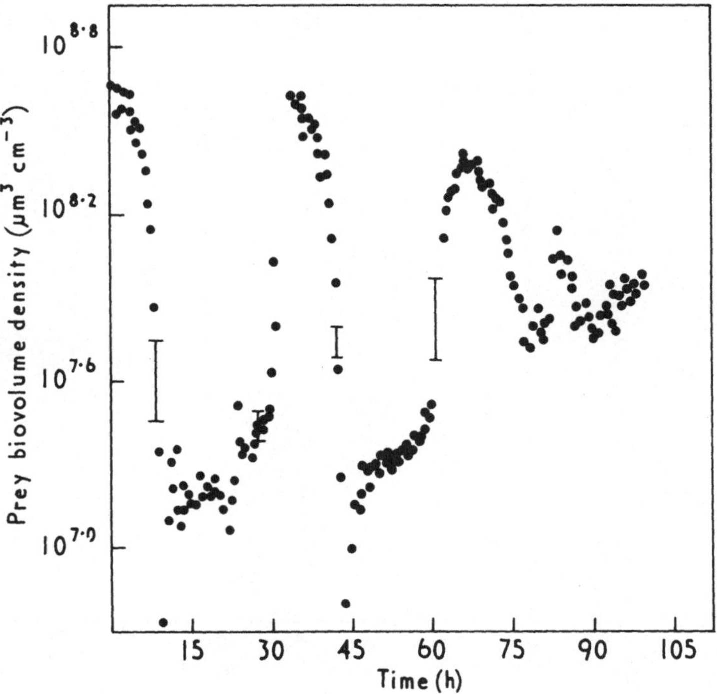

Fig. 10. Prey biovolume density as a function of time. Times at which sudden changes in λ occurred are indicated by vertical lines.

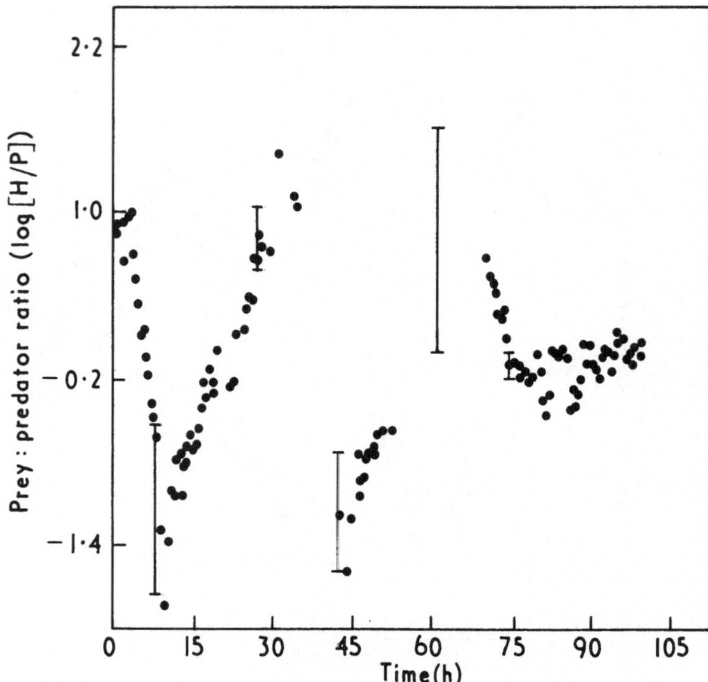

Fig. 11. As for Fig. 10 but with Prey:predator ratio as the ordinate. Even though it is difficult to locate the positions of the jumps in λ (the error bars are themselves estimates) it is clear that the configuration in Fig. 11 is consistent with the perfect delay convention, whereas that in Fig. 10 is not.

References:

d'Ancona, U. (1926). R. Comit. Talass. Ital. Mem. 126, 95

Curds, C.R. & Cockburn, A. (1971). J. gen. Microbiol. 54, 343-358.

Dent, V., Bazin, M.J. & Saunders, P.T. (1976). Arch. Microbiol. 109, 187-194

Owen, B.A. (1979). Ph.D. thesis, London University.

Poston, T. & Stewart, I. (1978). Catastrophe Theory and its Applications. Pitman, London.

Rosenzweig, M.D. (1971). Science, N.Y. 171, 385-387.

Saunders, P.T. & Bazin, M.J. (1975). J. theor. Biol. 52, 121-142.

Thom, R. (1972). Stabilité Structurelle et Morphogénèse. Benjamin, Reading, Mass.

Tsuchiya, H.M., Drake, J.F., Jost, J.L. and Fredrickson, A.G. (1972) J. Bact. 110, 1147-1153.

Volterra, V. (1926). Mem. R. Acad. Lincei, Ser. IV 2, 31.

Zeeman, E.C. (1976). Sci. Amer. 234, 65-83.

CHAPTER XXIII

J. RICHELLE (1)

Boolean approach of a prey-predator system.

Prey-predator systems have been extensively studied since their first statement by Volterra (1931) and Lotka (1956). Here I want to show the convergence of the boolean predictions with the well-known properties of the classical model. In the following, boolean equations are derived from a verbal description of the system and two models of different complexity are examined. This study gives the opportunity to outline some considerations on the meaning and the bearing of results obtained by a boolean analysis.

Models.

The system considered consists of two populations, the prey and the predators. The prey population grows when there are no or a few predators and the predator population grows when there are many prey.

x and y are boolean functions which describe the growing state of prey and predators respectively. The value 0 of a function means that the corresponding population is growing below an arbitrarily defined threshold rate, whereas the value 1 is attributed when the growing rate is above the threshold. The boolean variables ξ and υ describe in some sense the density of the prey and predator populations respectively. Their value, 0 or 1 , is discriminated by a density threshold. ξ and υ are the memorization variables of x and y (2). It is assumed that the density threshold for the prey population is the minimum density allowing the predator population to grow more rapidly than the threshold rate. On the other hand, the density threshold for the predator population is the maximum density allowing the prey population to grow more rapidly than the threshold rate. One notices that the values of the various thresholds are somehow implicitly present in the verbal description of the system.

---

(1) "Aspirant" of the Fonds National de la Recherche Scientifique (Belgium)

(2) In this case the relationship between function and associated memorization variable is obvious : if for example, the prey population is small but growing ($\xi = 0, x = 1$), sometime afterwards, it will be large ($\xi = 1$) .

From this verbal description the boolean equation of the system may be written

$$\begin{cases} x = \upsilon \\ y = \xi \end{cases} \qquad \text{(I)}$$

These equations constitute a negative loop as defined by Thomas (1978, and Chapter VII of this book) ; the cyclic behaviour expected is shown in the graph of states

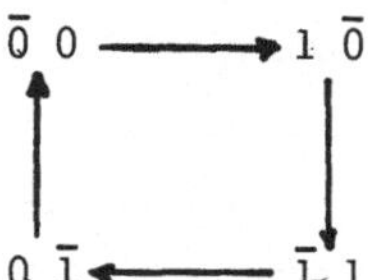

Let us consider the phase space ; the system trajectory is obviously restricted to the upper right quarter. One notes that the respective density thresholds divide it into four parts corresponding to the four boolean states. The boolean analysis predicts that the system travels cyclically through each of these space regions (figure 1).

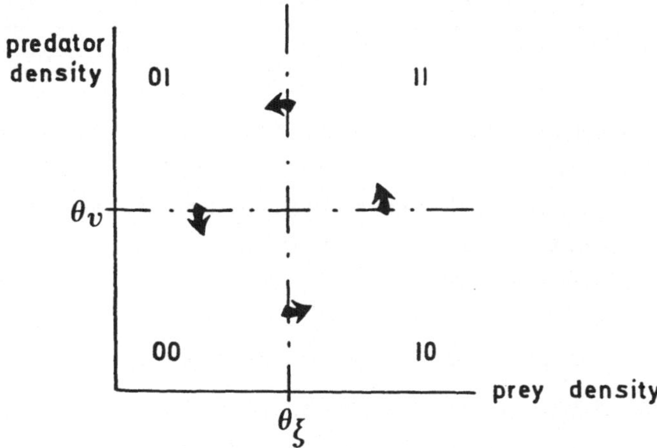

Figure 1. Representation of the boolean trajectory in the phase space. θ_ξ and θ_υ are the density thresholds of the prey and predator populations, the boolean equivalents of the four space parts defined by these thresholds are indicated.

Thus the well-known cyclical behaviour of the prey-predator system is already demonstrated by this elementary approach. A more precise analysis of the system requires a more refined description ; this may be obtained by using more than one boolean variable to describe a density state.

The density of the populations will now be described by couples of variables — $(\xi\xi_o)$ for the prey, $(\upsilon\upsilon_o)$ for the predators.

The first variable of the couple keeps the same meaning as in the preceding model, i.e. it describes the presence of individuals below (variable = 0) or above (variable = 1) the density threshold previously defined. As for the second variable of the couple, it discriminates between the (total) absence (variable = 0) and the presence (variable = 1) of individuals. The growing state of the populations will also be described by a couple of functions — (xx_o) for the prey, (yy_o) for the predators. As previously, the first function discriminates between growing at a rate less or greater than the threshold rate. The second function discriminates between non-growing and growing of the population. Table I shows the different meaningful states of the couples of variables or functions.

values	of the variables	of the functions
(0 0)	absence of individuals	absence of growth of the population
(0 1)	presence of few individuals	slow growth of the population
(1 1)	presence of many individuals	rapid growth of the population

Table I

One notes that the state (10) is "absurd" both for the functions (which would mean the growth of the population above the threshold rate and the absence of growth !) and for the variables (presence of more individuals than the density threshold and absence of individuals !). This coupling of functions is fundamentally equivalent to the use of multi-level boolean variables by Van Ham (this book, Chapter XV).

Let us now derive the set of equations describing the system

1° the growth of the prey population requires that prey be present

$$x_o = \xi_o \qquad \text{(II)}$$

2° the prey population grows rapidly if there are prey and no large density of predators (°)

$$x = \xi_o \cdot \bar{\upsilon} \qquad \text{(III)}$$

3° the predator population growth requires both prey and predators.

$$y_o = \xi_o \cdot \upsilon_o \qquad \text{(IV)}$$

4° the predator population grows at high rate if the predators are in the presence of a large density of prey.

$$y = \xi \cdot \upsilon_o \qquad \text{(V)}$$

(°) It is assumed that it is only on the rapid growing of the prey population that predators may have an effect ; this is why equation (II) is not $x_o = \xi_o \cdot \bar{\upsilon}_o$.

These equations (II-V) enable us to construct the matrix of the boolean model

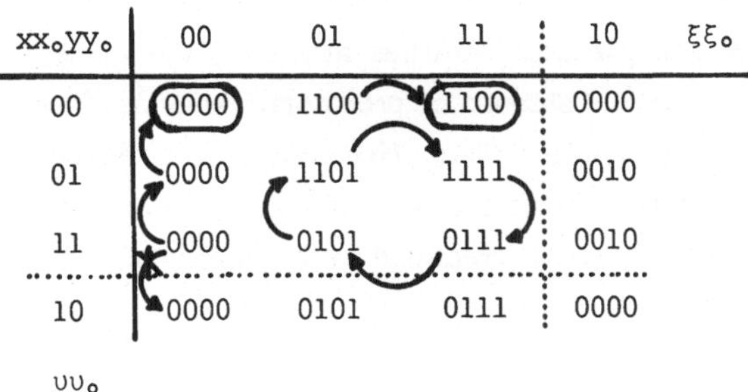

$xx_o yy_o$	00	01	11	10	$\xi\xi_o$
00	0000	1100	1100	0000	
01	0000	1101	1111	0010	
11	0000	0101	0111	0010	
10	0000	0101	0111	0000	
$\upsilon\upsilon_o$					

Table II : Boolean matrix of equations (II-V). The arrows between states indicate the possible transitions. (One notes that the transition $00\bar{1}\bar{1} \longrightarrow 00\dot{1}0$ is excluded because the target state 0010 is "absurd", and because the large density of predators will not reach zero before passing into a state of low density).

Let us now report the boolean transition in the phase space. As previously the upper right quarter of this space is divided in four parts by the density thresholds ; the states with either $\xi_o = 0$ or $\upsilon_o = 0$ corresponds to points on the axes (figure 2).

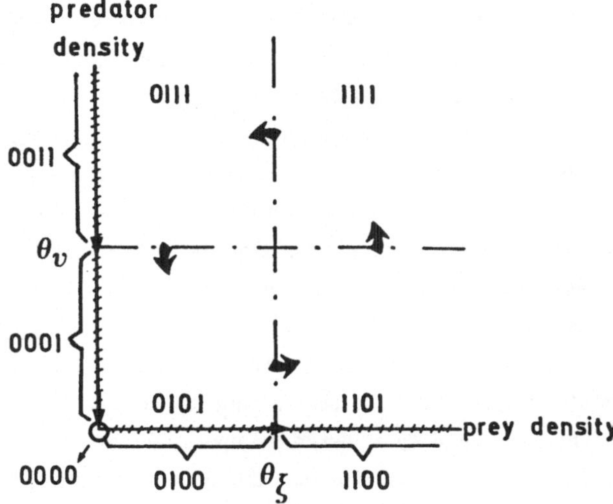

Figure 2. Representation of the boolean pathways of the matrix of Table II in the phase space. Θ_ξ and Θ_υ are the density thresholds of the prey and predator populations respectively ; the boolean equivalent of the various domains of the plane are indicated : 0101, 1101 , 1111 , 0111 , the four space parts ; 0001 , 0011 , 0100 , 1100 , the dashed axes segments ; 0000 , the axes origin.

The boolean approach now predicts two stables states, (0000) and (1100), and a cycle, $\bar{0}101 \rightarrow 11\bar{0}1 \rightarrow \bar{1}111 \longrightarrow 01\bar{1}1 \rightarrow \bar{0}101 \rightarrow$ etc. These behaviours correspond exactly to those revealed by the classical (continuous) approach : the trivial steady state, where prey and predator densities are zero, the cycle, and a continuous expansion of prey population in the absence of predators. This last (non-steady) behaviour corresponds to the boolean stable state 1100 : effectively $\xi = 1$ describes any prey density beyond the threshold, and thus this unique boolean state corresponds to an infinity of (continous) states for which the density is greater than the threshold.

Discussion.

A very simple boolean model has been shown to display the characteristic cyclical behaviour of a prey-predator system. When the boolean description is refined by using two pairs of function/memorization variables for each species, other characteristics of the system are recognized. Note that the two models are structurally equivalent ; indeed the first model may be derived from the second, by assuming that there are always some individuals of each species $\xi_o = \upsilon_o = 1$.

$$\begin{cases} x_o = 1 \\ x = 1 \cdot \bar{\upsilon} = \bar{\upsilon} \\ y_o = 1 \\ y = \xi \cdot 1 = \xi \end{cases} \qquad \text{(VI)}$$

In general, the Lotka-Volterra continuous model attributes to the system an infinite family of closed trajectories, with a (non trivial) steady state at their center. A given system will rotate around the center along a given trajectory, which depends only upon the initial condition or of the last perturbation. The system does not exhibit a limit cycle to which the system will tend whatever the initial condition or the perturbations.

When one tries to compare this continuous description to the boolean one presented here, it is reasonable to assume that the entirety of the trajectory family corresponds to the unique boolean cycle. It follows that the center is the same point (in the phase space) as the point defined by the density thresholds. At first glance, there is a discrepancy between the classical approach and this boolean analysis, the latter does not provide a stable state corresponding to the center while the former does. Note however that this steady state is immaterial since it would be impossible to locate a realistic system accuretly at the center. It is obvious then that the system will always rotate around the center, even if very close to it, in agreement with the boolean description.

It appears then that boolean states usually do not correspond to physically (or mathematically) punctual states, but rather to a finite or infinite set of states. This idea is consistent with the assimilation of the trajectory family to the boolean cycle, with the assimilation of the continuous prey growth to the unique state (1100), and also with the absence of a boolean state representing the trajectory center.

The boolean description may seem misleading when it associates a cyclical behaviour with an almost steady state. However this apparent contradiction may lead us to question the limitations of the technical or physical means of observation that may have reduced a microscopic cycle to an apparent steady behaviour. On the other hand, the boolean approach points out that the origin of a steady behaviour may be the infinitesimal limit of a fundamentally cyclic behaviour.

References

Lotka, A. (1956) Elements of Mathematical Biophysics. N.Y. Dover Publ. Inc.

Volterra, V. (1931) Leçons sur la Théorie Mathématique de la Lutte pour la Vie. Paris Gauthier - Villars.

Bio– mathematics

Managing Editors: K. Krickeberg, S. A. Levin

Editorial Board: H. J. Bremermann, J. Cowan, W. M. Hirsch, S. Karlin, J. Keller, R. C. Lewontin, R. M. May, J. Neyman, S. I. Rubinow, M. Schreiber, L. A. Segel

Volume 1:
Mathematical Topics in Population Genetics
Edited by K. Kojima
1970. 55 figures. IX, 400 pages
ISBN 3-540-05054-X

"...It is far and away the most solid product I have ever seen labelled biomathematics."
American Scientist

Volume 2: E. Batschelet
Introduction to Mathematics for Life Scientists
2nd edition. 1975. 227 figures. XV, 643 pages
ISBN 3-540-07293-4

"A sincere attempt to relate basic mathematics to the needs of the student of life sciences."
Mathematics Teacher

M. Iosifescu, P. Tăutu
Stochastic Processes and Applications in Biology and Medicine

Volume 3
Part 1: **Theory**
1973. 331 pages.
ISBN 3-540-06270-X

Volume 4
Part 2: **Models**
1973. 337 pages
ISBN 3-540-06271-8

Distributions Rights for the Socialist Countries: Romlibri, Bucharest

"... the two-volume set, with its very extensive bibliography, is a survey of recent work as well as a textbook. It is highly recommended by the reviewer."
American Scientist

Volume 5: A. Jacquard
The Genetic Structure of Populations
Translated by B. Charlesworth, D. Charlesworth
1974. 92 figures. XVIII, 569 pages
ISBN 3-540-06329-3

"...should take its place as a major reference work.."
Science

Volume 6: D. Smith, N. Keyfitz
Mathematical Demography
Selected Papers
1977. 31 figures. XI, 515 pages
ISBN 3-540-07899-1

This collection of readings brings together the major historical contributions that form the base of current population mathematics tracing the development of the field from the early explorations of Graunt and Halley in the seventeenth century to Lotka and his successors in the twentieth. The volume includes 55 articles and excerpts with introductory histories and mathematical notes by the editors.

Volume 7: E.R. Lewis
Network Models in Population Biology
1977. 187 figures. XII, 402 pages
ISBN 3-540-08214-X

Directed toward biologists who are looking for an introduction to biologically motivated systems theory, this book provides a simple, heuristic approach to quantitative and theoretical population biology.

A
Springer
Journal

Journal of

Mathematical Biology

Ecology and Population Biology
Epidemiology
Immunology
Neurobiology
Physiology
Artificial Intelligence
Developmental Biology
Chemical Kinetics

Edited by H.J. Bremermann, Berkeley, CA; F.A. Dodge, Yorktown Heights, NY; K.P. Hadeler, Tübingen; S.A. Levin, Ithaca, NY; D. Varjú, Tübingen.

Advisory Board: M.A. Arbib, Amherst, MA; E. Batschelet, Zürich; W. Bühler, Mainz; B.D. Coleman, Pittsburgh, PA; K. Dietz, Tübingen; W. Fleming, Providence, RI; D. Glaser, Berkeley, CA; N.S. Goel, Binghamton, NY; J.N.R. Grainger, Dublin; F. Heinmets, Natick, MA; H. Holzer, Freiburg i.Br.; W. Jäger, Heidelberg; K. Jänich, Regensburg; S. Karlin, Rehovot/Stanford CA; S. Kauffman, Philadelphia, PA; D.G. Kendall, Cambridge; N. Keyfitz, Cambridge, MA; B. Khodorov, Moscow; E.R. Lewis, Berkeley, CA; D. Ludwig, Vancouver; H. Mel, Berkeley, CA; H. Mohr, Freiburg i.Br.; E.W. Montroll, Rochester, NY; A. Oaten, Santa Barbara, CA; G.M. Odell, Troy, NY; G. Oster, Berkeley, CA; A.S. Perelson, Los Alamos, NM; T. Poggio, Tübingen; K.H. Pribram, Stanford, CA; S.I. Rubinow, New York, NY; W.v. Seelen, Mainz; L.A. Segel, Rehovot; W. Seyffert, Tübingen; H. Spekreijse, Amsterdam; R.B. Stein, Edmonton; R. Thom, Bures-sur-Yvette; Jun-ichi Toyoda, Tokyo; J.J. Tyson, Blacksbough, VA; J. Vandermeer, Ann Arbor, MI.

Springer-Verlag
Berlin
Heidelberg
New York

Journal of Mathematical Biology publishes papers in which mathematics leads to a better understanding of biological phenomena, mathematical papers inspired by biological research and papers which yield new experimental data bearing on mathematical models. The scope is broad, both mathematically and biologically and extends to relevant interfaces with medicine, chemistry, physics and sociology. The editors aim to reach an audience of both mathematicians and biologists.

Lecture Notes in Biomathematics

Vol. 26: M. B. Katz, Questions of Uniqueness and Resolution in Reconstruction Projections. IX, 175 pages. 1978.

Vol. 27: N. MacDonald, Time Lags in Biological Models. VII, 112 pages. 1978.

Vol. 28: P. C. Fife, Mathematical Aspects of Reacting and Diffusing Systems. IV, pages. 1979.

Vol. 29: Kinetic Logic – A Boolean Approach to the Analysis of Complex Regul Systems. Proceedings, 1977. Edited by R. Thomas. XIII, 507 pages. 1979.